VOLUME SEVENTY TWO

CURRENT TOPICS IN MEMBRANES

Functional Organization of Vertebrate Plasma Membrane

CURRENT TOPICS IN MEMBRANES, VOLUME 72

VOLUME SEVENTY TWO

CURRENT TOPICS IN MEMBRANES

Functional Organization of Vertebrate Plasma Membrane

Edited by

VANN BENNETT
HHMI, and Department of Biochemistry
Duke University Medical Center
Durham, North Carolina, USA

ELSEVIER

AMSTERDAM • BOSTON • HEIDELBERG • LONDON
NEW YORK • OXFORD • PARIS • SAN DIEGO
SAN FRANCISCO • SINGAPORE • SYDNEY • TOKYO
Academic Press is an imprint of Elsevier

Academic Press is an imprint of Elsevier
225 Wyman Street, Waltham, MA 02451, USA
525 B Street, Suite 1800, San Diego, CA 92101-4495, USA
The Boulevard, Langford Lane, Kidlington, Oxford, OX5 1GB, UK
32 Jamestown Road, London NW1 7BY, UK
Radarweg 29, PO Box 211, 1000 AE Amsterdam, The Netherlands

First edition 2013

ISBN: 978-0-12-417027-8
ISSN: 1063-5823

13 14 15 16 11 10 9 8 7 6 5 4 3 2 1

CONTENTS

CONTRIBUTORS

Francis J. Alenghat
Department of Biological Chemistry and Molecular Pharmacology, Harvard Medical School, Boston, Massachusetts, USA

Sheila A. Baker
Department of Biochemistry, Carver College of Medicine, University of Iowa, Iowa City, Iowa, USA

Vann Bennett
HHMI, and Department of Biochemistry, Duke University Medical Center, Durham, North Carolina, USA

Kae-Jiun Chang
Program in Developmental Biology, Baylor College of Medicine, Houston, Texas, USA

Donald N. Clarke
Department of Biology, Stanford University, Stanford, California, USA

Silvia Curado
Skirball Institute of Biomolecular Medicine; Department of Pathology, New York University School of Medicine, New York, USA, and Kennedy Institute of Rheumatology, NDORMS, University of Oxford, Headington, Oxfordshire, Oxford, UK

Michael L. Dustin
Skirball Institute of Biomolecular Medicine; Department of Pathology, New York University School of Medicine, New York, USA, and Kennedy Institute of Rheumatology, NDORMS, University of Oxford, Headington, Oxfordshire, Oxford, UK

Velia M. Fowler
Department of Cell and Molecular Biology, The Scripps Research Institute, La Jolla, California, USA

David E. Golan
Department of Biological Chemistry and Molecular Pharmacology, Harvard Medical School, and Hematology Division, Brigham and Women's Hospital, Boston, Massachusetts, USA

Reinhard Jahn
Department of Neurobiology, Max Planck Institute for Biophysical Chemistry, Göttingen, Germany

Vasily Kerov
Department of Biochemistry, Carver College of Medicine, University of Iowa, Iowa City, Iowa, USA

Crystal F. Kline
The Dorothy M. Davis Heart & Lung Research Institute, The Ohio State University Wexner Medical Center, Columbus, Ohio, USA

Sudha Kumari
Skirball Institute of Biomolecular Medicine; Department of Pathology, New York University School of Medicine, New York, USA, and Kennedy Institute of Rheumatology, NDORMS, University of Oxford, Headington, Oxfordshire, Oxford, UK

Thorsten Lang
Department of Membrane Biochemistry, Life & Medical Sciences (LIMES) Institute, University of Bonn, Bonn, Germany

Damaris N. Lorenzo
HHMI, and Department of Biochemistry, Duke University Medical Center, Durham, North Carolina, USA

Christopher J. Lowe
Department of Biology, Stanford University, Stanford, California, USA

Phillip W. Miller
Department of Molecular and Cellular Physiology, Stanford University School of Medicine, Stanford, California, USA

Peter J. Mohler
The Dorothy M. Davis Heart & Lung Research Institute; Division of Cardiovascular Medicine, Department of Internal Medicine, and Department of Physiology and Cell Biology, The Ohio State University Wexner Medical Center, Columbus, Ohio, USA

W. James Nelson
Department of Molecular and Cellular Physiology, Stanford University School of Medicine, and Department of Biology, Stanford University, Stanford, California, USA

Matthew N. Rasband
Program in Developmental Biology, and Department of Neuroscience, Baylor College of Medicine, Houston, Texas, USA

Geert van den Bogaart
Department of Tumor Immunology, Nijmegen Centre for Molecular Life Sciences, Radboud University Nijmegen Medical Centre, Nijmegen, The Netherlands

William I. Weis
Department of Molecular and Cellular Physiology, and Department of Structural Biology, Stanford University School of Medicine, Stanford, California, USA

PREFACE

Surfaces of most vertebrate cells are patterned into microdomains that were first appreciated by pioneering histologists employing light microscopes in the nineteenth century and later resolved in ultrastructural detail by electron microscopists. It has turned out that these structures play a central role in human physiology and are required for processes including signaling in the central nervous system, rhythmic beating and mechanical resilience of the heart, polarized transport of ions and water by epithelial tissues, detection of light by photoreceptors, and acquired immunity by lymphocytes. Many of the membrane domains that we will consider are vertebrate inventions and may not be familiar to those who have not been exposed to a histology course. Goals of this volume are to introduce new students to these structures and to supplement this description with molecular insights into their function, organization, and evolution.

Ironically the human erythrocyte, which is one of the few cells lacking membrane domains, has provided major insights into mechanisms for establishing long-range order in plasma membranes. The first chapter presents general principles of ankyrin- and spectrin-based domains, first discovered in erythrocytes, that now are directly applicable to excitable membranes in the nervous system and heart, lateral membranes of epithelial tissues, and outer and inner segments of photoreceptors (Bennett & Lorenzo, 2013). Next, we discuss the role of spectrin–actin interactions in formation of an extended membrane-associated network that was first appreciated in erythrocytes, but now is known to exist in axons and likely other cell domains (Fowler, 2013; Xu, Zhong, & Zhuang, 2013). We also provide a biophysical perspective, based initially on results from erythrocytes, on the general problem of how cells control mobility of membrane proteins embedded in a fluid phospholipid bilayer (Alenghat & Golan, 2013).

The next chapter presents intercalated discs and T-tubules of cardiomyocytes and their roles in both electrical and mechanical function of the heart (Kline & Mohler; 2013). Extending the theme of excitable membranes to the nervous system, we examine mechanisms for formation and maintenance of axon initial segments and nodes of Ranvier (Chang & Rasband, 2013), and discuss the molecular machinery responsible for exocytosis at synapses (van den Bogaart, Lang, & Jahn, 2013). We also present photoreceptor inner and outer segments and the adaptations that underlie their amazing ability to detect a single photon (Baker & Kerov, 2013).

In the final two chapters, we examine both the oldest and one of the most recently evolved membrane specializations. Mechanisms for cell adhesion of epithelial lateral membrane domains, which evolved in sponges and perhaps in nonmetazoans as well, are discussed from an evolutionary perspective (Miller, Clarke, Weis, Lowe, & Nelson, 2013). In contrast to ancient epithelial cells, we also present the immune synapse, which is a vertebrate adaptation, and present its organization and central role in the complex signaling of adaptive immunity (Curado, Kumari, & Dustin, 2013). While many of the domains discussed in this volume have been known since the nineteenth century, the immune synapse was only recently appreciated as a cellular structure (Grakoui et al., 1999). We hope readers will be motivated to not only contribute to further understanding of current plasma membrane domains but also to discover new examples where organization of membrane-spanning proteins optimizes their physiological function.

REFERENCES

Alenghat, F. J., & Golan, D. E. (2013). Membrane protein dynamics and functional implications in mammalian cells. *Current Topics in Membranes*, *72*, 91–222.

Baker, S. A., & Kerov, V. (2013). Photoreceptor inner and outer segments. *Current Topics in Membranes*, *72*, 233–267.

Chang, K.-J., & Rasband, M. N. (2013). Excitable domains of myelinated nerves: Axon initial segments and nodes of Ranvier. *Current Topics in Membranes*, *72*, 161–194.

Curado, S., Kumari, S., & Dustin, M. L. (2013). The immunological synapse. *Current Topics in Membranes*, *72*, 315–348.

Fowler, V. M. (2013). The human erythrocyte plasma membrane: A Rosetta stone for decoding membrane—Cytoskeleton structure. *Current Topics in Membranes*, *72*, 39–89.

Grakoui, A., Bromley, S. K., Sumen, C., Davis, M. M., Shaw, A. S., Allen, P. M., et al. (1999). The immunological synapse: A molecular machine controlling T cell activation. *Science*, *285*, 221–227.

Kline, C. F., & Mohler, P. J. (2013). Evolving form to fit function: Cardiomyocyte intercalated disc and transverse-tubule membranes. *Current Topics in Membranes*, *72*, 123–160.

Miller, P. W., Clarke, D. N., Weis, W. I., Lowe, C. J., & Nelson, W. J. (2013). The evolutionary origin of epithelial cell-cell adhesion mechanisms. *Current Topics in Membranes*, *72*, 269–313.

van den Bogaart, G., Lang, T., & Jahn, R. (2013). Microdomains of SNARE proteins in the plasma membrane. *Current Topics in Membranes*, *72*, 195–232.

Xu, K., Zhong, G., & Zhuang, X. (2013). Actin, spectrin, and associated proteins form a periodic cytoskeletal structure in axons. *Science*, *339*, 452–456.

PREVIOUS VOLUMES IN SERIES

Current Topics in Membranes and Transport

Volume 23 Genes and Membranes: Transport Proteins and Receptors* (1985)
Edited by Edward A. Adelberg and Carolyn W. Slayman

Volume 24 Membrane Protein Biosynthesis and Turnover (1985)
Edited by Philip A. Knauf and John S. Cook

Volume 25 Regulation of Calcium Transport across Muscle Membranes (1985)
Edited by Adil E. Shamoo

Volume 26 Na^+-H^+ Exchange, Intracellular pH, and Cell Function* (1986)
Edited by Peter S. Aronson and Walter F. Boron

Volume 27 The Role of Membranes in Cell Growth and Differentiation (1986)
Edited by Lazaro J. Mandel and Dale J. Benos

Volume 28 Potassium Transport: Physiology and Pathophysiology* (1987)
Edited by Gerhard Giebisch

Volume 29 Membrane Structure and Function (1987)
Edited by Richard D. Klausner, Christoph Kempf, and Jos van Renswoude

Volume 30 Cell Volume Control: Fundamental and Comparative Aspects in Animal Cells (1987)
Edited by R. Gilles, Arnost Kleinzeller, and L. Bolis

Volume 31 Molecular Neurobiology: Endocrine Approaches (1987)
Edited by Jerome F. Strauss, III, and Donald W. Pfaff

Volume 32 Membrane Fusion in Fertilization, Cellular Transport, and Viral Infection (1988)
Edited by Nejat Düzgünes and Felix Bronner

Volume 33 Molecular Biology of Ionic Channels* (1988)
Edited by William S. Agnew, Toni Claudio, and Frederick J. Sigworth

*Part of the series from the Yale Department of Cellular and Molecular Physiology

Volume 34 Cellular and Molecular Biology of Sodium Transport* (1989)
Edited by Stanley G. Schultz

Volume 35 Mechanisms of Leukocyte Activation (1990)
Edited by Sergio Grinstein and Ori D. Rotstein

Volume 36 Protein–Membrane Interactions* (1990)
Edited by Toni Claudio

Volume 37 Channels and Noise in Epithelial Tissues (1990)
Edited by Sandy I. Helman and Willy Van Driessche

Current Topics in Membranes

Volume 38 Ordering the Membrane Cytoskeleton Trilayer* (1991)
Edited by Mark S. Mooseker and Jon S. Morrow

Volume 39 Developmental Biology of Membrane Transport Systems (1991)
Edited by Dale J. Benos

Volume 40 Cell Lipids (1994)
Edited by Dick Hoekstra

Volume 41 Cell Biology and Membrane Transport Processes* (1994)
Edited by Michael Caplan

Volume 42 Chloride Channels (1994)
Edited by William B. Guggino

Volume 43 Membrane Protein–Cytoskeleton Interactions (1996)
Edited by W. James Nelson

Volume 44 Lipid Polymorphism and Membrane Properties (1997)
Edited by Richard Epand

Volume 45 The Eye's Aqueous Humor: From Secretion to Glaucoma (1998)
Edited by Mortimer M. Civan

Volume 46 Potassium Ion Channels: Molecular Structure, Function, and Diseases (1999)
Edited by Yoshihisa Kurachi, Lily Yeh Jan, and Michel Lazdunski

Volume 47 Amiloride-Sensitive Sodium Channels: Physiology and Functional Diversity (1999)
Edited by Dale J. Benos

CHAPTER ONE

Spectrin- and Ankyrin-Based Membrane Domains and the Evolution of Vertebrates

Vann Bennett[*,†,1], **Damaris N. Lorenzo**[*,†]
[*]HHMI, Duke University Medical Center, Durham, North Carolina, USA
[†]Department of Biochemistry, Duke University Medical Center, Durham, North Carolina, USA
[1]Corresponding author: e-mail address: benne012@mc.duke.edu

Contents

Abstract

Spectrin and ankyrin are membrane skeletal proteins that contribute to mechanical support of plasma membranes and micron-scale organization of diverse membrane-spanning proteins. This chapter provides a plausible scenario for the evolution of ankyrin- and spectrin-based membrane domains with a focus on vertebrates. The analysis integrates recent phylogenetic information with functional analyses of spectrin and ankyrin in erythrocytes, axon initial segments and nodes of Ranvier in neurons, T-tubules and intercalated disks of cardiomyocytes, lateral membrane domains of epithelial cells, and costameres of striated muscle. A core spectrin–ankyrin mechanism for coordinating membrane-spanning proteins and mechanically stabilizing membrane bilayers was expanded in vertebrates by gene duplication events, insertion of giant alternately spliced exons of axonal ankyrins, and a versatile peptide-binding fold of ANK repeats that facilitated acquisition of new protein partners. Cell adhesion molecules (CAM), including dystroglycan, L1 CAM family members, and cadherins, are the earliest examples of membrane-spanning proteins with ankyrin-binding motifs and were all present in urochordates. In contrast, ion channels have continued to evolve ankyrin-binding sites in vertebrates. These considerations suggest a model where proto-domains formed through interaction of ankyrin and spectrin with CAMs. These proto-domains then became populated with ion channels that developed ankyrin-binding

Current Topics in Membranes, Volume 72
ISSN 1063-5823
http://dx.doi.org/10.1016/B978-0-12-417027-8.00001-5

activity with selective pressure provided by optimization of physiological function. The best example is the axon initial segment where ankyrin-binding activity evolved sequentially and independently first in L1 CAMs, then in voltage-gated sodium channels, and finally in KCNQ2/3 channels, with the selective advantage of fast and precisely regulated signaling.

1. INTRODUCTION

First year medical students learn that plasma membranes of cells in human tissues are beautifully organized into functional micron-scale domains that are the basis for much of our physiology. What is less appreciated is the molecular novelty and evolutionary origin of these structures, especially those related to fast signaling in the heart and nervous system that exist only in vertebrates. Following the emergence of polarized epithelial cells in early metazoans beginning around 650 million years before present (mybp), animal cells rapidly diversified to include neurons and other sensory cells, as well as muscle cells. The first neurons and striated muscle in cnidarians (jellyfish, hydra, and corals) appeared by 580 mybp and were followed by multiple cell types organized with bilateral symmetry in the bilaterian lineage (nematodes, arthropods, flatworms, mollusks, etc.) around 550 mybp. A modern myelinated nervous system and closed cardiovascular system were likely present in the first jawed vertebrates by 420 mybp. Soon after these evolutionary developments, tetrapods invaded terrestrial environments, which required many adaptations, including new approaches to respiration and excretion. By 170 mybp, early eutherian mammals had developed homeothermy and high cardiac output, with cellular specializations including cardiomyocyte T-tubules and enucleated erythrocytes. Following the emergence of the first bilaterian, mammals had acquired a diverse set of newly configured assemblies of ion transporters and cell adhesion molecules (CAMs), each requiring new protein interactions and mechanisms for precise spatial patterning in the plasma membranes of multiple cell types.

Spectrin and ankyrin are membrane skeletal proteins present in their modern forms in bilaterians that contribute to mechanical support of plasma membranes and micron-scale organization of membrane-spanning proteins in many tissues (Bennett & Baines, 2001; Bennett & Healy, 2009). This chapter will consider the role of ankyrins and spectrins in the evolution of diverse vertebrate membrane domains. We will develop the thesis that members of the ankyrin and spectrin families were key substrates in adaptive evolution of diverse plasma membrane domains, including excitable membranes in neurons and heart, costameres of striated muscle, and lateral membrane domains of epithelial cells. We will review intrinsic features of spectrin and ankyrin that made these interacting proteins excellent starting points for establishing long-range order in an otherwise fluid phospholipid bilayer. We will consider the basis for diversity within the basic theme of a core spectrin–ankyrin mechanism for coordinating membrane-spanning proteins. We also will discuss gene duplication events in early vertebrates, vertebrate-specific alternately spliced isoforms of axonal ankyrins, and a versatile peptide-binding fold of ankyrins that facilitated acquisition of new protein partners. Lastly, we will review a parallel and perhaps ancient function for spectrins and ankyrins in directed intracellular trafficking of membrane organelles.

2. AN ANCIENT SPECTRIN–ANKYRIN PARTNERSHIP FOR COORDINATING MEMBRANE-SPANNING PROTEINS

Ankyrin and spectrin family members cooperate to provide a widely utilized mechanism for coordinating membrane-spanning proteins in the plane of the plasma membrane and coupling these proteins to an extended mechanically resilient submembrane network. The logic of this system is straightforward: membrane-spanning proteins, including cell adhesion proteins capable of interaction with the extracellular matrix and other cell surfaces as well as membrane transporters, are "anchored" by ankyrin to an extended spectrin network tightly associated with the plasma membrane (Fig. 1.1). Spectrin polymerizes into membrane-associated networks through association with specialized actin filaments, which interact with multiple spectrins and also are independently coupled to the membrane bilayer. Spectrin association with actin filaments is promoted by accessory proteins, including adducin, protein 4.1, and p55/MPP1 (described in more detail in the succeeding text) (Fig. 1.1). Ankyrin through its ANK repeats (Lee et al., 2006) and spectrin through its triple-helical repeats (Rief, Pascual, Saraste, & Gaub, 1999) both behave as elastic elements in single molecule atomic force microscopy measurements. Ankyrin and spectrin both experience stretching in erythrocytes under shear stress (Krieger et al., 2011). Moreover, deficiency of ankyrin and spectrin lead to fragile erythrocyte membranes (Eber & Lux, 2004) and axons (Hammarlund, Jorgensen, & Bastiani, 2007). The ankyrin–spectrin assembly thus provides mechanical stability to the lipid bilayer in addition to organization of protein assemblies.

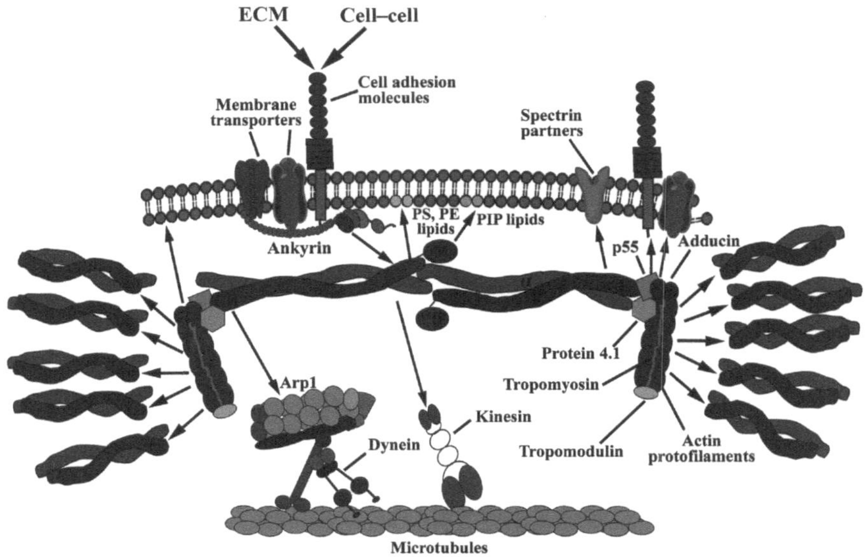

Figure 1.1 A conserved spectrin–ankyrin partnership coordinates membrane-spanning proteins within micron-scale plasma membrane domains responsive to extracellular cues. Membrane-spanning proteins, including cell adhesion proteins and membrane transporters, are "anchored" by ankyrin through its ANK repeats to an extended spectrin–actin network tightly associated with the plasma membrane. Spectrin polymerizes into membrane-associated networks through association with specialized actin filaments, which interact with multiple spectrins and also are independently coupled to the membrane bilayer. Spectrin association with actin filaments is promoted by accessory proteins, including adducin, protein 4.1, and p55/MPP1. The ankyrin–spectrin assembly provides mechanical stability to the lipid bilayer in addition to organization of membrane proteins. Parallel functions of spectrin and ankyrin in intracellular organelle transport are mediated by interactions with dynactin (shown for CH domains of beta-spectrin) and kinesin. (See Color Plate.)

The human erythrocyte provided the prototype where the elements of a spectrin–ankyrin-based assembly were first elucidated (see Chapter 2; Bennett & Baines, 2001). Ankyrin and/or spectrin has subsequently been implicated in organization and/or stabilization of multiple membrane domains, including axon initial segments and nodes of Ranvier (Dzhashiashvili et al., 2007; Galiano et al., 2012; Hedstrom, Ogawa, & Rasband, 2008; Hedstrom et al., 2007, Jenkins & Bennett, 2001; Sobotzik et al., 2009; Susuki et al., 2013; Zhou et al., 1998), unmyelinated axons (Scotland, Zhou, Benveniste, & Bennett, 1998), cardiomyocyte T-tubules and intercalated disks (Hund et al., 2010; Lowe et al., 2008;

Mohler, Davis, & Bennett, 2005; Mohler, Rivolta, et al., 2004), epithelial lateral membranes (Kizhatil & Bennett, 2004; Kizhatil, Davis, et al., 2007; Kizhatil, Yoon, et al., 2007), costameres, which are mechano-domains in heart and skeletal muscles (Ayalon, Davis, Scotland, & Bennett, 2008, Ayalon et al., 2011), and photoreceptor inner and outer segments (Kizhatil, Baker, Arshavsky, & Bennett, 2009, Kizhatil, Sandhu, Peachy, & Bennett, 2009; reviewed by Bennett & Healy, 2009) (see chapters 3–5; 7 in this volume).

We will next consider the molecular properties of spectrin and ankyrin that provide the basis for their function as membrane domain coordinators (Figs. 1.2 and 1.3). Spectrin is a flexible elongated tetramer nearly 200 nm in length that is comprised of alpha- and beta-subunits assembled side-to-side in an antiparallel orientation and head to head through association of alpha-spectrin with beta-spectrin (Figs. 1.1 and 1.2) (Shotton, Burke, & Branton, 1979). Alpha- and beta-spectrin subunits are both related to alpha-actinin (Djinovic-Carugo, Young, Gautel, & Saraste, 1999; Viel, 1999) but are extended in length from 40 to nearly 100 nm by multiple copies of a triple-helical repeat (Grum et al., 1999; Speicher & Marchesi, 1984; Yan et al., 1993). End-to-end association between alpha- and beta-spectrins results from noncovalent assembly of partial triple-helical repeats of each spectrin subunit (Ipsaro et al., 2010; Mehboob et al., 2010; Tse et al., 1990).

Beta-spectrin contributes to the principal interactions with other proteins and mediates interactions with F-actin (or Arp1 of the dynactin complex; see succeeding text) through N-terminal tandem calponin homology (CH) domains (Banuelos, Saraste, & Carugo, 1998; Carugo, Banuelos, & Saraste, 1997), with ankyrin through the 14th and 15th triple-helical repeats (Davis et al., 2009; Ipsaro, Huang, & Mondragón, 2009; Ipsaro & Mondragón, 2010; Kennedy, Warren, Forget, & Morrow, 1991; Stabach et al., 2009) and with $PI4,5P_2$ phosphatidylinositol lipids through a C-terminal pleckstrin homology (PH) domain (Hyvönen et al., 1995; Macias et al., 1994; Fig. 1.2). The beta-spectrin PH domain is absent as a result of alternative splicing in mammalian erythrocyte spectrin as well as some other spectrin isoforms (Hayes et al., 2000, Winkelmann, Chang, et al., 1990; Winkelmann, Costa, Linzie, & Forget, 1990).

Spectrin tetramers in erythrocyte membranes form a membrane-coupled polygonal network through association of five to seven spectrins at 50–70° angles with short 40 nm actin protofilaments (Byers & Branton, 1985). These actin protofilaments are capped on their fast-growing ends by adducin, which also promotes association with spectrin (Gardner & Bennett, 1987; Kuhlman,

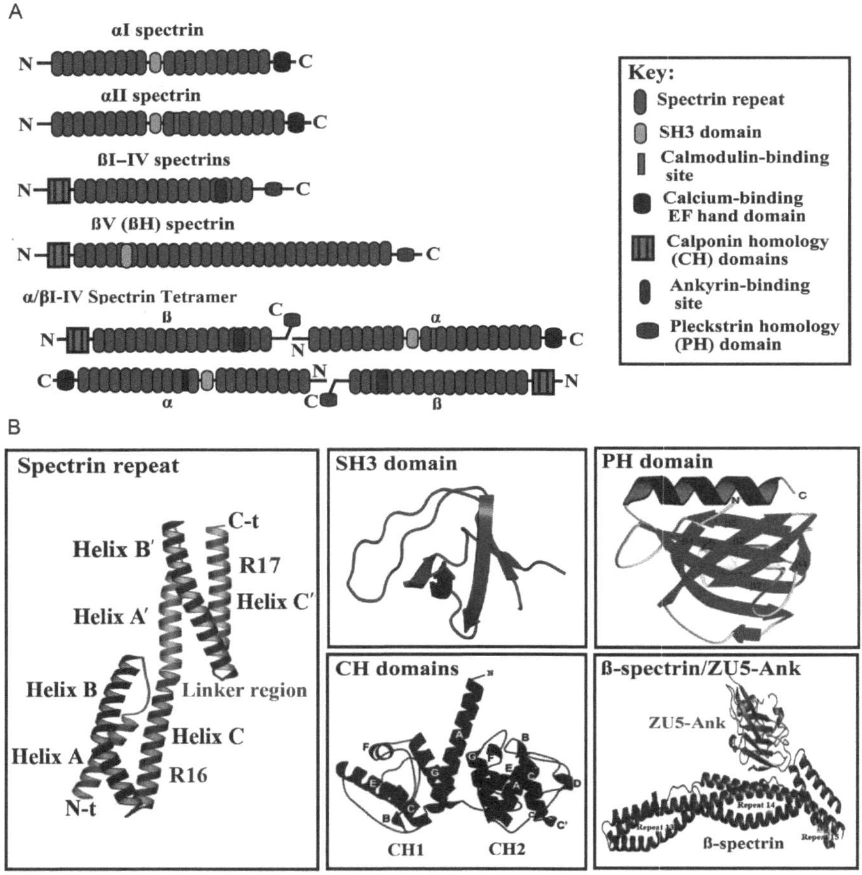

Figure 1.2 Domain structure of spectrins. (A) The domain organization of two α- and five β-spectrins is shown. Spectrins are comprised of modular spectrin repeat units (blue). Other functional domains include Src-homology domain (SH3, yellow), calcium-binding EF hand domain (red), and calmodulin-binding domain (gray). β-Spectrin proteins also have two in tandem calponin homology domains (CH, gray and red) and a pleckstrin homology domain (PH, green), and with the exception of β_V-spectrin, they contain an ankyrin-binding site (orange). The spectrin tetramer is the fundamental unit of the spectrin-based membrane skeleton. The N-terminus of each α-spectrin subunit associates with the C-terminal portion of β-spectrin to form a dimer. Tetramer formation depends on the lateral and antiparallel head-to-head association between two α/β-spectrin heterodimers. (B) Ribbon diagram representation of the crystal structure of two spectrin repeats (Grum, MacDonald, & Mondragón, 1999), SH3 domain (Musacchio, Noble, Pauptit, Wierenga, & Saraste, 1992), in tandem CH domains (Sjöblom, Ylänne, & Djinović-Carugo, 2008), PH domain (Lemmon, Ferguson, & Abrams, 2002), and the spectrin–ankyrin interaction binding domains (Ipsaro & Mondragón, 2010). (See Color Plate.)

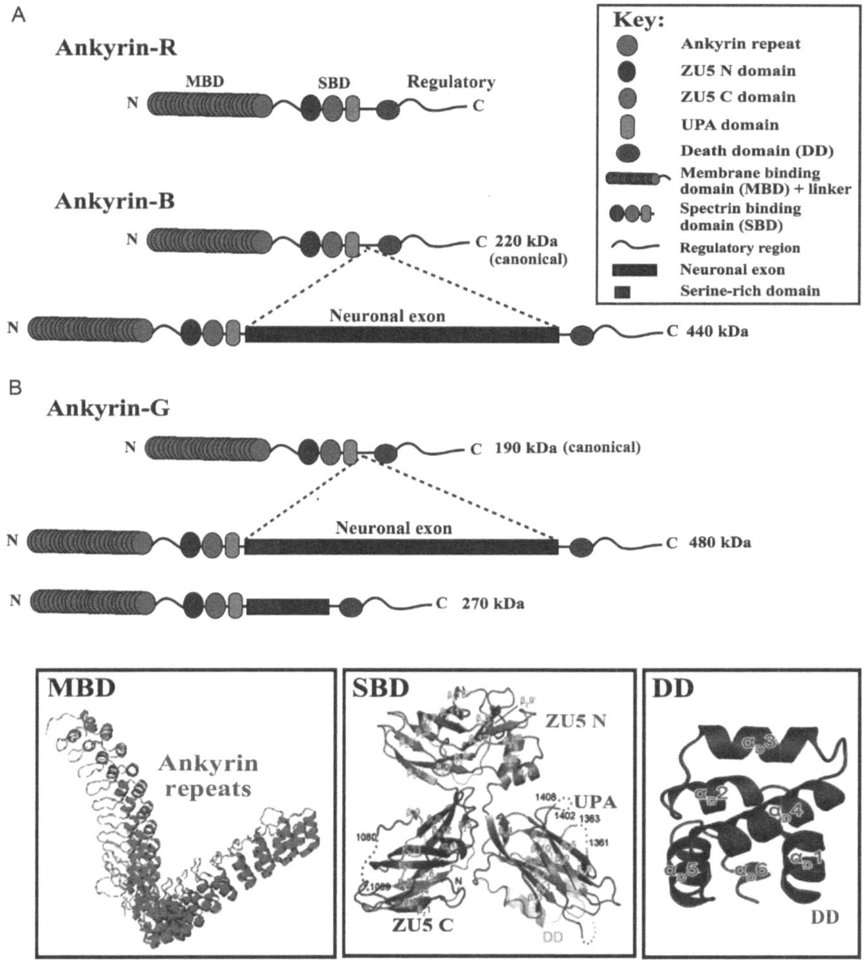

Figure 1.3 Domain structure of ankyrins. (A) The domain organization of canonical and neuronal giant ankyrins is shown. The membrane-binding domain of ankyrin (green) is comprised of 24 ankyrin repeats. The spectrin-binding domain supermodule contains two ZU5 domains (teal and blue) and a UPA domain (orange). Other functional domains include a death domain (pink) and a C-terminal unstructured regulatory domain (black line). The giant ankyrin isoforms have an insertion of a single exon (red) after the UPA domain. Neuronal giant 480 kDa and 270 kDa ankyrin-G proteins also contain a serine-rich domain at the beginning of the inserted region (black bar). (B) Ribbon representation of the crystal structure of the deduced 24 ANK repeat ankyrin membrane-binding domain (Michaely, Tomchick, Machius, & Anderson, 2002), ZU5n–ZU5c–UPA supermodule (Wang, Yu, Ye, Wei, & Zhang, 2012), and death domain (Wang et al., 2009). (See Color Plate.)

Hughes, Bennett, & Fowler, 1996; Li, Matsuoka, & Bennett, 1998), and on the slow-growing ends by tropomodulin (Weber, Pennise, Babcock, & Fowler, 1994). Spectrin interaction with actin in erythrocytes also is promoted by protein 4.1, a member of the FERM family (Baines, Lu, & Bennett, 2013; Pearson, Reczek, Bretscher, & Karplus, 2000; Ungewickell, Bennett, Calvert, Ohanian, & Gratzer, 1979). The spectrin–actin network is attached to erythrocyte membranes through a high-affinity association between beta-spectrin and ankyrin near the midregion of the spectrin tetramers (Bennett, 1978; Bennett & Stenbuck, 1979a; Ipsaro & Mondragón, 2010; Kennedy et al., 1991; Tyler, Hargreaves, & Branton, 1979). Additional protein-dependent membrane contacts at the spectrin–actin junction are provided by association of the MAGUK protein p55/MPP1 with protein 4.1 and the membrane-spanning protein glycophorin C (Marfatia, Leu, Branton, & Chishti, 1995; Marfatia, Lue, Branton, & Chishti, 1994, Ruff, Speicher, & Husain-Chishti, 1991) and of adducin with the anion exchanger (Anong et al., 2009). In addition to protein-based interactions, spectrin associates with membrane phosphatidylinositol lipids in nonerythroid cells through its PH domain (Das, Base, Dhulipala, & Dubreuil, 2006; Das, Base, Manna, Cho, & Dubreuil, 2008, Wang, Miller, Shaw, & Shaw, 1996). Spectrin also interacts with other membrane phospholipids such as phosphatidylserine at multiple sites identified in both alpha- and beta-subunits (An et al., 2004). Palmitoylation of ankyrin (He, Jenkins, & Bennett, 2012; Staufenbiel, 1987) and p55/MPP1 (membrane-palmitoylated protein-1) (Ruff et al., 1991) provides yet another mechanism for membrane attachment of the ankyrin–spectrin network.

Plasma membrane-associated spectrin–actin–adducin assemblies have been resolved in axons of cultured neurons using super-resolution light microscopy, although they exhibit a different organization from the polygonal erythrocyte configuration (Xu, Zhong, & Zhuang, 2013). Axonal actin and adducin are organized into periodic membrane-associated rings separated by 190 nm, which is precisely the length of brain spectrin tetramers (Bennett, Davis, & Fowler, 1982). Spectrin antibody labels between the actin–adducin rings suggesting that these structures are interconnected by spectrin tetramers attached at a 90° angle (Xu et al., 2013). Spectrin–actin–adducin networks, now resolved in both erythrocytes and axons, are likely to be a general feature of spectrin organization in epithelial lateral membranes as well as other membrane domains (Abdi & Bennett, 2008), although their precise geometry likely will depend on the cellular context.

Ankyrin provides a mechanism for utilizing extracellular cues to form plasma membrane domains through its ability to associate with CAMs as well as membrane transporters and to couple these membrane-spanning proteins to membrane-associated spectrin–actin networks. Ankyrin is a monomer with an N-terminal membrane-binding domain containing 24 tandem ANK repeats folded as a solenoid (Michaely et al., 2002), followed by a supermodule comprised of two ZU5 domains and a UPA domain (Wang et al., 2012), a death domain (Wang et al., 2009), and an unstructured regulatory domain (Fig. 1.3). The binding site for beta-spectrin is located in the first ZU5 domain (Ipsaro & Mondragón, 2010; Mohler, Yoon, & Bennett, 2004). The anion exchanger (band 3 in early literature) of erythrocytes provided the first example of an ankyrin-linked integral membrane protein (Bennett & Stenbuck, 1979a,1979b, 1980a). Membrane-spanning ankyrin-binding partners now include over 14 families of CAMs and membrane transporters (Bennett & Healy, 2009; Table 1.1).

Ankyrins interact with their membrane protein partners through ANK repeats that are folded into an extended solenoid with a 240-Å groove running along its length (Michaely et al., 2002). Interestingly, ankyrin can bind to more than one partner at a time and thus can form homo- and hetero-complexes (Michaely & Bennett, 1995a,1995b). ANK repeats in general perform a wide range of functions related to protein recognition and occur in tandem arrays throughout nature including viruses, bacteria, archaea, fungi, plants, and animals (Al-Khodor, Price, Kalia, & Abu Kwaik, 2010; Mosavi, Cammett, Desrosiers, & Peng, 2004). The versatility of ANK repeats in macromolecular recognition has been exploited using designed ankyrin repeats (DARPINS) expressed in bacteria, which provide an alternative to antibodies with biomedical applications, including diagnostics and drug delivery (Stumpp & Amstutz, 2007).

The ankyrin-binding motifs identified so far are relatively short peptide sequences (10–20 residues) that are distinct in their primary sequence but share a lack of secondary structure. For example, the ankyrin-binding activity of the erythrocyte anion exchanger is due to two loops evident in its crystal structure (Chang & Low, 2003; Grey et al., 2012), the cytoplasmic domains of E-cadherin and L1 CAMs are established to be natively unstructured by biophysical methods (Huber, Stewart, Laurents, Nelson, & Weis, 2001; Zhang et al., 1998), and sites of Nav channels, KCNQ2/3 channels, RhBG ammonium transporter, and beta-dystroglycan are predicted to be unstructured (Bennett & Healy, 2009). The ANK repeat groove can bind peptides based on an atomic structure of erythrocyte ankyrin where the

Table 1.1 Representative independently evolved ankyrin-binding sites of membrane-spanning proteins

Family	*Ankyrin Partner*	*Ankyrin Binding Motif*	*Present*	*~ mybp*
Cell adhesion molecules	Dystroglycan	GVPIIFADELDDSK	Cnidarian	580
	NF/L1CAMs	QFNEDGSFIGQY	Bilaterian	555
	E-cadherin	KEPLLPPEDDTRDNVYYYDEE	Urochordate	520
Membrane Transporters	NaV	VPIAVGESDFE	Cephalochordate	500
	KCNQ2	PYIAEGESDTDSD	Jawed fish	440
	Kir6.2	VPIVAEED	Jawed fish	440
	CNG beta	PEPGEQILSVKMPE	Jawed fish	440
	AE1	ELVMDEKNQEL...PAVLTRSGDPS	Mammals	200
	RhBG	KLPFLDSPP	Mammals	200

Ankyrin-binding sites in cell adhesion molecules and membrane transporters: dystroglycan (Ayalon et al., 2008); L1 CAMS (Zhang, Davis, Carpenter, & Bennett, 1998), E-cadherin (Jenkins et al., 2013; Kizhatil, Davis, et al., 2007), voltage-gated sodium channels (Garrido et al., 2003; Lemaillet, Walker, & Lambert, 2003), KCNQ2/3 channels (Pan et al., 2006), Kir6.2 (Kline et al., 2009), cyclic nucleotide-gated channel beta-subunit (Kizhatil, Baker, et al., 2009), anion exchanger 1 (Grey, Kodippili, Simon, & Low, 2012), and RhB/G ammonium transporter (Lopez et al., 2005). Residues that are critical for ankyrin binding are depicted in red. Residues required for interacting with other partners are illustrated in green. (See Color Plate.)

linker following the ANK repeats was resolved lying in the groove (Michaely et al., 2002). However, a structure of ANK repeats in a complex with a membrane-spanning protein is not yet available.

Natively unstructured motifs in general are widely utilized in protein recognition and are rapidly evolving in eukaryotic genomes (Dyson & Wright, 2005). Such a code offers multiple advantages, including ease of evolution of new partners as well as capacity for integrating interacting pathways through multiple partners. It is of interest with respect to multitasking that the ankyrin-binding motif of E-cadherin contains dileucine residues that are required for clathrin-dependent endocytosis, but do not participate in ankyrin binding (Jenkins et al., 2013). The E-cadherin ankyrin-binding motif thus is better described as a polarity motif that utilizes both ankyrin binding for retention and clathrin for editing to maintain E-cadherin apical–lateral polarity (Jenkins et al., 2013).

Efforts to determine epistatic relationships between spectrin and ankyrin have had mixed results. In *Drosophila* epithelial development, spectrin acts either upstream or independently from ankyrin (Das et al., 2006, 2008; Dubreuil, Wang, Dahl, Lee, & Goldstein, 2000). However, ankyrin-G recruits beta-4 spectrin to axon initial segments and nodes of Ranvier (Jenkins & Bennett, 2001; Komada & Soriano, 2002; Yang, Ogawa, Hedstrom, & Rasband, 2007), and ankyrin-B directs beta-2 spectrin to an intracellular compartment in cardiomyocytes (Mohler, Yoon, et al., 2004). Both ankyrin-G and beta-2 spectrin are required for biogenesis of epithelial lateral membranes, although ankyrin-G lacking beta-spectrin-binding activity still associates with lateral membranes (Kizhatil, Yoon, et al., 2007). These observations suggest that ankyrin and spectrins should be viewed as obligatory partners in an interactive network rather than individual components of a linear pathway.

Which came first, spectrin or ankyrin? The answer to this question is clearly spectrin and is facilitated by mapping binding sites of spectrin for ankyrin (Davis et al., 2009; Ipsaro et al., 2009; Ipsaro & Mondragón, 2010) and of ankyrin for spectrin (Ipsaro & Mondragón, 2010; Mohler, Yoon, et al., 2004) (Fig. 1.4). The most ancient beta-spectrins are evident in sponge and Placozoa genomes. These spectrins were of the same length as vertebrate spectrins, had tandem calponin homology domains and a PH domain, but lacked a recognizable ankyrin-binding site, which includes a critical tyrosine at position 1874 (Davis et al., 2009; Ipsaro & Mondragón, 2010). In addition, these organisms also express a much larger protein termed beta-H spectrin, with homologues in *Drosophila* (Dubreuil,

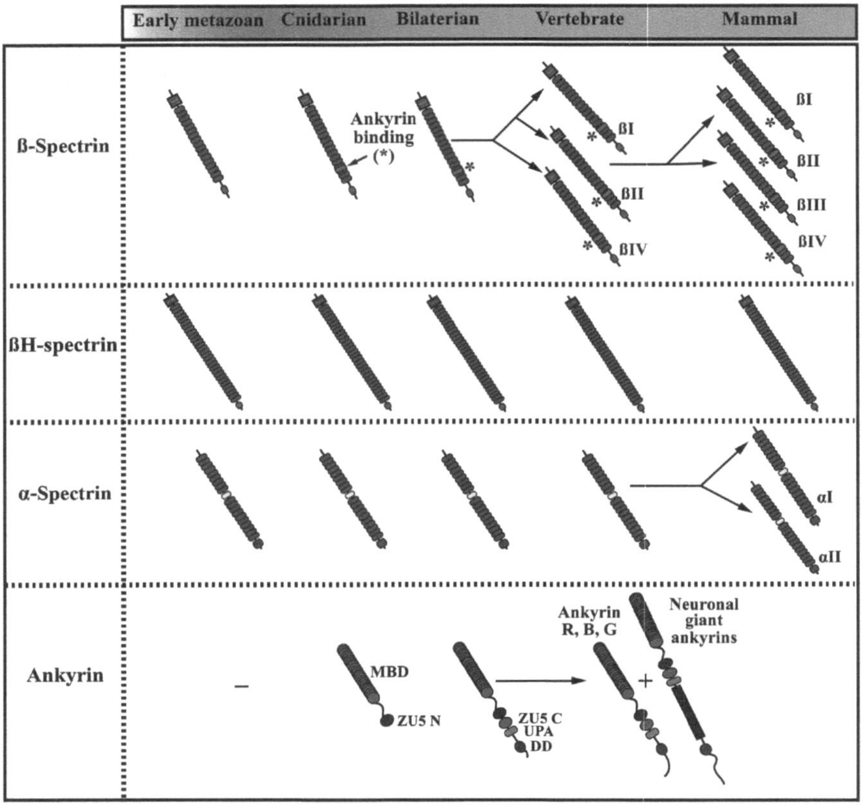

Figure 1.4 Evolution of ankyrin and spectrin protein diversity in vertebrates. Vertebrate ankyrin and spectrin proteins have diversified through evolution as a result of gene duplication, insertion of new sequences, and alternative splicing events. (See Color Plate.)

Byers, Stewart, & Kiehart, 1990), *C. elegans* (McKeown, Praitis, & Austin, 1998) and vertebrates (*SPTBN5* or beta-5 spectrin) (Stabach & Morrow, 2000). Beta-H spectrins lack ankyrin-binding activity and their functions include intracellular trafficking in mammalian photoreceptors (Papal et al., 2013) and *Drosophila* epithelial cells (Phillips & Thomas, 2006) as well as participation in apical domains of epithelial cells (Médina et al., 2002).

Cnidarians have a beta-spectrin with a potential ankyrin-binding site, which has not been evaluated experimentally (Fig. 1.4). In addition, cnidarian genomes also contain ankyrin-like sequences, including a ZU5 domain with a potential spectrin-binding site containing a characteristic DARGG motif. Ankyrins with all modern folded domains (ANK repeats, tandem

ZU5 domains, UPA domain, and death domain) are present throughout bilaterians (Fig. 1.4). Thus, a fully functional spectrin and ankyrin system with potential for lateral organization of membrane-spanning proteins and capable of cell–cell and cell–matrix interactions into micron-scale mechanically resilient domains likely was in place in the precambrian, by the time of kimberella, the first bilaterian fossil dated to 555 million years ago (Martin et al., 2000).

3. DIVERSIFICATION OF VERTEBRATE ANKYRINS AND SPECTRINS

The basic bilaterian repertoire of a single copy of alpha-spectrin, one beta-spectrin with ankyrin-binding activity and a single ankyrin, has been markedly expanded in vertebrates (Figs. 1.4 and 1.5). Ankyrins have provided a major source of diversification due to gene duplications resulting in three ankyrin genes, insertion of new nervous system-specific exons, divergence of regulatory exons, and alternative splicing. These events have resulted in the acquisition of new functions as well as the partition of old functions between the duplicated genes. An additional driver of diversity has resulted from ANK repeats and their ability to bind to intrinsically unstructured peptides, as discussed in the preceding text. The number of spectrin genes also has expanded to two alpha-spectrins (one expressed only in mammalian erythrocytes) and four beta-spectrins with ankyrin-binding activity. Together, these mechanisms have fueled a dramatic expansion of the roles of the ankyrin–spectrin partnership in vertebrate physiology.

Ohno's conjecture that vertebrates experienced two rounds of whole genome duplication has been validated based on genomic data (Dehal & Boore, 2005; Kasahara, 2013; Ohno, 1970, 1999). Many duplicated genes were lost and only a minority has persisted in modern genomes. Ankyrins and spectrins are among those gene families that expanded by whole genome duplications and are candidates to play special roles in vertebrate adaptations. In addition, genome duplications also expanded ankyrin-binding partners, including L1 CAMs and cadherins, that had evolved binding activity prior to duplication events, but these will not be discussed further.

The three vertebrate ankyrins originated from a single ankyrin gene present in urochordates (Cai & Zhang, 2006) and include *ANK1*, encoding ankyrin-R, first characterized in red blood cells (Bennett & Stenbuck, 1980b; Lux, John, & Bennett, 1990); *ANK2*, encoding ankyrin-B, first

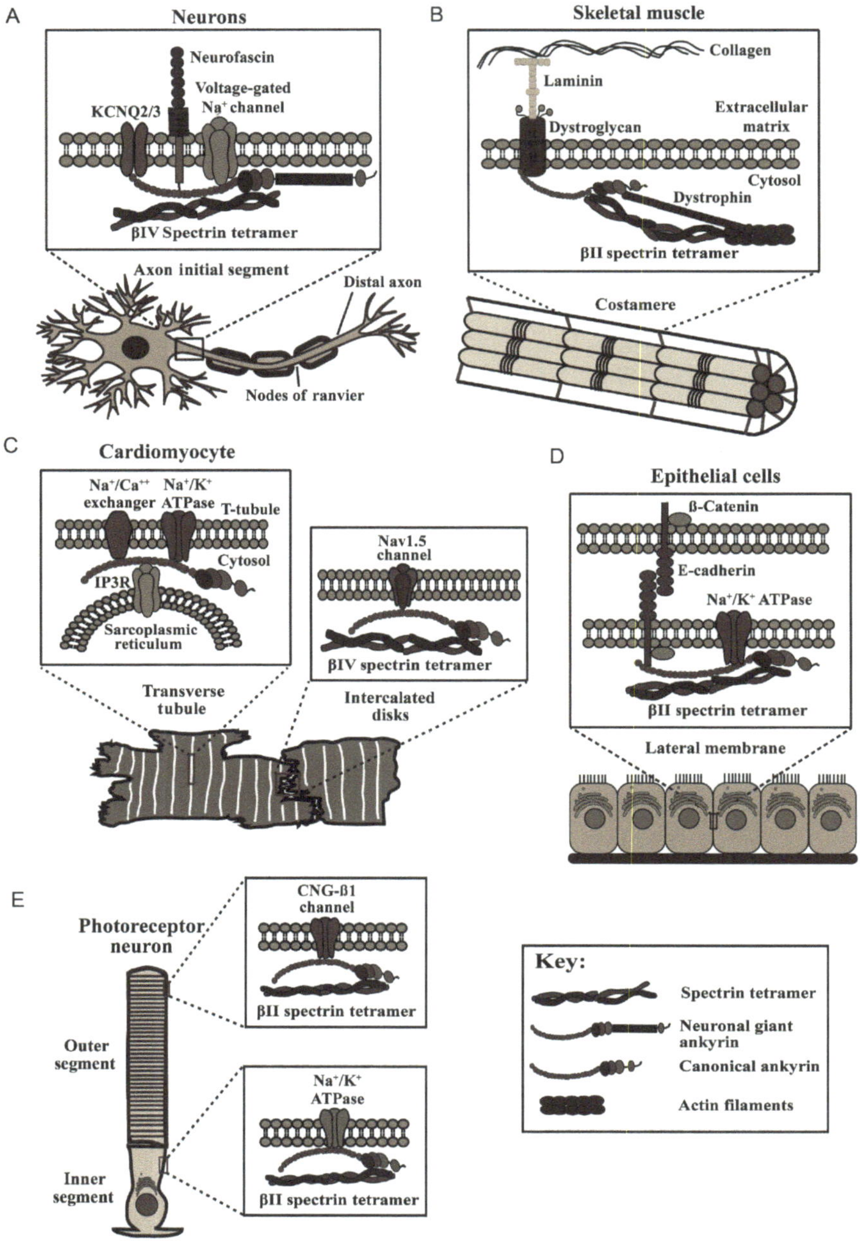

Figure 1.5 Vertebrate variations employing a core ankyrin–spectrin mechanism for coordinating membrane-spanning proteins into membrane domains. The ankyrin–spectrin network targets and stabilizes membrane transporters, cell adhesion molecules, and the dystrophin/dystroglycan complex to specialized membranes, including the axon initial segment of neurons (A), costameres of skeletal muscle (B), transverse tubules and intercalated disks of cardiomyocyte (C), lateral membranes of epithelial cells (D), and the outer and inner segments of photoreceptor neurons (E). (See Color Plate.)

characterized in the brain (Davis & Bennett, 1984; Otto, Kunimoto, McLaughlin, & Bennett, 1991); and *ANK 3,* encoding ankyrin-G, initially characterized in the brain (Kordeli, Lambert, & Bennett, 1995) as well as epithelial tissues (Peters et al., 1995; Thevananther, Kolli, & Devarajan, 1998). Phylogenetic tree analysis indicates that the first duplication event resulted in *ANK1* and the precursor of *ANK2* and *ANK3*, while the second event resulted in *ANK2* and *ANK3* but loss of the duplicate of *ANK1* (Cai & Zhang, 2006).

Vertebrate ankyrins retain extensive sequence similarity in their core folded domains but are divergent in intrinsically unstructured regulatory sequences. A major site of variation is in the unstructured C-terminal domain that modulates interactions with membrane proteins and spectrin through direct interactions with ANK repeats and the spectrin-binding domain (Abdi, Mohler, Davis, & Bennett, 2006; Davis, Davis, & Bennett, 1992; Hall & Bennett, 1987; Mohler, Gramolini, & Bennett, 2002). Another site of regulation and sequence divergence between ankyrin-G (encoded by *ANK3*) and ankyrin-B (encoded by *ANK2)* is located in the linker peptide connecting ANK repeats with the first ZU5 domain (He, Tseng, & Bennett, 2013). This peptide associates with ANK repeats and prevents binding of ankyrin-B with neurofascin and E-cadherin as well as association of ankyrin-B with the plasma membrane (He et al., 2013).

Alternative splicing adds a major source of functional diversity for vertebrate ankyrins (Cunha, Le Scouarnec, Schott, & Mohler, 2008; Cunha & Mohler, 2008; Hall & Bennett, 1987; Hopitzan, Baines, & Kordeli, 2006; Hopitzan, Baines, Ludosky, Recouvreur, & Kordeli, 2005, Lux et al., 1990; Otto et al., 1991; Peters et al., 1995). For example, an in-frame splice in the regulatory domain of ankyrin-R (band 2.2 in erythrocyte membranes) results in elimination of an acidic 186-residue segment and increased affinity for the anion exchanger as well as for spectrin (Davis et al., 1992; Hall & Bennett, 1987; Lux et al., 1990). Splicing within the regulatory domain of ankyrin-B regulates association with obscurin and the cochaperone Hsp40 (Cunha & Mohler, 2008). Ankyrin-G (and likely ankyrin-B) polypeptides include spliced variants lacking ANK repeats altogether (Hopitzan et al., 2005; Peters et al., 1995). These truncated polypeptides retain spectrin-binding domains and associate with intracellular organelles (Hoock, Peters, & Lux, 1997), although their functions are not known. The most extreme example of alternative splicing is small ankyrin1 that has lost both ANK repeats, as well as ZU5 and UPA domains, and retains only a C-terminal domain (Ackermann et al., 2011; Zhou et al., 1997).

Interestingly, vertebrate ankyrins share the unique feature that their ANK repeats are encoded by exons that begin and end at the same amino acid residue within the repeat sequence (Cai & Zhang, 2006; Cunha et al., 2008). In principle, it is thus possible, to generate transcripts encoding different numbers and linear combinations of ANK repeats that would still fold into extended although shorter solenoids. These variants would be predicted to both lose interactions and potentially to gain new partners due to juxtaposition of otherwise separated ANK repeats. Alternatively spliced variants lacking internal ANK repeats have been reported for ankyrin-B (Cunha et al., 2008), although the functional properties of the predicted polypeptides remain to be evaluated. A challenge in understanding the full scope of ankyrin diversity due to alternative splicing will be to determine actual exon usage in full-length transcripts. New sequencing technologies may be required to achieve this goal, given the diversity, low abundance, and large sizes of transcripts (4–13 kB), combined with incomplete annotation (Cunha et al., 2008; Otto et al., 1991; Peters et al., 1995).

Both *ANK2* and *ANK3* have a vertebrate-specific giant exon inserted at identical sites between their UPA and death domains that is subject to alternative splicing and is selectively expressed in the cells of the neuronal lineage (Fig. 1.3; Chan, Kordeli, & Bennett, 1993; Kordeli et al., 1995; Kunimoto, 1995; Kunimoto, Otto, & Bennett, 1991). This giant exon encodes 2085 amino acids in ankyrin-B and 2608 residues in ankyrin-G. The inserted exon of ankyrin-G also includes an N-terminal 40 kDa region absent from ankyrin-B that is enriched in serine and threonine residues and is modified by O-GlucNac monosaccharide residues (Vosseller et al., 2006; Zhang & Bennett, 1996). Ankyrin-G isoforms also include a 270 kDa polypeptide missing the C-terminal portion of this exon due to in-frame splicing (Kordeli et al., 1995). Acquisition of the giant exon likely occurred in the precursor to *ANK2* and *ANK3* genes prior to the second whole genome duplication event based on sequence similarity of inserted domains of ankyrin-B and ankyrin-G and precise maintenance of the insertion sites in the two genes.

480/270 kDa ankyrin-G and 440 kDa ankyrin-B polypeptides bearing giant exons are localized in axons (Chan et al., 1993; Kordeli et al., 1995; Kunimoto, 1995; Kunimoto et al., 1991). 440 kDa ankyrin-B is the predominant isoform in the neonatal brain prior to myelination but is largely replaced by 220 kDa ankyrin-B in adult rodents (Chan et al., 1993; Kunimoto, 1995). 480 and 270 kDa ankyrin-G variants containing the axonal exon are targeted to axon initial segments as well as nodes of Ranvier, although their specialized functions remain to be established (Kordeli et al., 1995; Zhang & Bennett, 1998). Interestingly, the closest mammalian matches to the giant axonal exons in Blast searches are proteins related to Titin, which have multiple Ig- and Fn3-like domains and perform mechanical roles in stabilizing sarcomeres of striated muscle. The functional importance of this domain is underscored by a report of a truncating mutation within the axonal exon of ankyrin-G that results in cognitive and behavioral deficits in humans (Iqbal et al., 2013).

Phenotypes of mice with knockout or deficiency of *ANK1*, *2*, and *3* indicate that the three vertebrate ankyrin genes have acquired distinct and nonoverlapping functions (Abdi et al., 2006; He et al., 2013; Mohler et al., 2002). For example, mice lacking ankyrin-R due to an ENU-induced null mutation in the *ANK1* gene die perinatally due to profound anemia and loss of the spectrin-based membrane skeleton (Rank et al., 2009). Ankyrin-G and ankyrin-B are coexpressed in neurons and striated muscle, where they have nonredundant functions, but also can collaborate in activities such as maintenance of costamere structure and organization of the axonal spectrin–actin skeleton (Ayalon et al., 2008; Galiano et al., 2012).

Alpha- and beta-spectrin genes have duplicated in parallel with the ankyrins (Fig. 1.4). Vertebrates, with the exception of mammals, have a single alpha-spectrin gene, *SPTAN1*, which is the generally expressed partner of beta-spectrins (Bennett et al., 1982, Davis & Bennett, 1983; Wasenius et al., 1989). In addition, mammals express a second alpha-spectrin gene, *SPTA1*, that is expressed primarily in erythrocytes and has reduced ability to assemble into tetramers (Mehboob et al., 2010; Salomao et al., 2006). Vertebrates, except for mammals, have three beta-spectrins genes: *SPTB*, encoding beta-1 spectrin, first characterized in red blood cells (Winkelmann, Chang, et al., 1990); *SPTBN1*, encoding beta-2 spectrin, first characterized in the brain (Bennett et al., 1982; Hu, Watanabe, & Bennett, 1992); and *SPTBN4,* encoding beta-4 spectrin, that is localized with 480 kDa ankyrin-G at nodes of Ranvier and axon initial segments (Berghs et al., 2000), as well as at intercalated disks of cardiomyocytes (Hund et al., 2010).

Mammals have an additional beta-spectrin gene, *SPTBN2*, encoding beta-3 spectrin, that was first identified in the brain (Ohara, Ohara, Yamakawa, Nakajima, & Nakayama, 1998; Stankewich et al., 1998) where it is most highly expressed in the cerebellum and is mutated in human spinocerebellar ataxia type 5 (Ikeda et al., 2006). Knockout of beta-3 spectrin in mice recapitulates human spinocerebellar ataxia symptoms and results in abnormal development of Purkinje cerebellar neurons (Gao et al., 2011; Perkins et al., 2010). In contrast, knockout of the generally expressed beta-2 spectrin markedly impairs development and is embryonic lethal (Tang et al., 2003).

While beta-1–3 spectrins are overall similar in sequence and domain organization, beta-4 spectrin has additional sequence between the final spectrin repeat and the PH domain (Berghs et al., 2000). Beta-4 spectrin also associates with calmodulin-dependent protein kinase 2 (CAM kinase 2) through this sequence and recruits CAM kinase 2 to cardiac intercalated disks and axon initial segments (Hund et al., 2010).

Alternative splicing increases the diversity of both alpha- (Zhang et al., 2010) and beta-spectrins (Berghs et al., 2000; Hayes et al., 2000; Winkelmann, Costa, et al., 1990). Variations include deletion of PH domains in beta-1 and beta-2 spectrins (Hayes et al., 2000; Winkelmann, Costa, et al., 1990) and deletion of either N-terminal or C-terminal portions of beta-4 spectrin (Berghs et al., 2000; Tse et al., 2001; Uemoto et al., 2007).

4. EVOLUTION OF SPECTRIN–ANKYRIN-BASED DOMAINS: LESSONS FROM THE AXON INITIAL SEGMENT

Axon initial segments with their dense clusters of ion channels, CAMs, and synaptic endings are responsible for both generation and modulation of action potentials (Chang & Rasband, 2013). This membrane domain and the closely related nodes of Ranvier of myelinated axons underlie the evolution of fast signaling in vertebrates and are the best understood ankyrin–spectrin-based membrane structures. Ankyrin-G is required for action potential initiation and for localization of all of the known initial segment components (Hedstrom et al., 2007; Jenkins & Bennett, 2001; Kordeli et al., 1995; Zhang & Bennett, 1998; Zhou et al., 1998). These ankyrin-G-dependent initial segment proteins include voltage-gated sodium channels (Hedstrom et al., 2007; Jenkins & Bennett, 2001), KCNQ2/3 channels that modulate sodium channel activity (Chung, Jan, & Jan, 2006; Cooper, 2011; Pan et al., 2006), 186 kDa neurofascin, a L1 CAM that directs GABAergic synapses to the initial segment (Ango et al., 2004; Jenkins & Bennett, 2001), and beta-4 spectrin, which stabilizes initial segments (Komada & Soriano, 2002; Lacas-Gervais et al., 2004; Yang et al., 2007). Moreover, ankyrin-G is also required to form microtubule bundles at the initial segment (Sobotzik et al., 2009). Consistent with the findings that multiple initial segment proteins depend on ankyrin-G, ankyrin-G-null axons acquire properties of dendrites both in cultured neurons and in mice (Hedstrom et al., 2008; Sobotzik et al., 2009).

The phylogenetic record of evolution of ankyrin-binding motifs in L1 CAMs and in ion channels provides a series of molecular "snapshots" of the events that ultimately resulted in emergence of axon initial segments and fast signaling in vertebrates. The L1 family of CAMs was the first among initial segment components to acquire ankyrin-binding activity. L1 CAM family members are expressed throughout modern bilaterian organisms and likely had evolved over 550 mybp in the Ediacaran era proceeding the Cambrian period. L1 CAMs are represented in *C. elegans* by LAD1, encoded by the *Sax-7* gene, and in *Drosophila* by neuroglian (Chen, Ong, & Bennett, 2001; Hortsch, Nagaraj, & Godenschwege, 2009). L1 CAMs all have a highly conserved ankyrin-binding motif, including the residues FIGQY (Chen et al., 2001; Garver, Ren, Tuvia, & Bennett, 1997; Zhang et al., 1998). Phosphorylation of the FIGQY tyrosine eliminates ankyrin binding, which is promoted in *C. elegans* by FGF receptor signaling (Chen et al., 2001; Garver et al., 1997). The *C. elegans* L1 CAM is localized with ankyrin (Unc44) at sites of cell–cell contact in multiple cell types (Chen et al., 2001). The adhesive functions of the *C. elegans* L1 CAM include correct positioning of neuronal cell bodies and axons, and the FIGQY ankyrin-binding motif is both necessary and sufficient for these activities (Pocock, Bénard, Shapiro, & Hobert, 2008). The *Drosophila* L1 CAM, neuroglian, has roles in coordinating synaptic connections and again these functions require ankyrin-binding activity (Enneking et al., 2013; Godenschwege, Kristiansen, Uthaman, Hortsch, & Murphey, 2006; Hortsch et al., 2009).

Voltage-gated sodium channels acquired an ankyrin-binding motif following evolution of the L1 CAM site, and this occurred early in chordate evolution, likely in the Cambrian period between 550 and 500 mybp (Hill et al., 2008). KCNQ2/3 channels were the next initial segment channel to gain ankyrin-binding activity, which occurred in jawed fish at the end of the Ordovician period around 450 mybp (Hill et al., 2008). Cooper and colleagues have reported an extensive phylogenetic analysis of ankyrin-binding motifs of both the voltage-gated sodium channels and KCNQ2/3 channels (Hill et al., 2008). The most distant organism with a sodium channel containing a recognizable ankyrin-binding motif was Amphioxus, which is a cephalochordate (Hill et al., 2008). Interestingly, even though the ankyrin-binding motifs of KCNQ2/3 and voltage-gated sodium channels are similar, these sequences most likely evolved independently (Hill et al., 2008).

The sequential and independent evolution of ankyrin-binding activities beginning with the L1 CAMs suggests a scenario where a proto-axon initial segment containing ankyrin and an L1 CAM evolved first and then was populated by ion channels, first the voltage-gated sodium channel and later by KCNQ2/3 channels. The selective pressure presumably was the advantages of faster signaling and its regulation. Higher concentration of sodium channels would have resulted in increased current density and greater amplitudes of depolarization that eventually became self-renewing action potentials. Evolutionary selection likely acted on ankyrin in parallel with its binding partners and thus development of the axon initial segment was an iterative as well as a sequential process. For example, urochordates have sodium channels with an ankyrin-binding motif but a single ankyrin gene that lacked the axonal-specific exons acquired later in vertebrate giant ankyrins. The giant exon of vertebrate 480 kDa ankyrin-G thus evolved in the context of axonal expression and clustering of sodium channels and now likely has axon initial segment-specific capabilities.

The axon initial segment with its clustered sodium channels likely was present in jawless fish, now represented by lampreys, and thus predated myelination and development of nodes of Ranvier, which appeared only in jawed fish (Hill et al., 2008). Interestingly, even though there are many parallels between the composition of nodes of Ranvier and initial segments (Davis, Lambert, & Bennett, 1996), these domains exhibit distinct mechanisms of assembly (Dzhashiashvili et al., 2007; Susuki et al., 2013). Thus, the basic ankyrin interactome of the axon initial segment was co-opted and modified during the process of myelination, resulting in closely related excitatory domains at nodes of Ranvier. A similar process of co-option likely resulted in utilization of ankyrin voltage-gated sodium channel interaction in intercalated disks and T-tubules in mammalian cardiomyocytes, where Nav1.5 requires ankyrin-binding activity for cell surface expression (Lowe et al., 2008; Mohler, Rivolta, et al., 2004). In this example of the heart, an ankyrin–sodium channel coupling occurs in T-tubules that lack cell–cell interactions and are present only in mammals. Thus, the initial connection of ankyrin with L1 CAMs that occurred in ancient axons and may have facilitated interaction with sodium channels no longer is required for ankyrin partners in the heart and possibly other tissues.

More generally, CAMs are the earliest examples of membrane-spanning proteins with ankyrin-binding motifs (Table 1.1). Dystroglycan has an ankyrin-binding site that retains critical residues in cnidarians, while the cadherin ankyrin-binding site is present in the urochordate *Ciona intestinalis* (Jenkins et al., 2013). In contrast, ion channel binding sites have continued to evolve in terrestrial vertebrates, where the RhB/G ammonium transporter has an ankyrin-binding motif present in mammals but is not conserved at critical residues in chickens (Lopez et al., 2005; Table 1.1). These considerations suggest a model for ankyrin-based membrane domains where the initiating event was the formation of a proto-domain through interaction of membrane-associated ankyrin and spectrin with membrane-spanning CAMs that in turn engaged in transcellular and cell–matrix interactions. These proto-domains then became populated with ion channels and likely other molecules that developed ankyrin-binding activity at sites determined by extracellular signals with selective pressure provided by optimization of physiological function.

We have focused on axon initial segments where the physiological adaptation would be efficient generation of action potentials and fast signaling. Ankyrin, along with L1 CAMs and cadherins, also is localized at sites of cell–cell contact in other cell types including epithelial tissues (Chen et al., 2001). Thus, it is conceivable that epithelial lateral membranes served as an "incubator" similar to the axon initial segment, where membrane transporters could have acquired ankyrin-binding activity with the advantage of increased efficiency through polarized fluxes of ions and nutrients. An apparent exception to the initiating role of CAMs in this scheme is the circulating erythrocyte, which contains only ion transporters as ankyrin-binding partners and does not engage in interactions with other cells. However, it is possible that an ankyrin–anion exchanger interaction evolved first in an epithelial cell type and later was co-opted in erythrocytes.

Ankyrins, with exception of the giant axonal variants, have remained relatively unchanged while their partners have evolved ankyrin-binding activity. An essential feature of this model is the ease of evolution of ankyrin-binding motifs, which require no folded structure and only 10–20 amino acids. It will be of interest in the future to understand the structural basis for ANK repeat association with peptides and to use this information to systematically identify potential partners through bioinformatics.

5. FUNCTIONS OF SPECTRIN AND ANKYRIN IN POLARIZED ORGANELLE TRANSPORT

An implication of the proposed role of CAMs in recruitment of a pre-existing ankyrin–spectrin skeleton is that ankyrin and spectrin may have other functions that occur in parallel and/or predated their association with the plasma membrane. In fact, multiple lines of evidence support roles of ankyrin and spectrin in polarized organelle transport. Brain spectrin was initially characterized by Willard and colleagues based on its association with multiple classes of axonally transported organelles (Levine & Willard, 1981). The kinesin KIF3 associates with alpha-spectrin through its light chain Kap3, and this interaction was implicated in organelle transport required for neurite extension (Takeda et al., 2000). Actin-related protein 1 (Arp1) of the dynactin complex, which interacts with both dynein and kinesin motors, associates with beta-spectrin through its calponin homology domains (Holleran et al., 2001, Holleran, Tokito, Karki, & Holzbaur, 1996). Arp1 shares sequence and folding similarity to actin, and calponin homology domains of beta-spectrin likely interact with residues conserved between these proteins. Interestingly, overexpression of beta-3 spectrin bearing mutations that cause spinocerebellar ataxia type 5 impairs axonal transport in *Drosophila* larvae (Lorenzo et al., 2010). Moreover, beta-spectrin together with dynactin and dynein is sufficient to reconstitute motility of liposomes lacking membrane proteins (Muresan et al., 2001). In addition to axonal transport, *Drosophila* spectrin, ankyrin, adducin, and tropomodulin are all components of an intracellular microtubule-based structure termed the fusome, which delivers organelles and RNAs from nurse cells to oocytes (de Cuevas, Lee, & Spradling, 1996; Lighthouse, Buszczak, & Spradling, 2008; Lin, Yue, & Spradling, 1994).

Ankyrin in *C. elegans* and *Drosophila* determines polarized organelle transport in dendrites and axons through organization of microtubules (Koch et al., 2008; Maniar et al., 2011; Pielage et al., 2008). A dual role of invertebrate ankyrins in both cell surface and intracellular functions appears to have largely been subdivided between ankyrin-B and ankyrin-G vertebrates. Ankyrin-G primarily is localized to plasma membrane domains, while ankyrin-B associates with intracellular membranes despite a high level of sequence similarity (He et al., 2013). This behavior results from loss of plasma membrane activity of ankyrin-B due to an ankyrin-B-specific linker peptide connecting the ankyrin repeat domain to the ZU52–UPA module that inhibits binding of ankyrin-B to membrane protein partners E-cadherin and neurofascin and prevents association of ankyrin-B with epithelial lateral membranes as well as the axon initial segment. The residues of the ankyrin-B linker required for autoinhibition are encoded by a small exon that is highly divergent between ankyrin family members but conserved in the ankyrin-B lineage. These considerations argue against neofunctionalization of ankyrin-B and rather support partition of the dual functions of the ancestral single ankyrin following gene duplication in the vertebrate lineage (He et al., 2013).

Further support for partition of plasma membrane and intracellular functions between ankyrin-B and ankyrin-G comes from the coordinated but distinct role of these proteins in organization of dystrophin and dystroglycan at costameres of striated muscle (Ayalon et al., 2008, 2011). Ankyrin-G binds directly to dystroglycan and dystrophin and stabilizes these proteins at costameres, but is not required for their accumulation at the plasma membrane. Ankyrin-B, in contrast, does not bind directly to dystroglycan, but is required for transport of dystroglycan from an intracellular compartment to the plasma membrane, perhaps through association with dystrophin. Ankyrin-B also directs a specialized population of microtubules to costameres through binding to p62/dynactin-4 subunit of the dynactin complex (Ayalon et al., 2008, 2011). Interestingly, p62/dynactin-4 and associated proteins at the minus end of the dynactin protofilament are proposed to couple the dynactin complex to membrane cargo (Yeh, Quintyne, Scipioni, Eckley, & Schroer, 2012). Dynactin engages both kinesin and dynein motors in transport of membrane cargo in axons and likely other examples of directed organelle movement (Schroer, 2004). It will be of interest to determine if ankyrin-B plays a general role in linking organelles to the dynactin complex and the evolutionary origins of this function.

Ankyrin-B also interacts with members of the family of Eps15 homology (EH) domain containing (EHD)/receptor-mediated endocytosis (Rme) proteins, which are ATPases related to dynamin that are involved in endosomal recycling (Daumke et al., 2007; Gudmundsson et al., 2010). EHD/Rme1–4 proteins exhibit increased cardiac expression in ankyrin-B haploinsufficient mice and in acquired ankyrin-B-deficiency that occurs in heart failure (Gudmundsson et al., 2010, 2012). These observations suggest that ankyrin-B, which has been viewed as a T-tubule-associated structural protein that stabilizes the Na/Ca exchanger and Na/K ATPase (Mohler et al., 2005), may also have a dynamic role in directing intracellular trafficking of these transporters. Live cell imaging of ankyrin-B, ankyrin-B membrane partners, and EHD/Rme proteins has not yet been reported but could give insight into the nature or their interaction.

6. SUMMARY AND PERSPECTIVES

This chapter provides a plausible scenario for evolution of ankyrin- and spectrin-based plasma membrane domains in vertebrates. The analysis integrates recent phylogenetic information with extensive functional analyses of spectrin and ankyrin beginning in human erythrocytes and more recently in excitable membranes in neurons and heart, lateral membrane domains of epithelial cells, and costameres of striated muscle. A membrane-associated spectrin–actin network likely was the starting point in early metazoans. Alpha-spectrin, and a beta-spectrin with the same length as modern beta-spectrins with calponin homology domains, and a PH domain are present in sponges and Placozoa. These spectrin subunits presumably are capable of forming tetramers, binding to phosphoinositide lipids, and of forming a network through association with actin, but lack an ankyrin-binding site, and precede appearance of ankyrin. In addition, a large beta-related spectrin lacking ankyrin-binding activity, termed beta-H, also was present and has a homologue in vertebrates. Both beta- and beta-H-spectrins also likely functioned in polarized organelle transport and this activity either preceded or has evolved in parallel with their association with plasma membranes.

The potential of spectrin–actin networks to recruit and coordinate membrane-spanning proteins was greatly amplified in cnidarians by the emergence of a partnership with ankyrins with their highly versatile ANK repeat domain configured as a solenoid with an extended peptide-binding groove (Fig. 1.5). The most ancient currently known membrane proteins with ankyrin-binding peptide motifs engage in cell–cell and cell–matrix interactions. These ankyrin partners include dystroglycan, present in cnidarians; the L1 family of CAMs, present in bilaterians; and cadherins, with an ankyrin-binding motif, evident in early chordates. Functions of these early plasma membrane-associated ankyrin–spectrin–actin networks likely included mechanical support for the membrane bilayer, an activity still evident in mammalian erythrocytes (Eber & Lux, 2004) and in *C. elegans* axons (Hammarlund et al., 2007).

The next phase in evolution of functional membrane domains was the independent acquisition of ankyrin-binding activity by diverse membrane transporters. This process was greatly facilitated by the ability of ankyrin to associate with unstructured peptides. The vertebrate axon initial segment with its ankyrin-G-dependent characteristics and clear physiological benefit of fast signaling with small diameter axons provides an instructive case history. Ankyrin is associated with L1 CAMs in axons of *C. elegans* and *Drosophila*, which either lack voltage-gated sodium channels altogether (*C. elegans*) or have sodium channels lacking an ankyrin-binding motif (*Drosophila*) (Hill et al., 2008). Sodium channels developed a recognizable ankyrin-binding motif in early chordates and were followed by KCNQ2/3 channels, which independently acquired their ankyrin-binding motif nearly 100 million years later, during the emergence of jawed fish (Hill et al., 2008). The principal ion channels clustered at vertebrate axon initial segments thus evolved ankyrin-binding activity sequentially and independently through rather modest changes over a short stretch of protein sequence. This adaptive process presumably was favored by selective advantages of clustering sodium channels to generate action potentials and modulating their excitability by KCNQ2/3 channels. We speculate that other proto-domains based on spectrin–ankyrin–CAM assemblies served a similar role as "incubators," where ion transporters and likely other yet to be identified membrane-spanning proteins acquired ankyrin-binding activity through convergent evolution. One candidate incubator is the lateral membrane domain of epithelial cells, where ankyrin and L1 CAMs are colocalized in *C. elegans* and *Drosophila*. Here, a selective advantage favoring gain of ankyrin-binding activity could have arisen from increased efficiency due to vectorial transport of ions and nutrients.

Two whole genome duplication events in the vertebrate lineage have markedly expanded the repertoire of ankyrin–spectrin assemblies and their membrane-spanning partners (Figs. 1.4 and 1.5). Vertebrate spectrins and ankyrins have retained core features including ANK repeats that are highly conserved among bilaterians. However, ankyrins in particular have diversified in other respects due to gain of protein interactions of regulatory domains (both intramolecular and intermolecular), gain of large neuronal-specific exons, and through alternative splicing. The scope of ankyrin partners and their membrane domains in vertebrates as well as other organisms thus likely is quite broad and only partially appreciated at this point.

A recurrent theme in this discussion is the acquisition of new protein function through mutation for ankyrin partners and ankyrin regulatory domains or insertion of a new exon in the case of the giant axonal ankyrins. The role of adaptive evolution of protein structure has been the subject of active debate among evolutionary biologists. The evolutionary developmental (evo devo) school favors the idea that the principal origin of variation lies in cis elements of regulatory DNA that determine levels of protein expression through regulatory networks, while the proteins themselves are essentially unchanged (Carroll, 2008). This view has been countered by arguments citing examples of positive functional consequences of variation in protein sequences (Hoekstra & Coyne, 2007; Linnen et al., 2013; Nery, González, & Opazo, 2013). A major concern has been that variation in protein sequence may lead to misfolding and is highly selected against due to pleiotropic expression of most genes. However, intrinsically unstructured protein sequences, which provide ankyrin-binding sites for membrane-spanning proteins as well as interactions of ankyrin regulatory domains, are relatively tolerant of mutation since they have no folded structure. Another general concern is the challenge in establishing a direct causal connection between evolutionary variation in either protein or cis-regulatory DNA sequences and actual morphological phenotype. However, a connection between protein and phenotype is clear in the example of the vertebrate axon initial segment, which depends on ankyrin-G for its defining characteristics both in cultured neurons and in animals.

Finally, it is important to not underestimate the creativity of adaptive evolution in response to selective pressures. Striking examples of this are the morphologically similar versions of striated muscle that evolved independently in cnidarians and bilaterians (Steinmetz et al., 2012). Cnidarian and bilaterian muscles both were based on the same ancient motility proteins and solved similar problems of generating coordinated contractile force but arrived at solutions with distinct molecular organization and composition. Extrapolation of this lesson to plasma membrane domains suggests that it is likely that there also are multiple independently evolved approaches, many utilizing spectrin and ankyrin, to address the core functional problem of establishing long-range organization and mechanical stability in a fluid membrane bilayer.

REFERENCES

Abdi, K. M., & Bennett, V. (2008). Adducin promotes micron-scale organization of beta 2-spectrin in lateral membranes of bronchial epithelial cells. *Molecular Biology of the Cell*, *19*, 536–545.

Abdi, K. M., Mohler, P. J., Davis, J. Q., & Bennett, V. (2006). Isoform specificity of ankyrin-B: A site in the divergent C-terminal domain is required for intramolecular association. *The Journal of Biological Chemistry*, *281*(9), 5741–5749.

Ackermann, M. A., Ziman, A. P., Strong, J., Zhang, Y., Hartford, A. K., Ward, C. W., et al. (2011). Integrity of the network sarcoplasmic reticulum in skeletal muscle requires small ankyrin 1. *Journal of Cell Science*, *124*, 3619–3630.

Al-Khodor, S., Price, C. T., Kalia, A., & Abu Kwaik, Y. (2010). Functional diversity of ankyrin repeats in microbial proteins. *Trends in Microbiology, 18*, 132–139.
An, X., Guo, X., Sum, H., Morrow, J., Gratzer, W., & Mohandas, N. (2004). Phosphatidylserine binding sites in erythroid spectrin: Location and implications for membrane stability. *Biochemistry, 43*, 310–315.
Ango, F., di Cristo, G., Higashiyama, H., Bennett, V., Wu, P., & Huang, Z. J. (2004). Ankyrin-based subcellular gradient of neurofascin, an immunoglobulin family protein, directs GABAergic innervation at purkinje axon initial segment. *Cell, 119*, 257–272.
Anong, W. A., Franco, T., Chu, H., Weis, T. L., Devlin, E. E., Bodine, D. M., et al. (2009). Adducin forms a bridge between the erythrocyte membrane and its cytoskeleton and regulates membrane cohesion. *Blood, 114*, 1904–1912.
Ayalon, G., Davis, J. Q., Scotland, P., & Bennett, V. (2008). An ankyrin-based mechanism for functional organization of dystrophin and dystroglycan. *Cell, 135*, 1189–1200.
Ayalon, G., Hostettler, J. D., Hoffman, J., Kizhatil, K., Davis, J. Q., & Bennett, V. (2011). Ankyrin-B interactions with spectrin and dynactin-4 are required for dystrophin- based protection of skeletal muscle from exercise injury. *The Journal of Biological Chemistry, 286*, 7370–7378.
Baines, A. J., Lu, H. C., & Bennett, P. M. (2013). The Protein 4.1 family: Hub proteins in animals for organizing membrane proteins. *Biochimica et Biophysica Acta,* pii: S0005-2736 (13)00183-1.
Banuelos, S., Saraste, M., & Carugo, K. D. (1998). Structural comparisons of calponin homology domains: Implications for actin binding. *Structure, 6*, 1419–1431.
Bennett, V. (1978). Purification of an active proteolytic fragment of the membrane attachment site for human erythrocyte spectrin. *The Journal of Biological Chemistry, 253*, 2292–2299.
Bennett, V., & Baines, A. J. (2001). Spectrin and ankyrin-based pathways: Metazoan inventions for integrating cells into tissues. *Physiological Reviews, 81*, 1353–1392.
Bennett, V., Davis, J., & Fowler, W. E. (1982). Brain spectrin, a membrane-associated protein related in structure and function to erythrocyte spectrin. *Nature, 299*, 126–131.
Bennett, V., & Healy, J. (2009). Membrane domains based on ankyrin and spectrin associated with cell-cell interactions. *Cold Spring Harbor Perspectives in Biology, 1*, a003012. http://dx.doi.org/10.1101/chsperpect.a003012.
Bennett, V., & Stenbuck, P. (1979a). Identification and partial purification of ankyrin, the high affinity membrane attachment site for human erythrocyte spectrin. *The Journal of Biological Chemistry, 254*, 2533–2541.
Bennett, V., & Stenbuck, P. (1979b). The membrane attachment site for spectrin is associated with band 3 in human erythrocyte membranes. *Nature, 280*, 468–473.
Bennett, V., & Stenbuck, P. J. (1980a). Association between ankyrin and the cytoplasmic domain of band 3 isolated from the human erythrocyte membrane. *The Journal of Biological Chemistry, 255*, 6424–6432.
Bennett, V., & Stenbuck, P. J. (1980b). Human erythrocyte ankyrin. Purification and properties. *The Journal of Biological Chemistry, 255*, 2540–2548.
Berghs, S., Aggujaro, D., Dirkx, R., Jr., Maksimova, E., Stabach, P., Hermel, J. M., et al. (2000). betaIV spectrin, a new spectrin localized at axon initial segments and nodes of ranvier in the central and peripheral nervous system. *The Journal of Cell Biology, 151*, 985–1002.
Byers, T. J., & Branton, D. (1985). Visualization of the protein associations in the erythrocyte membrane skeleton. *Proceedings of the National Academy of Sciences, 82*, 6153–6157.
Cai, X., & Zhang, Y. (2006). Molecular evolution of the ankyrin gene family. *Molecular Biology and Evolution, 23*(3), 550–558.
Carroll, S. B. (2008). Evo-devo and an expanding evolutionary synthesis: A genetic theory of morphological evolution. *Cell, 134*, 25–36.

Carugo, K. D., Banuelos, S., & Saraste, M. (1997). Crystal structure of a calponin homology domain. *Nature Structural Biology, 4*, 175–179.
Chan, W., Kordeli, E., & Bennett, V. (1993). 440-kD ankyrinB: Structure of the major developmentally regulated domain and selective localization in unmyelinated axons. *The Journal of Cell Biology, 123*, 1463–1473.
Chang, S. H., & Low, P. S. (2003). Identification of a critical ankyrin-binding loop on the cytoplasmic domain of erythrocyte membrane band 3 by crystal structure analysis and site-directed mutagenesis. *The Journal of Biological Chemistry, 278*, 6879–6884.
Chang, K. J., & Rasband, M. N. (2013). Excitable domains of myelinated nerves: Axon initial segments and nodes of Ranvier. *Current Topics in Membranes, 72.*
Chen, L., Ong, B., & Bennett, V. (2001). LAD-1, the Caenorhabditis elegans L1CAM homologue, participates in embryonic and gonadal morphogenesis and is a substrate for fibroblast growth factor receptor pathway-dependent phosphotyrosine-based signaling. *The Journal of Cell Biology, 154*, 841–855.
Chung, H. J., Jan, Y. N., & Jan, L. Y. (2006). Polarized axonal surface expression of neuronal KCNQ channels is mediated by multiple signals in the KCNQ2 and KCNQ3 C-terminal domains. *Proceedings of the National Academy of Sciences of the United States of America, 103*, 8870–8875.
Cooper, E. C. (2011). Made for "anchorin": Kv7.2/7.3 (KCNQ2/KCNQ3) channels and the modulation of neuronal excitability in vertebrate axons. *Seminars in Cell & Developmental Biology, 22*, 185–192.
Cunha, S. R., Le Scouarnec, S., Schott, J. J., & Mohler, P. J. (2008). Exon organization and novel alternative splicing of the human ANK2 gene: Implications for cardiac function and human cardiac disease. *Journal of Molecular and Cellular Cardiology, 45*, 724–734.
Cunha, S. R., & Mohler, P. J. (2008). Obscurin targets ankyrin-B and protein phosphatase 2A to the cardiac M-line. *The Journal of Biological Chemistry, 283*, 31968–31980.
Das, A., Base, C., Dhulipala, S., & Dubreuil, R. R. (2006). Spectrin functions upstream of ankyrin in a spectrin cytoskeleton assembly pathway. *The Journal of Cell Biology, 175*, 325–335.
Das, A., Base, C., Manna, D., Cho, W., & Dubreuil, R. R. (2008). Unexpected complexity in the mechanisms that target assembly of the spectrin cytoskeleton. *The Journal of Biological Chemistry, 283*, 12643–12653.
Daumke, O., Lundmark, R., Vallis, Y., Martens, S., Butler, P. J., & McMahon, H. T. (2007). Architectural and mechanistic insights into an EHD ATPase involved in membrane remodelling. *Nature, 449*, 923–927.
Davis, L., Abdi, K., Machius, M., Brautigam, C., Tomchick, D. R., Bennett, V., et al. (2009). Localization and structure of the ankyrin-binding site on beta 2-spectrin. *The Journal of Biological Chemistry, 284*, 6982–6987.
Davis, J., & Bennett, V. (1983). Brain spectrin. Isolation of subunits and formation of hybrids with erythrocyte spectrin subunits. *The Journa of Biological Chemistry, 258*, 7757–7766.
Davis, J. Q., & Bennett, V. (1984). Brain ankyrin. A membrane-associated protein with binding sites for spectrin, tubulin, and the cytoplasmic domain of the erythrocyte anion channel. *The Journal of Biological Chemistry, 259*, 13550–13559.
Davis, L. H., Davis, J. Q., & Bennett, V. (1992). Ankyrin regulation: An alternatively spliced segment of the regulatory domain functions as an intramolecular modulator. *The Journal of Biological Chemistry, 267*, 18966–18972.
Davis, J. Q., Lambert, S., & Bennett, V. (1996). Molecular composition of the node of Ranvier: Identification of ankyrin-binding cell adhesion molecules neurofascin (mucin+/third FNIII domain-) and NrCAM at nodal axon segments. *The Journal of Cell Biology, 135*, 1355–1367.
de Cuevas, M., Lee, J. K., & Spradling, A. C. (1996). Alpha-spectrin is required for germline cell division and differentiation in the Drosophila ovary. *Development, 122*, 3959–3968.

Dehal, P., & Boore, J. L. (2005). Two rounds of whole genome duplication in the ancestral vertebrate. *PLoS Biology*, *3*(10), e314.

Djinovic-Carugo, K., Young, P., Gautel, M., & Saraste, M. (1999). Structure of the alpha-actinin rod: Molecular basis for cross-linking of actin filaments. *Cell*, *98*, 537–546.

Dubreuil, R. R., Byers, T. J., Stewart, C. T., & Kiehart, D. P. (1990). A beta-spectrin isoform from Drosophila (beta H) is similar in size to vertebrate dystrophin. *The Journal of Cell Biology*, *111*, 1849–1858.

Dubreuil, R. R., Wang, P., Dahl, S., Lee, J., & Goldstein, L. S. (2000). Drosophila beta spectrin functions independently of alpha spectrin to polarize the Na,K ATPase in epithelial cells. *The Journal of Cell Biology*, *149*, 647–656.

Dyson, H. J., & Wright, P. E. (2005). Intrinsically unstructured proteins and their functions. *Nature Reviews Molecular Cell Biology*, *6*, 197–208.

Dzhashiashvili, Y., Zhang, Y., Galinska, J., Lam, I., Grumet, M., & Salzer, J. L. (2007). Nodes of Ranvier and axon initial segments are ankyrin G-dependent domains that assemble by distinct mechanisms. *The Journal of Cell Biology*, *177*, 857–870.

Eber, S., & Lux, S. E. (2004). Hereditary spherocytosis—Defects in proteins that connect the membrane skeleton to the lipid bilayer. *Seminars in Hematology*, *41*, 118–141.

Enneking, E. M., Kudumala, S. R., Moreno, E., Stephan, R., Boerner, J., Godenschwege, T. A., et al. (2013). Transsynaptic coordination of synaptic growth, function, and stability by the L1-type CAM Neuroglian. *PLoS Biology*, *11*, e1001537.

Galiano, M. R., Jha, S., Ho, T. S., Zhang, C., Ogawa, Y., Chang, K. J., et al. (2012). A distal axonal cytoskeleton forms an intra-axonal boundary that controls axon initial segment, assembly. *Cell*, *149*, 1125–1139.

Gao, Y., Perkins, E. M., Clarkson, Y. L., Tobia, S., Lyndon, A. R., Jackson, M., et al. (2011). β-III spectrin is critical for development of purkinje cell dendritic tree and spine morphogenesis. *The Journal of Neuroscience*, *31*, 16581–16590.

Gardner, K., & Bennett, V. (1987). Modulation of spectrin-actin assembly by erythrocyte adducin. *Nature*, *328*, 359–362.

Garrido, J. J., Giraud, P., Carlier, E., Fernandes, F., Moussif, A., Fache, M. P., et al. (2003). A targeting motif involved in sodium channel clustering at the axonal initial segment. *Science*, *300*, 2091–2094.

Garver, T. D., Ren, Q., Tuvia, S., & Bennett, V. (1997). Tyrosine phosphorylation at a site highly conserved in the L1 family of cell adhesion molecules abolishes ankyrin-binding and increases lateral mobility of neurofascin. *The Journal of Cell Biology*, *137*, 703–714.

Godenschwege, T. A., Kristiansen, L. V., Uthaman, S. B., Hortsch, M., & Murphey, R. K. (2006). A conserved role for Drosophila Neuroglian and human L1-CAM in central-synapse formation. *Current Biology*, *16*, 12–23.

Grey, J. L., Kodippili, G. C., Simon, K., & Low, P. S. (2012). Identification of contact sites between ankyrin and band 3 in the human erythrocyte membrane. *Biochemistry*, *51*, 6838–6846.

Grum, V. L., MacDonald, R. I., & Mondragón, A. (1999). Structures of two repeats of spectrin suggest models of flexibility. *Cell*, *98*, 523–535.

Gudmundsson, H., Curran, J., Kashef, F., Snyder, J. S., Smith, S. A., Vargas-Pinto, P., et al. (2012). Differential regulation of EHD3 in human and mammalian heart failure. *Journal of Molecular and Cellular Cardiology*, *52*, 1183–1190.

Gudmundsson, H., Hund, T. J., Wright, P. J., Kline, C. F., Snyder, J. S., Qian, L., et al. (2010). EH domain proteins regulate cardiac membrane protein targeting. *Circulation Research*, *107*, 84–95.

Hall, T. G., & Bennett, V. (1987). Regulatory domains of erythrocyte ankyrin. *The Journal of Biological Chemistry*, *262*, 10537–10545.

Hammarlund, M., Jorgensen, E. M., & Bastiani, M. J. (2007). Axons break in animals lacking beta-spectrin. *The Journal of Cell Biology*, *176*, 269–275.

Hayes, N. V., Scott, C., Heerkens, E., Ohanian, V., Maggs, A. M., Pinder, J. C., et al. (2000). Identification of a novel C-terminal variant of beta II spectrin: Two isoforms of beta II spectrin have distinct intracellular locations and activities. *Journal of Cell Science, J113*, 2023–2034.

He, M., Jenkins, P., & Bennett, V. (2012). Cysteine 70 of ankyrin-G is S-palmitoylated and is required for function of ankyrin-G in membrane domain assembly. *The Journal of Biological Chemistry, 287*, 43995–44005.

He, M., Tseng, W. C., & Bennett, V. (2013). A single divergent exon inhibits ankyrin-B association with the plasma membrane. *The Journal of Biological Chemistry, 288*, 14769–14779.

Hedstrom, K. L., Ogawa, Y., & Rasband, M. N. (2008). AnkyrinG is required for maintenance of the axon initial segment and neuronal polarity. *The Journal of Cell Biology, 183*, 635–640.

Hedstrom, K. L., Xu, X., Ogawa, Y., Frischknecht, R., Seidenbecher, C. I., Shrager, P., et al. (2007). Neurofascin assembles a specialized extracellular matrix at the axon initial segment. *The Journal of Cell Biology, 178*, 875–886.

Hill, A. S., Nishino, A., Nakajo, K., Zhang, G., Fineman, J. R., Selzer, M. E., et al. (2008). Ion channel clustering at the axon initial segment and node of Ranvier evolved sequentially in early chordates. *PLoS Genetics, 4*(12), e1000317.

Hoekstra, H. E., & Coyne, J. A. (2007). The locus of evolution: Evo devo and the genetics of adaptation. *Evolution, 6*, 995–1016.

Holleran, E. A., Ligon, L. A., Tokito, M., Stankewich, M. C., Morrow, J. S., & Holzbaur, E. L. (2001). Beta III spectrin binds to the Arp1 subunit of dynactin. *The Journal of Biological Chemistry, 276*, 36598–36605.

Holleran, E. A., Tokito, M. K., Karki, S., & Holzbaur, E. L. (1996). Centractin (ARP1) associates with spectrin revealing a potential mechanism to link dynactin to intracellular organelles. *The Journal of Cell Biology, 135*, 1815–1829.

Hoock, T. C., Peters, L. L., & Lux, S. E. (1997). Isoforms of ankyrin-3 that lack the NH2-terminal repeats associate with mouse macrophage lysosomes. *The Journal of Cell Biology, 136*, 1059–1070.

Hopitzan, A. A., Baines, A. J., & Kordeli, E. (2006). Molecular evolution of ankyrin: Gain of function in vertebrates by acquisition of an obscurin/titin-binding-related domain. *Molecular Biology and Evolution, 23*, 46–55.

Hopitzan, A. A., Baines, A. J., Ludosky, M. A., Recouvreur, M., & Kordeli, E. (2005). Ankyrin-G in skeletal muscle: Tissue-specific alternative splicing contributes to the complexity of the sarcolemmal cytoskeleton. *Experimental Cell Research, S309*, 86–98.

Hortsch, M., Nagaraj, K., & Godenschwege, T. A. (2009). The interaction between L1-type proteins and ankyrins—A master switch for L1-type CAM function. *Cellular & Molecular Biology Letters, 14*, 57–69.

Hu, R. J., Watanabe, M., & Bennett, V. (1992). Characterization of human brain cDNA encoding the general isoform of beta-spectrin. *The Journal of Biological Chemistry, 267*, 18715–18722.

Huber, A. H., Stewart, D. B., Laurents, D. V., Nelson, W. J., & Weis, W. I. (2001). The cadherin cytoplasmic domain is unstructured in the absence of beta-catenin. A possible mechanism for regulating cadherin turnover. *The Journal of Biological Chemistry, 276*, 12301–12309.

Hund, T. J., Koval, O. M., Li, J., Wright, P. J., Qian, L., Snyder, J. S., et al. (2010). A β(IV)-spectrin/CaMKII signaling complex is essential for membrane excitability in mice. *The Journal of Clinical Investigation, 120*, 3508–3519.

Hyvönen, M., Macias, M. J., Nilges, M., Oschkinat, H., Saraste, M., & Wilmanns, M. (1995). Structure of the binding site for inositol phosphates in a PH domain. *The EMBO Journal, 14*, 4676–4685.

Ikeda, Y., Dick, K. A., Weatherspoon, M. R., Gincel, D., Armbrust, K. R., Dalton, J. C., et al. (2006). Spectrin mutations cause spinocerebellar ataxia type 5. *Nature Genetics*, *38*, 184–190.

Ipsaro, J. J., Harper, S. L., Messick, T. E., Marmorstein, R., Mondragón, A., & Speicher, D. W. (2010). Crystal structure and functional interpretation of the erythrocyte spectrin tetramerization domain, complex. *Blood*, *115*, 4843–4852.

Ipsaro, J. J., Huang, L., & Mondragón, A. (2009). Structures of the spectrin-ankyrin interaction binding domains. *Blood*, *113*, 5385–5393.

Ipsaro, J. J., & Mondragón, A. (2010). Structural basis for spectrin recognition by ankyrin. *Blood*, *115*, 4093–4101.

Iqbal, Z., Vandeweyer, G., van der Voet, M., Waryah, A. M., Zahoor, M. Y., Besseling, J. A., et al. (2013). Homozygous and heterozygous disruptions of ANK3: At the crossroads of neurodevelopmental and psychiatric disorders. *Human Molecular Genetics*, *22*, 1960–1970.

Jenkins, S. M., & Bennett, V. (2001). Ankyrin-G coordinates assembly of the spectrin-based membrane skeleton, voltage-gated sodium channels, and L1 CAMs at Purkinje neuron initial segments. *The Journal of Cell Biology*, *155*, 739–746.

Jenkins, P. M., Vasavda, C., Hostettler, J., Davis, J. Q., Abdi, K., & Bennett, V. (2013). E-cadherin polarity is determined by a multi-function motif mediating lateral membrane retention through ankyrin-G and apical-lateral transcytosis through clathrin. *The Journal of Biological Chemistry*, *288*, 14018–14031.

Kasahara, M. (2013). Impact of whole-genome duplication on vertebrate development and evolution. *Seminars in Cell & Developmental Biology*, *24*, 81–82.

Kennedy, S. P., Warren, S. L., Forget, B. G., & Morrow, J. S. (1991). Ankyrin binds to the 15th repetitive unit of erythroid and nonerythroid beta-spectrin. *The Journal of Cell Biology*, *115*, 267–277.

Kizhatil, K., Baker, S. A., Arshavsky, V. Y., & Bennett, V. (2009). Ankyrin-G promotes cyclic nucleotide-gated channel transport to rod photoreceptor sensory cilia. *Science*, *323*, 1614–1617.

Kizhatil, K., & Bennett, V. (2004). Lateral membrane biogenesis in human bronchial epithelial cells requires 190 kDa ankyrin-G. *The Journal of Biological Chemistry*, *279*, 16706–16714.

Kizhatil, K., Davis, J. Q., Davis, L., Hoffman, J., Hogan, B. L., & Bennett, V. (2007). Ankyrin-G is a molecular partner of E-cadherin in epithelial cells and early embryos. *The Journal of Biological Chemistry*, *282*, 26552–26561.

Kizhatil, K., Sandhu, N., Peachy, N. S., & Bennett, V. (2009). Ankyrin-B is required for coordinated expression of beta-2 spectrin, the Na/K ATPase and the Na/Ca exchanger in the inner segment of rod photoreceptors. *Experimental Eye Research*, *88*, 59–64.

Kizhatil, K., Yoon, W., Mohler, P. J., Davis, L. H., Hoffman, J. A., & Bennett, V. (2007). Ankyrin-G and beta2-spectrin collaborate in biogenesis of lateral membrane of human bronchial epithelial cells. *The Journal of Biological Chemistry*, *282*, 2029–2037.

Kline, C. F., Kurata, H. T., Hund, T. J., Cunha, S. R., Koval, O. M., Wright, P. J., et al. (2009). Dual role of K ATP channel C-terminal motif in membrane targeting and metabolic regulation. *Proceedings of the National Academy of Sciences of the United States of America*, *106*, 16669–16674.

Koch, I., Schwarz, H., Beuchle, D., Goellner, B., Langegger, M., & Aberle, H. (2008). Drosophila ankyrin 2 is required for synaptic stability. *Neuron*, *58*, 210–222.

Komada, M., & Soriano, P. (2002). [Beta] IV-spectrin regulates sodium channel clustering through ankyrin-G at axon initial segments and nodes of Ranvier. *The Journal of Cell Biology*, *156*, 337–348.

Kordeli, E., Lambert, S., & Bennett, V. (1995). AnkyrinG. A new ankyrin gene with neural-specific isoforms localized at the axonal initial segment and node of Ranvier. *The Journal of Biological Chemistry*, *270*, 2352–2359.

Krieger, C. C., An, X., Tang, H. Y., Mohandas, N., Speicher, D. W., & Discher, D. E. (2011). Cysteine shotgun-mass spectrometry (CS-MS) reveals dynamic sequence of protein structure changes within mutant and stressed cells. *Proceedings of the National Academy of Sciences of the United States of America*, *108*, 8269–8274.

Kuhlman, P. A., Hughes, C. A., Bennett, V., & Fowler, V. M. (1996). A new function for adducin. Calcium/calmodulin-regulated capping of the barbed ends of actin filaments. *The Journal of Biological Chemistry*, *271*, 7986–7991.

Kunimoto, M. (1995). A neuron-specific isoform of brain ankyrin, 440-kD ankyrinB, is targeted to the axons of rat cerebellar neurons. *The Journal of Cell Biology*, *131*, 1821–1829.

Kunimoto, M., Otto, E., & Bennett, V. (1991). A new 440-kD isoform is the major ankyrin in neonatal rat brain. *The Journal of Cell Biology*, *115*, 1319–1331.

Lacas-Gervais, S., Guo, J., Strenzke, N., Scarfone, E., Kolpe, M., Jahkel, M., et al. (2004). BetaIVSigma1 spectrin stabilizes the nodes of Ranvier and axon initial segments. *The Journal of Cell Biology*, *166*, 983–990.

Lee, G., Abdi, K., Jiang, Y., Michaely, P., Bennett, V., & Marszalek, P. E. (2006). Nanospring behaviour of ankyrin repeats. *Nature*, *440*, 246–249.

Lemaillet, G., Walker, B., & Lambert, S. (2003). Identification of a conserved ankyrin-binding motif in the family of sodium channel alpha subunits. *The Journal of Biological Chemistry*, *278*, 27333–27339.

Lemmon, M. A., Ferguson, K. M., & Abrams, C. S. (2002). Pleckstrin homology domains and the cytoskeleton. *FEBS Letters*, *513*, 71–76.

Levine, J., & Willard, M. (1981). Fodrin: Axonally transported polypeptides associated with the internal periphery of many cells. *The Journal of Cell Biology*, *90*, 631–642.

Li, X., Matsuoka, Y., & Bennett, V. (1998). Adducin preferentially recruits spectrin to the fast growing ends of actin filaments in a complex requiring the MARCKS-related domain and a newly defined oligomerization domain. *The Journal of Biological Chemistry*, *273*, 19329–19338.

Lighthouse, D. V., Buszczak, M., & Spradling, A. C. (2008). New components of the Drosophila fusome suggest it plays novel roles in signaling and transport. *Developmental Biology*, *317*, 59–71.

Lin, H., Yue, L., & Spradling, A. C. (1994). The Drosophila fusome, a germline-specific organelle, contains membrane skeletal proteins and functions in cyst formation. *Development*, *120*, 947–956.

Linnen, C. R., Poh, Y. P., Peterson, B. K., Barrett, R. D., Larson, J. G., Jensen, J. D., et al. (2013). Adaptive evolution of multiple traits through multiple mutations at a single gene. *Science*, *339*, 1312–1316.

Lopez, C., Metral, S., Eladari, D., Drevensek, S., Gane, P., Chambrey, R., et al. (2005). The ammonium transporter RhBG requirement of a tyrosine-based signal and ankyrin-G for basolateral targeting and membrane anchorage in polarized kidney epithelial cells. *The Journal of Biological Chemistry*, *280*, 8221–8228.

Lorenzo, D. N., Li, M. G., Mische, S. E., Armbrust, K. R., Ranum, L. P., & Hays, T. S. (2010). Spectrin mutations that cause spinocerebellar ataxia type 5 impair axonal transport and induce neurodegeneration in Drosophila. *The Journal of Cell Biology*, *189*, 143–158.

Lowe, J. S., Palygin, O., Bhasin, N., Hund, T. J., Boyden, P. A., Shibata, E., et al. (2008). Voltage-gated Nav channel targeting in the heart requires an ankyrin-G dependent cellular pathway. *The Journal of Cell Biology*, *180*, 173–186.

Lux, S. E., John, K. M., & Bennett, V. (1990). Analysis of cDNA for human erythrocyte ankyrin indicates a repeated structure with homology to tissue- differentiation and cell-cycle control proteins. *Nature*, *344*, 36–42.

Macias, M. J., Musacchio, A., Ponstingl, H., Nilges, M., Saraste, M., & Oschkinat, H. (1994). Structure of the pleckstrin homology domain from beta-spectrin. *Nature, 369*, 675–677.

Maniar, T. A., Kaplan, M., Wang, G. J., Shen, K., Wei, L., Shaw, J. E., et al. (2011). UNC-33 (CRMP) and ankyrin organize microtubules and localize kinesin to polarize axon-dendrite sorting. *Nature Neuroscience, 15*, 48–56.

Marfatia, S. M., Leu, R. A., Branton, D., & Chishti, A. H. (1995). Identification of the protein 4.1 binding interface on glycophorin C and p55, a homologue of the Drosophila discs-large tumor suppressor protein. *The Journal of Biological Chemistry, 270*, 715–719.

Marfatia, S. M., Lue, R. A., Branton, D., & Chishti, A. H. (1994). In vitro binding studies suggest a membrane-associated complex between erythroid p55, protein 4.1, and glycophorin C. *The Journal of Biological Chemistry, 269*, 8631–8634.

Martin, M. W., Grazhdankin, D. V., Bowring, S. A., Evans, D. A., Fedonkin, M. A., & Kirschvink, J. L. (2000). Age of Neoproterozoic bilatarian body and trace fossils, White Sea, Russia: Implications for metazoan evolution. *Science, 288*, 841–845.

McKeown, C., Praitis, V., & Austin, J. (1998). sma-1 encodes a betaH-spectrin homolog required for Caenorhabditis elegans morphogenesis. *Development, 125*, 2087–2098.

Médina, E., Williams, J., Klipfell, E., Zarnescu, D., Thomas, G., & Le Bivic, A. (2002). Crumbs interacts with moesin and beta(Heavy)-spectrin in the apical membrane skeleton of Drosophila. *The Journal of Cell Biology, 158*, 941–951.

Mehboob, S., Song, Y., Witek, M., Long, F., Santarsiero, B. D., Johnson, M. E., et al. (2010). Crystal structure of the nonerythroid alpha-spectrin tetramerization site reveals differences between erythroid and nonerythroid spectrin tetramer formation. *The Journal of Biological Chemistry, 285*, 14572–14584.

Michaely, P., & Bennett, V. (1995a). The ANK repeats of erythrocyte ankyrin form two distinct but cooperative binding sites for the erythrocyte anion exchanger. *The Journal of Biological Chemistry, 270*, 22050–22057.

Michaely, P., & Bennett, V. (1995b). Mechanism for binding site diversity on ankyrin. Comparison of binding sites on ankyrin for neurofascin and the Cl2/HCO32 anion exchanger. *The Journal of Biological Chemistry, 270*, 31298–31302.

Michaely, P., Tomchick, D. R., Machius, M., & Anderson, R. G. (2002). Crystal structure of a 12 ANK repeat stack from human ankyrinR. *The EMBO Journal, 21*, 6387–6396.

Mohler, P. J., Davis, J. Q., & Bennett, V. (2005). Ankyrin-B coordinates the Na/K ATPase, Na/Ca exchanger, and InsP3 receptor in a cardiac T-tubule/SR microdomain. *PLoS Biology, 3*, e423.

Mohler, P. J., Gramolini, A. O., & Bennett, V. (2002). The ankyrin-B C-terminal domain determines activity of ankyrin-B/G chimeras in rescue of abnormal inositol 1,4,5-trisphosphate and ryanodine receptor distribution in ankyrin-B (-/-) neonatal cardiomyocytes. *The Journal of Biological Chemistry, 277*, 10599–10607.

Mohler, P. J., Rivolta, I., Napolitano, C., LeMaillet, G., Lambert, S., Priori, S. G., et al. (2004). Nav1.5 E1053K mutation causing Brugada syndrome blocks binding to ankyrin-G and expression of Nav1.5 on the surface of cardiomyocytes. *Proceedings of the National Academy of Sciences of the United States of America, 101*, 17533–17538.

Mohler, P. J., Yoon, W., & Bennett, V. (2004). Ankyrin-B targets beta2-spectrin to an intracellular compartment in neonatal cardiomyocytes. *The Journal of Biological Chemistry, 279*, 40185–40193.

Mosavi, L. K., Cammett, T. J., Desrosiers, D. C., & Peng, Z. Y. (2004). The ankyrin repeat as molecular architecture for protein recognition. *Protein Science, 13*, 1435–1448.

Muresan, V., Stankewich, M. C., Steffen, W., Morrow, J. S., Holzbaur, E. L., & Schnapp, B. J. (2001). Dynactin-dependent, dynein-driven vesicle transport in the absence of membrane proteins: A role for spectrin and acidic phospholipids. *Molecular Cell, 7*, 173–183.

Musacchio, A., Noble, M., Pauptit, R., Wierenga, R., & Saraste, M. (1992). Crystal structure of a Src-homology 3 (SH3) domain. *Nature, 359*, 851–855.

Nery, M. F., González, D. J., & Opazo, J. C. (2013). How to make a dolphin: Molecular signature of positive selection in cetacean genome. *PLoS One*, *8*, e65491.

Ohara, O., Ohara, R., Yamakawa, H., Nakajima, D., & Nakayama, M. (1998). Characterization of a new beta-spectrin gene which is predominantly expressed in brain. *Brain Research Molecular Brain Research*, *57*, 181–192.

Ohno, S. (1970). *Evolution by gene duplication*. Berlin: Springer Verlag.

Ohno, S. (1999). Gene duplication and the uniqueness of vertebrate genomes circa 1970–1999. *Seminars in Cell & Developmental Biology*, *10*, 517–522, Review.

Otto, E., Kunimoto, M., McLaughlin, T., & Bennett, V. (1991). Isolation and characterization of cDNAs encoding human brain ankyrins reveal a family of alternatively-spliced genes. *The Journal of Cell Biology*, *114*, 241–253.

Pan, Z., Kao, T., Horvath, Z., Lemos, J., Sul, J. Y., Cranstoun, S. D., et al. (2006). A common ankyrin-G-based mechanism retains KCNQ and NaV channels at electrically active domains of the axon. *The Journal of Neuroscience*, *26*, 2599–2613.

Papal, S., Cortese, M., Legendre, K., Sorusch, N., Dragavon, J., Sahly, I., et al. (2013). The giant spectrin βV couples the molecular motors to phototransduction and Usher syndrome type I proteins along their trafficking route. *Human Molecular Genetics*, *22*, 3773–3788.

Pearson, M. A., Reczek, D., Bretscher, A., & Karplus, P. A. (2000). Structure of the ERM protein moesin reveals the FERM domain fold masked by an extended actin binding tail domain. *Cell*, *101*, 259–270.

Perkins, E. M., Clarkson, Y. L., Sabatier, N., Longhurst, D. M., Millward, C. P., Jack, J., et al. (2010). Loss of beta-III spectrin leads to Purkinje cell dysfunction recapitulating the behavior and neuropathology of spinocerebellar ataxia type 5 in humans. *The Journal of Neuroscience*, *30*, 4857–4867.

Peters, L. L., John, K. M., Lu, F. M., Eicher, E. M., Higgins, A., Yialamas, M., et al. (1995). Ank3 (epithelial ankyrin), a widely distributed new member of the ankyrin gene family and the major ankyrin in kidney, is expressed in alternatively spliced forms, including forms that lack the repeat domain. *The Journal of Cell Biology*, *130*, 313–330.

Phillips, M. D., & Thomas, G. H. (2006). Brush border spectrin is required for early endosome recycling in Drosophila. *Journal of Cell Science*, *119*, 1361–1370.

Pielage, J., Cheng, L., Fetter, R. D., Carlton, P. M., Sedat, J. W., & Davis, G. W. (2008). A presynaptic giant ankyrin stabilizes the NMJ through regulation of presynaptic microtubules and transsynaptic cell adhesion. *Neuron*, *58*, 195–209.

Pocock, R., Bénard, C. Y., Shapiro, L., & Hobert, O. (2008). Functional dissection of the C. elegans cell adhesion molecule SAX-7, a homologue of human L1. *Molecular and Cellular Neuroscience*, *37*, 56–68.

Rank, G., Sutton, R., Marshall, V., Lundie, R. J., Caddy, J., Romeo, T., et al. (2009). Novel roles for erythroid Ankyrin-1 revealed through an ENU-induced null mouse mutant. *Blood*, *113*, 3352–3362.

Rief, M., Pascual, J., Saraste, M., & Gaub, H. E. (1999). Single molecule force spectroscopy of spectrin repeats: Low unfolding forces in helix bundles. *Journal of Molecular Biology*, *286*, 553–561.

Ruff, P., Speicher, D. W., & Husain-Chishti, A. (1991). Molecular identification of a major palmitoylated erythrocyte membrane protein containing the src homology 3 motif. *Proceedings of the National Academy of Sciences of the United States of America*, *88*, 6595–6599.

Salomao, M., An, X., Guo, X., Gratzer, W. B., Mohandas, N., & Baines, A. J. (2006). Mammalian alpha I-spectrin is a neofunctionalized polypeptide adapted to small highly deformable erythrocytes. *Proceedings of the National Academy of Sciences of the United States of America*, *103*, 643–648.

Schroer, T. A. (2004). Dynactin. *Annual Review of Cell and Developmental Biology*, *20*, 759–779.

Scotland, P., Zhou, D., Benveniste, H., & Bennett, V. (1998). Nervous system defects of ankyrin-B (-/-) mice suggest functional overlap between the cell adhesion molecule L1 and 440 kD ankyrin-B in premyelinated axons. *The Journal of Cell Biology*, *143*, 1305–1315.

Shotton, D. M., Burke, B. E., & Branton, D. (1979). The molecular structure of human erythrocyte spectrin. Biophysical and electron microscopic studies. *Journal of Molecular Biology*, *131*, 303–329.

Sjöblom, B., Ylänne, J., & Djinović-Carugo, K. (2008). Novel structural insights into F-actin-binding and novel functions of calponin homology domains. *Current Opinion in Structural Biology*, *18*, 702–708.

Sobotzik, J. M., Sie, J. M., Politi, C., Del Turco, D., Bennett, V., Deller, T., et al. (2009). AnkyrinG is required to maintain axo-dendritic polarity in vivo. *Proceedings of the National Academy of Sciences of the United States of America*, *106*, 17564–17569.

Speicher, D. W., & Marchesi, V. T. (1984). Erythrocyte spectrin is comprised of many homologous triple helical segments. *Nature*, *311*, 177–180.

Stabach, P. R., & Morrow, J. S. (2000). Identification and characterization of beta V spectrin, a mammalian ortholog of Drosophila beta H spectrin. *The Journal of Biological Chemistry*, *275*, 21385–21395.

Stabach, P. R., Simonović, I., Ranieri, M. A., Aboodi, M. S., Steitz, T. A., Simonović, M., et al. (2009). The structure of the ankyrin-binding site of beta-spectrin reveals how tandem spectrin-repeats generate unique ligand-binding properties. *Blood*, *113*, 5377–5384.

Stankewich, M. C., Tse, W. T., Peters, L. L., Ch'ng, Y., John, K. M., Stabach, P. R., et al. (1998). A widely expressed betaIII spectrin associated with Golgi and cytoplasmic vesicles. *Proceedings of the National Academy of Sciences of the United States of America*, *95*, 14158–14163.

Staufenbiel, M. (1987). Ankyrin-bound fatty acid turns over rapidly at the erythrocyte plasma membrane. *Molecular and Cellular Biology*, *7*, 2981–2984.

Steinmetz, P. R., Kraus, J. E., Larroux, C., Hammel, J. U., Amon-Hassenzahl, A., Houliston, E., et al. (2012). Independent evolution of striated muscles in cnidarians and bilaterians. *Nature*, *487*, 231–234.

Stumpp, M. T., & Amstutz, P. (2007). DARPins: A true alternative to antibodies. *Current Opinion in Drug Discovery & Development*, *10*, 153–159.

Susuki, K., Chang, K. J., Zollinger, D. R., Liu, Y., Ogawa, Y., Eshed-Eisenbach, Y., et al. (2013). Three mechanisms assemble central nervous system nodes of Ranvier. *Neuron*, *78*, 469–482.

Takeda, S., Yamazaki, H., Seog, D. H., Kanai, Y., Terada, S., & Hirokawa, N. (2000). Kinesin superfamily protein 3 (KIF3) motor transports fodrin-associating vesicles important for neurite building. *The Journal of Cell Biology*, *148*, 1255–1265.

Tang, Y., Katuri, V., Dillner, A., Mishra, B., Deng, C. X., & Mishra, L. (2003). Disruption of transforming growth factor-beta signaling in ELF beta-spectrin-deficient mice. *Science*, *299*, 574–777.

Thevananther, S., Kolli, A. H., & Devarajan, P. (1998). Identification of a novel ankyrin isoform (AnkG190) in kidney and lung that associates with the plasma membrane and binds alpha-Na, K-ATPase. *The Journal of Biological Chemistry*, *273*, 23952–23958.

Tse, W. T., Lecomte, M. C., Costa, F. F., Garbarz, M., Feo, C., Boivin, P., et al. (1990). Point mutation in the beta-spectrin gene associated with alpha I/74 hereditary elliptocytosis. Implications for the mechanism of spectrin dimer self-association. *The Journal of Clinical Investigation*, *86*, 909–916.

Tse, W. T., Tang, J., Jin, O., Korsgren, C., John, K. M., Kung, A. L., et al. (2001). A new spectrin, beta IV, has a major truncated isoform that associates with promyelocytic leukemia protein nuclear bodies and the nuclear matrix. *The Journal of Biological Chemistry*, *276*, 23974–23985.

Tyler, J. M., Hargreaves, W. R., & Branton, D. (1979). Purification of two spectrin-binding proteins: Biochemical and electron microscopic evidence for site-specific reassociation between spectrin and bands 2.1 and 4.1. *Proceedings of the National Academy of Sciences of the United States of America, 76*, 5192–5196.

Uemoto, Y., Suzuki, S., Terada, N., Ohno, N., Ohno, S., Yamanaka, S., et al. (2007). Specific role of the truncated betaIV-spectrin Sigma6 in sodium channel clustering at axon initial segments and nodes of Ranvier. *The Journal of Biological Chemistry, 282*, 6548–6555.

Ungewickell, E., Bennett, P. M., Calvert, R., Ohanian, V., & Gratzer, W. B. (1979). In vitro formation of a complex between cytoskeletal proteins of the human erythrocyte. *Nature, 280*, 811–814.

Viel, A. (1999). Alpha-actinin and spectrin structures: An unfolding family story. *FEBS Letters, 460*, 391–394.

Vosseller, K., Trinidad, J. C., Chalkley, R. J., Specht, C. G., Thalhammer, A., Lynn, A. J., et al. (2006). O-linked N-acetylglucosamine proteomics of postsynaptic density preparations using lectin weak affinity chromatography and mass spectrometry. *Molecular & Cellular Proteomics, 5*, 923–934.

Wang, D. S., Miller, R., Shaw, R., & Shaw, G. (1996). The pleckstrin homology domain of human beta I sigma II spectrin is targeted to the plasma membrane in vivo. *Biochemical and Biophysical Research Communications, 225*, 420–426.

Wang, R., Wei, Z., Jin, H., Wu, H., Yu, C., Wen, W., et al. (2009). Autoinhibition of UNC5b revealed by the cytoplasmic domain structure of the receptor. *Molecular Cell, 33*, 692–703.

Wang, C., Yu, C., Ye, F., Wei, Z., & Zhang, M. (2012). Structure of the ZU5-ZU5-UPA-DD tandem of ankyrin-B reveals interaction surfaces necessary for ankyrin function. *Proceedings of the National Academy of Sciences of the United States of America, 109*, 4822–4827.

Wasenius, V. M., Saraste, M., Salvén, P., Erämaa, M., Holm, L., & Lehto, V. P. (1989). Primary structure of the brain alpha-spectrin. *The Journal of Cell Biology, 108*, 79–93.

Weber, A., Pennise, C. R., Babcock, G. G., & Fowler, V. M. (1994). Tropomodulin caps the pointed ends of actin filaments. *The Journal of Cell Biology, 127*, 1627–1635.

Winkelmann, J. C., Chang, J. G., Tse, W. T., Scarpa, A. L., Marchesi, V. T., & Forget, B. G. (1990). Full-length sequence of the cDNA for human erythroid beta-spectrin. *The Journal of Biological Chemistry, 265*, 11827–11832.

Winkelmann, J. C., Costa, F. F., Linzie, B. L., & Forget, B. G. (1990). Beta spectrin in human skeletal muscle. Tissue-specific differential processing of 3' beta spectrin pre-mRNA generates a beta spectrin isoform with a unique carboxyl terminus. *The Journal of Biological Chemistry, 265*, 20449–20454.

Xu, K., Zhong, G., & Zhuang, X. (2013). Actin, spectrin, and associated proteins form a periodic cytoskeletal structure in axons. *Science, 339*, 452–456.

Yan, Y., Winograd, E., Viel, A., Cronin, T., Harrison, S. C., & Branton, D. (1993). Crystal structure of the repetitive segments of spectrin. *Science, 262*, 2027–2030.

Yang, Y., Ogawa, Y., Hedstrom, K. L., & Rasband, M. N. (2007). betaIV spectrin is recruited to axon initial segments and nodes of Ranvier by ankyrinG. *The Journal of Cell Biology, 176*, 509–519.

Yeh, T. Y., Quintyne, N. J., Scipioni, B. R., Eckley, D. M., & Schroer, T. A. (2012). Dynactin's pointed-end complex is a cargo-targeting module. *Molecular Biology of the Cell, 23*, 3827–3837.

Zhang, X., & Bennett, V. (1996). Identification of O-linked N-acetyl glucosamine modification of ankyrin G isoforms targeted to nodes of Ranvier. *The Journal of Biological Chemistry, 271*, 31391–31398.

Zhang, X., & Bennett, V. (1998). Restriction of 480/270kD ankyrin G to axon proximal segments requires multiple ankyrin G-specific domains. *The Journal of Cell Biology, 142*, 1671–1682.

Zhang, X., Davis, J. Q., Carpenter, S., & Bennett, V. (1998). Structural requirements for association of neurofascin with ankyrin. *The Journal of Biological Chemistry, 273*, 30785–30794.
Zhang, Y., Resneck, W. G., Lee, P. C., Randall, W. R., Bloch, R. J., & Ursitti, J. A. (2010). Characterization and expression of a heart-selective alternatively spliced variant of alpha II-spectrin, cardi+, during development in the rat. *Journal of Molecular and Cellular Cardiology, 48*, 1050–1059.
Zhou, D., Birkenmeier, C. S., Williams, M. W., Sharp, J. J., Barker, J. E., & Bloch, R. J. (1997). Small, membrane-bound, alternatively spliced forms of ankyrin 1 associated with the sarcoplasmic reticulum of mammalian skeletal muscle. *The Journal of Cell Biology, 136*, 621–631.
Zhou, D., Lambert, S., Malen, P. L., Carpenter, S., Boland, L. M., & Bennett, V. (1998). Ankyrin G is required for clustering of voltage-gated Na channels at axon initial segments and for normal action potential firing. *The Journal of Cell Biology, 143*, 1295–1304.

CHAPTER TWO

The Human Erythrocyte Plasma Membrane: A Rosetta Stone for Decoding Membrane–Cytoskeleton Structure

Velia M. Fowler[1]
Department of Cell and Molecular Biology, The Scripps Research Institute, La Jolla, California, USA
[1]Corresponding author: e-mail address: velia@scripps.edu

Contents

Abstract

The mammalian erythrocyte, or red blood cell (RBC), is a unique experiment of nature: a cell with no intracellular organelles, nucleus or transcellular cytoskeleton, and a plasma membrane with uniform structure across its entire surface. By virtue of these specialized properties, the RBC membrane has provided a template for discovery of the fundamental actin filament network machine of the membrane skeleton, now known to confer mechanical resilience, anchor membrane proteins, and organize membrane domains

Current Topics in Membranes, Volume 72
ISSN 1063-5823
http://dx.doi.org/10.1016/B978-0-12-417027-8.00002-7

in all cells. This chapter provides a historical perspective and critical analysis of the biochemistry, structure, and physiological functions of this actin filament network in RBCs. The core units of this network are nodes of ~35–37 nm-long actin filaments, interconnected by long strands of $(\alpha 1\beta 1)_2$-spectrin tetramers, forming a 2D isotropic lattice with quasi-hexagonal symmetry. Actin filament length and stability is critical for network formation, relying upon filament capping at both ends: tropomodulin-1 at pointed ends and αβ-adducin at barbed ends. Tropomodulin-1 capping is essential for precise filament lengths, and is enhanced by tropomyosin, which binds along the short actin filaments. αβ-adducin capping recruits spectrins to sites near barbed ends, promoting network formation. Accessory proteins, 4.1R and dematin, also promote spectrin binding to actin and, with αβ-adducin, link to membrane proteins, targeting actin nodes to the membrane. Dissection of the molecular organization within the RBC membrane skeleton is one of the paramount achievements of cell biological research in the past century. Future studies will reveal the structure and dynamics of actin filament capping, mechanisms of precise length regulation, and spectrin–actin lattice symmetry.

1. INTRODUCTION

Mature human erythrocytes, or red blood cells (RBCs), are biconcave disk-shaped cells ~8 μm in diameter and 2 μm thick at their rim, containing no nucleus or intracellular organelles, and packed with ~450 mg/ml hemoglobin in their cytoplasm for O_2 delivery and CO_2 removal. RBCs are remarkably deformable and amazingly stable, repeatedly traversing capillaries smaller than their diameter in the peripheral tissues, and withstanding the shear stresses in the large arteries, with a lifespan of ~120 days in humans (~40 days in mice) (An, Lecomte, Chasis, Mohandas, & Gratzer, 2002; Handin, Lux, & Stossel, 2003; Mohandas & Gallagher, 2008). To perform its circulatory function, the RBC membrane contains abundant and specialized ion and gas transporters to regulate O_2/CO_2 exchange, intracellular pH, ion and water homeostasis, as well as glycosylated proteins that form the basis of the blood group antigen system. The membrane proteins are anchored to a thin cytoskeleton layer (~100 nm thick), termed the membrane skeleton, a micron-scale network of long spectrin strands connecting short actin filaments, extending across the cytoplasmic surface of the entire RBC membrane (Fig. 2.1). RBC membrane assembly, integrity, and mechanics rely exclusively on the membrane skeleton, such that defects in the membrane skeleton lead to abnormal RBC shapes, reduced deformability, and decreased stability. This impairs RBC survival in the circulation, leading to hemolytic anemias in mice and humans (Gallagher, 2004; Mohandas & Evans, 1994; Mohandas & Gallagher, 2008; Palek, 1985; Perrotta, Gallagher, & Mohandas, 2008).

The mammalian RBC membrane is a unique experiment of nature that has created a uniform and specialized membrane domain. At the last stage of erythroid differentiation when the nucleus is expelled (Fig. 2.1A), a subset of plasma membrane components are segregated to the membrane of the nascent reticulocyte, leaving behind unwanted membrane proteins, such as integrins, on the plasma membrane surrounding the ejected nucleus (Ji, Murata-Hori, & Lodish, 2011; Keerthivasan, Wickrema, & Crispino, 2011; Mohandas & Gallagher, 2008). In a further cellular simplification, intracellular organelles and transcellular cytoskeletal structures (microtubules, intermediate filaments, and cytoplasmic actin filaments) are also removed during enucleation, leaving the membrane skeleton as the sole cytoskeletal structure in mature RBCs. Remnants of unwanted membrane and cytoskeletal proteins continue to be removed during maturation of reticulocytes to RBCs over several days, via complex membrane vesicular trafficking, remodeling, autophagy, and other degradation processes (Blanc & Vidal, 2010; Chasis, Prenant, Leung, & Mohandas, 1989; Johnstone, 2005; Liu, Mohandas, & An, 2011; Ney, 2011). The end result is a plasma membrane domain with a homogenous molecular composition and structural organization across the entire RBC surface. When hemoglobin is removed by osmotic lysis and washing to make membrane "ghosts," grams of this pure plasma membrane domain are available for biochemical, biophysical, structural, and functional analysis.

Due to these unique biological features, studies of the human RBC membrane have historically assumed a central role in the elucidation of basic concepts in membrane biology and medicine, some of which have been recognized by a series of Nobel prizes. Landsteiner's identification of the blood group antigen system in RBCs in 1901 had a huge impact on safe blood transfusions and effective treatment for Rh-antigen-induced hemolytic anemias in newborns, for which Landsteiner received the 1930 Nobel Prize in Physiology and Medicine. Pioneering biophysical studies by Gorter and Grendel in the 1920s (Gorter & Grendel, 1925), Danielli and Davson in the 1930s (Danielli & Davson, 1935), and Robertson in the 1950s led to the fundamental concept of the lipid bilayer (Robertson, 1959). Analysis of RBC membrane proteins provided key insights into the topology of membrane-spanning glycoproteins and concepts of peripheral and integral proteins, using selective extraction and chemical labeling (Marchesi, 1979; Steck, 1974; Fig. 2.1B and C). Freeze-fracture electron microscopy of RBCs also demonstrated that membrane proteins traversed the bilayer and were laterally mobile

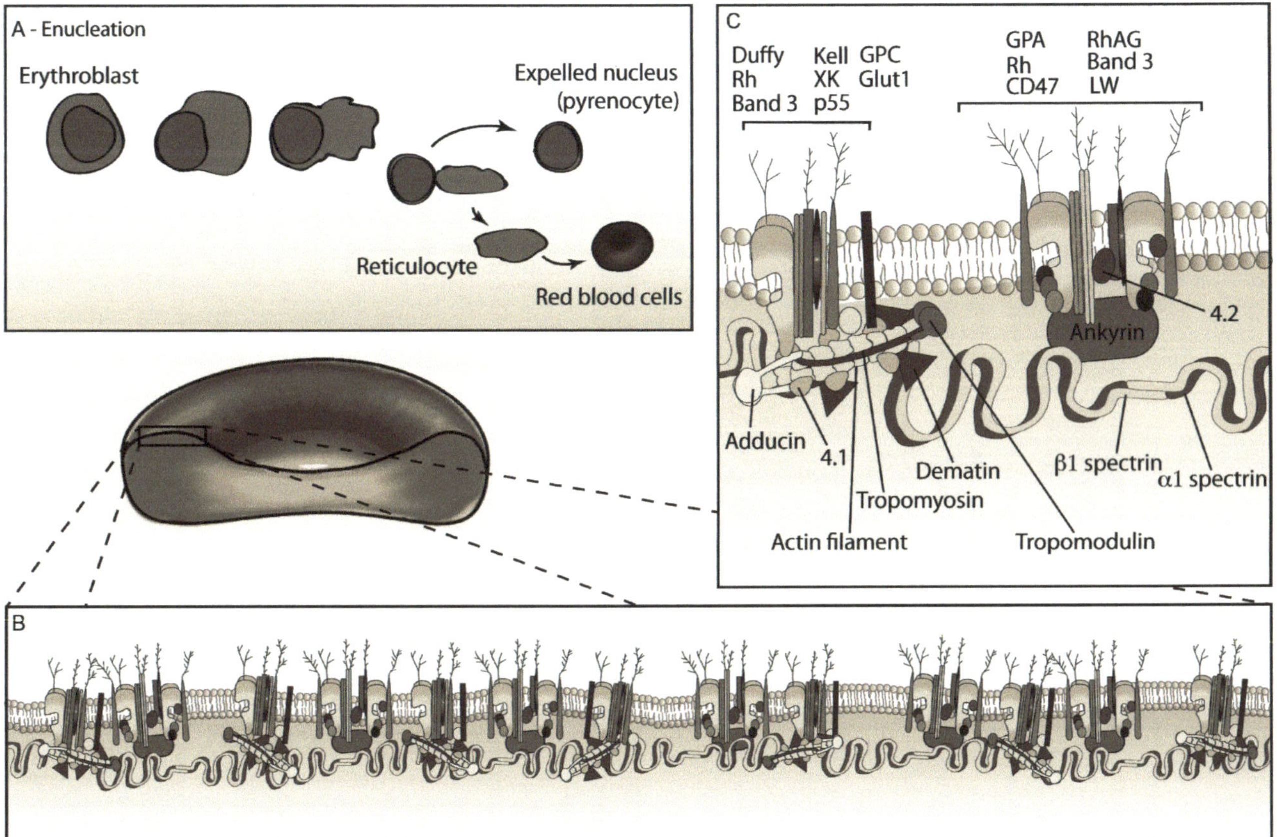

A - Enucleation
Erythroblast
Expelled nucleus
(pyrenocyte)
Reticulocyte
Red blood cells
C
Duffy
Rh
Band 3
Kell
XK
p55
GPC
Glut1
GPA
Rh
CD47
RhAG
Band 3
LW
4.2
Ankyrin
Adducin
4.1
Dematin
Tropomyosin
Actin filament
Tropomodulin
β1 spectrin
α1 spectrin
B

Figure 2.1 (A) Red blood cells (RBCs) arise from nucleated progenitors (erythroblasts), which terminally differentiate and expel their nucleus (pyrenocyte) to yield reticulocytes. Reticulocytes continue to synthesize proteins and contain intracellular organelles, which are eliminated over several days by complex membrane remodeling and degradation processes to yield mature biconcave RBCs with no intracellular organelles or transcellular cytoskeleton. (B) Schematic representation of RBC membrane structure depicting abundant transmembrane multiprotein complexes spanning the lipid bilayer, with the associated membrane skeleton forming a thin layer attached to the cytoplasmic domains of membrane proteins. The membrane skeleton is a 2D network of long flexible spectrin tetramers that cross-link short actin filaments into a micron-scale cytoskeletal domain that extends uniformly across the entire surface of the RBC membrane. (C) Components in the transmembrane multiprotein complexes and on the short actin filaments. There are two types of transmembrane complexes with overlapping components, one is anchored at the short actin filaments (junctional complexes, JCs) and the other is anchored via ankyrin near the middle of the $(\alpha 1\beta 1)_2$-spectrin tetramer. In addition to $\alpha 1\beta 1$-spectrin, the short actin filaments are associated with five actin-binding proteins, tropomodulin (Tmod1), $\alpha\beta$ adducin, protein 4.1R, tropomyosin (TM), and dematin, each with distinct actin-regulatory functions (Table 2.1; Fig. 2.3). *Panel (B) adapted from Salomao et al. (2008) and Yamashiro, Gokhin, Kimura, Nowak, and Fowler (2012).* (See color plate.)

(Pinto da Silva & Branton, 1970), contributing to the seminal "fluid-mosaic model" of membranes (Pinto da Silva & Branton, 1972; Singer & Nicolson, 1972).

The membrane water channel (aquaporin1) was discovered in RBCs (Benga, Popescu, Borza, et al., 1986; Benga, Popescu, Pop, & Holmes, 1986; Denker, Smith, Kuhajda, & Agre, 1988; Preston, Carroll, Guggino, & Agre, 1992), launching a revolution in the field of water regulation and ion homeostasis in the kidney and other tissues, for which Peter Agre received the 2003 Nobel Prize in Chemistry. The spectrin–actin membrane skeleton that supports the membrane via binding to ankyrin and other adaptors was discovered in RBCs (Bennett & Stenbuck, 1979; Branton, Cohen, & Tyler, 1981; Lux, 1979; Marchesi & Steers, 1968) and subsequently shown to be critical for membrane domain biogenesis and stability in metazoans, with mutations in its components leading to human diseases of hemolytic anemias, cardiac arrhythmias, and cerebellar ataxias (Bennett & Baines, 2001; Bennett & Healy, 2008; Mohandas & Evans, 1994; see Chapter 1). The only known actin filament pointed end capping proteins, the tropomodulins (Tmods), were discovered in RBCs (Fowler, 1987; Weber, Pennise, Babcock, & Fowler, 1994) and demonstrated to regulate precise thin filament lengths and sarcomere contraction in striated muscle (Gokhin & Fowler, 2011; Gregorio, Weber, Bondad, Pennise, & Fowler, 1995) and micron-scale domain organization of the spectrin–actin network in differentiated cells (Yamashiro et al., 2012).

The RBC membrane skeleton is the paradigmatic membrane-associated actin cytoskeleton, defined by a long-range isotropic filament network associated with the cytoplasmic surface of membranes via multipoint connections to transmembrane proteins (Fig. 2.1B and C). In this chapter, I will discuss the historical basis for our current understanding of RBC actin filament assembly and structural organization, the properties of RBC actin-binding proteins and their functions in RBC biology, and highlight some unsolved questions. This chapter is not meant to be comprehensive, and the reader is directed to previous reviews for more details on many of the topics discussed. In this area, as in so many others, the RBC membrane has been a powerful model system, enabling discovery of the properties of a specialized membrane-associated actin cytoskeleton with broad significance to other cells. It is hoped that this chapter will motivate continuing studies of RBC actin filaments as a valuable paradigm for actin assembly and associations with plasma membranes.

2. OVERVIEW OF SPECTRIN–ACTIN LATTICE STRUCTURE IN THE MEMBRANE SKELETON

The RBC membrane skeleton consists of a 2D lattice of long $(\alpha1\beta1)_2$-spectrin tetramers attached by their ends to short actin filaments at junctional complexes (JCs; Fig. 2.1B and C; for reviews, see Gilligan & Bennett, 1993; Fowler, 1996). α1β1-Spectrin binds to the actin filaments using two calponin homology (CH1 and CH2) domains at the N-terminal end of the β1 subunit and EF-hand domains at the C-terminal end of the α1 subunit, similar to the homologous actin-binding protein, α-actinin (Korsgren & Lux, 2010) [for a review, see Bennett and Baines (2001)]. The RBC actin filaments are all the same length, ~35–37 nm long, capped by Tmod1 at their pointed ends and αβ-adducin at their barbed ends (Fig. 2.1C). Two tropomyosin (TM) dimers bind to the sides of each short actin filament, spanning their length and binding to Tmod1 at the pointed filament end. Tmod1 binds actin and TMs, stabilizing TMs on the filament, and αβ-adducin binds actin and β1-spectrin, similarly helping to stabilize spectrin binding to the filaments. Caldesmon is a TM-binding and actin filament stabilizing protein that may also be associated with each actin filament (der Terrossian, Deprette, & Cassoly, 1989). Protein 4.1R is also bound to the actin filaments and to the β1-spectrin, playing an important role in enhancing β1-spectrin binding to actin (Takakuwa, 2000). Finally, dematin (protein 4.9) is an actin filament-bundling protein associated with the JCs, which also enhances α1β1-spectrin binding to actin filaments (Koshino, Mohandas, & Takakuwa, 2012). Thus, in total, there are six different flavors of actin-binding proteins (barbed or pointed end capping, side-binding, cross-linking, and bundling) stoichiometrically associated with each short actin filament at the JCs (Table 2.1)! In addition to their network linkage function, some of the actin-binding proteins (αβ-adducin, dematin, and protein 4.1R) also serve as adaptors to link the JCs in the membrane skeleton to transmembrane proteins (band 3, glycophorin C, Rh, Duffy, Kell, XK, and Glut1; Fig. 2.1), which will not be discussed here (Mohandas & Gallagher, 2008; Salomao et al., 2008). This chapter will also not discuss the molecular basis and functions of $(\alpha1\beta1)_2$-spectrin and protein 4.1R interactions with actin filaments, which have been covered extensively in prior reviews (e.g., Branton et al., 1981; Takakuwa, 2000; Bennett & Baines, 2001). Instead, I will focus on the properties of the actin filament linkers and regulation of their polymerization and dynamics by Tmod1, αβ-adducin, TMs, and dematin.

Table 2.1 RBC membrane skeleton actin-binding proteins

Protein Molecular weight (Da)	**Copies/actin filament**[a]	**Actin binding function**
β Actin 42,000	12–17 Subunits (30–40,000 filaments/cell)	Polymerizes to ∼35–37 nm long filaments at nodes of spectrin–actin hexagonal lattice.
Capping proteins		
Tropomodulin (Tmod1) 40,000	2 Monomers	Caps pointed ends of actin filaments in membrane skeleton (K_{cap} ∼100 nM for pure actin). Binds TM which promotes capping of TM-actin filaments (K_{cap} ∼2 nm). Specifies precise actin filament lengths.
Adducin[b] α 103,000 β 97,000	1–2 αβ Heterodimers	Caps barbed ends of actin filaments in membrane skeleton (K_{cap} ∼100 nm). Recruits β1-spectrin to actin filaments near barbed ends (K_d ∼15 nm). Bundles actin filaments. Ca^{++}-calmodulin binding or PKA phosphorylation inhibits adducin binding to actin. Links actin to membrane by binding band 3 or glucose transporter.
Capping protein (EcapZ) α 36,000 $β_2$ 32,000	∼2 $α1β_2$ Heterodimers (in cytosol)	In absence of adducin, caps actin filament barbed ends in membrane skeleton (K_{cap} ∼1 nM).
Side-binding proteins		
Tropomyosin (TM) TM5b ∼29,000 TM5NM1 ∼27,000	2 TM5b or TM5NM1 Homodimers	Stabilizes actin filaments in membrane skeleton. May help specify precise actin filament lengths with Tmod1. Mg^{++}-dependent association with actin filaments in membrane.
Caldesmon 71,000	2 Monomers	May strengthen TM binding to actin and stabilize filaments. May also regulate actomyosin ATPase. No *in vivo* data. Mg^{++}-dependent association with actin filaments in membrane.

Table 2.1 RBC membrane skeleton actin-binding proteins—cont'd

Protein Molecular weight (Da)	Copies/actin filament	Actin binding function
Dematin	3 Monomers	Bundles actin filaments.
48,000	48 kDa (3): 52 KDa (1)	Promotes α1β1-spectrin binding to actin filaments.
52,000		Links actin to membrane by binding glucose transporter.
Spectrin α1 260,000	5–7 α1β1 Heterodimers	Cross-links actin filaments into a hexagonal lattice via β1 subunit tail binding to actin.
β1 225,000		Links actin to membrane via binding to ankyrin, which links to band 3.
Protein 4.1 R	5–6 Monomers	Strengthens α1β1-spectrin binding to actin. Binds to β1-spectrin and actin forming a ternary complex.
78,000		Linsks actin to membrane via binding to band 3 or glycophorin C.

[a]This value is experimentally determined for each component (see text for references).
[b]Adducin is also an actin filament side-binding protein, as indicated by its actin bundling function.
References for information in this table are provided in the relevant sections of the text and see Fowler (1996).

3. HISTORY

3.1. Discovery of actin filaments as linkers in the spectrin–actin lattice

Actin was identified in human RBCs by Ohnishi in 1962 based on its filamentous structure and ability to activate muscle myosin ATPase (Ohnishi, 1962, 1977), and later, it was purified and its polymerization properties were characterized by several groups (Nakashima & Beutler, 1979; Sheetz, Painter, & Singer, 1976; Tilley & Ralston, 1984; Tilney & Detmers, 1975). Human RBC actin consists exclusively of the β-actin isoform, providing a useful source for studies of β-actin's biochemical properties (V.M. Fowler, unpublished data; Pinder, Ungewickell, Bray, & Gratzer, 1978). Improved purification methods have been developed, but have so far not been taken advantage of for studies of β-actin properties (Pinder, Sleep, Bennett, & Gratzer, 1995; Schafer, Jennings, & Cooper, 1998).

The first evidence that actin was a linking element in a spectrin network on the cytoplasmic surface of the RBC membranes was obtained by Tilney and Detmers (1975), who concluded from transmission electron microscopy (TEM) studies of membranes that actin and spectrin formed an "anastomosing framework like a net woven by a myopic fisherman (not too well-ordered)." Subsequent elegant studies of membrane skeleton ultrastructure by TEM revealed a horizontally organized network of thin (~9 nm) spectrin strands linked to the lipid bilayer via vertical connectors, most likely consisting of ankyrin attached to the cytoplasmic domain of band 3 (Tsukita, Tsukita, & Ishikawa, 1980; Tsukita, Tsukita, Ishikawa, Sato, & Nakao, 1981). In these preparations, the actin filaments themselves could not be directly visualized *in situ*, leading to early proposals that spectrins were linked into a network via interactions with actin monomers (Pinder et al., 1978; Sheetz, 1979; Tilney & Detmers, 1975). The difficulty of observing actin filaments *in situ*, together with spectrin's abundance, elongated shape, and ability to self-associate, also led to an alternative idea that spectrin strands formed a self-associating polymeric network (without actin) directly attached to the lipid bilayer.

The concept that α1β1-spectrin was associated with short actin "protofilaments" in RBCs emerged at this time, based on the stoichiometry in cells of actin and filament ends and their polymerizing activities (Pinder, Clark, Baines, Morris, & Gratzer, 1981). For example, large complexes of spectrin, 4.1R, and actin were isolated from membranes that behaved functionally like actin filament seeds (short filaments), stimulating polymerization of exogenous actin from their barbed ends (Brenner & Korn, 1980; Cohen & Branton, 1979; Lin & Lin, 1979; Pinder, Bray, & Gratzer, 1975; Pinder, Ohanian, & Gratzer, 1984; Pinder, Ungewickell, Calvert, Morris, & Gratzer, 1979; Sato, Yanagida, Maruyama, & Ohnishi, 1979). Ultrastructural examination of these oligomeric spectrin–4.1R–actin complexes revealed spiderlike structures with several 200 nm-long spectrin molecules attached to central nodes; extended networks were observed under conditions promoting spectrin tetramer formation (Beaven et al., 1985; Matsuzaki, Sutoh, & Ikai, 1985; Shen, Josephs, & Steck, 1984). The strong actin nucleating activity of the actin seeds in these oligomeric spectrin–4.1R–actin complexes explained previous observations that partially purified preparations of spectrin-stimulated actin polymerization, which had confused the field for some time into thinking that spectrin itself was an actin nucleator or could itself polymerize into long filaments (Marchesi & Steers, 1968; Pinder et al., 1975). Evidence for the existence of short actin "protofilaments" associated with the RBC membrane also derived from quantitative cytochalasin binding assays for barbed filament ends (Lin, 1981; Lin & Lin, 1978, 1979) and DNAseI binding assays for pointed filament ends in membranes (Podolski & Steck, 1988). Based on the numbers of filament ends and the total numbers of actin monomers per cell, a number average of 30–40,000 short filaments containing 12–17 subunits each were predicted to be associated with the membrane of each RBC (Pinder et al., 1981; Pinder & Gratzer, 1983).

A definitive role for actin filaments in long-range spectrin network formation was finally established, based on reconstitution experiments with purified proteins in the late 1970s and early 1980s, which showed that a spectrin–actin network only formed from actin filaments cross-linked by spectrin tetramers and not by self-association of spectrin itself (Brenner & Korn, 1979; Cohen, Tyler, & Branton, 1980; Fowler & Taylor, 1980; Ungewickell, Bennett, Calvert, Ohanian, & Gratzer, 1979). The ingredients for α1β1-spectrin–actin network formation are (1) actin filaments with spectrin attachment sites; (2) $(\alpha1\beta1)_2$-spectrin tetramers with two actin binding sites, one at each end, allowing cross-linking of one actin filament to another; and (3) protein 4.1R binding to spectrin and actin, enhancing α1β1-spectrin's binding affinity for actin filaments. Interestingly, protein 4.1R is not required for actin filament network formation with $(\alpha1\beta1)_2$-spectrin from sheep RBCs or with nonerythroid $(\alpha2\beta2)_2$-spectrin (fodrin), as these spectrin tetramers bind actin with sufficient affinity to cross-link actin filaments effectively on their own (Bennett, Davis, & Fowler, 1982; Brenner & Korn, 1979; Coleman et al., 1989). The biochemistry and structure of spectrin and protein 4.1R interactions with actin filaments has been the topic of other reviews and will not be covered here (Bennett & Baines, 2001; Cohen, 1983; Lux & Palek, 1995; Takakuwa, 2000).

3.2. Actin filaments are nodes in a quasi-hexagonal symmetric spectrin–actin lattice

A wealth of biochemical studies measuring the stoichiometries of actin, actin-binding proteins, and numbers of filament ends per cell provided compelling evidence for the existence of short actin filaments connecting the spectrin strands in the membrane skeleton, as depicted in several reviews in the 1980s (Branton et al., 1981; Cohen, 1983; Lux, 1979; Pinder et al., 1981). Nevertheless, direct visualization of the structural organization of the spectrin–actin network *in situ* on the membrane remained elusive, due to the amazing density of spectrin and associated proteins, making it impossible to visualize the actin filaments clearly (Pinder et al., 1981; Tilney & Detmers, 1975; Tsukita et al., 1980). A breakthrough in the field came when membrane skeletons were visualized by negative staining electron microscopy after expansion at low ionic strength and mechanical stretching while spreading on grids (Fig. 2.2A; Byers & Branton, 1985; Liu et al., 1987; Shen et al., 1986; Terada, Fujii, & Ohno, 1996). These studies revealed that the membrane skeleton network consists of long spectrin strands attached to central nodes of morphologically recognizable short actin filaments, forming the strands and vertices of a quasi-hexagonal symmetric lattice, as diagrammed schematically in Fig. 2.3B. Measurements from electron micrographs revealed that the short actin filaments were quite uniform in their lengths (33 ± 5 nm), with five to seven ~200 nm-long $(\alpha1\beta1)_2$-spectrin tetramers attached by their distal ends to each short filament. The head-to-head self-association sites of the α1β1-spectrin dimers were located in the middle of the 200 nm strands, with a globular particle corresponding to ankyrin attached to the spectrin strands about 30 nm from the middle (Byers & Branton, 1985; Liu et al., 1987), consistent with the location of the ankyrin binding site on β1-spectrin (Bennett & Baines, 2001; Branton et al., 1981).

Immunogold labeling of spread membrane skeletons further demonstrated conclusively that protein 4.1R, Tmod1, TMs, dematin, and α-adducin are all located at the central nodes of the hexagonal lattice with the actin filaments (Fig. 2.3B and C; Derick, Liu, Chishti, & Palek, 1992; Ursitti & Fowler, 1994; Ursitti & Wade, 1993). However, the relatively low resolution of this labeling approach did not provide any information about the exact locations and structural associations of the spectrin or the other actin-binding proteins in the JCs. Thus, models for the molecular organization of the short actin filaments in the JCs (Figs. 2.1C and 2.3B, C) were derived from biochemical and morphological investigations of protein–protein interactions and determinations of the numbers of actin and each actin-binding protein per cell (Table 2.1; Bennett & Baines, 2001; Branton et al., 1981; Cohen, 1983; Fowler, 1996; Mohandas & Gallagher, 2008; Salomao et al., 2008). While spectrins are typically depicted as attached randomly along the length of the short actin filaments (Fig. 2.3C), other locations for spectrin binding sites have been proposed

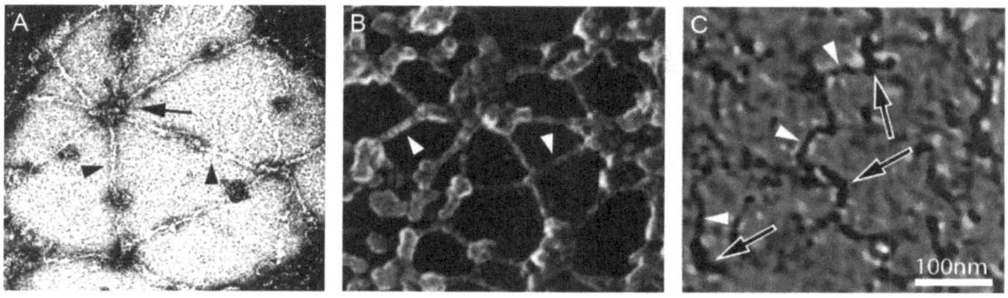

Figure 2.2 Electron microscopy images of the RBC membrane skeleton. (A) Image of the expanded spectrin–actin lattice visualized *en face* by negative staining TEM. Short actin filaments (~35–37 nm; black arrows) are located at the vertices of a quasi-symmetric hexagonal lattice whose strands are ~200 nm-long spectrin tetramers (arrowheads). Between 4 and 7 spectrin strands are attached to each actin filament. (B) Image of the membrane skeleton *in situ*, visualized in replicas of unexpanded membrane skeletons prepared by Triton permeabilization and fixation followed by rapid freezing, freeze-drying, and platinum/carbon shadowing. Connecting strands of varying thicknesses and lengths are evident, formed by self-association of spectrins (white arrowheads), which intersect at 3- and 4-way junctions, as previously described (Ohno, Terada, Fujii, & Ueda, 1994; Ursitti, Pumplin, Wade, & Bloch, 1991; Ursitti & Wade, 1993), but actin filaments are not visible, likely obscured by the numerous globular particles. (C) Image of the unexpanded membrane skeleton visualized in cryo-electron tomograms of Triton-extracted membranes quick-frozen in low ionic strength buffer. Convoluted spectrin strands of varying thickness and length are evident (white arrowheads), intersecting with one another as in B. Denser, thick rodlike structures from which many thin spectrin strands emanate are also evident, likely representing actin filaments (black arrowheads). These actin filaments are shorter than expected (~27 nm), possibly due to some actin dissociation during preparation, and some are distinctly bent, which is unexpected. *Panel (A) reproduced from Fig. 3 in Byers and Branton (1985); panel (B) reproduced from Fig. 4A in Moyer et al. (2010); and panel (C) individual slice of a tomogram, reproduced from Fig. 4A in Nans, Mohandas, and Stokes (2011).*

(Fig. 2.3D–F). For example, based on the ability of RBC TMs to inhibit α1β1-spectrin binding to actin in cosedimentation assays, spectrins were proposed to attach to actin subunits not covered by TMs and located near filament ends (Fowler & Bennett, 1984b; Fig. 2.3D). Later, based on Tmod1 ability to bind TM and cap actin pointed ends and adducin's ability to recruit spectrin and cap actin barbed ends (Sections 4.1.1 and 4.2.1), the spectrin attachment sites were relocated to TM-free actin subunits near the barbed filament end (Fig. 2.3E; Fowler, 1996; Kuhlman, Hughes, Bennett, & Fowler, 1996). Fluorescence polarization microscopy of actin filament orientations using rhodamine phalloidin labeling of RBC membranes under deformation indicates that filaments have a random azimuthal orientation tangential to the bilayer (Discher, 2000; Picart, Dalhaimer, & Discher,

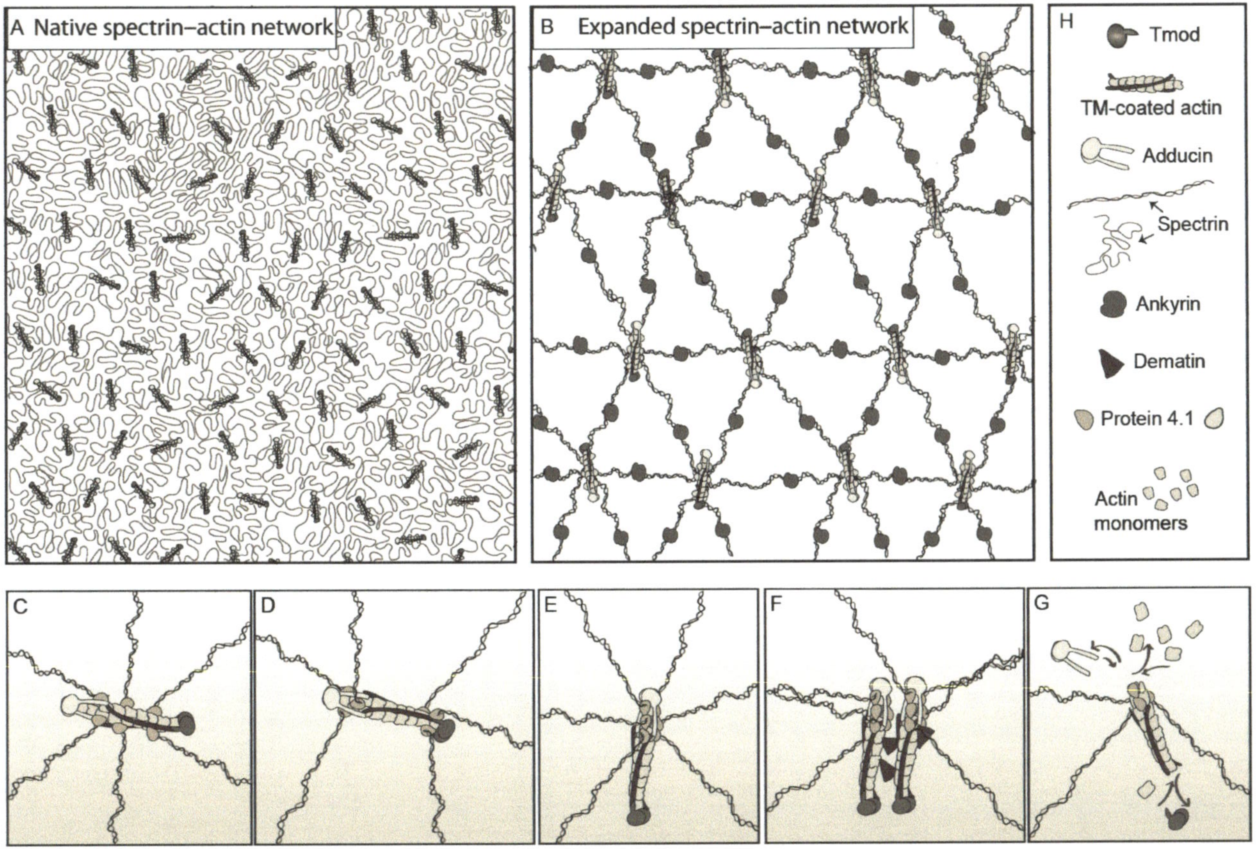

Figure 2.3 Spectrin–actin lattice organization viewed *en face* at the cytoplasmic surface of the RBC membrane. (A) Schematic of the density of the spectrin–actin lattice *in situ*, depicting long, convoluted spectrin strands attached to short actin filaments approximately ~60 nm apart. (B) Schematic of the symmetric (quasi-)hexagonal organization of the spectrin–actin lattice in well-spread preparations of the membrane skeleton, based on images of specimens visualized by negative staining TEM. The distances between adjacent actin filaments in the extended

lattice are ~200 nm, that of a fully extended $(\alpha 1\beta 1)_2$-spectrin tetramer (Byers & Branton, 1985; Liu, Derick, & Palek, 1987; Shen, Josephs, & Steck, 1986). (C–G) Enlargement of an actin filament, depicting alternative molecular configurations. Each actin filament is 12–17 subunits long (~35–37 nm), associated with 5–7 α1β1-spectrin dimers and 4.1R molecules (spectrin:4.1R = 1:1), two Tmod1s, two TM homodimers (TM5b and TM5NM1), one αβ-adducin heterodimer, and three dematin monomers (Table 2.1; Fowler, 1996; Gilligan & Bennett, 1993). Protein 4.1R binds to the end of the α1β1-spectrin dimer near α1β1-spectrin's actin binding site and to the actin filament, promoting spectrin binding along the side of the actin filament. Tmod1s cap the pointed filament end where they also bind to the end of each TM rod, which span the actin filament, and may restrict spectrin binding to TM-free actin subunits, as depicted in D and E. An αβ-adducin heterodimer caps the actin filament barbed end, likely recruiting spectrins to sites on actin near the barbed end, as depicted in E. The location of dematin is less certain and may gather filaments into bundles, as depicted in F. αβ-Adducin and/or Tmod1 capping may be dynamic under some conditions, allowing actin subunit association and dissociation with filament ends, as depicted in G. See text for details regarding each protein's interactions with actin filaments. *Panel (A) drawn from a quick-freeze deep-etch TEM image in Fig. 2b from Coleman, Fishkind, Mooseker, and Morrow (1989) and panel (B) schematic adapted from Moyer et al. (2010).* (For color version of this figure, the reader is referred to the online version of this chapter.)

2000; Picart & Discher, 1999), which may be accommodated by filament structures in Fig. 2.3C or E, with Fig. 2.3D less likely. Such considerations of mechanics of actin filaments suspended in a spectrin network attached to the membrane also led to models with spectrins attached periodically along the short actin filament, projecting radially due to the helical symmetry of the filament (e.g., Fig. 2.3C; radial disposition not shown; Sche, Vera, & Sung, 2011; Zhu, Vera, Asaro, Sche, & Sung, 2007).

3.3. Actin filament structures in the membrane skeleton *in situ*

What is known about the structural basis for actin filament associations in the membrane skeleton in unspread RBC membranes *in situ*? A tantalizing image from John Heuser at Washington University showed 67 nm actin filaments (nuggets) connected by spaghettilike spectrin strands in NP40/NaCl-extracted membrane skeletons (depicted schematically in Fig. 2.3A), but this was not followed up (see Fig. 2b in Coleman et al., 1989). In the 1990s, several investigators used quick-freezing, deep etching, and rotary shadowing–TEM to visualize native membranes, revealing a highly interconnected, complex network topography with numerous associated globular particles (e.g., Fig. 2.2B; Ohno et al., 1994; Terada et al., 1996; Ursitti & Wade, 1993; Ursitti et al., 1991). Many strand intersections in the network were evident, some due to spectrin connections with JCs containing actin (as expected from the spread images), but many others were ascribed to spectrin–spectrin lateral contacts at non-actin junctions based on immunogold labeling and measurements of strand thicknesses (Ursitti et al., 1991; Ursitti & Wade, 1993). Atomic force microscopy (AFM) was also used to visualize network topology on the cytoplasmic surface of the RBC membrane, but again, actin filaments were not identifiable (Liu, Burgess, Mizukami, & Ostafin, 2003; Swihart, Mikrut, Ketterson, & Macdonald, 2001; Takeuchi, Miyamoto, Sako, Komizu, & Kusumi, 1998). Recently, cryo-electron tomography has succeeded at identifying actin filaments in intact human RBCs preserved by plunge-freezing, revealing short actin filaments, 30–40 nm long and 6.8 ± 0.5 nm thick, satisfyingly confirming previous TEM data from the negatively stained spread membrane skeleton preparations (Cyrklaff et al., 2011). However, the thin spectrin strands could not be detected in the tomograms of the frozen intact cells, nor was the resolution sufficient to visualize actin filament subunit structure and associated proteins. The presence of high cytosolic concentrations of electron-dense hemoglobin undoubtedly interfered with the visualization of spectrin or actin filament features in these tomograms.

To get around this, Nans and colleagues used cryo-electron tomography to visualize the membrane skeleton of ghosts from which hemoglobin had been removed by osmotic lysis followed by extraction by Triton (Nans et al., 2011). These preparations also revealed a complex and variable topology of the spectrin–actin network, with strands converging at a variety of junctions formed by short actin filaments (JCs), or spectrin–spectrin intersections, remarkably similar to the results of the prior quick-freeze deep-etch studies (Fig. 2.2C; Ursitti et al., 1991; Ursitti & Wade, 1993). Curiously, Nans et al. (2011) observed that the short actin filaments often appeared to be bent in the middle (Fig. 2.2C). Regrettably, the resolution of the tomogram images was insufficient to identify the structural features of the actin filaments and their associated proteins. Future progress towards elucidating the structure of actin filaments in JCs, and their disposition in the native membrane skeleton, likely awaits improved sample preparations along with higher-resolution electron microscopy and computational image averaging approaches across many JCs. Such investigations would be expected to provide insights into the structural basis for actin filament end capping (not well understood in any system) and the structural basis for the quasi-hexagonal symmetry of the spectrin–actin lattice (i.e., what determines the binding of 5–7 spectrins to each filament?).

3.4. Actin filament capping restricts filament lengths in RBCs

Actin filaments are polarized polymers of actin subunits, with one filament end that polymerizes and depolymerizes at about 10 × the rate of the other; the former is referred to as the fast-growing (barbed) end, while the latter is referred to as the slow-growing (pointed) end. During assembly, actin filaments can elongate up to many microns in length, but the RBC actin filaments are less than 40 nm long (Section 3.2). At steady state, actin monomers continue to associate and dissociate from filament ends, so that over time, purified actin filaments achieve an exponential length distribution with filaments of varying lengths (Littlefield & Fowler, 1998). Thus, the uniform (Gaussian) length distribution of the short RBC actin filaments suggests that they are capped tightly at both ends to prevent subunit loss or gain that would otherwise lead to filament length changes over the RBC lifetime (~120 days in humans, ~40 days in mice). In the 1990s, I and my colleagues identified RBC Tmod1 and αβ-adducin as the pointed and barbed end actin filament capping proteins, respectively, supporting the idea that actin capping restricts RBC actin filament length (Section 4). This is a nice example of how the unique properties of the RBC membrane (short filaments with abundant numbers of filament ends) enabled discovery of novel actin capping proteins and provided insights into the important problem of actin filament length regulation in all cells.

Despite the *a priori* necessity for actin capping proteins to restrict actin filament lengths, the idea that RBC actin filaments were capped at both ends was under dispute for some time before the Tmod1 and αβ-adducin cappers were discovered. For example, under some conditions, exogenous actin was observed to elongate from the ends of the short red cell actin filaments, indicating that filament ends are not always capped (Byers & Branton, 1985; Pinder & Gratzer, 1983; Pinder, Weeds, & Gratzer, 1986; Podolski & Steck, 1988; Tsukita, Tsukita, Tsukita, Hosoya, & Mabuchi, 1985; Tsukita, Tsukita, & Ishikawa, 1984). In some investigators' experiments, incubation of the exposed cytoplasmic surface of ghosts with actin monomer concentrations above the barbed but below the pointed end critical concentration led to elongation only from barbed ends, while incubation at concentrations above the pointed end critical concentration led to elongation from both ends—results similar to experiments with purified uncapped filaments (Tsukita et al., 1984, 1985). In others, elongation was only observed from barbed, but not pointed ends (Pinder & Gratzer, 1983; Pinder et al., 1986; Podolski & Steck, 1988). Experiments measuring binding of dihydrocytochalasin B (binds specifically to barbed ends) or DNAseI (binds specifically to pointed ends) to ghost membranes were also consistent with the existence of many short, uncapped red cell actin filaments (Lin & Lin, 1978; Podolski & Steck, 1988).

Subsequent investigations revealed that the low ionic strength conditions typically used to purify RBC membranes most likely led to filament uncapping. For pointed ends, low ionic strength conditions without magnesium extract RBC TMs (Fowler & Bennett, 1984a, 1984b), Tmod1's binding partner, and would be expected to convert Tmod1 to a low-affinity cap (Section 4.1.1), thus allowing actin subunit addition and filament elongation from pointed ends or DNAseI binding by displacement of the weak Tmod1 cap from the pointed ends. For the barbed ends, osmotic lysis and washing of ghosts in low ionic strength buffers without divalent cations leads to extraction or uncapping by αβ-adducin, allowing actin subunit addition and filament elongation, or binding of EcapZ (a barbed end capping protein) to the free barbed ends (DiNubile, 1999; Kuhlman, 2000; Kuhlman & Fowler, 1997; Section 4.3).

Actin filament breakage during osmotic lysis and centrifugal shearing of RBCs to prepare ghosts may also have accounted for the appearance of new filament ends, based on the presence of fewer EcapZ binding sites on membranes when the filament stabilizer, phallacidin, was included in the osmotic lysis buffers (Kuhlman & Fowler, 1997). This raises the possibility that at least some of the short actin filaments observed at nodes of the quasi-hexagonal spectrin–actin lattice prepared by low ionic strength expansion and mechanical stretching may have been created by filament breakage (Byers & Branton, 1985; Liu et al., 1987; Shen et al., 1986). The idea that some RBC actin filaments may be longer than is commonly accepted was originally proposed by Atkinson and colleagues, from observations of ~100 nm-long actin filaments in extracts prepared from membranes by phalloidin stabilization, mild proteolysis, and gel filtration (Atkinson, Morrow, & Marchesi, 1982). Long actin filaments have also been observed in spectrin–actin networks prepared by nonionic detergent extraction followed by high salt extraction (Shen et al., 1984). However, proteolysis or extraction of filament caps, followed by end-to-end annealing of the short filaments, cannot be ruled out in these preparations.

4. RBC ACTIN FILAMENT CAPPING PROTEINS: PROPERTIES AND FUNCTIONS

4.1. Tropomodulin1 (Tmod1) is the pointed end capper

4.1.1 Tmod1 binds TM and actin to cap filament pointed ends

The abundance of capped actin filament ends in the RBC membrane skeleton (short filaments have high numbers of ends with respect to total actin) enabled the serendipitous discovery of Tmod1, the founding member of the Tmod family of pointed end capping proteins (Table 2.1) (for a review, see Yamashiro et al., 2012). Tmod1 was identified and purified from ghost membranes on the basis of its ability to bind RBC TMs, for which it was initially termed a "TMBP" (TM-binding protein; Fowler, 1987, 1990). At the time, I had been looking for a TM-binding protein with troponin-like properties that might regulate actomyosin and RBC shape (Section 5.1.2) but instead turned up a completely different molecule (Fowler, 1987), which bound to the end of RBC TMs and prevented cooperative binding of the TMs along actin filaments (Fowler, 1990). This led to the idea that Tmod1 might regulate RBC actin filament length via preventing TMs' head-to-tail polymerization along actin filaments. We only suspected Tmod1 to be an actin filament pointed end cap after immunofluorescence staining of skeletal muscle myofibrils showed Tmod1 localization at the thin filament pointed ends (Fowler, Sussmann, Miller, Flucher, & Daniels, 1993). This motivated us to directly Tmod1 for pointed end capping in pyrene–actin polymerization assays, using actin seeds capped at their barbed ends by gelsolin—the method required to detect subunit association/dissociation from the 10× slower polymerizing pointed ends (Weber et al., 1994). In these assays, Tmod1 specifically inhibited actin association and dissociation rates at pointed ends without binding monomers, barbed ends, or filament sides, and Tmod1's pointed end capping activity was enhanced by TM (Weber et al., 1994; Weber, Pennise, & Fowler, 1999). Note that previous attempts to identify RBC pointed end capping factors were hindered by poor assay design and interference by barbed end events, leading to the mistaken attribution of pointed end capping activity to spectrin and protein 4.1 (e.g., Pinder et al., 1984).

Tight capping of actin filaments by Tmod1 depends on cooperative protein–protein associations at the filament pointed end. Tmod1 is an asymmetric monomer in solution (Fowler, 1987) and, on its own, has a relatively weak affinity ($K_{cap} \sim 100$–200 nm) for the actin filament pointed end, insufficient to prevent actin association/dissociation and filament length changes (Weber et al., 1994). Tmod1 is converted to a high-affinity cap via binding to TM, a rodlike protein (Section 5.1) that binds along the sides of actin filaments (Kostyukova & Hitchcock-DeGregori, 2004; Weber et al., 1994, 1999). High-affinity capping requires direct binding of Tmod1's N-terminal domain to TM, together with binding of two sites in Tmod1's N-terminal and C-terminal domains to actin. The C-terminal actin capping site does not require TM (K_{cap} ~0.2–0.4 μM; Fowler, Greenfield, & Moyer, 2003), while the second, weaker, actin binding site in the N-terminal domain depends on TM binding to an adjacent region for capping activity (K_{cap} ~0.02–0.2 nM; Fowler et al., 2003; Kong & Kedes, 2006; Kostyukova, Choy, & Rapp, 2006; Kostyukova, Rapp, Choy, Greenfield, & Hitchcock-DeGregori, 2005). Based on these multiple interactions, Tmod1's affinity for TM–actin pointed ends is enhanced by several orders of magnitude as compared to filaments without TMs (K_{cap} ~2 nM for RBC TM5b and 50 pM for skeletal muscle α/β-TM; Weber et al., 1999; S. Yamashiro and V.M. Fowler, unpublished data). TM associations with actin filaments are also stabilized by Tmod1 capping, since the terminal

TMs at the end of the filament can interact with both actin and Tmod1 (Mudry, Perry, Richards, Fowler, & Gregorio, 2003). Thus, ternary associations of Tmod1, TMs, and actin at the pointed filament end can cap the filament pointed end tightly to prevent RBC actin filament growth or shrinkage. While only 1 Tmod1 molecule is required to cap TM–actin filament pointed ends *in vitro* (Weber et al., 1999), there are two Tmod1 molecules associated with each short actin filament in the RBC membrane (Moyer et al., 2010). A comprehensive review of Tmod structure, properties, and functions was published recently (Yamashiro et al., 2012).

4.1.2 *Tmod1 regulates RBC actin filament lengths and membrane skeleton integrity* in vivo

Tmods are ~40kD monomeric proteins encoded by four closely related genes in mammals, *Tmods 1–4*. Tmod1 is expressed in postmitotic, differentiated cells such as striated muscle, lens fiber cells, neurons, epithelial cells, and mature mammalian RBCs, while Tmod3 is expressed in erythroid progenitors as well as in many other cell types (Sui, Nowak, Bacconi, et al. 2013; Yamashiro et al., 2012). Global deletion of Tmod1 in mice is embryonic lethal at E8.5–9.5 due to defects in cardiac development and contractile function (Chu et al., 2003; Fritz-Six et al., 2003). In addition, the primitive nucleated RBCs circulating at this stage of embryonic development display mechanical instability in the absence of Tmod1 (Chu et al., 2003). The embryonic lethality and development can be rescued by introduction of a Tmod1 transgene under the control of the cardiac-restricted, α-myosin heavy chain promoter, allowing studies of Tmod1-null RBCs in adult mice (McKeown, Nowak, Moyer, Sussman, & Fowler, 2008). Tmod1-null mouse RBCs are sphero-elliptocytic in shape and osmotically fragile with reduced deformability, leading to a mild, compensated anemia resembling human hereditary sphero-elliptocytosis (Table 2.2; Moyer et al., 2010). The Tmod1-null mouse hematological phenotype is characteristic of RBC defects with mutations or deficiencies in membrane skeleton components. Such defects compromise the stability of the membrane skeleton, resulting in reduced RBC survival and life span (Mohandas & Evans, 1994; Mohandas & Gallagher, 2008).

Does Tmod1 regulate actin filament assembly, length, or stability *in vivo*? Negative staining electron microscopy of spread membrane skeletons reveals abnormally variable filament lengths, ranging from 19 to 56 nm in Tmod1-null RBCs, as compared to the expected narrow range of 32–42 nm in wild-type RBCs (Moyer et al., 2010). Moreover, electron microscopy of critical

Table 2.2 Phenotypes of actin regulatory protein knockouts

Mutation	RBC phenotype	Isoform compensation	Altered membrane skeleton proteins	Fold change	Actin and membrane skeleton structure
Tmod1 −/−[a]	Mild hemolytic anemia Sphero-elliptocytosis Osmotically fragile Reduced deformability	Tmod3 present at 1/5th wild-type Tmod1 levels	No changes		Actin filament numbers similar, but lengths variable (TEM) Skeleton network pore sizes larger (TEM)
β-Adducin −/−[b]	Mild hemolytic anemia Sphero-elliptocytosis Osmotically fragile Reduced deformability	α-Adducin—0.2–0.3 × γ-Adducin—4–5 ×	EcapZ Tmod1 Actin TM(CH1) Dematin 52 kD	~9 × ~1.65 × ~0.85 × ~0.35 × ~1.8 ×	Skeleton network elements damaged and aggregated (AFM)
γ-Adducin −/−[c]	Normal	α- and β-adducin levels normal	Normal		ND
γ,β-Adducin −/−[d]	Mild hemolytic anemia Sphero-elliptocytosis Osmotically fragile Reduced deformability	α-Adducin—<1%	EcapZ TM(CH1)	>10 × Slightly reduced	ND

α-Adducin−/−[e]	Mild hemolytic anemia Sphero-elliptocytosis Osmotically fragile Reduced deformability	No β-adducin No γ-adducin	EcapZ α EcapZ β TM(CH1)	Increased Unchanged ~0.20×	ND
Dematin Headpiece−/−[f]	Mild hemolytic anemia Sphero-elliptocytosis Osmotically fragile Reduced deformability	Truncated 40 kD dematin at 30% wild-type dematin levels	Actin Actin, spectrin more extractable in TX-100	~0.35×	Skeleton network elements damaged and aggregated (AFM)
Dematin Headpiece−/−; β-adducin−/−[g]	Severe hemolytic anemia Spherocytosis Microcytosis Osmotically fragile	Truncated 40 kD dematin at 30% WT levels γ-Adducin present	Spectrin Actin 4.1R Actin, spectrin more extractable in TX-100	~0.85× ~0.85× Reduced	Actin aggregates (IF) Skeleton network elements damaged and aggregated (AFM)

(Continued)

Table 2.2 Phenotypes of actin regulatory protein knockouts—cont'd

Mutation	RBC phenotype	Isoform compensation	Altered membrane skeleton proteins	Fold change	Actin and membrane skeleton structure
Rac1 −/−; *Rac2* −/−[h]	Mild hemolytic anemia	None	Actin	~2.6×	Actin aggregates (IF)
	Microcytosis		Adducin, dematin	Reduced	
	Fragmentation		Adducin P-Ser724	Increased	
	Osmotically fragile		Actin, P-adducin more extractable in TX-100		Skeleton network irregular and aggregated (TEM)
	Reduced deformability				
Hem-1 −/− *(WAVE-family member)*[i]	Mild hemolytic anemia	WAVE1, WAVE2, Abi2 in WT and KO	Adducin, dematin, Tmod1, β-spectrin, ankyrin, 4.1R, band 3, p55	~0.2–0.5×	Actin aggregates (IF)
	Microcytosis		Phospho-adducin	~2.6×	
	Fragmentation		Tmod3	~2.6×	
	Osmotically fragile				

[a]Moyer et al. (2010)
[b]Gilligan et al. (1999), Muro et al. (2000), Porro et al. (2004), Chen et al. (2007)
[c]Sahr et al. (2009)
[d]Sahr et al. (2009)
[e]Robledo et al. (2008)
[f]Khanna et al. (2002)
[g]Chen et al. (2007), Liu, Khan et al. (2011)
[h]Kalfa et al. (2006)
[i]Chan et al. (2013)

point dried, rotary shadowed preparations of unspread skeletons reveals an attenuated network with larger and more variable pore sizes, indicating that the long-range organization of the membrane skeleton is also abnormal. These filament length changes and network architectural abnormalities are likely due to molecular rearrangements, since the total levels of actin, TMs, α- and β-adducins, dematin, and α1- and β1-spectrin are normal in the absence of Tmod1 (Table 2.2). Thus, exactly how such relatively small changes in actin filament lengths lead to perturbations in the overall architecture of the membrane skeleton is unclear. This highlights the uncertain structural relationship between the quasi-hexagonal symmetry of the spectrin–actin lattice in spread preparations (Fig. 2.2A) and the dense and irregular membrane skeleton network visualized in unspread preparations (Fig. 2.2B and C), as discussed earlier (Section 3.3).

The mild phenotype likely results from the appearance of Tmod3, an isoform not normally found in wild-type mouse (or human) mature RBCs. Since Tmod3 message and protein is present in RBC progenitors during terminal differentiation (Sui et al., 2013), Tmod3 protein likely persists in mature Tmod1-null RBCs by binding to vacant Tmod1 binding sites at actin filament pointed ends. However, Tmod3 is present in the Tmod1-null RBCs at only 1/5 of Tmod1 levels normally present in wild-type RBCs, indicating that the misregulated and variable actin filament lengths in Tmod1-null RBCs can be explained by capping of some but not all filaments by Tmod3 (Moyer et al., 2010). For some uncapped filaments, actin and TM may dissociate and filaments shorten, while others may lengthen by addition of the previously dissociated actin subunits and their stabilization with another pair of TMs (see Fig. 9 in Moyer et al., 2010). Actin monomer binding by Tmod3 (a function specific to Tmod3) may further destabilize the actin filaments (Fischer et al., 2006; Yamashiro, Speicher, Speicher, & Fowler, 2010). It is not known whether initial assembly of short actin filaments into the membrane skeleton is abnormal in the absence of Tmod1 or whether the observed length variability results from length redistribution during RBC passage through the circulation, possibly as a consequence of mechanical stresses resulting in filament instability and subunit loss. To date, Tmod1 is the only protein shown to regulate the precise lengths of the short actin filaments in the RBC membrane skeleton.

4.1.3 Significance

Tmods, first discovered in RBCs in 1987, are the only known proteins to cap actin filament pointed ends and are now established as a unique and conserved family of TM-regulated, actin capping proteins present in all metazoans (Yamashiro et al., 2012). Biochemical, cell biological, and molecular genetic approaches have shown that Tmods regulate the precise actin filament lengths in the RBC spectrin–actin network (as discussed here) as well as in the sarcomeres of striated muscle, both examples of highly organized actin filament architectures (Gokhin & Fowler, 2011). Tmods also control actin assembly and stability in the spectrin-based membrane skeletons of nonerythroid cells, and regulate actin turnover and dynamics in more dynamic cellular contexts (Fischer & Fowler, 2003). In these capacities, Tmods are essential for embryonic development, differentiated cell architectures, tissue mechanics, and physiology [for recent reviews, see Gokhin & Fowler, 2011; Yamashiro et al., 2012].

4.2. Adducin is the barbed end capper

4.2.1 Adducin caps barbed ends and recruits spectrin to actin

RBC adducin was first characterized as a calmodulin-binding, PKC- and PKA-phosphorylated protein in RBCs that could bind to spectrin–actin complexes and promote spectrin binding to actin (Gardner & Bennett, 1986, 1987; Ling, Gardner, & Bennett, 1986; Mische, Mooseker, & Morrow, 1987; Waseem & Palfrey, 1988). Adducin was also shown to bind along the sides of actin filaments and bundle them in a calmodulin-regulated fashion (Mische et al., 1987). Subsequently, two considerations led me and my colleagues to test whether adducin capped the barbed ends of RBC actin filaments (Kuhlman et al., 1996). First, adducin was the only RBC membrane-associated actin-binding protein (other than Tmod1) present at stoichiometric levels with respect to the actin filaments, the right number to be a filament cap (Table 2.1; Fowler, 1996). Second, the other RBC actin-binding proteins (spectrin, protein 4.1R, and dematin) all bound along the sides of actin filaments (Branton et al., 1981; Lux, 1979), leaving adducin as the only likely candidate for a filament end capper. Indeed, we found that purified αβ-adducin inhibited elongation and depolymerization from the free barbed ends of spectrin–actin nuclei (seeds) in pyrene–actin elongation assays, with a K_{cap} ~100 nM (Kuhlman et al., 1996). This then led to the discovery that adducin preferentially recruits spectrin to actin binding sites near barbed ends (Li, Matsuoka, & Bennett, 1998), as had been predicted in a model for the RBC actin filament (Fowler, 1996; Kuhlman et al., 1996). Adducin's barbed end capping activity and ability to recruit spectrin to actin filaments are contained in a basic MARCKS-related tail domain plus a neck domain (Hughes & Bennett, 1995; Kuhlman et al., 1996; Li et al., 1998). Based on a half-maximal concentration of 15 nM for the β-adducin tail + neck domain to recruit β-spectrin to gelsolin-sensitive sites on actin filaments (i.e., barbed ends), it was proposed that adducin's capping affinity may be increased ~10-fold by also binding to β-spectrin on actin (Li et al., 1998). Nevertheless, the capping affinity of adducin remains considerably weaker than that of Tmod1 for TM-coated actin filaments (K_{cap} of ~2 nM for RBC TM5b; S. Yamashiro and V.M. Fowler, unpublished data), suggesting that RBC barbed ends are more likely to be uncapped than are pointed ends *in vivo* (Sections 4.2.2 and 4.3). Indeed, adducin's ability to cap actin or recruit spectrin to actin filaments is inhibited by calmodulin binding to the MARCKS-related tail domain (Gardner & Bennett, 1987; Kuhlman et al., 1996; Mische et al., 1987) or by phosphorylation by PKC and PKA (Matsuoka, Hughes, & Bennett, 1996; Matsuoka, Li, & Bennett, 1998). Conversely, adducin–actin interactions are enhanced by Rho-kinase phosphorylation of two sites in the adducin neck domain (Fukata et al., 1999; Kimura et al., 1998) [for a review on adducin, see Matsuoka, Li, & Bennett, 2000].

4.2.2 Adducin stabilizes the RBC membrane skeleton in vivo

Does adducin regulate RBC actin filament assembly and length *in vivo*? RBC adducin consists of obligate heterodimers (and heterotetramers) of α- and β-subunits (726 and 713 amino acids, respectively) encoded by closely related genes. There is a third, closely related γ-adducin gene not normally expressed in human RBCs (and at very low levels in mouse RBCs), encoding a 674-amino acid polypeptide (Matsuoka et al., 2000). Targeted deletion of the β-adducin gene in mice results in a mild compensated hemolytic anemia, in which RBCs are abnormally shaped and osmotically fragile with reduced deformability (Table 2.2; Gilligan et al., 1999; Muro et al., 2000). The mild phenotype is undoubtedly due to the compensatory upregulation of the γ-adducin gene, which likely forms heterodimers with the α-subunit, but in insufficient levels to completely restore function, since α-adducin levels are reduced to only 20–30% normal. The overall architecture of the membrane skeleton is abnormal, based on atomic force microscopy, which reveals aggregation and damage to network elements (Chen et al., 2007; Liu, Khan, Chishti, & Ostafin, 2011). Unfortunately, no information is available about actin filament lengths, since negatively stained spread membrane skeleton preparations were not studied. However, striking changes in levels of some of the actin-binding proteins associated with the membrane skeleton may provide some clues (Table 2.2; Porro et al., 2004). First, levels of the normally cytosolic barbed end capping protein, EcapZ, are increased nearly 10-fold on the membrane, likely compensating for reduced αβ-adducin by capping the barbed ends of the RBC actin filaments (Kuhlman & Fowler, 1997) (Section 4.3). Second, TM levels are reduced to 1/3 normal and actin is slightly reduced, but Tmod1 levels are unchanged or even slightly increased. Since RBC TMs must span 34 nm along the length of an actin filament to bind (Fowler, 1990), RBC actin filaments may be shorter, which would impair TM binding, leading to loss of TM. Alternatively, filament numbers could be reduced to 1/3. Quantification of the numbers of EcapZ and remaining α- and γ-adducin molecules in the membranes of the β-adducin-null RBCs would be a biochemical approach to address these possibilities.

Targeted deletion of γ-adducin in mice had no RBC phenotype (as expected, due to low γ-adducin expression), and the combined deletion of β- and γ-adducin, which led to <1% normal levels of α-adducin, did not exacerbate the phenotype of the β-adducin-null RBCs (Table 2.2; Sahr et al., 2009). Moreover, deletion of α-adducin led to complete absence of both β- and γ-adducin in RBCs but only a mild compensated hemolytic anemia, similar to the β-adducin nulls, with >10× increased levels of EcapZ on the membrane and some loss of TM (Table 2.2; Robledo et al., 2008). An interesting implication of EcapZ upregulation in absence of adducins (thus capping the filament barbed ends) is that the relatively mild anemia and spherocytic RBC phenotypes may be due principally to loss of the αβ-adducin-mediated attachment of JCs to band 3 (Anong et al., 2009), as well as loss of αβ-adducin-mediated recruitment of spectrin to actin (Gardner & Bennett, 1987; Mische et al., 1987), since presumably EcapZ cannot perform either of these functions. Thus, to further explore the role of barbed end capping in actin filament length regulation, it will be necessary to also interfere with EcapZ function (Section 4.3).

Another line of evidence supports the idea that αβ-adducin–actin interactions are critical for RBC actin assembly and stability. Targeted combined deletion of Rac1 and Rac2 GTPases from RBCs using an inducible Mx-Cre approach resulted in a mild microcytic hemolytic anemia with smaller RBCs displaying abnormal shapes, increased fragmentation, and reduced deformability (Table 2.2; Kalfa et al., 2006). The Rac1/2-null RBC membranes had reduced levels of adducin (isoforms were not determined) and dematin, as well as a ~two- to threefold increased ratio of actin to spectrin. Phosphorylation of adducin at Ser 724, a PKC and PKA site in the adducin MARCKS domain, was increased, and the phosphorylated adducin and actin were more readily extracted from membranes by nonionic detergents at low ionic strength. This indicates reduced interactions of phosphorylated adducin with actin and spectrin, consistent with *in vitro* studies discussed earlier. Fluorescence confocal microscopy of actin filament staining in Rac1/2-null RBCs and TEM of rotary shadowed replicas of membrane skeletons suggest abnormal aggregation of network elements. Yet, since individual actin filaments were not evident in these specimens, no information was obtained regarding filament lengths or numbers or how the spectrin strands were attached to each filament. It is tempting to speculate that Rac-regulated pathways leading to Ser 724 phosphorylation of adducin may result in reduced actin filament capping and impaired recruitment of spectrin to actin, permitting abnormal actin filament growth and misspecification of spectrin attachments to actin, leading to lattice asymmetry and disorganization.

4.2.3 Significance

Adducins, also discovered in RBCs like Tmods, are a unique family of actin filament barbed end capping proteins that recruit spectrin to actin filaments, promoting formation of an extended spectrin–actin network. A fascinating feature of RBC adducin, whose implications have not yet been extended to other cells, is its ability to bind the cytoplasmic domain of the anion channel (band 3, AEI; Anong et al., 2009) and the glucose transporter, Glut1, in human RBCs (Khan et al., 2008), thus directly linking actin filament barbed ends to the membrane. Thus, adducins comprise a novel membrane-associated class of actin filament barbed end capping and network-forming proteins [for reviews, see Gilligan & Bennett, 1993; Matsuoka et al., 2000].

4.3. Capping protein (EcapZ) also caps barbed ends in RBCs

RBCs also contain another actin filament barbed end capping protein, so-called capping protein, a nonmuscle isoform of the striated muscle thin filament capping protein, capZ (Table 2.1; Fowler, 1996). Erythrocyte capZ (EcapZ) is an obligate $\alpha1\beta2$ heterodimer and is fully functional in blocking actin elongation from barbed ends (K_{cap} ~1–5 nM) and in nucleating actin polymerization (Kuhlman & Fowler, 1997). However, EcapZ is present exclusively in the cytosol of mature human RBCs and is only present in the membrane skeleton in the absence of adducin. As discussed above, exogenous EcapZ binds to membrane skeletons from which $\alpha\beta$-adducin has been dissociated by washing at low ionic strength in the absence of magnesium, with binding saturating at levels corresponding to expected numbers of actin filament barbed ends (Kuhlman & Fowler, 1997). Increased amounts of

EcapZ subunits are also detected on membranes of mouse RBCs in which adducins have been genetically deleted (Table 2.2; Section 4.2.2; Porro et al., 2004; Robledo et al., 2008; Sahr et al., 2009). Whether EcapZ has a function in normal RBC biology is not known, but it is possible that EcapZ may play a role in initiating assembly of actin filaments into the membrane skeleton during RBC biogenesis. Studies of EcapZ function *in vivo* may be challenging as the α2 isoform may compensate for absence of the α1, and the β1 isoform may compensate for absence of the β2 (Hart, Korshunova, & Cooper, 1997).

5. RBC ACTIN FILAMENT SIDE-BINDING PROTEINS

5.1. Tropomyosin (TM) stabilizes actin filaments

5.1.1 TM regulation of actin filament length and stability

TMs are coiled-coil, rodlike dimers that bind along the length of actin filaments, stabilizing filaments from disassembly, severing, or mechanical breakage (Gunning, O'Neill, & Hardeman, 2008). In striated muscle, TMs also regulate actomyosin contractile activity via Ca^{++} regulation of the troponin–TM complex. I discovered TMs serendipitously in RBCs as ~30 kD proteins that copurified with RBC actin, cosedimenting with the RBC actin in polymerization assays (Table 2.1; Fowler & Bennett, 1984a, 1984b). A key observation enabling the discovery that these ~30 kD proteins were TMs was based on previous studies that TM–actin interactions are magnesium-dependent; thus, inclusion of magnesium in osmotic lysis and washing buffers was required to retain the TMs on the RBC membranes, resulting in "pink" ghosts (Fowler & Bennett, 1984a, 1984b). Standard procedures for preparation of RBC membranes in low ionic strength and EDTA to generate "white" ghosts led to selective depletion of over 50–80% of the TMs from the RBC membranes.

Two TM isoforms are present in mouse and human RBCs, TM5b, a short TM product of the α-TM (*TPM1*) gene, and TM5NM1, a short TM product of the γ-TM (*TPM3*) gene (Dunn, Mohteshamzadeh, Daly, & Thomas, 2003; Sung et al., 2000; Sung & Lin, 1994). The TM5NM1 (~29 kDa) and TM5b (~27 kDa) proteins in human RBCs are present in an equimolar ratio and associate to homodimers rather than heterodimers, based on oxidative cross-linking (V.M. Fowler, unpublished data). As expected from studies with other TMs, binding of RBC TMs to actin filaments is strongly magnesium-dependent. The RBC TMs bind cooperatively along actin filaments, saturating at a molar ratio of 1 TM for every

6–7 actin subunits, with a Hill coefficient of ~2.8 (Fowler & Bennett, 1984a; Mak, Roseborough, & Baker, 1987), with TM5b one of the tightest actin filament binding TMs described (Maytum, Konrad, Lehrer, & Geeves, 2001). Despite their cooperative binding to actin filaments, RBC TMs self-associate poorly in solution, unlike striated muscle TMs (Mak et al., 1987). In addition, Tmod1 binds to the N-terminal end of RBC TMs (Vera et al., 2000) and effectively blocks TM head-to-tail self-association along actin filaments (Fowler, 1990). Measurement of TM–actin stoichiometry reveals 1 TM for every 7–8 actin subunits or 2 TMs per short RBC filament, one on each actin filament strand (Fowler & Bennett, 1984).

The close correspondence in length of RBC TMs (~34 nm; Fowler, 1990) with the lengths of the RBC actin filaments (~35–37 nm; Byers & Branton, 1985; Shen et al., 1986) led to the idea that RBC TM may function as a molecular ruler to determine the lengths of the short filaments (Fowler, 1996). However, RBC TMs span 6–7 actin subunits along an actin filament strand, while the stoichiometry for TM to actin on the membrane is 1 TM:7–8 actin subunits, suggesting that RBC actin filaments have a few TM-free subunits extending beyond the ends of the TM rods (Fig. 2.3D–F; Fowler, 1996; Fowler & Bennett, 1984b). Thus, since Tmod1 can bind simultaneously to the actin filament pointed end and to the N-terminal end of TM (Fowler, 1990; Vera et al., 2000), Tmod1 could anchor the end of TM precisely at the actin filament pointed end, thus setting the minimum filament length to that of TM.

A puzzle is how lengths are set at the barbed filament end (i.e., maximum length). The following observations suggest a possible mechanism. First, RBC TMs inhibit erythrocyte α1β1-spectrin binding to actin filaments (Fowler & Bennett, 1984b; Mak et al., 1987). Second, TM levels are reduced substantially in both α- and β-adducin-null RBCs (Table 2.2; Porro et al., 2004; Robledo et al., 2008; Sahr et al., 2009), suggesting that adducin may bind to TM and stabilize TM association to actin. Third, the adducin neck and extended tail domain caps barbed ends and recruits spectrin to actin subunits near barbed ends (Matsuoka et al., 2000). Thus, the extreme end of each extended αβ-adducin tail might bind to the C-terminal end of each TM, setting the location of the barbed end at several actin subunits past the end of the TM (Fig. 2.3D–F). This model can be tested by biochemical and structural studies with isolated proteins.

What is the function of the TMs in regulating RBC actin filament length and stability? There is one study that addresses the function of TMs in RBC, taking advantage of TM depletion from white ghosts prepared in the absence of magnesium (Fowler & Bennett, 1984a, 1984b). An and colleagues compared membrane mechanical stability in pink ghosts (with TM) and white ghosts (TM-depleted), using a shear-based method to measure membrane fragmentation (ektacytometry; An, Salomao, Guo, Gratzer, & Mohandas, 2007). These experiments showed that TM-depleted white ghosts were considerably more fragile than pink ghosts containing TMs. In addition, normal mechanical stability to shear-induced fragmentation could be restored by reconstitution of ghosts with purified RBC TMs, but not skeletal muscle α/β-TMs. Thus, RBC TMs may stabilize the short RBC actin filaments to mechanical breakage induced by shear stress, fortifying the membrane to withstand repetitive passages through the circulation *in vivo*. However, this idea is difficult to evaluate, as RBC actin filament lengths were not determined after shear stress. Future studies with RBCs from mice with targeted deletions in TMs will also be necessary to understand RBC TM function *in vivo*; but this will be challenging due to the multiple splicing of TMs, with compensation by other genes or by alternatively spliced exons often observed (Gunning et al., 2008).

5.1.2 TM regulation of RBC actomyosin ATPase

In addition to stabilizing actin filaments in RBCs, TMs were hypothesized to play a role in regulation of RBC actomyosin ATPase (Fowler & Bennett, 1984a, 1984b). Human RBCs contain a nonmuscle myosin II, which is mostly present in the cytosol (Table 2.1; Fowler, Davis, & Bennett, 1985; Wong, Kiehart, & Pollard, 1985). The RBC myosin has a 200 kDa heavy chain with 26 kDa and 19.5 kDa light chains, forms typical dimers with two globular heads and a long rodlike tail, self-associates to typical bipolar filaments, and has a characteristic pattern of ATPase activity activated by actin (Fowler et al., 1985; Higashihara, Hartshorne, Craig, & Ikebe, 1989; Wong et al., 1985). The myosin is present in RBCs at about 6000 copies per cell, at 1 myosin:80 actins, which is similar to other nonmuscle cells. Myosin is localized in a punctate pattern in RBCs (Fowler et al., 1985), suggesting that the RBC actin filaments may not be uniformly distributed in the membrane skeleton *in situ*. I have speculated that RBC myosin controls RBC shape and deformability (Fowler, 1986), but in the absence of *in vivo* functional evidence, the prevailing view is that myosin in mature RBCs is a remnant of a prior stage of RBC biogenesis, for example, functioning in enucleation (Colin & Schrier, 1991; Ubukawa et al., 2012).

Nevertheless, the possibility that myosin may have a functional role in mature RBCs was also supported by the identification in pig RBCs of caldesmon, a well-established TM-binding and actomyosin regulatory protein (Table 2.1; der Terrossian, Deprette, & Cassoly, 1989). Caldesmon is an actin filament and calmodulin-binding protein that is associated with actin filaments in smooth muscle and nonmuscle cells (Lin, Li, Eppinga, Wang, & Jin, 2009). Caldesmon stabilizes actin filaments and participates with TMs in the inhibition of actomyosin ATPase activity, which can be reversed by phosphorylation of caldesmon or by Ca^{++}–calmodulin binding to caldesmon. Similar to RBC TMs, an immunoreactive ~71kD caldesmon polypeptide is only present in pink ghosts isolated by lysis in magnesium-containing buffers (der Terrossian et al., 1989). RBC caldesmon was purified and found to have the expected properties, including Ca^{++}-sensitive calmodulin binding, actin filament binding, and the ability to inhibit actin-activated myosin ATPase in the presence of erythrocyte TMs, which was reversed by Ca^{++}–calmodulin (der Terrossian, Deprette, Lebbar, & Cassoly, 1994). The ratio of caldesmon–TM–actin was determined to be 1:1:7–8, consistent with two caldesmons per short actin filament, so that each TM could be associated with one caldesmon.

Moreover, immunofluorescence staining of human RBCs revealed punctate patterns of caldesmon, TM, actin, and myosin, in contrast to the smooth pattern of spectrin staining along the membrane, again suggesting a nonuniform organization of actin and its associated proteins in the membrane skeleton (der Terrossian et al., 1994). It may also be significant that an alternative transcript of the β1-spectrin gene, β1E2 expressed in muscle and brain, has been identified in human RBCs and localized in patches along the membrane (Pradhan, Tseng, Cianci, & Morrow, 2004). A nonuniform organization of the membrane skeleton is also suggested by the actin-bundling properties of dematin (Section 5.2). To date, these intriguing observations for regional specialization of the membrane skeleton in RBCs or an *in vivo* function for caldesmon in regulating RBC actomyosin or other RBC functions have not been followed up.

5.2. Dematin: A role for actin filament bundling?

5.2.1 Dematin bundles actin filaments

Dematin, originally referred to as band 4.9, is a set of related 48 kDa and 52 kDa polypeptides (ratio 3:1) that were initially identified as prominent substrates for phosphorylation by cAMP-dependent kinase (PKA) in RBC membranes [for a review, see Cohen and Gascard (1992)]. Protein 4.9 was purified by Siegel and Branton (1985) based on the idea that it might interact with spectrin and regulate spectrin–membrane associations to control ATP-dependent RBC shape changes, which was a hot topic of investigation at the time (Chishti, A., personal communication). Instead, Siegel discovered that 4.9 was a potent actin filament-bundling protein, forming tight parallel bundles of actin filaments with a ~36 nm banding pattern, similar to actin bundles formed by fimbrin or villin (Siegel & Branton, 1985). While these preparations of 4.9 also reduced the rate of actin elongation at barbed ends, there was no effect on the actin critical concentration, suggesting that elongation rates were slower due to steric hindrance in bundles, rather than due to barbed end capping. Husain-Chishti and colleagues then showed that PKA phosphorylation of 4.9 completely eliminated its actin filament-bundling activity (Husain-Chishti, Levin, & Branton, 1988). At the time, this was the first demonstration that phosphorylation of any actin-binding protein regulated its functional activity, another first for the RBC.

Protein 4.9 was renamed dematin in 1989 (Husain-Chishti, Faquin, Wu, & Branton, 1989), and cDNA cloning revealed that dematin was a member of a class of actin-binding proteins with a "headpiece" domain, similar to villin (Azim, Knoll, Beggs, & Chishti, 1995; Rana, Ruff, Maalouf, Speicher,& Chishti, 1993). Both dematin polypeptides are derived from the same gene, with the 52 kDa differing from the 48 kDa by the presence of a 22-amino acid insertion in the headpiece domain (Azim et al., 1995). While originally thought to be a trimer (Husain-Chishti et al., 1988; Siegel & Branton, 1985), analytical ultracentrifugation now indicates that native dematin is monomeric (Chen, Brown, Mok, Hatters, & McKnight, 2013). Dematin monomers contain two actin filament binding sites, one in the folded "headpiece" domain (Vardar et al., 2002) and one in an unstructured "core" domain (Chen et al., 2013). PKA phosphorylation of Ser 381 in the headpiece domain leads to interaction of headpiece with the unstructured region, sterically eliminating one of the actin filament binding sites and eliminating filament-bundling but not binding activity (Chen et al., 2013).

5.2.2 Dematin stabilizes the RBC membrane skeleton in vivo

It remains a puzzle what the function of an actin filament-bundling protein such as dematin might be in the spectrin–actin lattice, as actin filament bundles have never been observed. With three dematin monomers present per short actin filament (Husain-Chishti et al., 1988), it seems possible that dematin could gather RBC actin filaments into small bundles (Fig. 2.3F). Such bundles may partly explain the irregular actin filament distribution patterns observed by der Terrossian et al. (1994) for phalloidin staining of intact human RBCs (Section 5.1.2). However, two recent studies indicate that dematin may have other functions than actin bundling in RBCs. First, dematin binds to the cytoplasmic domain of the Glut1 glucose transporter in human RBC membranes, indicating it can link the short actin filaments to the membrane (Khan et al., 2008). The effect of PKA phosphorylation of dematin on this function has not been examined. Second, dematin binds to the actin-binding tail region of α1β1-spectrin and facilitates spectrin interactions with actin filaments (Koshino et al., 2012). This interaction is inhibited by PKA phosphorylation of dematin and is proposed to account for the decreased membrane mechanical stability observed for cAMP-treated RBC membranes (Koshino et al., 2012). Thus, dematin displays remarkable functional similarities to adducin—they can both bundle actin filaments, promote spectrin binding to actin, and provide a linkage for the JCs to the membrane (via Glut1 or band 3; Section 4.2.1).

To investigate an *in vivo* function for dematin's actin-bundling activity, knockout mice were created with a targeted deletion of the C-terminal headpiece domain (Table 2.2; Khanna et al., 2002). RBCs from these mice contained a truncated ~40 kDa dematin polypeptide but no full-length 52 or 48 kDa polypeptides. Similar to the Tmod1-null and adducin-null mice described earlier, dematin headpiece-null mice had a mild compensated hemolytic anemia with abnormally shaped and smaller spherocytic RBCs that were osmotically fragile and less deformable. The RBC membrane appeared to be unstable, due to somewhat reduced levels of spectrin and actin and an increased propensity for spectrin and actin to be extracted in the presence of nonionic detergent (TX-100). Examination of membrane skeleton structure *in situ* by atomic force microscopy (AFM) revealed that skeleton network elements were damaged and partially aggregated (Chen et al., 2007; Liu, Khan, et al., 2011), although the actin filaments themselves could not be visualized. Thus, the dematin headpiece domain appears to be important for actin stability and for the architecture of the membrane skeleton, possibly by linking to actin and promoting actin–actin filament associations, or by mediating JC linkage to the membrane.

Interestingly, mice deficient for both dematin headpiece and β-adducin demonstrated a more severe spherocytic hemolytic anemia than either single knockout alone, in terms of hematological parameters, RBC shapes, osmotic fragility, dissociation of spectrin and actin, and disruption and aggregation of the skeletal network visualized by AFM (Chen et al., 2007; Liu, Khan, et al., 2011). Based on analogous membrane linkages of dematin and adducin (Section 4.2.1), the absence of dematin- and/or adducin-mediated JC linkages to transmembrane proteins could contribute to the RBC phenotypes, similar to other spherocytic phenotypes (Mohandas & Evans, 1994; Mohandas & Gallagher, 2008; Perrotta et al., 2008). Clearly, a complication in assigning structural defects in the spectrin–actin network *per se*, to RBC physiological phenotypes *in vivo*, is that loss of some proteins (dematin and adducin) can affect both lattice integrity (so-called horizontal interactions, leading to elliptocytosis) and linkages to the membrane (so-called vertical interactions, leading to spherocytosis; Gallagher, 2004; Mohandas & Gallagher, 2008; Perrotta et al., 2008).

6. ARE RBC ACTIN FILAMENTS DYNAMIC?

Dynamic actin filament capping is not incompatible with precise length regulation of filaments. In striated muscle cells with precisely regulated actin filament lengths and Tmod1 and capZ caps at pointed and barbed filament ends, respectively, it is well established that both the cappers and the terminal actin subunits transiently associate and dissociate from filament ends, indicating dynamic mechanisms of length regulation (Littlefield, Almenar-Queralt, & Fowler, 2001; Littlefield & Fowler, 2008). Tmod1 regulation of actin dynamics at pointed ends controls thin filament lengths in striated muscle, while regulation of barbed end dynamics does not influence lengths, instead likely regulating initial filament assembly (Gokhin & Fowler, 2011; Littlefield & Fowler, 1998).

Indeed, there are tantalizing hints indicating that a simple tight capping mechanism for actin filament length regulation in RBCs is likely oversimplified. Pinder and Gratzer showed in 1983 that human RBC cytosol contains ~10 μg/ml free actin monomer, equal to the barbed end critical concentration, which is similar to the concentration of free monomer in other cells (Pinder & Gratzer, 1983). The presence of this concentration of free actin monomer implies that the barbed ends of RBC actin filaments are at steady state with the free monomer pool and are thus not permanently capped (Fig. 2.3G; Zigmond, 2004). Pointed ends may also be dynamic since the lower critical concentration of the barbed filament end sets the free monomer level. Moreover, the presence of actin in the cytoplasm of normal human RBCs is supported by immunogold labeling of β-actin throughout the cytoplasm of intact human RBCs prepared by high-pressure freezing, freeze-substitution, and thin sectioning (Supplemental Fig. S3 in Cyrklaff et al., 2011). Western blotting of actin in cytosol and isolated membrane fractions after osmotic lysis of mouse RBCs indicates that about 5–10% of the actin may be in the cytosol, although the exact amount was not quantified carefully (Moyer et al., 2010). Actin subunits in the cytosol could potentially serve as a reservoir for actin exchange or elongation at filament ends of preexisting membrane-associated actin filaments or for new actin polymerization.

The idea that the capping state of RBC actin filaments may be dynamically regulated *in vivo* is supported by several additional intriguing observations. First, Cyrklaff and colleagues (Cyrklaff et al., 2011) reported that infection of human RBCs by the malaria parasite, *Plasmodium falciparum,* led to dramatic remodeling of RBC actin. The short filaments near the membrane were completely disassembled and replaced by a branching actin filament network in the cytosol that may facilitate vesicular trafficking of parasite proteins to the RBC membrane. These branched actin filaments are strikingly reminiscent of Arp2/3-nucleated branching actin filament networks in lamellipodia (Pollard & Borisy, 2003) and suggest that RBC cytosol may contain Arp2/3 or related actin nucleators, as well as their upstream regulators (see succeeding text). The malaria parasite could hijack an endogenous (but normally silent) pathway in RBCs to dismantle the actin filaments in the spectrin–actin lattice and reassemble them in the cytosol for its own purposes (Zuccala & Baum, 2011). In support of this idea, sickle RBCs containing hemoglobins S and C were observed to be resistant to malaria parasite-induced actin filament remodeling, which was proposed to be due to prevalence of oxidized forms of hemoglobin that interfere with actin polymerization (Cyrklaff et al., 2011). The β-actin in irreversibly sickled RBCs was previously shown to be oxidized at cysteines 284 and 373, forming a disulfide bridge, which interferes with β-actin polymerization and with disassembly of spectrin–4.1R–actin complexes (Abraham, Bencsath, Shartava, Kakhniashvili, & Goodman, 2002; Bencsath, Shartava, Monteiro, & Goodman, 1996; Shartava et al., 1995).

Another recent study provides evidence that RBCs contain a signaling pathway that could regulate Arp2/3-induced actin nucleation. Namely, Hem-1, a hematopoietic-specific component of the Rac-regulated WAVE complex, which activates Arp2/3 to nucleate actin filament assembly (Park, Chan, & Iritani, 2010), was identified in mouse RBCs (Chan et al., 2013). A nonsense mutation in *Hem1* leading to Hem-1 deficiency resulted in defective actin regulation in immune cells, including defective migration and phagocytosis of neutrophils, and defects in T cell development and function (Park et al., 2008). The *Hem1* mutant mice also displayed a mild compensated anemia with abnormally shaped and osmotically fragile RBCs with reduced life span, resembling a microcytic, hypochromic hemolytic anemia (Table 2.2; Chan et al., 2013; Park et al., 2008), similar to Rac-deficient mice (Kalfa et al., 2006). RBCs of *Hem1* mice are also similar to those of Rac-deficient mice, with reduced levels of many membrane skeleton proteins relative to actin, including spectrin, ankyrin, Tmod1, and dematin. Most strikingly, levels of phospho-adducin were elevated in the Hem-1-deficient RBCs, and abnormal aggregates of actin filaments were evident by fluorescent microscopy, similar to the Rac-deficient RBCs. These data suggest that Rac signaling pathways may modulate actin filament remodeling in RBCs, both by controlling phosphorylation of adducin to regulate adducin–actin interactions and by activating Hem-1 in the WAVE complex to stimulate Arp2/3-mediated actin assembly. However, it is conceivable that both the Hem-1 and Rac-deficiency phenotypes are a consequence of defects in actin remodeling and assembly during the process of reticulocyte maturation into RBCs, rather than defects in dynamic actin homeostasis in mature RBCs.

In mature RBCs, mechanical stresses on cells as they pass through the circulation may potentially enhance actin subunit dynamics at filament ends or lead to filament breakage and reannealing, as observed for purified actin filaments. Nakashima and Beutler showed many years ago that phalloidin (an actin filament stabilizer) reduced the deformability of resealed ghosts as measured by ektacytometry (Nakashima & Beutler, 1979). Cytochalasin B (an actin barbed end capping molecule) treatment of intact cells increased RBC resistance to osmotic lysis and reduced deformability in a micropipette aspiration assay (Beck, Jay, & Saari, 1972), although cytochalasin B effects on the glucose transporter and cell volume cannot be excluded (Jung & Rampal, 1977; Lin & Lin, 1978). It is well established that spectrin dimer–dimer interaction sites, spectrin–4.1R, and adducin–band 3 interactions "breathe" when RBC membranes are subjected to shear stresses, allowing incorporation of their specific binding peptides into the membrane skeleton (An et al., 2002; Anong et al., 2009; Discher et al., 1995). Thus, it would be interesting to examine actin subunit incorporation (or exchange) into the membrane skeleton during shear stresses, which may promote Tmod1 or adducin dissociation from filament ends or lead to filament breakage, allowing new actin subunit binding and incorporation. This would imply active mechanisms of actin filament length control in human RBCs, requiring ongoing regulation of dynamics by Tmod1 at pointed ends and adducins at barbed ends (Fig. 2.3G). So far, our ideas of RBC actin filament capping and filament stability are derived principally from studies of isolated membranes or detergent-extracted membrane skeletons, preparations in which factors regulating dynamic capping may have been removed. It will be important to perform direct studies of dynamics in RBCs using fluorescence microscopy approaches in living cells into which fluorescent-labeled actin, Tmod1, or αβ-adducin probes have been introduced. Actin filament dynamics represents a relatively unexplored control point for RBC membrane skeleton assembly and stability.

7. CONCLUSIONS AND FUTURE DIRECTIONS

The RBC membrane skeleton remains a powerful model system for structure/function studies due to the accessibility and purity of RBCs. Studies of RBC spectrins and ankyrins and their linkages to the membrane have provided a jumping-off point for many other biological problems, leading to novel insights into basic science and human diseases (Ayalon, Davis, Scotland, & Bennett, 2008; Bennett & Healy, 2008), as is evident from the topics of several chapters in this volume. The insights into RBC actin assembly and organization discussed here can serve as a paradigm to elucidate the roles of actin dynamics and filament capping in the spectrin-based membrane skeletons of all cells. Not only has the RBC provided a fertile ground for discovery of new families of actin-binding proteins (Tmods, adducins, spectrins, 4.1R, and dematin), but also the rigorous exploration of their biochemical, structural, and functional interactions has been enabled by the purity and homogeneity of the RBC membrane. While much has been learned, there remain some mysteries. For example, how are all the proteins arranged on each actin filament, and are all the filaments the same? Are the actin filaments uniformly distributed along the RBC membrane? How do dematin and its bundling activity contribute to actin filament organization in the spectrin–actin lattice? What are the *in vivo* functions of TMs, caldesmon, and the contractile protein, myosin II, in mature RBCs? Are RBC actin filaments dynamic, and how are their dynamics regulated? Answering these questions may reveal new principles for actin dynamics and stability on membranes, and how actomyosin contractile activity might be coupled to plasma membranes to transmit forces. This may also lead to insights into the impact of mechanical stresses on actin assembly at plasma membranes and could be important for RBC pathologies such as sickle-cell disease and malaria parasite invasion.

Finally, the stretched spectrin–actin lattice with its repeating nodes of short actin filaments may yet provide a unique opportunity to obtain a structural understanding of actin filament capping and filament length regulation at a molecular level. A major unsolved problem also remains the structural relationship between the stretched quasi-hexagonal spectrin–actin lattice and the complex topography of the *in situ* membrane skeleton—its resolution can be expected to have implications for the membrane skeletons of other cell types. A recent super-resolution examination of spectrin and actin filament organization in neuronal axons revealed periodic rings of actin filaments associated with adducin, located at ~180–190 nm intervals with spectrin in between; 200 nm is the distance expected for fully extended spectrin tetramers (Xu, Zhong, & Zhuang, 2013). Thus, the basic organizational unit of short actin filaments attached by long spectrin tetramers first visualized in the RBC may be a fundamental feature of the plasma membrane skeleton! Future super-resolution studies with the other RBC actin-binding proteins, both in RBCs and in other cells, will define key conserved features, or reveal divergent features allowing plasticity of the spectrin–actin lattice in different cellular contexts. In many ways, we may be entering an exciting era for study of the membrane skeleton; now that the principal actors are well understood individually and in combinations, we can tease apart how this complex supramolecular network-forming machine is assembled and functions at a cellular and tissue scale.

ACKNOWLEDGMENTS

I am grateful to Roberta Nowak for the preparation of the artwork, figures, and tables, and to David Gokhin for help with writing the Abstract. This work was supported by a grant from the NIH (HL083464 to V. M. F.).

REFERENCES

Abraham, A., Bencsath, F. A., Shartava, A., Kakhniashvili, D. G., & Goodman, S. R. (2002). Preparation of irreversibly sickled cell beta-actin from normal red blood cell beta-actin. *Biochemistry*, *41*(1), 292–296.

An, X., Lecomte, M. C., Chasis, J. A., Mohandas, N., & Gratzer, W. (2002). Shear-response of the spectrin dimer-tetramer equilibrium in the red blood cell membrane. *The Journal of Biological Chemistry*, *277*(35), 31796–31800.

An, X., Salomao, M., Guo, X., Gratzer, W., & Mohandas, N. (2007). Tropomyosin modulates erythrocyte membrane stability. *Blood*, *109*(3), 1284–1288.

Anong, W. A., Franco, T., Chu, H., Weis, T. L., Devlin, E. E., Bodine, D. M., et al. (2009). Adducin forms a bridge between the erythrocyte membrane and its cytoskeleton and regulates membrane cohesion. *Blood*, *114*(9), 1904–1912.

Atkinson, M. A., Morrow, J. S., & Marchesi, V. T. (1982). The polymeric state of actin in the human erythrocyte cytoskeleton. *Journal of Cellular Biochemistry*, *18*(4), 493–505.

Ayalon, G., Davis, J. Q., Scotland, P. B., & Bennett, V. (2008). An ankyrin-based mechanism for functional organization of dystrophin and dystroglycan. *Cell*, *135*(7), 1189–1200.

Azim, A. C., Knoll, J. H., Beggs, A. H., & Chishti, A. H. (1995). Isoform cloning, actin binding, and chromosomal localization of human erythroid dematin, a member of the villin superfamily. *The Journal of Biological Chemistry*, *270*(29), 17407–17413.

Beaven, G. H., Jean-Baptiste, L., Ungewickell, E., Baines, A. J., Shahbakhti, F., Pinder, J. C., et al. (1985). An examination of the soluble oligomeric complexes extracted from the red cell membrane and their relation to the membrane cytoskeleton. *European Journal of Cell Biology, 36*(2), 299–306.

Beck, J. S., Jay, A. W. L., & Saari, J. T. (1972). Effects of cytochalasin B on osmotic fragility and deformability of human erythrocytes. *Canadian Journal of Physiology and Pharmacology, 50*, 684–688.

Bencsath, F. A., Shartava, A., Monteiro, C. A., & Goodman, S. R. (1996). Identification of the disulfide-linked peptide in irreversibly sickled cell beta-actin. *Biochemistry, 35*(14), 4403–4408.

Benga, G., Popescu, O., Borza, V., Pop, V. I., Muresan, A., Mocsy, I., et al. (1986). Water permeability in human erythrocytes: Identification of membrane proteins involved in water transport. *European Journal of Cell Biology, 41*(2), 252–262.

Benga, G., Popescu, O., Pop, V. I., & Holmes, R. P. (1986). p-(Chloromercuri) benzenesulfonate binding by membrane proteins and the inhibition of water transport in human erythrocytes. *Biochemistry, 25*(7), 1535–1538.

Bennett, V., & Baines, A. J. (2001). Spectrin and ankyrin-based pathways: Metazoan inventions for integrating cells into tissues. *Physiological Reviews, 81*(3), 1353–1392.

Bennett, V., Davis, J., & Fowler, W. E. (1982). Brain spectrin, a membrane-associated protein related in structure and function to erythrocyte spectrin. *Nature, 299*(5879), 126–131.

Bennett, V., & Healy, J. (2008). Organizing the fluid membrane bilayer: Diseases linked to spectrin and ankyrin. *Trends in Molecular Medicine, 14*(1), 28–36.

Bennett, V., & Stenbuck, P. J. (1979). Identification and partial purification of ankyrin, the high affinity membrane attachment site for human erythrocyte spectrin. *The Journal of Biological Chemistry, 254*(7), 2533–2541.

Blanc, L., & Vidal, M. (2010). Reticulocyte membrane remodeling: Contribution of the exosome pathway. *Current Opinion in Hematology, 17*(3), 177–183.

Branton, D., Cohen, C. M., & Tyler, J. (1981). Interaction of cytoskeletal proteins on the human erythrocyte membrane. *Cell, 24*(1), 24–32.

Brenner, S. L., & Korn, E. D. (1979). Spectrin-actin interaction. Phosphorylated and dephosphorylated spectrin tetramer cross-link F-actin. *The Journal of Biological Chemistry, 254*(17), 8620–8627.

Brenner, S. L., & Korn, E. D. (1980). Spectrin/actin complex isolated from sheep erythrocytes accelerates actin polymerization by simple nucleation. Evidence for oligomeric actin in the erythrocyte cytoskeleton. *The Journal of Biological Chemistry, 255*(4), 1670–1676.

Byers, T. J., & Branton, D. (1985). Visualization of the protein associations in the erythrocyte membrane skeleton. *Proceedings of the National Academy of Sciences of the United States of America, 82*(18), 6153–6157.

Chan, M. M., Wooden, J. M., Tsang, M., Gilligan, D. M., Hirenallur, S. D., Finney, G. L., et al. (2013). Hematopoietic protein-1 regulates the actin membrane skeleton and membrane stability in murine erythrocytes. *PLoS One, 8*(2), e54902.

Chasis, J. A., Prenant, M., Leung, A., & Mohandas, N. (1989). Membrane assembly and remodeling during reticulocyte maturation. *Blood, 74*(3), 1112–1120.

Chen, L., Brown, J. W., Mok, Y. F., Hatters, D. M., & McKnight, C. J. (2013). The allosteric mechanism induced by protein kinase A (PKA) phosphorylation of dematin (band 4.9). *The Journal of Biological Chemistry, 288*(12), 8313–8320.

Chen, H., Khan, A. A., Liu, F., Gilligan, D. M., Peters, L. L., Messick, J., et al. (2007). Combined deletion of mouse dematin-headpiece and beta-adducin exerts a novel effect on the spectrin-actin junctions leading to erythrocyte fragility and hemolytic anemia. *The Journal of Biological Chemistry, 282*(6), 4124–4135.

Chu, X., Chen, J., Reedy, M. C., Vera, C., Sung, K. L., & Sung, L. A. (2003). E-Tmod capping of actin filaments at the slow-growing end is required to establish mouse embryonic circulation. *American Journal of Physiology Heart and Circulatory Physiology*, *284*(5), H1827–H1838.

Cohen, C. M. (1983). The molecular organization of the red cell membrane skeleton. *Seminars in Hematology*, *20*(3), 141–158.

Cohen, C. M., & Branton, D. (1979). The role of spectrin in erythrocyte membrane-stimulated actin polymerisation. *Nature*, *279*(5709), 163–165.

Cohen, C. M., & Gascard, P. (1992). Regulation and post-translational modification of erythrocyte membrane and membrane-skeletal proteins. *Seminars in Hematology*, *29*(4), 244–292.

Cohen, C. M., Tyler, J. M., & Branton, D. (1980). Spectrin-actin associations studied by electron microscopy of shadowed preparations. *Cell*, *21*(3), 875–883.

Coleman, T. R., Fishkind, D. J., Mooseker, M. S., & Morrow, J. S. (1989). Functional diversity among spectrin isoforms. *Cell Motility and the Cytoskeleton*, *12*(4), 225–247.

Colin, F. C., & Schrier, S. L. (1991). Myosin content and distribution in human neonatal erythrocytes are different from adult erythrocytes. *Blood*, *78*(11), 3052–3055.

Cyrklaff, M., Sanchez, C. P., Kilian, N., Bisseye, C., Simpore, J., Frischknecht, F., et al. (2011). Hemoglobins S and C interfere with actin remodeling in Plasmodium falciparum-infected erythrocytes. *Science*, *334*(6060), 1283–1286.

Danielli, J. F., & Davson, H. (1935). A contribution to the theory of permeability of thin films. *Journal of Cellular and Comparative Physiology*, *5*(4), 495–508.

Denker, B. M., Smith, B. L., Kuhajda, F. P., & Agre, P. (1988). Identification, purification, and partial characterization of a novel Mr 28,000 integral membrane protein from erythrocytes and renal tubules. *The Journal of Biological Chemistry*, *263*(30), 15634–15642.

Derick, L. H., Liu, S. C., Chishti, A. H., & Palek, J. (1992). Protein immunolocalization in the spread erythrocyte membrane skeleton. *European Journal of Cell Biology*, *57*(2), 317–320.

der Terrossian, E., Deprette, C., & Cassoly, R. (1989). Caldesmon is present in human and pig erythrocytes. *Biochemical and Biophysical Research Communications*, *159*(2), 395–401.

der Terrossian, E., Deprette, C., Lebbar, I., & Cassoly, R. (1994). Purification and characterization of erythrocyte caldesmon. Hypothesis for an actin-linked regulation of a contractile activity in the red blood cell membrane. *European Journal of Biochemistry*, *219*(1–2), 503–511.

DiNubile, M. J. (1999). Erythrocyte membrane fractions contain free barbed filament ends despite sufficient concentrations of retained capper(s) to prevent barbed end growth. *Cell Motility and the Cytoskeleton*, *43*(1), 10–22.

Discher, D. E. (2000). New insights into erythrocyte membrane organization and microelasticity. *Current Opinion in Hematology*, 7(2), 117–122.

Discher, D. E., Winardi, R., Schischmanoff, P. O., Parra, M., Conboy, J. G., & Mohandas, N. (1995). Mechanochemistry of protein 4.1's spectrin-actin-binding domain: Ternary complex interactions, membrane binding, network integration, structural strengthening. *The Journal of Cell Biology*, *130*(4), 897–907.

Dunn, S. A., Mohteshamzadeh, M., Daly, A. K., & Thomas, T. H. (2003). Altered tropomyosin expression in essential hypertension. *Hypertension*, *41*(2), 347–354.

Fischer, R. S., & Fowler, V. M. (2003). Tropomodulins: Life at the slow end. *Trends in Cell Biology*, *13*(11), 593–601.

Fischer, R. S., Yarmola, E. G., Weber, K. L., Speicher, K. D., Speicher, D. W., Bubb, M. R., et al. (2006). Tropomodulin 3 binds to actin monomers. *The Journal of Biological Chemistry*, *281*(47), 36454–36465.

Fowler, V. M. (1986). An actomyosin contractile mechanism for erythrocyte shape transformations. *Journal of Cellular Biochemistry*, *31*(1), 1–9.

Fowler, V. M. (1987). Identification and purification of a novel Mr 43,000 tropomyosin-binding protein from human erythrocyte membranes. *The Journal of Biological Chemistry, 262*(26), 12792–12800.
Fowler, V. M. (1990). Tropomodulin: A cytoskeletal protein that binds to the end of erythrocyte tropomyosin and inhibits tropomyosin binding to actin. *The Journal of Cell Biology, 111*(2), 471–481.
Fowler, V. M. (1996). Regulation of actin filament length in erythrocytes and striated muscle. *Current Opinion in Cell Biology, 8*(1), 86–96.
Fowler, V. M., & Bennett, V. (1984a). Erythrocyte membrane tropomyosin. Purification and properties. *The Journal of Biological Chemistry, 259*(9), 5978–5989.
Fowler, V. M., & Bennett, V. (1984b). Tropomyosin: A new component of the erythrocyte membrane skeleton. *Progress in Clinical and Biological Research, 159*, 57–71.
Fowler, V. M., Davis, J. Q., & Bennett, V. (1985). Human erythrocyte myosin: Identification and purification. *The Journal of Cell Biology, 100*(1), 47–55.
Fowler, V. M., Greenfield, N. J., & Moyer, J. (2003). Tropomodulin contains two actin filament pointed end-capping domains. *The Journal of Biological Chemistry, 278*(41), 40000–40009.
Fowler, V. M., Sussmann, M. A., Miller, P. G., Flucher, B. E., & Daniels, M. P. (1993). Tropomodulin is associated with the free (pointed) ends of the thin filaments in rat skeletal muscle. *The Journal of Cell Biology, 120*(2), 411–420.
Fowler, V., & Taylor, D. L. (1980). Spectrin plus band 4.1 cross-link actin. Regulation by micromolar calcium. *The Journal of Cell Biology, 85*(2), 361–376.
Fritz-Six, K. L., Cox, P. R., Fischer, R. S., Xu, B., Gregorio, C. C., Zoghbi, H. Y., et al. (2003). Aberrant myofibril assembly in tropomodulin1 null mice leads to aborted heart development and embryonic lethality. *The Journal of Cell Biology, 163*(5), 1033–1044.
Fukata, Y., Oshiro, N., Kinoshita, N., Kawano, Y., Matsuoka, Y., Bennett, V., et al. (1999). Phosphorylation of adducin by Rho-kinase plays a crucial role in cell motility. *The Journal of Cell Biology, 145*(2), 347–361.
Gallagher, P. G. (2004). Hereditary elliptocytosis: Spectrin and protein 4.1R. *Seminars in Hematology, 41*(2), 142–164.
Gardner, K., & Bennett, V. (1986). A new erythrocyte membrane-associated protein with calmodulin binding activity. Identification and purification. *The Journal of Biological Chemistry, 261*(3), 1339–1348.
Gardner, K., & Bennett, V. (1987). Modulation of spectrin-actin assembly by erythrocyte adducin. *Nature, 328*(6128), 359–362.
Gilligan, D. M., & Bennett, V. (1993). The junctional complex of the membrane skeleton. *Seminars in Hematology, 30*(1), 74–83.
Gilligan, D. M., Lozovatsky, L., Gwynn, B., Brugnara, C., Mohandas, N., & Peters, L. L. (1999). Targeted disruption of the beta adducin gene (Add2) causes red blood cell spherocytosis in mice. *Proceedings of the National Academy of Sciences of the United States of America, 96*(19), 10717–10722.
Gokhin, D. S., & Fowler, V. M. (2011). Tropomodulin capping of actin filaments in striated muscle development and physiology. *Journal of Biomedicine & Biotechnology, 2011*, 103069.
Gorter, E., & Grendel, F. (1925). On bimolecular layers of lipoids on the chromocytes of the blood. *The Journal of Experimental Medicine, 41*(4), 439–443.
Gregorio, C. C., Weber, A., Bondad, M., Pennise, C. R., & Fowler, V. M. (1995). Requirement of pointed-end capping by tropomodulin to maintain actin filament length in embryonic chick cardiac myocytes. *Nature, 377*(6544), 83–86.
Gunning, P., O'Neill, G., & Hardeman, E. (2008). Tropomyosin-based regulation of the actin cytoskeleton in time and space. *Physiological Reviews, 88*(1), 1–35.

Handin, R. I., Lux, S. E., & Stossel, T. P. (2003). *Blood: Principles and Practice of Hematology*. Philadelphia, PA: Lippincott Williams & Wilkins.

Hart, M. C., Korshunova, Y. O., & Cooper, J. A. (1997). Vertebrates have conserved capping protein alpha isoforms with specific expression patterns. *Cell Motility and the Cytoskeleton*, *38*(2), 120–132.

Higashihara, M., Hartshorne, D. J., Craig, R., & Ikebe, M. (1989). Correlation of enzymatic properties and conformation of bovine erythrocyte myosin. *Biochemistry*, *28*(4), 1642–1649.

Hughes, C. A., & Bennett, V. (1995). Adducin: A physical model with implications for function in assembly of spectrin-actin complexes. *The Journal of Biological Chemistry*, *270*(32), 18990–18996.

Husain-Chishti, A., Faquin, W., Wu, C. C., & Branton, D. (1989). Purification of erythrocyte dematin (protein 4.9) reveals an endogenous protein kinase that modulates actin-bundling activity. *The Journal of Biological Chemistry*, *264*(15), 8985–8991.

Husain-Chishti, A., Levin, A., & Branton, D. (1988). Abolition of actin-bundling by phosphorylation of human erythrocyte protein 4.9. *Nature*, *334*(6184), 718–721.

Ji, P., Murata-Hori, M., & Lodish, H. F. (2011). Formation of mammalian erythrocytes: Chromatin condensation and enucleation. *Trends in Cell Biology*, *21*(7), 409–415.

Johnstone, R. M. (2005). Revisiting the road to the discovery of exosomes. *Blood Cells, Molecules & Diseases*, *34*(3), 214–219.

Jung, C. Y., & Rampal, A. L. (1977). Cytochalasin B binding sites and glucose transport carrier in human erythrocyte ghosts. *The Journal of Biological Chemistry*, *252*(15), 5456–5463.

Kalfa, T. A., Pushkaran, S., Mohandas, N., Hartwig, J. H., Fowler, V. M., Johnson, J. F., et al. (2006). Rac GTPases regulate the morphology and deformability of the erythrocyte cytoskeleton. *Blood*, *108*(12), 3637–3645.

Keerthivasan, G., Wickrema, A., & Crispino, J. D. (2011). Erythroblast enucleation. *Stem Cells International*, *2011*, 139851.

Khan, A. A., Hanada, T., Mohseni, M., Jeong, J. J., Zeng, L., Gaetani, M., et al. (2008). Dematin and adducin provide a novel link between the spectrin cytoskeleton and human erythrocyte membrane by directly interacting with glucose transporter-1. *The Journal of Biological Chemistry*, *283*(21), 14600–14609.

Khanna, R., Chang, S. H., Andrabi, S., Azam, M., Kim, A., Rivera, A., et al. (2002). Headpiece domain of dematin is required for the stability of the erythrocyte membrane. *Proceedings of the National Academy of Sciences of the United States of America*, *99*(10), 6637–6642.

Kimura, K., Fukata, Y., Matsuoka, Y., Bennett, V., Matsuura, Y., Okawa, K., et al. (1998). Regulation of the association of adducin with actin filaments by Rho-associated kinase (Rho-kinase) and myosin phosphatase. *The Journal of Biological Chemistry*, *273*(10), 5542–5548.

Kong, K. Y., & Kedes, L. (2006). Leucine 135 of tropomodulin-1 regulates its association with tropomyosin, its cellular localization, and the integrity of sarcomeres. *The Journal of Biological Chemistry*, *281*(14), 9589–9599.

Korsgren, C., & Lux, S. E. (2010). The carboxyterminal EF domain of erythroid alpha-spectrin is necessary for optimal spectrin-actin binding. *Blood*, *116*(14), 2600–2607.

Koshino, I., Mohandas, N., & Takakuwa, Y. (2012). Identification of a novel role for dematin in regulating red cell membrane function by modulating spectrin-actin interaction. *The Journal of Biological Chemistry*, *287*(42), 35244–35250.

Kostyukova, A. S., Choy, A., & Rapp, B. A. (2006). Tropomodulin binds two tropomyosins: A novel model for actin filament capping. *Biochemistry*, *45*(39), 12068–12075.

Kostyukova, A. S., & Hitchcock-DeGregori, S. E. (2004). Effect of the structure of the N terminus of tropomyosin on tropomodulin function. *The Journal of Biological Chemistry*, *279*(7), 5066–5071.

Kostyukova, A. S., Rapp, B. A., Choy, A., Greenfield, N. J., & Hitchcock-DeGregori, S. E. (2005). Structural requirements of tropomodulin for tropomyosin binding and actin filament capping. *Biochemistry*, *44*(12), 4905–4910.

Kuhlman, P. A. (2000). Characterization of the actin filament capping state in human erythrocyte ghost and cytoskeletal preparations. *The Biochemical Journal*, *349*(Pt 1), 105–111.

Kuhlman, P. A., & Fowler, V. M. (1997). Purification and characterization of an alpha 1 beta 2 isoform of CapZ from human erythrocytes: Cytosolic location and inability to bind to Mg2+ ghosts suggest that erythrocyte actin filaments are capped by adducin. *Biochemistry*, *36*(44), 13461–13472.

Kuhlman, P. A., Hughes, C. A., Bennett, V., & Fowler, V. M. (1996). A new function for adducin. Calcium/calmodulin-regulated capping of the barbed ends of actin filaments. *The Journal of Biological Chemistry*, *271*(14), 7986–7991.

Li, X., Matsuoka, Y., & Bennett, V. (1998). Adducin preferentially recruits spectrin to the fast growing ends of actin filaments in a complex requiring the MARCKS-related domain and a newly defined oligomerization domain. *The Journal of Biological Chemistry*, *273*(30), 19329–19338.

Lin, D. C. (1981). Spectrin-4.1-actin complex of the human erythrocyte: Molecular basis of its ability to bind cytochalasins with high-affinity and to accelerate actin polymerization in vitro. *Journal of Supramolecular Structure and Cellular Biochemistry*, *15*(2), 129–138.

Lin, J. J., Li, Y., Eppinga, R. D., Wang, Q., & Jin, J. P. (2009). Chapter 1: Roles of caldesmon in cell motility and actin cytoskeleton remodeling. *International Review of Cell and Molecular Biology*, *274*, 1–68.

Lin, D. C., & Lin, S. (1978). High affinity binding of [3H]dihydrocytochalasin B to peripheral membrane proteins related to the control of cell shape in the human red cell. *The Journal of Biological Chemistry*, *253*(5), 1415–1419.

Lin, D. C., & Lin, S. (1979). Actin polymerization induced by a motility-related high-affinity cytochalasin binding complex from human erythrocyte membrane. *Proceedings of the National Academy of Sciences of the United States of America*, *76*(5), 2345–2349.

Ling, E., Gardner, K., & Bennett, V. (1986). Protein kinase C phosphorylates a recently identified membrane skeleton-associated calmodulin-binding protein in human erythrocytes. *The Journal of Biological Chemistry*, *261*(30), 13875–13878.

Littlefield, R., Almenar-Queralt, A., & Fowler, V. M. (2001). Actin dynamics at pointed ends regulates thin filament length in striated muscle. *Nature Cell Biology*, *3*(6), 544–551.

Littlefield, R., & Fowler, V. M. (1998). Defining actin filament length in striated muscle: Rulers and caps or dynamic stability? *Annual Review of Cell and Developmental Biology*, *14*, 487–525.

Littlefield, R. S., & Fowler, V. M. (2008). Thin filament length regulation in striated muscle sarcomeres: Pointed-end dynamics go beyond a nebulin ruler. *Seminars in Cell & Developmental Biology*, *19*(6), 511–519.

Liu, F., Burgess, J., Mizukami, H., & Ostafin, A. (2003). Sample preparation and imaging of erythrocyte cytoskeleton with the atomic force microscopy. *Cell Biochemistry and Biophysics*, *38*(3), 251–270.

Liu, S. C., Derick, L. H., & Palek, J. (1987). Visualization of the hexagonal lattice in the erythrocyte membrane skeleton. *The Journal of Cell Biology*, *104*(3), 527–536.

Liu, F., Khan, A. A., Chishti, A. H., & Ostafin, A. E. (2011). Atomic force microscopy demonstration of cytoskeleton instability in mouse erythrocytes with dematin-headpiece and beta-adducin deficiency. *Scanning*, *33*(6), 426–436.

Liu, J., Mohandas, N., & An, X. (2011). Membrane assembly during erythropoiesis. *Current Opinion in Hematology*, *18*(3), 133–138.

Lux, S. E. (1979). Dissecting the red cell membrane skeleton. *Nature*, *281*(5731), 426–429.

Lux, S. E., & Palek, J. (1995). Disorders of the red cell membrane. In R. I. Handin, S. E. Lux, & T. P. Stossel (Eds.), *Blood: Principles and Practice of Hematology* (pp. 1701–1818). Philadelphia, PA: J.B. Lippincott Company.

Mak, A. S., Roseborough, G., & Baker, H. (1987). Tropomyosin from human erythrocyte membrane polymerizes poorly but binds F-actin effectively in the presence and absence of spectrin. *Biochimica et Biophysica Acta, 912*(2), 157–166.

Marchesi, V. T. (1979). Functional proteins of the human red blood cell membrane. *Seminars in Hematology, 16*(1), 3–20.

Marchesi, V. T., & Steers, E., Jr. (1968). Selective solubilization of a protein component of the red cell membrane. *Science, 159*(3811), 203–204.

Matsuoka, Y., Hughes, C. A., & Bennett, V. (1996). Adducin regulation. Definition of the calmodulin-binding domain and sites of phosphorylation by protein kinases A and C. *The Journal of Biological Chemistry, 271*(41), 25157–25166.

Matsuoka, Y., Li, X., & Bennett, V. (1998). Adducin is an in vivo substrate for protein kinase C: Phosphorylation in the MARCKS-related domain inhibits activity in promoting spectrin-actin complexes and occurs in many cells, including dendritic spines of neurons. *The Journal of Cell Biology, 142*(2), 485–497.

Matsuoka, Y., Li, X., & Bennett, V. (2000). Adducin: Structure, function and regulation. *Cellular and Molecular Life Sciences, 57*(6), 884–895.

Matsuzaki, F., Sutoh, K., & Ikai, A. (1985). Structural unit of the erythrocyte cytoskeleton. Isolation and electron microscopic examination. *European Journal of Cell Biology, 39*(1), 153–160.

Maytum, R., Konrad, M., Lehrer, S. S., & Geeves, M. A. (2001). Regulatory properties of tropomyosin effects of length, isoform, and N-terminal sequence. *Biochemistry, 40*(24), 7334–7341.

McKeown, C. R., Nowak, R. B., Moyer, J., Sussman, M. A., & Fowler, V. M. (2008). Tropomodulin1 is required in the heart but not the yolk sac for mouse embryonic development. *Circulation Research, 103*(11), 1241–1248.

Mische, S. M., Mooseker, M. S., & Morrow, J. S. (1987). Erythrocyte adducin: A calmodulin-regulated actin-bundling protein that stimulates spectrin-actin binding. *The Journal of Cell Biology, 105*(6 Pt 1), 2837–2845.

Mohandas, N., & Evans, E. (1994). Mechanical properties of the red cell membrane in relation to molecular structure and genetic defects. *Annual Review of Biophysics and Biomolecular Structure, 23*, 787–818.

Mohandas, N., & Gallagher, P. G. (2008). Red cell membrane: Past, present, and future. *Blood, 112*(10), 3939–3948.

Moyer, J. D., Nowak, R. B., Kim, N. E., Larkin, S. K., Peters, L. L., Hartwig, J., et al. (2010). Tropomodulin 1-null mice have a mild spherocytic elliptocytosis with appearance of tropomodulin 3 in red blood cells and disruption of the membrane skeleton. *Blood, 116*(14), 2590–2599.

Mudry, R. E., Perry, C. N., Richards, M., Fowler, V. M., & Gregorio, C. C. (2003). The interaction of tropomodulin with tropomyosin stabilizes thin filaments in cardiac myocytes. *The Journal of Cell Biology, 162*(6), 1057–1068.

Muro, A. F., Marro, M. L., Gajovic, S., Porro, F., Luzzatto, L., & Baralle, F. E. (2000). Mild spherocytic hereditary elliptocytosis and altered levels of alpha- and gamma-adducins in beta-adducin-deficient mice. *Blood, 95*(12), 3978–3985.

Nakashima, K., & Beutler, E. (1979). Comparison of structure and function of human erythrocyte and human muscle actin. *Proceedings of the National Academy of Sciences of the United States of America, 76*(2), 935–938.

Nans, A., Mohandas, N., & Stokes, D. L. (2011). Native ultrastructure of the red cell cytoskeleton by cryo-electron tomography. *Biophysical Journal, 101*(10), 2341–2350.

Ney, P. A. (2011). Normal and disordered reticulocyte maturation. *Current Opinion in Hematology, 18*(3), 152–157.
Ohnishi, T. (1962). Extraction of actin- and myosin-like proteins from erythrocyte membrane. *Journal of Biochemistry, 52*, 307–308.
Ohnishi, T. (1977). Isolation and characterization of an actin-like protein from membranes of human red cells. *British Journal of Haematology, 35*(3), 453–458.
Ohno, S., Terada, N., Fujii, Y., & Ueda, H. (1994). Membrane skeleton in fresh unfixed erythrocytes as revealed by a rapid-freezing and deep-etching method. *Journal of Anatomy, 185*(Pt 2), 415–420.
Palek, J. (1985). Hereditary elliptocytosis and related disorders. *Clinics in Haematology, 14*(1), 45–87.
Park, H., Chan, M. M., & Iritani, B. M. (2010). Hem-1: Putting the "WAVE" into actin polymerization during an immune response. *FEBS Letters, 584*(24), 4923–4932.
Park, H., Staehling-Hampton, K., Appleby, M. W., Brunkow, M. E., Habib, T., Zhang, Y., et al. (2008). A point mutation in the murine Hem1 gene reveals an essential role for Hematopoietic protein 1 in lymphopoiesis and innate immunity. *The Journal of Experimental Medicine, 205*(12), 2899–2913.
Perrotta, S., Gallagher, P. G., & Mohandas, N. (2008). Hereditary spherocytosis. *Lancet, 372*(9647), 1411–1426.
Picart, C., Dalhaimer, P., & Discher, D. E. (2000). Actin protofilament orientation in deformation of the erythrocyte membrane skeleton. *Biophysical Journal, 79*(6), 2987–3000.
Picart, C., & Discher, D. E. (1999). Actin protofilament orientation at the erythrocyte membrane. *Biophysical Journal, 77*(2), 865–878.
Pinder, J. C., Bray, D., & Gratzer, W. B. (1975). Actin polymerisation induced by spectrin. *Nature, 258*(5537), 765–766.
Pinder, J. C., Clark, S. E., Baines, A. J., Morris, E., & Gratzer, W. B. (1981). The construction of the red cell cytoskeleton. *Progress in Clinical and Biological Research, 55*, 343–361.
Pinder, J. C., & Gratzer, W. B. (1983). Structural and dynamic states of actin in the erythrocyte. *The Journal of Cell Biology, 96*(3), 768–775.
Pinder, J. C., Ohanian, V., & Gratzer, W. B. (1984). Spectrin and protein 4.1 as an actin filament capping complex. *FEBS Letters, 169*(2), 161–164.
Pinder, J. C., Sleep, J. A., Bennett, P. M., & Gratzer, W. B. (1995). Concentrated Tris solutions for the preparation, depolymerization, and assay of actin: Application to erythroid actin. *Analytical Biochemistry, 225*(2), 291–295.
Pinder, J. C., Ungewickell, E., Bray, D., & Gratzer, W. B. (1978). The spectrin-actin complex and erythrocyte shape. *Journal of Supramolecular Structure, 8*(4), 439–445.
Pinder, J. C., Ungewickell, E., Calvert, R., Morris, E., & Gratzer, W. B. (1979). Polymerisation of G-actin by spectrin preparations: Identification of the active constituent. *FEBS Letters, 104*(2), 396–400.
Pinder, J. C., Weeds, A. G., & Gratzer, W. B. (1986). Study of actin filament ends in the human red cell membrane. *Journal of Molecular Biology, 191*(3), 461–468.
Pinto da Silva, P., & Branton, D. (1970). Membrane splitting in freeze-ethching. Covalently bound ferritin as a membrane marker. *The Journal of Cell Biology, 45*(3), 598–605.
Pinto da Silva, P., & Branton, D. (1972). Membrane intercalated particles: The plasma membrane as a planar fluid domain. *Chemistry and Physics of Lipids, 8*(4), 265–278.
Podolski, J. L., & Steck, T. L. (1988). Association of deoxyribonuclease I with the pointed ends of actin filaments in human red blood cell membrane skeletons. *The Journal of Biological Chemistry, 263*(2), 638–645.
Pollard, T. D., & Borisy, G. G. (2003). Cellular motility driven by assembly and disassembly of actin filaments. *Cell, 112*(4), 453–465.
Porro, F., Costessi, L., Marro, M. L., Baralle, F. E., & Muro, A. F. (2004). The erythrocyte skeletons of beta-adducin deficient mice have altered levels of tropomyosin, tropomodulin and EcapZ. *FEBS Letters, 576*(1–2), 36–40.

Pradhan, D., Tseng, K., Cianci, C. D., & Morrow, J. S. (2004). Antibodies to betaISigma2 spectrin identify in-homogeneities in the erythrocyte membrane skeleton. *Blood Cells, Molecules & Diseases*, *32*(3), 408–410.
Preston, G. M., Carroll, T. P., Guggino, W. B., & Agre, P. (1992). Appearance of water channels in Xenopus oocytes expressing red cell CHIP28 protein. *Science*, *256*(5055), 385–387.
Rana, A. P., Ruff, P., Maalouf, G. J., Speicher, D. W., & Chishti, A. H. (1993). Cloning of human erythroid dematin reveals another member of the villin family. *Proceedings of the National Academy of Sciences of the United States of America*, *90*(14), 6651–6655.
Robertson, J. D. (1959). The ultrastructure of cell membranes and their derivatives. *Biochemical Society Symposium*, *16*, 3–43.
Robledo, R. F., Ciciotte, S. L., Gwynn, B., Sahr, K. E., Gilligan, D. M., Mohandas, N., et al. (2008). Targeted deletion of alpha-adducin results in absent beta- and gamma-adducin, compensated hemolytic anemia, and lethal hydrocephalus in mice. *Blood*, *112*(10), 4298–4307.
Sahr, K. E., Lambert, A. J., Ciciotte, S. L., Mohandas, N., & Peters, L. L. (2009). Targeted deletion of the gamma-adducin gene (Add3) in mice reveals differences in alpha-adducin interactions in erythroid and nonerythroid cells. *American Journal of Hematology*, *84*(6), 354–361.
Salomao, M., Zhang, X., Yang, Y., Lee, S., Hartwig, J. H., Chasis, J. A., et al. (2008). Protein 4.1R-dependent multiprotein complex: New insights into the structural organization of the red blood cell membrane. *Proceedings of the National Academy of Sciences of the United States of America*, *105*(23), 8026–8031.
Sato, S. B., Yanagida, M., Maruyama, K., & Ohnishi, S. (1979). Seeding role of spectrin in polymerization of skeletal muscle actin. *Biochimica et Biophysica Acta*, *578*(2), 436–444.
Schafer, D. A., Jennings, P. B., & Cooper, J. A. (1998). Rapid and efficient purification of actin from nonmuscle sources. *Cell Motility and the Cytoskeleton*, *39*(2), 166–171.
Sche, P., Vera, C., & Sung, L. A. (2011). Intertwined alphabeta spectrin meeting helical actin protofilament in the erythrocyte membrane skeleton: Wrap-around vs. point-attachment. *Annals of Biomedical Engineering*, *39*(7), 1984–1993.
Shartava, A., Monteiro, C. A., Bencsath, F. A., Schneider, K., Chait, B. T., Gussio, R., et al. (1995). A posttranslational modification of beta-actin contributes to the slow dissociation of the spectrin-protein 4.1-actin complex of irreversibly sickled cells. *The Journal of Cell Biology*, *128*(5), 805–818.
Sheetz, M. P. (1979). DNase-I-dependent dissociation of erythrocyte cytoskeletons. *The Journal of Cell Biology*, *81*(1), 266–270.
Sheetz, M. P., Painter, R. G., & Singer, S. J. (1976). Relationships of the spectrin complex of human erythrocyte membranes to the actomyosins of muscle cells. *Biochemistry*, *15*(20), 4486–4492.
Shen, B. W., Josephs, R., & Steck, T. L. (1984). Ultrastructure of unit fragments of the skeleton of the human erythrocyte membrane. *The Journal of Cell Biology*, *99*(3), 810–821.
Shen, B. W., Josephs, R., & Steck, T. H. (1986). Ultrastructure of the intact skeleton of the human erythrocyte. *The Journal of Cell Biology*, *102*, 997–1006.
Siegel, D. L., & Branton, D. (1985). Partial purification and characterization of an actin-bundling protein, band 4.9, from human erythrocytes. *The Journal of Cell Biology*, *100*(3), 775–785.
Singer, S. J., & Nicolson, G. L. (1972). The fluid mosaic model of the structure of cell membranes. *Science*, *175*(4023), 720–731.
Steck, T. L. (1974). The organization of proteins in the human red blood cell membrane. A review. *The Journal of Cell Biology*, *62*(1), 1–19.
Sui, Z., Nowak, R.B., Bacconi, A., Kim, N.E., Lui, H., Li, J., Wickrema, A., An, X., & Fowler, V.M. (2013). Tropomodulin3-null mice are embryonic lethal with anemia due to impaired erythroid terminal differentiation in the fetal liver. Blood, in press.

Sung, L. A., Gao, K. M., Yee, L. J., Temm-Grove, C. J., Helfman, D. M., Lin, J. J., et al. (2000). Tropomyosin isoform 5b is expressed in human erythrocytes: Implications of tropomodulin-TM5 or tropomodulin-TM5b complexes in the protofilament and hexagonal organization of membrane skeletons. *Blood*, *95*(4), 1473–1480.

Sung, L. A., & Lin, J. J. (1994). Erythrocyte tropomodulin binds to the N-terminus of hTM5, a tropomyosin isoform encoded by the gamma-tropomyosin gene. *Biochemical and Biophysical Research Communications*, *201*(2), 627–634.

Swihart, A. H., Mikrut, J. M., Ketterson, J. B., & Macdonald, R. C. (2001). Atomic force microscopy of the erythrocyte membrane skeleton. *Journal of Microscopy*, *204*(Pt 3), 212–225.

Takakuwa, Y. (2000). Protein 4.1, a multifunctional protein of the erythrocyte membrane skeleton: Structure and functions in erythrocytes and nonerythroid cells. *International Journal of Hematology*, *72*(3), 298–309.

Takeuchi, M., Miyamoto, H., Sako, Y., Komizu, H., & Kusumi, A. (1998). Structure of the erythrocyte membrane skeleton as observed by atomic force microscopy. *Biophysical Journal*, *74*(5), 2171–2183.

Terada, N., Fujii, Y., & Ohno, S. (1996). Three-dimensional ultrastructure of in situ membrane skeletons in human erythrocytes by quick-freezing and deep-etching method. *Histology and Histopathology*, *11*(3), 787–800.

Tilley, L., & Ralston, G. (1984). Purification and kinetic characterisation of human erythrocyte actin. *Biochimica et Biophysica Acta*, *790*(1), 46–52.

Tilney, L. G., & Detmers, P. (1975). Actin in erythrocyte ghosts and its association with spectrin. Evidence for a nonfilamentous form of these two molecules in situ. *The Journal of Cell Biology*, *66*(3), 508–520.

Tsukita, S., Tsukita, S., Hosoya, H., & Mabuchi, I. (1985). Barbed end-capping protein regulates polarity of actin filaments from the human erythrocyte membrane. *Experimental Cell Research*, *158*(1), 280–285.

Tsukita, S., Tsukita, S., & Ishikawa, H. (1980). Cytoskeletal network underlying the human erythrocyte membrane. Thin-section electron microscopy. *The Journal of Cell Biology*, *85*(3), 567–576.

Tsukita, S., Tsukita, S., & Ishikawa, H. (1984). Bidirectional polymerization of G-actin on the human erythrocyte membrane. *The Journal of Cell Biology*, *98*(3), 1102–1110.

Tsukita, S., Tsukita, S., Ishikawa, H., Sato, S., & Nakao, M. (1981). Electron microscopic study of reassociation of spectrin and actin with the human erythrocyte membrane. *The Journal of Cell Biology*, *90*(1), 70–77.

Ubukawa, K., Guo, Y. M., Takahashi, M., Hirokawa, M., Michishita, Y., Nara, M., et al. (2012). Enucleation of human erythroblasts involves non-muscle myosin IIB. *Blood*, *119*(4), 1036–1044.

Ungewickell, E., Bennett, P. M., Calvert, R., Ohanian, V., & Gratzer, W. B. (1979). In vitro formation of a complex between cytoskeletal proteins of the human erythrocyte. *Nature*, *280*(5725), 811–814.

Ursitti, J. A., & Fowler, V. M. (1994). Immunolocalization of tropomodulin, tropomyosin and actin in spread human erythrocyte skeletons. *Journal of Cell Science*, *107*(Pt 6), 1633–1639.

Ursitti, J. A., Pumplin, D. W., Wade, J. B., & Bloch, R. J. (1991). Ultrastructure of the human erythrocyte cytoskeleton and its attachment to the membrane. *Cell Motility and the Cytoskeleton*, *19*(4), 227–243.

Ursitti, J. A., & Wade, J. B. (1993). Ultrastructure and immunocytochemistry of the isolated human erythrocyte membrane skeleton. *Cell Motility and the Cytoskeleton*, *25*(1), 30–42.

Vardar, D., Chishti, A. H., Frank, B. S., Luna, E. J., Noegel, A. A., Oh, S. W., et al. (2002). Villin-type headpiece domains show a wide range of F-actin-binding affinities. *Cell Motility and the Cytoskeleton*, *52*(1), 9–21.

Vera, C., Sood, A., Gao, K. M., Yee, L. J., Lin, J. J., & Sung, L. A. (2000). Tropomodulin-binding site mapped to residues 7-14 at the N-terminal heptad repeats of tropomyosin isoform 5. *Archives of Biochemistry and Biophysics*, *378*(1), 16–24.

Waseem, A., & Palfrey, H. C. (1988). Erythrocyte adducin. Comparison of the alpha- and beta-subunits and multiple-site phosphorylation by protein kinase C and cAMP-dependent protein kinase. *European Journal of Biochemistry*, *178*(2), 563–573.

Weber, A., Pennise, C. R., Babcock, G. G., & Fowler, V. M. (1994). Tropomodulin caps the pointed ends of actin filaments. *The Journal of Cell Biology*, *127*(6 Pt 1), 1627–1635.

Weber, A., Pennise, C. R., & Fowler, V. M. (1999). Tropomodulin increases the critical concentration of barbed end-capped actin filaments by converting ADP.P(i)-actin to ADP-actin at all pointed filament ends. *The Journal of Biological Chemistry*, *274*(49), 34637–34645.

Wong, A. J., Kiehart, D. P., & Pollard, T. D. (1985). Myosin from human erythrocytes. *The Journal of Biological Chemistry*, *260*(1), 46–49.

Xu, K., Zhong, G., & Zhuang, X. (2013). Actin, spectrin, and associated proteins form a periodic cytoskeletal structure in axons. *Science*, *339*(6118), 452–456.

Yamashiro, S., Gokhin, D. S., Kimura, S., Nowak, R. B., & Fowler, V. M. (2012). Tropomodulins: Pointed-end capping proteins that regulate actin filament architecture in diverse cell types. *Cytoskeleton (Hoboken)*, *69*(6), 337–370.

Yamashiro, S., Speicher, K. D., Speicher, D. W., & Fowler, V. M. (2010). Mammalian tropomodulins nucleate actin polymerization via their actin monomer binding and filament pointed end-capping activities. *The Journal of Biological Chemistry*, *285*(43), 33265–33280.

Zhu, Q., Vera, C., Asaro, R. J., Sche, P., & Sung, L. A. (2007). A hybrid model for erythrocyte membrane: A single unit of protein network coupled with lipid bilayer. *Biophysical Journal*, *93*(2), 386–400.

Zigmond, S. H. (2004). Beginning and ending an actin filament: Control at the barbed end. *Current Topics in Developmental Biology*, *63*, 145–188.

Zuccala, E. S., & Baum, J. (2011). Cytoskeletal and membrane remodelling during malaria parasite invasion of the human erythrocyte. *British Journal of Haematology*, *154*(6), 680–689.

CHAPTER THREE

Membrane Protein Dynamics and Functional Implications in Mammalian Cells

Francis J. Alenghat[*,1,2], **David E. Golan**[*,†,1]
[*]Department of Biological Chemistry and Molecular Pharmacology, Harvard Medical School, Boston, Massachusetts, USA
[†]Hematology Division, Brigham and Women's Hospital, Boston, Massachusetts, USA
[1]Corresponding authors: e-mail address: alenghat@post.harvard.edu; david_golan@hms.harvard.edu

Contents

Abstract

The organization of the plasma membrane is both highly complex and highly dynamic. One manifestation of this dynamic complexity is the lateral mobility of proteins within the plane of the membrane, which is often an important determinant of intermolecular protein-binding interactions, downstream signal transduction, and local membrane mechanics. The mode of membrane protein mobility can range from random Brownian motion to immobility and from confined or restricted motion to actively directed motion. Several methods can be used to distinguish among the various modes of protein mobility, including fluorescence recovery after photobleaching, single-particle tracking, fluorescence correlation spectroscopy, and variations of these techniques. Here, we present both a brief overview of these methods and examples of their use to elucidate the dynamics of membrane proteins in mammalian cells—first in erythrocytes, then in erythroblasts and other cells in the hematopoietic lineage, and finally in non-hematopoietic cells. This multisystem analysis shows that the cytoskeleton frequently governs modes of membrane

[2] Present address: Cardiology Section, University of Chicago, Chicago, Illinois, USA.

Current Topics in Membranes, Volume 72
ISSN 1063-5823
http://dx.doi.org/10.1016/B978-0-12-417027-8.00003-9

protein motion by stably anchoring the proteins through direct-binding interactions, by restricting protein diffusion through steric interactions, or by facilitating directed protein motion. Together, these studies have begun to delineate mechanisms by which membrane protein dynamics influence signaling sequelae and membrane mechanical properties, which, in turn, govern cell function.

1. INTRODUCTION

The composition and organization of the plasma membrane are both highly complex and ever-changing. The modes by which proteins move in the plane of the membrane provide insights into the molecular interactions between these proteins and neighboring membrane proteins, membrane lipids, the underlying cytoskeleton, and counter-receptors on cells or other structures in the extracellular environment. Over the past several decades, investigators have developed increasingly sophisticated methods for probing the dynamics of membrane proteins, and, in so doing, have helped to define structure–function relationships among membrane receptors, counter-receptors, and structural proteins.

Membrane-associated proteins comprise a large subset of all proteins synthesized by mammalian cells. Understanding how these proteins move and interact with one another in their native environment is essential to understanding their cellular function. For example, regardless of their intrinsic affinity for binding to each other, two proteins expressed on the plasma membrane may have little chance for interaction if their positions are fixed at random locations, unless the protein concentration is very high. With slow random diffusion of both proteins, the opportunity for interaction increases. With directed movement of one or both proteins toward a membrane landmark, or with confinement or corralling of the proteins in specific sites (such as focal adhesions or membrane microdomains), the opportunities for interaction increase further. It is in this context that the diffusion modes and kinetics of membrane proteins direct their function.

2. THE FLUID MOSAIC MODEL AND BEYOND

The fluid mosaic model was a critical insight that has guided all subsequent refinements of our understanding of membrane protein dynamics. Developed by Singer and Nicolson in 1972 in the face of limited knowledge of the complexity of the plasma membrane's composition and organization, this model attempted to characterize biological membranes in a unifying manner (Singer & Nicolson, 1972). One main tenet of the model was that biological membranes consist of a phospholipid bilayer in which globular proteins are embedded. The nonpolar portions of membrane proteins were thought to be sequestered from contact with the aqueous extracellular and cytoplasmic environments, whereas the polar portions of the proteins were thought to be relatively exposed to the extracellular or cytoplasmic environment in order to minimize the free energy of the membrane. These concepts may seem obvious to cell biologists now, but, at the time, the fluid mosaic model was at odds with other ideas, such as the possibility that membrane proteins are in some way tethered to the membrane and are extended into the extracellular environment without any consideration for thermodynamic stability. The fluid mosaic model further predicted that proteins embedded in a lipid bilayer would be free to undergo translational diffusion at rates determined by the viscosity of the lipid bilayer. Despite this ability to undergo lateral diffusion, the proteins would maintain their membrane-embedded status; that is, their "degree of intercalation" with the membrane would not change (Singer & Nicolson, 1972). These predictions about mobility came with an important caveat—the lateral diffusion of a membrane protein would occur freely *unless* the protein interacts specifically with other proteins or lipids.

We now know that such interactions are the rule rather than the exception. In fact, interactions of membrane proteins with other membrane proteins, membrane lipids, and intracellular and extracellular structures are so common that measured diffusion rates of membrane proteins rarely approach the rate or pattern of free Brownian diffusion. Below is a useful categorization of the major classes of nonrandom interactions involving membrane proteins:

(A) *Interactions with membrane lipids*. The lipid bilayer surrounding any given membrane protein is typically inhomogeneous. This is evident on the molecular scale, in that adjacent phospholipid molecules have different molecular structures such that the interface of the protein with the lipid bilayer is not radially symmetric. The inhomogeneity of the lipid bilayer is also manifested on the nanometer and micrometer scale, in that the plasma membrane contains areas that are enriched for certain lipids such as cholesterol and glycosphingolipids (Simons & Ikonen, 1997). These cholesterol-enriched membrane microdomains (CEMMs; sometimes called lipid rafts) are present in most cell types, and many membrane proteins have demonstrated a clear preference for localizing to these microdomains (Brown & London, 1998; Galbiati, Razani, & Lisanti, 2001). Indeed, compartmentalization of such proteins to CEMMs allows the proteins to serve as organizing centers for signal transduction at the membrane (Lingwood & Simons, 2010; Patel, Murray, & Insel, 2008). Observations of the lateral diffusion of membrane proteins that are preferentially localized to CEMMs reflect this relative confinement (see "*Confinement*" below).

(B) *Steric interactions and complex formation with other membrane proteins and with intracellular proteins.* Any interaction with another protein may affect the lateral diffusion of a membrane protein. Relatively immobile protein obstacles that are either within the membrane or just below or above the membrane can present a corralling effect to confine the diffusion of membrane proteins. Binding interactions with other membrane proteins and with cytoskeletal or adaptor proteins can also constrain membrane protein diffusion. The affinity of the binding interaction, the degree of immobilization of the binding partner, and the number of proteins interacting with the membrane protein combine to determine the magnitude of the restriction on lateral diffusion (Frick, Schmidt, & Nichols, 2007). For example, a membrane protein anchored to the underlying cytoskeleton through a large number of other proteins may exhibit more restricted diffusion than a membrane protein that forms a heterodimer with another membrane protein but has no other significant protein–protein interactions.

(C) *Interactions with extracellular structures.* Many membrane proteins are receptors or ligands for receptors on other cells or for extracellular matrix (ECM) proteins. These interactions can affect membrane protein diffusion through two mechanisms: first, the physical restriction of lateral diffusion through the receptor–counter-receptor binding interaction; and second, the binding-induced conformational change in the membrane protein that may alter downstream signaling and protein associations to further modulate diffusion.

(D) *Self-organization, self-assembly, and hierarchical ordering of membrane proteins.* There is a growing understanding that all cellular proteins—including proteins found at the plasma membrane—are highly organized and dynamic. In order to function properly, membrane proteins must bind to and release from other proteins, form and dissolve protein complexes, and traffic to and from intracellular and membrane-localized sites. These dynamic interactions result in membrane protein mobility that is not consistent with free diffusion (Kusumi, Suzuki,

Kasai, Ritchie, & Fujiwara, 2011; Watkins, Miller, Majewski, & Kuhl, 2011). In one extreme form, directed motion along a submembranous scaffold can be observed as a protein travels from a site of insertion in the membrane to a distant site of action (Serge, Fourgeaud, Hemar, & Choquet, 2003). Small-scale directed motion can also be observed as proteins cluster at specific sites to form supramolecular complexes (Sander, Arora, & Smith, 2012; Wehrle-Haller, 2007).

Given the prevalence of these complex interactions involving proteins in the plasma membranes of mammalian cells, there is surprisingly little room for free diffusion of membrane proteins such as that predicted by the fluid mosaic model. Free diffusion may occur within some membrane subdomains or with some mild restriction for certain proteins, but it is the exception rather than the rule. Moreover, the existence of interactions involving lipid rafts, protein complexes, extracellular structures, and self-organization all suggest that the distribution of membrane proteins is nonrandom across the plasma membrane.

Over the past 30–40 years, the dynamics of many different proteins in mammalian plasma membranes have been described and several distinct models for the regulation of these dynamics have been advanced. In general, these models have been developed in concert with the elucidation of the complex membrane protein interactions described above. The major current models for the regulation of membrane protein dynamics include (Fig. 3.1):

(A) *Random diffusion.* This mode of motion is based on a model of independent particles (such as proteins) that are randomly distributed in a homogenous, two-dimensional fluid (such as the plasma membrane). In such a model, the particles diffuse freely (also called Brownian motion) and the rate of diffusion is characterized by the particle's diffusion coefficient. The distance traveled by the particle as a function of time is expressed as its mean square displacement (MSD), which is related to the diffusion coefficient. In two dimensions, $\mathrm{MSD}=4Dt^{\alpha}$, where $\alpha=1$ for Brownian motion (Mirchev & Golan, 2001). Therefore, for random diffusion, the diffusion coefficient $D=\mathrm{MSD}/4t$. As an example, if a membrane protein randomly diffuses in two dimensions and travels an average of 0.2 μm in 1 s, then $D=(2\times10^{-5}\,\mathrm{cm})^2/(4\times1\,\mathrm{s})=1\times10^{-10}\,\mathrm{cm}^2/\mathrm{s}$.

(B) *Confinement.* This mode of motion occurs when a membrane protein is preferentially localized to specific subdomains of the membrane (such as lipid rafts), or when a membrane protein encounters obstacles in its

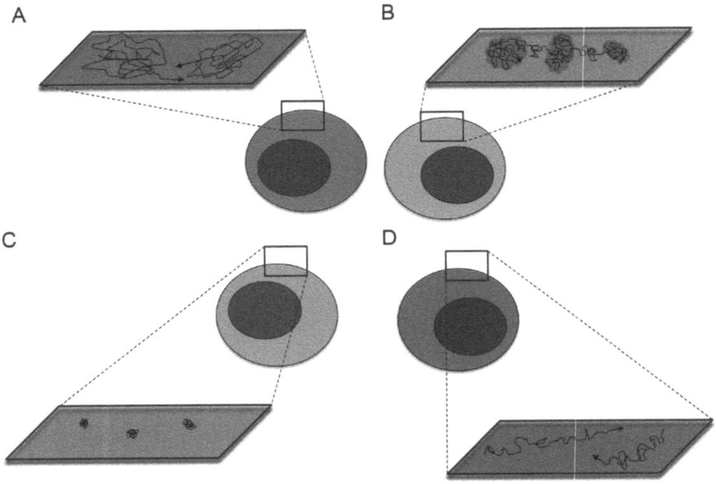

Figure 3.1 *A general framework for the major modes of membrane protein mobility. Random diffusion* (A) is most closely approximated by the Brownian motion of a protein embedded in a homogeneous lipid bilayer. Protein mobility is relatively free from encumbrances due to interactions with extracellular ligands or receptors, other membrane proteins, cytoskeletal elements, or cytoplasmic proteins. *Confinement* (B) occurs when a protein diffuses preferentially within specific domains in the membrane. These can be cholesterol-enriched membrane microdomains (CEMMs) that stabilize the protein or contain docking partners for the protein, patches bounded by less mobile membrane-embedded molecules that corral the diffusing protein, or areas overlying a meshwork of cytoskeletal or other structural proteins that corral or otherwise interact sterically with the cytoplasmic domain of the protein. Proteins experiencing *restriction* (C) have diffusion coefficients substantially lower than those predicted by Brownian motion. Such proteins may be restrained through direct or indirect binding to relatively immobile structures such as extracellular binding partners or macromolecular complexes anchored to the cytoskeleton. *Directed motion* (D) is characterized by either channeled movement or active transport of the protein and is typically guided by linearly polarized cytoskeletal elements or driven by subjacent motor mechanisms. (For color version of this figure, the reader is referred to the online version of this chapter.)

lateral motion (such as relatively immobile protein complexes or cytoskeleton-anchored membrane proteins) so that it is effectively corralled. In both scenarios, the probability that the membrane protein will move away from the local area is less than the probability that it will stay within the local area. In many cases, the lateral mobility of the protein within the area of confinement is consistent with random diffusion, and the escape of the protein from the local area results in its relocation to a new area elsewhere on the membrane, where random diffusion resumes. If the predominant behavior is confined diffusion with rare relocation events, this is known as *hop diffusion*. If the immobile obstacles or preferred membrane environments are sparse, then the confinement is transient and the larger-scale movements predominate. In either case, the lateral mobility (characterized by the diffusion coefficient D) differs during periods of confinement and nonconfinement. For this reason, it is useful to characterize the lateral mobility of membrane proteins both on small time and distance scales (D_{μ}, on the ms and nm scale) and on large time and distance scales (D_M, on the s and μm scale). When D_{μ} exceeds D_M—a common situation for many proteins in mammalian plasma membranes—short time-scale diffusion is not constrained, but obstacles or areas of confinement become apparent over longer periods. When D_M and D_{μ} are equal (and not extremely low), confinement is not present.

(C) *Restriction*. This mode of motion can be considered an extreme form of confinement. It occurs when a membrane protein is bound to a significantly less mobile element (such as a large protein complex that is anchored to the cytoskeleton) such that its diffusion is far less than what would be predicted by Brownian motion, even on small time and distance scales (see "*Confinement*" above). In the extreme example, fixation of cells results in the complete immobilization of membrane proteins (Umenishi, Verbavatz, & Verkman, 2000). When D_M and D_{μ} are both extremely low, restriction is present.

(D) *Directed Motion*. Active transport of a membrane protein results in its directed motion in the plane of the membrane. In such a situation, the calculated diffusion coefficient may be less informative than the velocity of the transport. The equation $MSD = 4Dt^{\alpha}$ may also be used to characterize directed motion since the parameter α describes the time dependence of D and indicates the mode of motion. As noted above, in Brownian diffusion $\alpha = 1$ and the equation reduces to $MSD = 4Dt$. In directed motion, $\alpha > 1$.

3. TECHNIQUES FOR MEASURING LATERAL MOBILITY OF MEMBRANE PROTEINS

Several different methods are used to measure membrane protein dynamics. The primary techniques currently in use include:

(A) *Fluorescence Recovery After Photobleaching (FRAP)* [also known as Fluorescence Photobleaching Recovery (FPR)]. In this technique

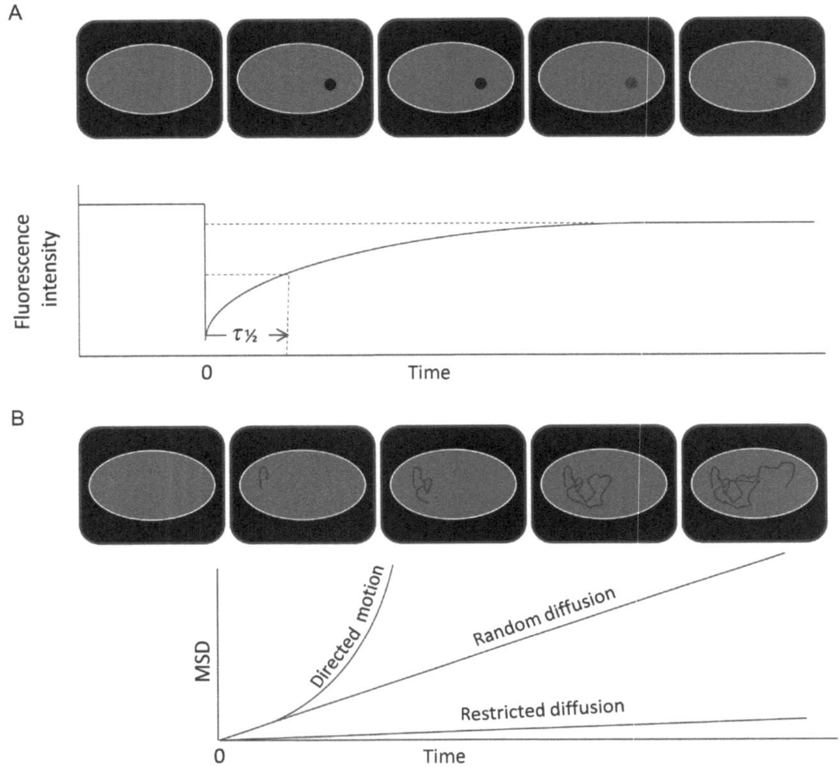

Figure 3.2 *Fluorescence recovery after photobleaching (FRAP) and single-particle tracking (SPT) measurements are the basis for the majority of studies of membrane protein dynamics.* In *FRAP* (A), a small area of membrane containing a fluorescently labeled protein is photobleached using a high-intensity laser. As labeled proteins from the surrounding (unbleached) area migrate laterally into the bleached area, fluorescence recovers in the bleached area. The half-time for maximal fluorescence recovery is used to calculate the lateral diffusion coefficient of the protein. The ratio of the maximal fluorescence recovery to the prebleach fluorescence corresponds to the fractional mobility of the protein. In *SPT* (B), the membrane protein is tagged at very low density with a microparticle, nanoparticle, or fluorescent molecule, and this tag is tracked using video microscopy, often at high frame rates. The mean square displacement (MSD) from the origin is plotted versus time. The lateral diffusion coefficient is calculated based on the MSD versus time relationship. Certain modes of motion have characteristic MSD versus time relationships (e.g., directed motion, random diffusion, restricted diffusion). Because each tracked particle is monitored individually, SPT can be used to elucidate molecular heterogeneity among lateral diffusion coefficients and modes of motion. (See color plate.)

(Fig. 3.2A), membrane proteins of interest are fluorescently labeled and a small area of the plasma membrane is rapidly photobleached by an intense laser (Axelrod, Koppel, Schlessinger, Elson, & Webb, 1976). Over time, fluorescently labeled proteins diffuse laterally from outside the bleached region into the bleached region. The rate of lateral diffusion can be quantified using the equation $D = r^2/4\tau_{1/2}$ (a modified version of $D = \mathrm{MSD}/4t$), where r is the radius of the bleached spot and $\tau_{1/2}$ is the time for recovery of half the maximal postbleach fluorescence (Axelrod et al., 1976; Kang, Day, Kenworthy, & DiBenedetto, 2012; Lippincott-Schwartz, Snapp, & Kenworthy, 2001). The fractional mobility f is defined as the ratio of the maximal postbleach fluorescence intensity to the prebleach intensity. A value of $f < 1$ suggests that a fraction of the labeled membrane proteins is relatively immobile. One advantage of FRAP is its ability to characterize the average dynamics of the labeled membrane proteins, since the observed fluorescence signal represents the sum of the signals from many copies of the protein. At the same time, it is difficult to define the heterogeneity of dynamics of individual membrane proteins in a population using this technique.

(B) *Single-Particle Tracking* (*SPT*). In SPT (Fig. 3.2B), video microscopy is used to track the motion of labeled proteins within the plasma membrane (Qian, Sheetz, & Elson, 1991). SPT is often performed by binding 20–1000 nm-diameter gold, latex, or fluorescent beads to the protein of interest, and recording the position of the bead in x- and y-coordinates at high frame rates. Larger beads (~1000 nm) can be tracked using light microscopy. Smaller (~20 nm) particles include colloidal gold, which can be tracked using differential interference contrast microscopy, and fluorescent nanoparticles or quantum dots, which can be tracked using high-speed fluorescence microscopy (fluorescent SPT). In all cases, high spatial resolution and reasonable time resolution are achieved by using short exposure times (Gonda, Watanabe, Ohuchi, & Higuchi, 2010). The primary advantage of SPT and its variations is, as the name indicates, the visualization of individual membrane proteins. Trajectories and MSD are recorded for each tracked particle, and, therefore, the heterogeneity of motion among a population of protein molecules can be better appreciated (Saxton, 1997; Saxton & Jacobson, 1997). SPT also provides information about the mode of motion of individual protein molecules, including confinement, restriction, and directed motion. As described above, this information can help elucidate the biological structures responsible for the nonrandom motion, such as membrane microdomains, cytoskeletal anchorage, hierarchical organization, and active transport. Fluorescent SPT can also be used to detect clustering of tracked particles, which is visualized as an increase in fluorescence intensity (Jaqaman et al., 2008). Fluorescence speckle microscopy (FSM) is used to analyze macromolecular assemblies. FSM has been used primarily to study the dynamics of cytoskeletal and cytoskeleton-binding proteins (Mendoza, Besson, & Danuser, 2012). In this technique, a small fraction of a protein of interest is labeled, often by fusion with a fluorescent protein. Because the protein concentration is heterogeneous in two and three dimensions and because the label is not saturating, the protein labeling appears speckled. In the context of membrane protein dynamics, speckles can represent transient clustering of the protein of interest. The motion, appearance, and disappearance of such speckles can be tracked and used to quantitatively describe local protein concentrations and dynamics. Although used primarily as a technique to study the dynamics of cytoskeletal proteins, FSM could also serve as a complement to SPT in the study of membrane proteins, particularly adhesion molecules (Hu, Ji, Applegate, Danuser, & Waterman-Storer, 2007; Mendoza et al., 2012).

(C) *Fluorescence Correlation Spectroscopy* (*FCS*). FCS is a third technique to measure membrane protein diffusion. In this method, membrane proteins of interest are fluorescently labeled and a small area of the plasma membrane is illuminated using a confocal microscope at the excitation wavelength of the fluorophore (Bacia, Kim, & Schwille, 2006; Haustein & Schwille, 2007). As the labeled proteins diffuse into the area of illumination, fluorescence emission is detected. The longer the protein remains in the illuminated area, the longer is the recorded burst of fluorescence emission. Analysis of the multiple transits of the labeled protein molecules yields a characteristic transit time, and this transit time, together with the size of the illumination spot, allows for calculation of the diffusion coefficient (Haustein & Schwille, 2007). The transit time can also be expressed as a function of the size of the illumination spot; the larger the spot is, the longer is the transit time. In cases where there is active partitioning of proteins among different membrane subdomains, such as confinements conferred by lipid rafts, the transit time is less dependent on the size of the illumination spot (Bacia, Scherfeld, Kahya, & Schwille, 2004; Billaudeau et al.,

2013; Kahya, Scherfeld, Bacia, Poolman, & Schwille, 2003). For example, if a small area of illumination happens to contain one or more lipid rafts in their entirety, then a raft-dependent protein that enters the area would likely be confined preferentially to those rafts, and shrinking the illumination area further would not change the transit time significantly if the rafts were still within the illumination area. Advantages of FCS are its ability to quantify particle density in the area of illumination and its use of low-intensity laser illumination (unlike FRAP). All of the above techniques can be combined with other experimental modalities to measure protein–protein interactions (Lippincott-Schwartz et al., 2001). One such method is fluorescence resonance energy transfer (FRET). For example, FRET can be combined with fluorescent SPT to determine whether two fluorescently tagged proteins are clustering together or are otherwise associated at the plasma membrane (Rolfe et al., 2011). Similar analysis is also enabled using a combination of FCS and FRET (Haustein & Schwille, 2007). Protein diffusion before and after such protein–protein association can be measured and compared. FRAP, fluorescent SPT, and FCS can also be combined with total internal reflection fluorescence (TIRF) microscopy to gain better resolution of the plasma membrane and to measure the effect of protein–ligand interactions on membrane protein lateral mobility (Nechyporuk-Zloy, Dieterich, Oberleithner, Stock, & Schwab, 2008).

4. MEMBRANE PROTEIN DYNAMICS IN MAMMALIAN CELLS

Much of the early work on membrane protein dynamics was carried out on erythrocytes. Since that work, studies have expanded to include the dynamics of membrane proteins on all major hematopoietic lineages and many other nucleated mammalian cells. In this section, we review some of the studies on major membrane proteins in these various cell types.

4.1. Red cell membrane protein dynamics

The human red blood cell (RBC) membrane was an early model of choice for studying membrane protein dynamics. The reasons for this choice included: relative ease of obtaining the cells for study; relative uniformity of the cells; relatively abundant knowledge about the content, organization, and characterization of RBC membrane proteins and lipids (Bennett, 1985); and the availability of RBCs with abnormal membrane proteins that were known to be involved in several RBC diseases (Iolascon, Perrotta, & Stewart, 2003; Palek, 1987). Like the plasma membrane of other mammalian cells, the human RBC membrane consists of a lipid bilayer with embedded proteins. In the case of the RBC membrane, the lipid bilayer is biochemically and biophysically coupled to an underlying membrane skeleton. The major proteins in the membrane skeleton are spectrin tetramers and short actin filaments that, together, form a triangular lattice arrangement. The linking protein ankyrin has high-affinity binding sites for both spectrin and band 3 (see Section 4.1.1 below), and thereby serves as an important connection between the membrane skeleton and the lipid bilayer (Bennett, 1985). Another important connection is provided by adducin, which links the spectrin–actin junctional complex to band 3 (Anong et al., 2009). In this section, we review the dynamics of some of the major RBC membrane proteins.

4.1.1 Band 3

Each human RBC contains ~1million copies of band 3 protein (Steck, 1974). This protein has two important roles in RBC structure and function: (1) to serve as the major membrane anchor for the protein–protein interactions that couple the lipid bilayer to the underlying membrane skeleton and (2) to maintain anion exchange (HCO_3^- and Cl^-) between the interior of the cell and its extracellular environment. Molecular defects in band 3 are responsible for a fraction of cases of hereditary spherocytosis (Corbett, Agre, Palek, & Golan, 1994) and for all cases of Southeast Asian ovalocytosis (Liu et al., 1990; Mirchev, Lam, & Golan, 2011). Early studies of band 3 lateral mobility utilized FRAP on RBC ghost membranes (Golan & Veatch, 1980). Band 3 was found to be immobilized under conditions of low temperature (21 °C) and moderate ionic strength, whereas increasing temperature (toward 37 °C) and reducing ionic strength first increased the fractional mobility of band 3 and then significantly increased the diffusion coefficient (Golan & Veatch, 1980). Subsequent experiments suggested that this increase in mobility was not attributable to proteolysis of any major membrane protein. Rather, the progressive severing of the connections between the spectrin-based membrane skeleton and the overlying lipid bilayer, mediated by the release of ankyrin (and possibly adducin) from its binding site on spectrin, was responsible for the large changes in the fractional mobility and diffusion coefficient of band 3 (Cho, Eber, Liu, Lux, & Golan, 1998; Tsuji & Ohnishi, 1986). Indeed, RBCs from patients with ankyrin-deficient hereditary spherocytosis had markedly increased band 3 diffusion (Cho et al., 1998). Given that the ankyrin link to spectrin is critical for limiting band 3 diffusion, it followed that disruption of the spectrin-based membrane skeleton would also lead to higher diffusion coefficients for band 3. As expected, the diffusion coefficient of band 3 was eightfold higher on RBCs from patients with spectrin-deficient hereditary spherocytosis (12×10^{-11} cm^2/s) than on control RBCs (1.6×10^{-11} cm^2/s), without a change in band 3 fractional mobility (Corbett, Agre, et al., 1994).

Band 3 mobility has also been measured using SPT. The first SPT analyses of band 3 suggested that one third of the protein was relatively tightly confined, likely tethered to spectrin, whereas the remainder demonstrated confinement on a larger time and distance scale with episodic hops between the zones of confinement (Tomishige, Sako, & Kusumi, 1998). These findings, together with measurements of band 3 mobility under conditions of membrane skeletal disruption, led to the hypothesis that corralling of the cytoplasmic domain of band 3 by the underlying membrane skeleton was the mechanism most likely responsible for band 3 confinement and hopping (Tomishige et al., 1998). In another SPT study on normal RBCs, about half of the band 3 molecules displayed confinement and the other half were not confined (Mirchev et al., 2011). Interestingly, RBCs with the Southeast Asian ovalocytosis (SAO) band 3 mutation demonstrated even tighter areas of band 3 confinement. The SAO mutation causes band 3 to form linear oligomers in the plane of the RBC membrane, and this tighter confinement likely reflected a greater degree of mobility constraint on the larger protein complex. A third set of SPT studies also confirmed that a fraction of band 3 is quite confined—this is likely to represent the molecules that are associated with spectrin through ankyrin or adducin linkages—whereas another fraction is corralled in larger confinement zones (Kodippili et al., 2012).

4.1.2 Glycophorins

Glycophorins are the major sialoglycoproteins that span the membrane of the RBC. The predominant RBC glycophorin is glycophorin A (0.5 million copies per RBC). Several early studies measured the lateral mobility of nonspecifically labeled RBC membrane glycoproteins (including glycophorin A) in RBCs and RBC ghosts (Schindler, Koppel, & Sheetz, 1980) and the lateral mobility of purified glycophorin in synthetic lipid bilayers (Kapitza, Ruppel, Galla, & Sackmann, 1984). Our laboratory used fluorescein thiosemicarbazide to label RBC glycophorins specifically and FRAP to measure glycophorin lateral mobility in intact RBCs. The measured diffusion coefficient for the glycophorins was $2–5 \times 10^{-11}$ cm^2/s, similar to that of band 3 (Corbett, Cho, & Golan, 1994). The lateral mobility of both glycophorin and band 3 decreased as a function of cell density in RBCs from patients with sickle cell disease, likely due to the irreversible damage that characterizes the membrane of dense sickle RBCs (Corbett & Golan, 1993). These and other data led to the hypothesis that a fraction of band 3 and glycophorin molecules are physically linked in the RBC membrane. Consistent with this hypothesis, FRAP studies showed that extracellular engagement of glycophorin A induced lateral immobilization both of this protein and of band 3, probably through a mechanism involving rigidification of the underlying membrane skeleton (Knowles, Chasis, Evans, & Mohandas, 1994).

4.1.3 Complement components

Decay-accelerating factor (DAF) is a membrane protein that inhibits activation of the C3 complement component and thereby protects RBCs from complement-mediated pore formation and eventual cell lysis (Brodbeck, Mold, Atkinson, & Medof, 2000; Michaels, Abramovitz, Hammer, & Mayer, 1976). As a glycosylphosphatidylinositol (GPI)-anchored protein, DAF is well suited for this function, since the lateral mobility of GPI-linked proteins is generally high in mammalian plasma membranes. High lateral mobility allows DAF to "patrol" the RBC membrane and inactivate small amounts of newly deposited complement proteins wherever they are located on the cell. Using SPT, our laboratory found that the majority (~80%) of DAF molecules exhibited Brownian lateral motion with a diffusion coefficient of 41×10^{-11} cm^2/s in untreated (noncomplement activated) cells (Karnchanaphanurach et al., 2009).

In contrast, cells treated with abundant amounts of complement showed relative immobilization of DAF at sites of complement deposition. Under these conditions, 70% of DAF molecules exhibited a diffusion coefficient of 6×10^{-11} cm^2/s. Moreover, this immobilization of DAF was accompanied by immobilization of complement component C3b and of the major RBC membrane proteins glycophorin A (which is also the RBC receptor for C3b) and band 3. Mass spectrometry analysis showed that, in such complement-treated cells, band 3, glycophorin A, and C3b formed a macromolecular complex that was immobilized via links to the underlying spectrin–ankyrin membrane skeleton. Finally, as measured using laser optical tweezers, the mechanical properties of the complement-treated RBCs were affected in the form of increased stiffness and membrane viscosity. Together, these data suggested that complement activation caused a membrane skeleton-linked DAF-C3b-glycophorin A–band 3 complex to form on the RBC surface, and that formation of this complex increased RBC stiffness and membrane viscosity (Karnchanaphanurach et al., 2009). These changes in membrane stiffness may facilitate removal of senescent RBCs from the circulation.

4.1.4 Aquaporins

Approximately 50,000 tetramers of the aquaporin water transport protein (AQP1) are expressed on the human RBC membrane (Smith & Agre, 1991). FRAP-based measurements showed that fluorescently labeled AQP1 had a relatively high mobile fraction (66%) and a relatively low lateral diffusion coefficient (3.1×10^{-11} cm^2/s) in the membrane of intact RBCs. The fractional mobility of AQP1 was not significantly altered by immobilization of band 3 or glycophorin A, indicating that AQP1 is not tightly associated with either of these membrane proteins. Interestingly, stretching the membrane (and thereby dilating the underlying membrane skeleton) caused a 20-fold increase in the lateral diffusion coefficient of AQP1. Membrane deformation did not have the same effect on the lateral mobility of band 3 and glycophorin A, suggesting that aquaporin dynamics may be uniquely responsive to cell deformation. These findings also suggested that, unlike other RBC membrane proteins, AQP1 is not tightly associated with the underlying membrane skeleton; instead, the spectrin-based lattice may serve to corral the aquaporin molecules in confinement zones that are susceptible to enlargement and/or rupture upon stretching of the membrane (Cho et al., 1999).

4.1.5 Toward a model of functional organization

Measurements of the lateral mobility of erythrocyte membrane proteins have provided insights into the organization of the RBC membrane and the functional implications of RBC membrane structure. For example, the lateral mobility of two of the most abundant RBC membrane proteins, band 3 and glycophorin A, is markedly confined in normal RBC membranes. Both of these proteins are intimately associated with the underlying membrane skeleton, the former through ankyrin and adducin linkages and the latter through interaction with band 3. These direct-binding interactions are responsible for the restricted (immobile) fraction of band 3 and glycophorin A molecules observed in FRAP and SPT experiments. The band 3 and glycophorin A molecules that are not directly bound to the membrane skeleton show confined, not free, lateral mobility, due to the corralling effect of the spectrin-based membrane skeleton. Disruption of the membrane skeleton in diseases such as hereditary spherocytosis leads to both release of band 3 and glycophorin A from these mobility constraints and to increased fragility of the RBCs. In other diseases such as sickle cell anemia, band 3 and glycophorin A are immobilized in the RBC membrane due to oxidative crosslinking reactions catalyzed by membrane-associated aggregates of hemoglobin S, and the immobilization (clustering) of band 3 contributes to the premature removal of these cells.

Lateral mobility measurements on less abundant membrane proteins have also been informative. For example, the complement inhibitory protein DAF is highly mobile on the RBC membrane until it encounters its target (deposited complement components), at which point it becomes immobile and associates with a number of other RBC membrane proteins to alter cell stiffness. AQP1, in contrast, is relatively confined at baseline and becomes increasingly mobile with deformation of the spectrin-based membrane skeleton. The functional consequence of this change in AQP1 mobility remains an open question. The high mobility of AQP1 under membrane deformation may allow for rapid redistribution of this important homeostatic channel, and this redistribution may be prevented in stiff, senescent RBCs that are rapidly cleared from the circulation (Yip et al., 1983).

4.2. Membrane protein dynamics in other hematopoietic cells

The erythrocyte provided an early model for studying membrane protein dynamics, and the field then turned to other hematopoietic cells. The larger number and variety of membrane proteins on these cells, many of which are receptors or counter-receptors, have yielded new insights into the functional consequences of membrane protein mobility regulation. In this section, we review the dynamics of membrane proteins in erythroblasts, lymphocytes, and other cells derived from the bone marrow.

4.2.1 Erythroblasts

Erythroblasts undergo massive cytoskeletal changes during their maturation into erythrocytes, eventually losing much of the actin cytoskeleton and assembling a spectrin-based membrane skeleton (Liu, Guo, Mohandas, Chasis, & An, 2010). The signals and mechanisms that regulate these changes are critical for erythroid maturation. Along with these cytoskeletal changes, receptors that are important for maintaining adhesive interactions between the erythroblast and its microenvironment become downregulated and disappear at the end of erythroid maturation. Integrins, for example, mediate cellular adhesion between the erythroblast and the ECM and neighboring cells (Mohandas & Chasis, 2010). Integrins signal bidirectionally across the plasma membrane, physically linking the extracellular environment to intracellular signaling molecules and to the cytoskeleton (Giancotti & Ruoslahti, 1999). Understanding the dynamics of critical erythroblast membrane proteins, such as integrins, could provide insights into the relative importance for cellular function of different membrane proteins at each stage of erythroblast maturation, and could allow for identification of structural and/or signaling complexes at the membrane at each stage of development.

Despite this promise, the field of erythroblast membrane protein dynamics is still in its infancy. Band 3 dynamics have recently been described in erythroblasts (Kodippili et al., 2012). Using SPT, a decrease in the lateral mobility of band 3 was found to occur with progressive stages of erythroblast development (Kodippili et al., 2012). The gradual immobilization of band 3 was thought to be due to the integration of the protein in a cooperative membrane assembly process such that, by the end of terminal erythroid maturation, band 3 was in its proper place and fully functional. A potential model is as follows. Band 3 is present on the membrane in early-stage erythroblasts, but the protein is not cytoskeletally anchored in these cells. Indeed, the spectrin and ankyrin required for membrane skeletal attachment are not yet present in these early cells. With erythroblast maturation, as the requisite membrane skeletal elements accumulate, band 3 becomes progressively anchored. This model may also apply to other erythroblast membrane proteins that increase in expression during erythroid maturation, that is, such proteins may become more tightly anchored and/or confined as they become more functional. It remains to be determined whether the lateral mobility of downregulated membrane proteins, such as integrins, the transferrin receptor, and the erythropoietin receptor, also changes with erythroid maturation. This is an area of active investigation.

4.2.2 Lymphocytes

The interactions of lymphocytes with other cells are highly regulated. For example, T cells interact with endothelial cells and with antigen-presenting cells (APCs) in order to home to sites of inflammation and to modulate inflammatory responses. Lymphocyte function-associated antigen 1 (LFA-1, also called CD11a/CD18 and $\alpha_L\beta_2$) is an integrin heterodimer expressed on T cells that interacts with ICAM-1 on endothelial cells and APCs. The affinity of LFA-1 for its counter-receptors and its extent of clustering on the membrane influence both T cell adhesion and activation

(Kim, Carman, & Springer, 2003; van Kooyk & Figdor, 2000). The topological alignment of LFA-1 and ICAM-1 is necessary for optimal receptor–counter-receptor binding and clustering to take place (Dustin, 1998). Therefore, the lateral mobility of LFA-1 has great functional consequence for determining its ability to engage ICAM-1 effectively.

Substantial work has been done to characterize the lateral mobility of lymphocyte adhesion and activation receptors. An early study showed that T-cell activation with PMA caused a 10-fold increase in the diffusion coefficient of LFA-1, from 2.3×10^{-11} cm^2/s in native cells to 2.9×10^{-10} cm^2/s after 10 min of PMA treatment. Cytochalasin D also increased the rate of LFA-1 diffusion in resting T cells, suggesting that the mobility of nonactivated LFA-1 was constrained by the underlying actin cytoskeleton (Kucik, Dustin, Miller, & Brown, 1996). In another study, removing the portion of LFA-1 that is known to interact with the actin cytoskeleton did not completely release LFA-1 from its mobility constraints. Instead, it appeared that disruption of the entire actin network was required to substantially increase LFA-1 diffusion. These findings suggested that tight binding of LFA-1 to the cytoskeleton was not the only determinant of confined mobility, and that LFA-1 confinement was, at least in part, due to corralling of the integrin by the underlying actin network (Peters et al., 1999). More recently, the observation that treatment with cytochalasin D increases the lateral mobility of LFA-1 was replicated in neutrophils (Gaborski, Clark, Waugh, & McGrath, 2008). An important study showed that the increased lateral mobility of LFA-1 induced by disruption of the actin cytoskeleton was capable of driving integrin clustering and cell adhesion, and suggested that such clustering is especially important in adhesion strengthening after integrin binding to multivalent ligands (Kim, Carman, Yang, Salas, & Springer, 2004).

Our group has used FRAP and SPT to measure the lateral mobility of active and inactive conformations of LFA-1 in the plasma membrane of the T cell. Using a conformation-nonspecific antibody to label LFA-1, FRAP studies showed that exogenous activation of T cells with PMA significantly increased the average lateral diffusion coefficient of this receptor, and SPT measurements indicated that activation with PMA caused a higher fraction of LFA-1 molecules to become laterally mobile. Interestingly, however, using an antibody that was specific for the active, high-affinity conformation of LFA-1, we found that activation of T cells with PMA increased the *immobile* fraction of LFA-1 molecules that were in the active (open) conformation (Cairo, Mirchev, & Golan, 2006). Conversely, PMA increased the *mobile* fraction of LFA-1 molecules that were in the inactive (closed) conformation. These results were consistent with a model in which T cell activation mobilizes non-ligated, closed-conformation LFA-1 in order to maximize the opportunity for receptor–counter-receptor interactions, whereas the same stimulus acts to anchor the ligated, open-conformation LFA-1 in order to reinforce the receptor–counter-receptor interaction and strengthen T-cell adhesion. FRAP studies were used by other investigators to demonstrate immobilization of high-affinity LFA-1 in focal zones on the membranes of rapidly migrating T cells, and talin was shown to be the critical adaptor protein mediating the interaction of activated LFA-1 with the underlying actin cytoskeleton (Smith et al., 2005).

The T cell receptor (TCR) is the centerpiece of the immunological synapse that forms between a T cell and an APC (Grakoui et al., 1999). Investigators used SPT to show that T-cell activation by an agonist signal from an APC caused relatively fast directed movement of unligated TCR molecules to the interface between the two cells, allowing these TCR molecules to participate in the growing synapse (Moss, Irvine, Davis, & Krummel, 2002). The directed movement of the TCR to the growing synapse was in stark contrast to the baseline motion of the TCR in nonactivated T cells, which was confined (diffusion coefficient $\sim 1 \times 10^{-10}$ cm^2/s) but not directed (Sloan-Lancaster et al., 1998). The mechanism underlying this switch from confined motion to directed motion was unclear, but interaction of the TCR with the cortical actin cytoskeleton was likely to be responsible since the cortical actin cytoskeleton is required for immunological synapse formation between T cell and APC (Valitutti, Dessing, Aktories, Gallati, & Lanzavecchia, 1995; Varma, Campi, Yokosuka, Saito, & Dustin, 2006; Wulfing & Davis, 1998). Calcium may also be involved in TCR mobility regulation, as T-cell activation by ionomycin (which increases intracellular calcium) reduced TCR mobility in an actin-dependent manner (Dushek et al., 2008). Future measurements of membrane protein dynamics will likely contribute to elucidating the mechanism by which the cytoskeleton guides directed motion of the TCR.

The diffusion properties of other participants in the immunological synapse have also been characterized. In resting T cells, the adhesion molecule CD2 had a relatively high lateral diffusion coefficient (7.9×10^{-10} cm^2/s) and fractional mobility (75%). Binding of CD2 to its counter-receptor CD58 did not alter either the diffusion coefficient or the fractional mobility of CD2 (Zhu, Dustin, Cairo, & Golan, 2007). In combination with T-cell activation, however, binding of CD2 to CD58 decreased the fractional mobility of CD2. These results were consistent with the role of the CD2–CD58 interaction at the immunological synapse, since high lateral mobility would facilitate rapid diffusion of CD2 to the nascent synapse, whereas, upon cell activation, immobilization of bound CD2–CD58 complexes at the mature synapse would strengthen adhesion (Zhu, Dustin, Cairo, Thatte, & Golan, 2006). Measurements of CD2 fractional mobility were combined with measurements of the number of CD2 molecules per cell, the surface area of the cell, the size of the contact area, and the bound and free counter-receptor (CD58) densities in the contact area to calculate the two-dimensional affinity of binding (2D K_d) between CD2 and CD58. The 2D K_d was shown to represent a quantitative measure of the mechanisms that regulate cell–cell adhesion (Zhu et al., 2007).

CD45 is a membrane protein expressed on T cells and certain other hematopoietic cells. The cytoplasmic domain of CD45 has phosphatase activity that is critical to T-cell activation. FRAP measurements showed that CD45 had a lateral diffusion coefficient of 4×10^{-10} cm^2/s in T-cell plasma membranes (Goldman et al., 1992). SPT measurements showed that CD45 mobility decreased upon cell activation, and this decrease in mobility was attenuated by disruption of the actin cytoskeleton. Peptide fragments of β1 spectrin were used to show that CD45 utilized spectrin to maintain cytoskeletal linkages with actin and with ankyrin (Cairo et al., 2010). These spectrin–ankyrin and spectrin–actin links stabilized CD45 in the membrane. SPT and time-resolved TIRF have been used to measure the lateral mobility of several other T-cell membrane proteins that are important for immunological synapse formation, including ZAP70, SLP76, CD3, and CD4 (Hsu et al., 2012; Mascalchi, Lamort, Salome, & Dumas, 2012).

Investigators have measured the lateral mobility of major histocompatibility complex (MHC) class I and class II molecules on APCs with the goal of understanding how such mobility could affect the interaction of APCs with T cells (Bierer, Herrmann, Brown, Burakoff, & Golan, 1987; Wilson, Morrison, Smith, Fernandez, & Cherry, 1996). Fluorescent SPT was used to track the motion of transfected HLA-DR molecules on fibroblasts. Multiple modes of motion were observed, including random diffusion, confined diffusion, and directed motion (Wilson et al., 1996). Reported lateral diffusion coefficients of MHC class II molecules have varied widely, from 1×10^{-13} cm^2/s to 32×10^{-10} cm^2/s, depending on cell type and experimental conditions (Umemura et al., 2008; Wilson et al., 1996; Yang, Kohler, Davis, & Burroughs, 2010). MHC class II molecules in which the transmembrane and cytoplasmic domains were replaced by

GPI linkers exhibited random diffusion with higher diffusion coefficients. This finding, coupled with experiments using latrunculin to perturb the underlying actin cytoskeleton, suggested that the cytoplasmic domain of MHC class II molecules interacted sterically with the actin-based membrane skeleton. In this model, membrane skeletal proteins do not bind the MHC class II molecules directly but rather serve as boundaries to confine and compartmentalize MHC class II diffusion. The MHC class II molecules periodically hop to adjacent regions of confinement, facilitated by transient perturbation of the membrane or transient disruption of the underlying actin lattice (Umemura et al., 2008; Vrljic, Nishimura, Brasselet, Moerner, & McConnell, 2002).

Membrane protein dynamics have been less well studied in hematopoietic cells of the myeloid lineage (Flannagan, Harrison, Yip, Jaqaman, & Grinstein, 2010). Such studies are at least partially complicated by high rates of membrane turnover associated with endocytosis and phagocytosis. One outstanding study used fluorescent SPT to measure the dynamics of CD36, the receptor for oxidized LDL, on macrophage membranes. This work showed that cortical F-actin, in cooperation with microtubules, organized to form linear troughs that confined CD36 lateral mobility and thereby promoted its clustering and priming for endocytosis of oxidized LDL (Jaqaman et al., 2011).

4.2.3 Common themes of functional organization in hematopoietic cells

Extending from lateral mobility measurements in erythrocytes, which focused primarily on the structural organization and architecture of the membrane, studies of membrane protein dynamics in other hematopoietic cells have helped to explain the mechanisms by which membrane protein lateral mobility affect cell–cell interactions. Characterization of the membrane protein dynamics of receptor–counter-receptor interactions involving lymphocytes has been especially illuminating. The functional consequence of changes in protein mobility in response to APC contact, and the dependence of these changes on underlying cytoskeletal elements, have become key tenets of immune system physiology. As initially shown in erythrocytes, the role of the underlying cytoskeleton in regulating membrane protein dynamics is now a common theme and has recently been reviewed in detail (Jaqaman & Grinstein, 2012). The potential for future studies to move beyond the standard three-dimensional K_d, as measured in solution chemistry, to the more physiologically relevant two-dimensional K_d, derived in part from measurements of membrane protein dynamics, has important physiologic and pharmacologic implications.

4.3. Membrane protein dynamics in non-hematopoietic cells

Measurements of membrane protein lateral diffusion have been performed in a wide array of non-hematopoietic cell types to address substantial questions about membrane physiology and cell function. The current discussion cannot cover every cell type that has been studied; instead, we illustrate the range by discussing one representative cell type derived from each of the three germ layers. In the neuron, the lateral diffusion and active transport of neurotransmitter receptors govern the cell's response to synaptic activity. In the endothelium, the availability and lateral mobility of counter-receptors for interacting leukocytes and of receptors for ECM proteins govern how the endothelial cells interact with their environment to direct inflammatory processes and vascular remodeling. In the epithelial parenchyma of most major organs, such as the alveolar epithelium, changes in receptor lateral mobility can determine cell-specific physiologic responses to drugs and natural ligands.

4.3.1 Neurons

SPT has been a useful tool for measuring the dynamics of neurotransmitter receptors as they move into and out of synapses (Alcor, Gouzer, & Triller, 2009). The lateral motion of these receptors is in addition to, and distinguished from, receptor turnover mediated by endocytosis and exocytosis. In many of the experiments on neurons, the beads used for SPT were brought into controlled contact with the cultured neuron of interest using optical tweezers. An early study of neurotransmitter receptor diffusion found that the 2-amino-3-(3-hydroxy-5-methyl-isoxazol-4-yl)propanoic acid (AMPA) receptor in rat hippocampal neurons switched rapidly from fast, relatively unconstrained diffusion to relative immobilization, and that the fraction of time spent immobilized at postsynaptic terminals increased as the neurons matured. Increases in intracellular calcium also promoted AMPA receptor immobilization (Borgdorff & Choquet, 2002). A similar rapid switching phenomenon was observed for the neuronal glycine receptor (Meier, Vannier, Serge, Triller, & Choquet, 2001). Switching between modes of lateral mobility required the scaffolding protein gephryin and was regulated by connections to the actin and microtubule cytoskeletons (Calamai et al., 2009; Charrier, Ehrensperger, Dahan, Levi, & Triller, 2006; Meier et al., 2001). Gephyrin was also implicated in controlling the motion of the GABA receptor during inhibitory clustering activity, but not during excitatory declustering (Mukherjee et al., 2011; Niwa et al., 2012). Using both SPT and FRAP, investigators found that ECM proteins acted in part as restrictive barriers to the lateral diffusion of AMPA receptors, and the interaction between the ECM proteins and the receptors may have modulated synaptic plasticity (Frischknecht et al., 2009).

Regulation of membrane protein lateral mobility appears to be important in axonal growth. For example, the ECM receptor β1 integrin exhibited restricted diffusion in axonal membranes in the absence of nerve growth factor (NGF), and treatment with NGF induced both threefold faster receptor diffusion (the apparent diffusion coefficient increased from 4.4×10^{-10} cm^2/s to 14.5×10^{-10} cm^2/s) as well as rapid directed receptor motion toward the end of the growth cone (at a typical rate of 37 μm/min, with brief periods of sustained forward excursions at rates of 75–150 μm/min). The directed motion of β1 integrin required an intact actin cytoskeleton and active actin–myosin coupling, since treatment with cytochalasin D or butanedione monoxime (a myosin ATPase inhibitor) blocked this response (Grabham, Foley, Umeojiako, & Goldberg, 2000). A similar pattern of response to NGF was observed with GABA receptors; in this case, the directed motion was dependent on microtubules (Bouzigues, Morel, Triller, & Dahan, 2007).

Recent work has probed the lateral mobility of ion channels in neuronal membranes. Sodium channels at the axonal initial segment, which are critical for action potential initiation, were restricted in their lateral diffusion by ankyrin G. Overexpression of ankyrin reduced the fractional mobility of the channels from 95% to 65% and decreased the lateral diffusion coefficient by 50% (Brachet et al., 2010). FRAP studies on cells cotransfected with neurofascin, a neuronal membrane protein, and fluorescently tagged ankyrin G have helped to define the domains of ankyrin G that are responsible for immobilizing the neurofascin–ankyrin G complex in the membrane. Neurofascin immobilization is important for the organization of sodium channels in the axonal initial segment. The lateral diffusion coefficient of neurofascin decreased by approximately 10-fold, and its fractional mobility decreased from 70% to 10%, upon formation of the neurofascin–ankyrin G complex (Zhang & Bennett, 1998). Ankyrin G appears to have a critical role in regulating the lateral mobility of both sodium channels and neurofascin in neuronal membranes, likely due to its function as a linking protein that mediates interactions between membrane proteins and the underlying cytoskeleton (Bennett & Baines, 2001; Galiano et al., 2012).

4.3.2 Endothelial cells

Given the myriad of cell–cell interactions involving the endothelium, it is likely that the dynamics of proteins in the endothelial cell membrane are highly regulated. Somewhat surprisingly, however, these cells have not been extensively studied in this context. For example, while the lateral mobility of LFA-1 on circulating lymphocytes has been probed in detail, the mobility of LFA-1's counter-receptor ICAM-1 on endothelial cells has only recently received attention. FRAP was used to measure the lateral mobility of GFP–ICAM-1 fusion proteins in the plasma membrane of human umbilical vein endothelial cells (HUVEC), and the lateral diffusion coefficient was 2.9×10^{-10} cm^2/s. A variant of ICAM-1 that lacked the cytoplasmic tail had a diffusion coefficient three- to fourfold higher than this value. Crosslinking of ICAM-1 (which mimics ICAM-1 engagement) increased the density of actin stress fibers in the cells in a cortactin-dependent manner while concomitantly restricting the lateral diffusion of the wild-type ICAM-1 but not the ICAM-1 variant lacking the cytoplasmic tail. Thus, the lateral mobility of ICAM-1 was dynamically linked to the actin cytoskeleton (Yang et al., 2006). A second FRAP study investigated the regulation of ICAM-1 clustering at the apical (lumenal) surface of endothelial cells upon leukocyte engagement. Such clustering was associated with a decrease in ICAM-1 fractional mobility, and the relative immobilization of the clustered molecules was reversed by disruption of the actin cytoskeleton (van Buul et al., 2010).

FRAP has been used to measure the lateral mobility of integrins in the plasma membrane of endothelial cells in the context of focal adhesions. Here, disruption of the actin cytoskeleton with cytochalasin D or latrunculin B *decreased* the fractional mobility of the integrins (Tsuruta et al., 2002). This result suggested a model in which integrins require the actin cytoskeleton in order to translocate within focal adhesion sites, where the endothelial cell engages with the underlying ECM. Integrin translocation likely occurs with the cooperation of other focal adhesion complex proteins; more work is needed to dissect the interactions that determine integrin trafficking in this context (Lele, Thodeti, Pendse, & Ingber, 2008). FRAP has also been used to measure the lateral mobility of GFP-tagged VE-cadherin in endothelial cell–cell junctions. Inhibition of the activity of tyrosine phosphatase, specifically SHP2, significantly decreased the fractional mobility and diffusion coefficient of VE-cadherin. This result suggested that SHP2-containing pathways may be important for promoting increased diffusion of VE-cadherin, and thereby controlling the recovery of cell–cell junctions in response to inflammation (Timmerman et al., 2012).

4.3.3 Epithelial cells

Few studies have been conducted to date on membrane protein dynamics in epithelial cells. One interesting set of observations involved the dynamics of the G protein-coupled β-adrenergic receptor in the plasma membrane of alveolar epithelial cells and model cell lines. FCS experiments showed that epinephrine binding caused β2 receptors to exhibit very rapid lateral diffusion for several minutes (diffusion coefficient, 288×10^{-10} cm^2/s), after which the diffusion rate slowed to more typical values (10×10^{-10} cm^2/s). This behavior could correspond to ligand binding and free diffusion of the ligand-bound receptor within CEMMs, followed by migration of the ligand-bound receptor out of CEMMs in preparation for receptor internalization (Hegener et al., 2004). FRAP experiments showed that the lateral diffusion coefficient of unligated, GFP-tagged β2 receptor was 40×10^{-10} cm^2/s in the plasma membrane of a transfected cell line (Barak et al., 1997). Initial SPT studies in a model cell line showed that ligated β2 receptors demonstrated a heterogeneous mobility profile. Global stimulation of β2 receptors by terbutaline markedly immobilized the labeled receptor through regulatory mechanisms downstream of the cAMP pathway (Sieben, Kaminski, Kubitscheck, & Haberlein, 2011). Future SPT studies in which CEMMs and cAMP pathways are disrupted may further elucidate the mechanisms underlying these agonist-induced changes in membrane protein dynamics.

4.3.4 Common themes of functional organization in non-hematopoietic cells

In the future, we expect that the dynamics of a much broader array of membrane proteins will be investigated in many different non-hematopoietic cell types. The techniques of FRAP, SPT, FCS, variations thereof, and complementary methods such as FRET are widely applicable. Much like the systems discussed here—that is, neurotransmitter receptor trafficking in neurons, cell–cell and cell–ECM interactions in endothelium, and β2-adrenergic responses in lung—the elucidation of cell physiology and pathophysiology will be advanced in other mammalian and model systems through studies of membrane protein dynamics. In these systems, as in the non-hematopoietic cells described above, we expect that a common theme will be the role of interactions between membrane proteins and the underlying cytoskeleton in regulating protein lateral mobility and cell function.

5. MEMBRANE DIFFUSION, PHYSIOLOGY, AND PHARMACOLOGIC IMPLICATIONS

Studies of membrane protein dynamics provide insight into the real-time behavior of receptors, counter-receptors, and structural proteins in living mammalian cells. When coupled with interventions such as ligand binding, crosslinking, or disruption of downstream signaling pathways or cytoskeletal linkages, a rich picture can emerge of the mechanisms by which membrane protein dynamics influence signaling sequelae and membrane mechanical properties. Ultimately, the regulation of membrane protein lateral mobility translates into the local availability of these proteins for interactions with the intracellular and extracellular environment, which includes not only other cells and ECM components but also pharmacologic agents that are designed to modulate some aspect of the protein's activity or binding capacity. Studies of membrane protein dynamics are, therefore, one important component of a complete depiction of the role of membrane proteins in cell function.

REFERENCES

Alcor, D., Gouzer, G., & Triller, A. (2009). Single-particle tracking methods for the study of membrane receptors dynamics. *The European Journal of Neuroscience, 30*, 987–997.

Anong, W. A., Franco, T., Chu, H., Weis, T. L., Devlin, E. E., Bodine, D. M., et al. (2009). Adducin forms a bridge between the erythrocyte membrane and its cytoskeleton and regulates membrane cohesion. *Blood, 114*, 1904–1912.

Axelrod, D., Koppel, D. E., Schlessinger, J., Elson, E., & Webb, W. W. (1976). Mobility measurement by analysis of fluorescence photobleaching recovery kinetics. *Biophysical Journal, 16*, 1055–1069.

Bacia, K., Kim, S. A., & Schwille, P. (2006). Fluorescence cross-correlation spectroscopy in living cells. *Nature Methods, 3*, 83–89.

Bacia, K., Scherfeld, D., Kahya, N., & Schwille, P. (2004). Fluorescence correlation spectroscopy relates rafts in model and native membranes. *Biophysical Journal, 87*, 1034–1043.

Barak, L. S., Ferguson, S. S., Zhang, J., Martenson, C., Meyer, T., & Caron, M. G. (1997). Internal trafficking and surface mobility of a functionally intact beta2-adrenergic receptor-green fluorescent protein conjugate. *Molecular Pharmacology, 51*, 177–184.

Bennett, V. (1985). The membrane skeleton of human erythrocytes and its implications for more complex cells. *Annual Review of Biochemistry, 54*, 273–304.

Bennett, V., & Baines, A. J. (2001). Spectrin and ankyrin-based pathways: Metazoan inventions for integrating cells into tissues. *Physiological Reviews, 81*, 1353–1392.

Bierer, B. E., Herrmann, S. H., Brown, C. S., Burakoff, S. J., & Golan, D. E. (1987). Lateral mobility of class I histocompatibility antigens in B lymphoblastoid cell membranes: Modulation by cross-linking and effect of cell density. *The Journal of Cell Biology, 105*, 1147–1152.

Billaudeau, C., Mailfert, S., Trombik, T., Bertaux, N., Rouger, V., Hamon, Y., et al. (2013). Probing the plasma membrane organization in living cells by spot variation fluorescence correlation spectroscopy. *Methods in Enzymology, 519*, 277–302.
Borgdorff, A. J., & Choquet, D. (2002). Regulation of AMPA receptor lateral movements. *Nature, 417*, 649–653.
Bouzigues, C., Morel, M., Triller, A., & Dahan, M. (2007). Asymmetric redistribution of GABA receptors during GABA gradient sensing by nerve growth cones analyzed by single quantum dot imaging. *Proceedings of the National Academy of Sciences of the United States of America, 104*, 11251–11256.
Brachet, A., Leterrier, C., Irondelle, M., Fache, M. P., Racine, V., Sibarita, J. B., et al. (2010). Ankyrin G restricts ion channel diffusion at the axonal initial segment before the establishment of the diffusion barrier. *The Journal of Cell Biology, 191*, 383–395.
Brodbeck, W. G., Mold, C., Atkinson, J. P., & Medof, M. E. (2000). Cooperation between decay-accelerating factor and membrane cofactor protein in protecting cells from autologous complement attack. *The Journal of Immunology, 165*, 3999–4006.
Brown, D. A., & London, E. (1998). Functions of lipid rafts in biological membranes. *Annual Review of Cell and Developmental Biology, 14*, 111–136.
Cairo, C. W., Das, R., Albohy, A., Baca, Q. J., Pradhan, D., Morrow, J. S., et al. (2010). Dynamic regulation of CD45 lateral mobility by the spectrin-ankyrin cytoskeleton of T cells. *The Journal of Biological Chemistry, 285*, 11392–11401.
Cairo, C. W., Mirchev, R., & Golan, D. E. (2006). Cytoskeletal regulation couples LFA-1 conformational changes to receptor lateral mobility and clustering. *Immunity, 25*, 297–308.
Calamai, M., Specht, C. G., Heller, J., Alcor, D., Machado, P., Vannier, C., et al. (2009). Gephyrin oligomerization controls GlyR mobility and synaptic clustering. *The Journal of Neuroscience, 29*, 7639–7648.
Charrier, C., Ehrensperger, M. V., Dahan, M., Levi, S., & Triller, A. (2006). Cytoskeleton regulation of glycine receptor number at synapses and diffusion in the plasma membrane. *The Journal of Neuroscience, 26*, 8502–8511.
Cho, M. R., Eber, S. W., Liu, S. C., Lux, S. E., & Golan, D. E. (1998). Regulation of band 3 rotational mobility by ankyrin in intact human red cells. *Biochemistry, 37*, 17828–17835.
Cho, M. R., Knowles, D. W., Smith, B. L., Moulds, J. J., Agre, P., Mohandas, N., et al. (1999). Membrane dynamics of the water transport protein aquaporin-1 in intact human red cells. *Biophysical Journal, 76*, 1136–1144.
Corbett, J. D., Agre, P., Palek, J., & Golan, D. E. (1994). Differential control of band 3 lateral and rotational mobility in intact red cells. *The Journal of Clinical Investigation, 94*, 683–688.
Corbett, J. D., Cho, M. R., & Golan, D. E. (1994). Deoxygenation affects fluorescence photobleaching recovery measurements of red cell membrane protein lateral mobility. *Biophysical Journal, 66*, 25–30.
Corbett, J. D., & Golan, D. E. (1993). Band 3 and glycophorin are progressively aggregated in density-fractionated sickle and normal red blood cells. Evidence from rotational and lateral mobility studies. *The Journal of Clinical Investigation, 91*, 208–217.
Dushek, O., Mueller, S., Soubies, S., Depoil, D., Caramalho, I., Coombs, D., et al. (2008). Effects of intracellular calcium and actin cytoskeleton on TCR mobility measured by fluorescence recovery. *PLoS One, 3*, e3913.
Dustin, M. L. (1998). Making a little affinity go a long way: A topological view of LFA-1 regulation. *Cell Adhesion and Communication, 6*, 255–262.
Flannagan, R. S., Harrison, R. E., Yip, C. M., Jaqaman, K., & Grinstein, S. (2010). Dynamic macrophage "probing" is required for the efficient capture of phagocytic targets. *The Journal of Cell Biology, 191*, 1205–1218.
Frick, M., Schmidt, K., & Nichols, B. J. (2007). Modulation of lateral diffusion in the plasma membrane by protein density. *Current Biology, 17*, 462–467.

Frischknecht, R., Heine, M., Perrais, D., Seidenbecher, C. I., Choquet, D., & Gundelfinger, E. D. (2009). Brain extracellular matrix affects AMPA receptor lateral mobility and short-term synaptic plasticity. *Nature Neuroscience*, *12*, 897–904.

Gaborski, T. R., Clark, A., Jr., Waugh, R. E., & McGrath, J. L. (2008). Membrane mobility of beta2 integrins and rolling associated adhesion molecules in resting neutrophils. *Biophysical Journal*, *95*, 4934–4947.

Galbiati, F., Razani, B., & Lisanti, M. P. (2001). Emerging themes in lipid rafts and caveolae. *Cell*, *106*, 403–411.

Galiano, M. R., Jha, S., Ho, T. S., Zhang, C., Ogawa, Y., Chang, K. J., et al. (2012). A distal axonal cytoskeleton forms an intra-axonal boundary that controls axon initial segment assembly. *Cell*, *149*, 1125–1139.

Giancotti, F. G., & Ruoslahti, E. (1999). Integrin signaling. *Science*, *285*, 1028–1032.

Golan, D. E., & Veatch, W. (1980). Lateral mobility of band 3 in the human erythrocyte membrane studied by fluorescence photobleaching recovery: Evidence for control by cytoskeletal interactions. *Proceedings of the National Academy of Sciences of the United States of America*, *77*, 2537–2541.

Goldman, S. J., Uniyal, S., Ferguson, L. M., Golan, D. E., Burakoff, S. J., & Kiener, P. A. (1992). Differential activation of phosphotyrosine protein phosphatase activity in a murine T cell hybridoma by monoclonal antibodies to CD45. *The Journal of Biological Chemistry*, *267*, 6197–6204.

Gonda, K., Watanabe, T. M., Ohuchi, N., & Higuchi, H. (2010). In vivo nano-imaging of membrane dynamics in metastatic tumor cells using quantum dots. *The Journal of Biological Chemistry*, *285*, 2750–2757.

Grabham, P. W., Foley, M., Umeojiako, A., & Goldberg, D. J. (2000). Nerve growth factor stimulates coupling of beta1 integrin to distinct transport mechanisms in the filopodia of growth cones. *Journal of Cell Science*, *113*(Pt 17), 3003–3012.

Grakoui, A., Bromley, S. K., Sumen, C., Davis, M. M., Shaw, A. S., Allen, P. M., et al. (1999). The immunological synapse: A molecular machine controlling T cell activation. *Science*, *285*, 221–227.

Haustein, E., & Schwille, P. (2007). Fluorescence correlation spectroscopy: Novel variations of an established technique. *Annual Review of Biophysics and Biomolecular Structure*, *36*, 151–169.

Hegener, O., Prenner, L., Runkel, F., Baader, S. L., Kappler, J., & Haberlein, H. (2004). Dynamics of beta2-adrenergic receptor-ligand complexes on living cells. *Biochemistry*, *43*, 6190–6199.

Hsu, C. J., Hsieh, W. T., Waldman, A., Clarke, F., Huseby, E. S., Burkhardt, J. K., et al. (2012). Ligand mobility modulates immunological synapse formation and T cell activation. *PLoS One*, *7*, e32398.

Hu, K., Ji, L., Applegate, K. T., Danuser, G., & Waterman-Storer, C. M. (2007). Differential transmission of actin motion within focal adhesions. *Science*, *315*, 111–115.

Iolascon, A., Perrotta, S., & Stewart, G. W. (2003). Red blood cell membrane defects. *Reviews in Clinical and Experimental Hematology*, *7*, 22–56.

Jaqaman, K., & Grinstein, S. (2012). Regulation from within: The cytoskeleton in transmembrane signaling. *Trends in Cell Biology*, *22*, 515–526.

Jaqaman, K., Kuwata, H., Touret, N., Collins, R., Trimble, W. S., Danuser, G., et al. (2011). Cytoskeletal control of CD36 diffusion promotes its receptor and signaling function. *Cell*, *146*, 593–606.

Jaqaman, K., Loerke, D., Mettlen, M., Kuwata, H., Grinstein, S., Schmid, S. L., et al. (2008). Robust single-particle tracking in live-cell time-lapse sequences. *Nature Methods*, *5*, 695–702.

Kahya, N., Scherfeld, D., Bacia, K., Poolman, B., & Schwille, P. (2003). Probing lipid mobility of raft-exhibiting model membranes by fluorescence correlation spectroscopy. *The Journal of Biological Chemistry*, *278*, 28109–28115.

Kang, M., Day, C. A., Kenworthy, A. K., & DiBenedetto, E. (2012). Simplified equation to extract diffusion coefficients from confocal FRAP data. *Traffic, 13*, 1589–1600.

Kapitza, H. G., Ruppel, D. A., Galla, H. J., & Sackmann, E. (1984). Lateral diffusion of lipids and glycophorin in solid phosphatidylcholine bilayers. The role of structural defects. *Biophysical Journal, 45*, 577–587.

Karnchanaphanurach, P., Mirchev, R., Ghiran, I., Asara, J. M., Papahadjopoulos-Sternberg, B., Nicholson-Weller, A., et al. (2009). C3b deposition on human erythrocytes induces the formation of a membrane skeleton-linked protein complex. *The Journal of Clinical Investigation, 119*, 788–801.

Kim, M., Carman, C. V., & Springer, T. A. (2003). Bidirectional transmembrane signaling by cytoplasmic domain separation in integrins. *Science, 301*, 1720–1725.

Kim, M., Carman, C. V., Yang, W., Salas, A., & Springer, T. A. (2004). The primacy of affinity over clustering in regulation of adhesiveness of the integrin {alpha}L{beta}2. *The Journal of Cell Biology, 167*, 1241–1253.

Knowles, D. W., Chasis, J. A., Evans, E. A., & Mohandas, N. (1994). Cooperative action between band 3 and glycophorin A in human erythrocytes: Immobilization of band 3 induced by antibodies to glycophorin A. *Biophysical Journal, 66*, 1726–1732.

Kodippili, G. C., Spector, J., Hale, J., Giger, K., Hughes, M. R., McNagny, K. M., et al. (2012). Analysis of the mobilities of band 3 populations associated with ankyrin protein and junctional complexes in intact murine erythrocytes. *The Journal of Biological Chemistry, 287*, 4129–4138.

Kucik, D. F., Dustin, M. L., Miller, J. M., & Brown, E. J. (1996). Adhesion-activating phorbol ester increases the mobility of leukocyte integrin LFA-1 in cultured lymphocytes. *The Journal of Clinical Investigation, 97*, 2139–2144.

Kusumi, A., Suzuki, K. G., Kasai, R. S., Ritchie, K., & Fujiwara, T. K. (2011). Hierarchical mesoscale domain organization of the plasma membrane. *Trends in Biochemical Sciences, 36*, 604–615.

Lele, T. P., Thodeti, C. K., Pendse, J., & Ingber, D. E. (2008). Investigating complexity of protein–protein interactions in focal adhesions. *Biochemical and Biophysical Research Communications, 369*, 929–934.

Lingwood, D., & Simons, K. (2010). Lipid rafts as a membrane-organizing principle. *Science, 327*, 46–50.

Lippincott-Schwartz, J., Snapp, E., & Kenworthy, A. (2001). Studying protein dynamics in living cells. *Nature Reviews Molecular Cell Biology, 2*, 444–456.

Liu, J., Guo, X., Mohandas, N., Chasis, J. A., & An, X. (2010). Membrane remodeling during reticulocyte maturation. *Blood, 115*, 2021–2027.

Liu, S. C., Zhai, S., Palek, J., Golan, D. E., Amato, D., Hassan, K., et al. (1990). Molecular defect of the band 3 protein in southeast Asian ovalocytosis. *The New England Journal of Medicine, 323*, 1530–1538.

Mascalchi, P., Lamort, A. S., Salome, L., & Dumas, F. (2012). Single particle tracking reveals two distinct environments for CD4 receptors at the surface of living T lymphocytes. *Biochemical and Biophysical Research Communications, 417*, 409–413.

Meier, J., Vannier, C., Serge, A., Triller, A., & Choquet, D. (2001). Fast and reversible trapping of surface glycine receptors by gephyrin. *Nature Neuroscience, 4*, 253–260.

Mendoza, M. C., Besson, S., & Danuser, G. (2012). Quantitative fluorescent speckle microscopy (QFSM) to measure actin dynamics. *Current Protocols in Cytometry, 62*, 2.18.1–2.18.26.

Michaels, D. W., Abramovitz, A. S., Hammer, C. H., & Mayer, M. M. (1976). Increased ion permeability of planar lipid bilayer membranes after treatment with the C5b-9 cytolytic attack mechanism of complement. *Proceedings of the National Academy of Sciences of the United States of America, 73*, 2852–2856.

Mirchev, R., & Golan, D. E. (2001). Single-particle tracking and laser optical tweezers studies of the dynamics of individual protein molecules in membranes of intact human and mouse red cells. *Blood Cells, Molecules & Diseases*, *27*, 143–147.

Mirchev, R., Lam, A., & Golan, D. E. (2011). Membrane compartmentalization in Southeast Asian ovalocytosis red blood cells. *British Journal of Haematology*, *155*, 111–121.

Mohandas, N., & Chasis, J. A. (2010). The erythroid niche: Molecular processes occurring within erythroblastic islands. *Transfusion Clinique et Biologique*, *17*, 110–111.

Moss, W. C., Irvine, D. J., Davis, M. M., & Krummel, M. F. (2002). Quantifying signaling-induced reorientation of T cell receptors during immunological synapse formation. *Proceedings of the National Academy of Sciences of the United States of America*, *99*, 15024–15029.

Mukherjee, J., Kretschmannova, K., Gouzer, G., Maric, H. M., Ramsden, S., Tretter, V., et al. (2011). The residence time of GABA(A)Rs at inhibitory synapses is determined by direct binding of the receptor alpha1 subunit to gephyrin. *The Journal of Neuroscience*, *31*, 14677–14687.

Nechyporuk-Zloy, V., Dieterich, P., Oberleithner, H., Stock, C., & Schwab, A. (2008). Dynamics of single potassium channel proteins in the plasma membrane of migrating cells. *American Journal of Physiology Cell Physiology*, *294*, C1096–C1102.

Niwa, F., Bannai, H., Arizono, M., Fukatsu, K., Triller, A., & Mikoshiba, K. (2012). Gephyrin-independent GABA(A)R mobility and clustering during plasticity. *PLoS One*, *7*, e36148.

Palek, J. (1987). Hereditary elliptocytosis, spherocytosis and related disorders: Consequences of a deficiency or a mutation of membrane skeletal proteins. *Blood Reviews*, *1*, 147–168.

Patel, H. H., Murray, F., & Insel, P. A. (2008). Caveolae as organizers of pharmacologically relevant signal transduction molecules. *Annual Review of Pharmacology and Toxicology*, *48*, 359–391.

Peters, I. M., van Kooyk, Y., van Vliet, S. J., de Grooth, B. G., Figdor, C. G., & Greve, J. (1999). 3D single-particle tracking and optical trap measurements on adhesion proteins. *Cytometry*, *36*, 189–194.

Qian, H., Sheetz, M. P., & Elson, E. L. (1991). Single particle tracking. Analysis of diffusion and flow in two-dimensional systems. *Biophysical Journal*, *60*, 910–921.

Rolfe, D. J., McLachlan, C. I., Hirsch, M., Needham, S. R., Tynan, C. J., Webb, S. E., et al. (2011). Automated multidimensional single molecule fluorescence microscopy feature detection and tracking. *European Biophysics Journal*, *40*, 1167–1186.

Sander, S., Arora, N., & Smith, E. A. (2012). Elucidating the role of select cytoplasmic proteins in altering diffusion of integrin receptors. *Analytical and Bioanalytical Chemistry*, *403*, 2327–2337.

Saxton, M. J. (1997). Single-particle tracking: The distribution of diffusion coefficients. *Biophysical Journal*, *72*, 1744–1753.

Saxton, M. J., & Jacobson, K. (1997). Single-particle tracking: Applications to membrane dynamics. *Annual Review of Biophysics and Biomolecular Structure*, *26*, 373–399.

Schindler, M., Koppel, D. E., & Sheetz, M. P. (1980). Modulation of membrane protein lateral mobility by polyphosphates and polyamines. *Proceedings of the National Academy of Sciences of the United States of America*, *77*, 1457–1461.

Serge, A., Fourgeaud, L., Hemar, A., & Choquet, D. (2003). Active surface transport of metabotropic glutamate receptors through binding to microtubules and actin flow. *Journal of Cell Science*, *116*, 5015–5022.

Sieben, A., Kaminski, T., Kubitscheck, U., & Haberlein, H. (2011). Terbutaline causes immobilization of single beta2-adrenergic receptor-ligand complexes in the plasma membrane of living A549 cells as revealed by single-molecule microscopy. *Journal of Biomedical Optics*, *16*, 026013.

Simons, K., & Ikonen, E. (1997). Functional rafts in cell membranes. *Nature*, *387*, 569–572.

Singer, S. J., & Nicolson, G. L. (1972). The fluid mosaic model of the structure of cell membranes. *Science, 175*, 720–731.

Sloan-Lancaster, J., Presley, J., Ellenberg, J., Yamazaki, T., Lippincott-Schwartz, J., & Samelson, L. E. (1998). ZAP-70 association with T cell receptor zeta (TCRzeta): Fluorescence imaging of dynamic changes upon cellular stimulation. *The Journal of Cell Biology, 143*, 613–624.

Smith, B. L., & Agre, P. (1991). Erythrocyte Mr 28,000 transmembrane protein exists as a multisubunit oligomer similar to channel proteins. *The Journal of Biological Chemistry, 266*, 6407–6415.

Smith, A., Carrasco, Y. R., Stanley, P., Kieffer, N., Batista, F. D., & Hogg, N. (2005). A talin-dependent LFA-1 focal zone is formed by rapidly migrating T lymphocytes. *The Journal of Cell Biology, 170*, 141–151.

Steck, T. L. (1974). The organization of proteins in the human red blood cell membrane. A review. *The Journal of Cell Biology, 62*, 1–19.

Timmerman, I., Hoogenboezem, M., Bennett, A. M., Geerts, D., Hordijk, P. L., & van Buul, J. D. (2012). The tyrosine phosphatase SHP2 regulates recovery of endothelial adherens junctions through control of beta-catenin phosphorylation. *Molecular Biology of the Cell, 23*, 4212–4225.

Tomishige, M., Sako, Y., & Kusumi, A. (1998). Regulation mechanism of the lateral diffusion of band 3 in erythrocyte membranes by the membrane skeleton. *The Journal of Cell Biology, 142*, 989–1000.

Tsuji, A., & Ohnishi, S. (1986). Restriction of the lateral motion of band 3 in the erythrocyte membrane by the cytoskeletal network: Dependence on spectrin association state. *Biochemistry, 25*, 6133–6139.

Tsuruta, D., Gonzales, M., Hopkinson, S. B., Otey, C., Khuon, S., Goldman, R. D., et al. (2002). Microfilament-dependent movement of the beta3 integrin subunit within focal contacts of endothelial cells. *The FASEB Journal, 16*, 866–868.

Umemura, Y. M., Vrljic, M., Nishimura, S. Y., Fujiwara, T. K., Suzuki, K. G., & Kusumi, A. (2008). Both MHC class II and its GPI-anchored form undergo hop diffusion as observed by single-molecule tracking. *Biophysical Journal, 95*, 435–450.

Umenishi, F., Verbavatz, J. M., & Verkman, A. S. (2000). cAMP regulated membrane diffusion of a green fluorescent protein-aquaporin 2 chimera. *Biophysical Journal, 78*, 1024–1035.

Valitutti, S., Dessing, M., Aktories, K., Gallati, H., & Lanzavecchia, A. (1995). Sustained signaling leading to T cell activation results from prolonged T cell receptor occupancy. Role of T cell actin cytoskeleton. *The Journal of Experimental Medicine, 181*, 577–584.

van Buul, J. D., van Rijssel, J., van Alphen, F. P., Hoogenboezem, M., Tol, S., Hoeben, K. A., et al. (2010). Inside-out regulation of ICAM-1 dynamics in TNF-alpha-activated endothelium. *PLoS One, 5*, e11336.

van Kooyk, Y., & Figdor, C. G. (2000). Avidity regulation of integrins: The driving force in leukocyte adhesion. *Current Opinion in Cell Biology, 12*, 542–547.

Varma, R., Campi, G., Yokosuka, T., Saito, T., & Dustin, M. L. (2006). T cell receptor-proximal signals are sustained in peripheral microclusters and terminated in the central supramolecular activation cluster. *Immunity, 25*, 117–127.

Vrljic, M., Nishimura, S. Y., Brasselet, S., Moerner, W. E., & McConnell, H. M. (2002). Translational diffusion of individual class II MHC membrane proteins in cells. *Biophysical Journal, 83*, 2681–2692.

Watkins, E. B., Miller, C. E., Majewski, J., & Kuhl, T. L. (2011). Membrane texture induced by specific protein binding and receptor clustering: Active roles for lipids in cellular function. *Proceedings of the National Academy of Sciences of the United States of America, 108*, 6975–6980.

Wehrle-Haller, B. (2007). Analysis of integrin dynamics by fluorescence recovery after photobleaching. *Methods in Molecular Biology*, *370*, 173–202.

Wilson, K. M., Morrison, I. E., Smith, P. R., Fernandez, N., & Cherry, R. J. (1996). Single particle tracking of cell-surface HLA-DR molecules using R-phycoerythrin labeled monoclonal antibodies and fluorescence digital imaging. *Journal of Cell Science*, *109*(Pt 8), 2101–2109.

Wulfing, C., & Davis, M. M. (1998). A receptor/cytoskeletal movement triggered by costimulation during T cell activation. *Science*, *282*, 2266–2269.

Yang, J., Kohler, K., Davis, D. M., & Burroughs, N. J. (2010). An improved strip FRAP method for estimating diffusion coefficients: Correcting for the degree of photobleaching. *Journal of Microscopy*, *238*, 240–253.

Yang, L., Kowalski, J. R., Yacono, P., Bajmoczi, M., Shaw, S. K., Froio, R. M., et al. (2006). Endothelial cell cortactin coordinates intercellular adhesion molecule-1 clustering and actin cytoskeleton remodeling during polymorphonuclear leukocyte adhesion and transmigration. *The Journal of Immunology*, *177*, 6440–6449.

Yip, R., Mohandas, N., Clark, M. R., Jain, S., Shohet, S. B., & Dallman, P. R. (1983). Red cell membrane stiffness in iron deficiency. *Blood*, *62*, 99–106.

Zhang, X., & Bennett, V. (1998). Restriction of 480/270-kD ankyrin G to axon proximal segments requires multiple ankyrin G-specific domains. *The Journal of Cell Biology*, *142*, 1571–1581.

Zhu, D. M., Dustin, M. L., Cairo, C. W., & Golan, D. E. (2007). Analysis of two-dimensional dissociation constant of laterally mobile cell adhesion molecules. *Biophysical Journal*, *92*, 1022–1034.

Zhu, D. M., Dustin, M. L., Cairo, C. W., Thatte, H. S., & Golan, D. E. (2006). Mechanisms of cellular avidity regulation in CD2-CD58-mediated T cell adhesion. *ACS Chemical Biology*, *1*, 649–658.

CHAPTER FOUR

Evolving Form to Fit Function: Cardiomyocyte Intercalated Disc and Transverse-Tubule Membranes

Crystal F. Kline*, Peter J. Mohler*,†,‡,1

*The Dorothy M. Davis Heart & Lung Research Institute, The Ohio State University Wexner Medical Center, Columbus, Ohio, USA

†Division of Cardiovascular Medicine, Department of Internal Medicine, The Ohio State University Wexner Medical Center, Columbus, Ohio, USA

‡Department of Physiology and Cell Biology, The Ohio State University Wexner Medical Center, Columbus, Ohio, USA

[1]Corresponding author: e-mail address: peter.mohler@osumc.edu1

Contents

Abstract

The vertebrate cardiac myocyte has evolved a highly organized cellular membrane architecture and cell–cell contacts in order to effectively transmit precisely timed and homogeneous depolarizing waves without failure (>2 billion times/human life span). Two unique specialized membrane domains, the intercalated disc and the transverse tubule (T-tubule), function to ensure the rapid and coordinated propagation of the action potential throughout the heart. Based on their critical roles in structure, signaling, and electric inter- and intracellular communication, it is not surprising that dysfunction in these membrane structures is associated with aberrant vertebrate physiology, resulting in potentially fatal congenital and acquired disease. This chapter will review the fundamental components of cardiomyocyte intercalated disc and transverse-tubule membranes with a focus on linking dysfunction in these membranes with human cardiovascular disease.

Current Topics in Membranes, Volume 72
ISSN 1063-5823
http://dx.doi.org/10.1016/B978-0-12-417027-8.00004-0

1. THE MULTIFUNCTIONAL MYOCYTE INTERCALATED DISC

Contractility and synchronicity of the heartbeat requires resilient mechanical and electric coupling to maintain the functional and structural stability of the heart. The heart endures constant mechanical stress, necessitating the need for anchoring cell–cell junctions. Anchoring cell–cell junctions are particularly critical as they provide stability to cardiac myocytes in the face of severe stress by mechanically anchoring cells during cardiac systolic and diastolic cycles. In vertebrate ventricular and atrial myocytes, these cell junctions are localized at the longitudinal end of the cell in a stair-like profile at lateral ends, forming a structure classified as the intercalated disc (Fig. 4.1). Hematoxylin/eosin staining of heart tissue sections reveals intercalated discs as eosinophilic bands, suggesting a structure that is highly proteinaceous in nature. On the molecular level, the intercalated disc is a highly organized triad of junctions between cardiac myocytes, including adhesive and communicating macromolecular complexes containing adhesive/anchoring (desmosomes and *fascia adherens* junctions) and communicating junctions (gap junctions), respectively (Fig. 4.2).

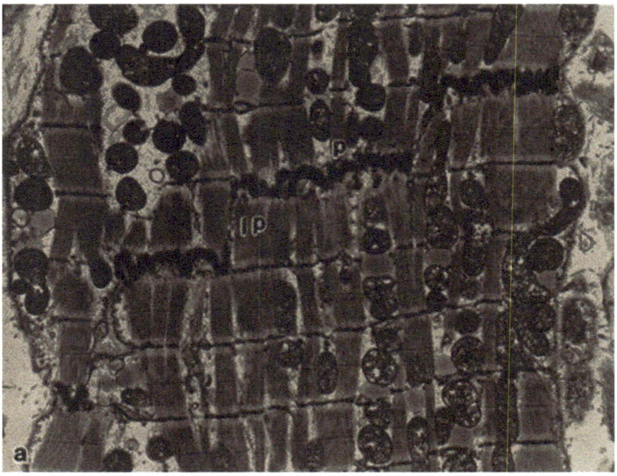

Figure 4.1 Vertebrate intercalated disc. Transmission electron microscope image of right ventricle longitudinal section of myocytes of monkey. This image depicts the plicate segments (P) and interplicate segments (IP) of the electron-dense intercalated disc structures (6300 ×). *Reprinted from Shimada, Kawazato, Yasuda, Ono, and Sueda (2004) with permission from Wiley.*

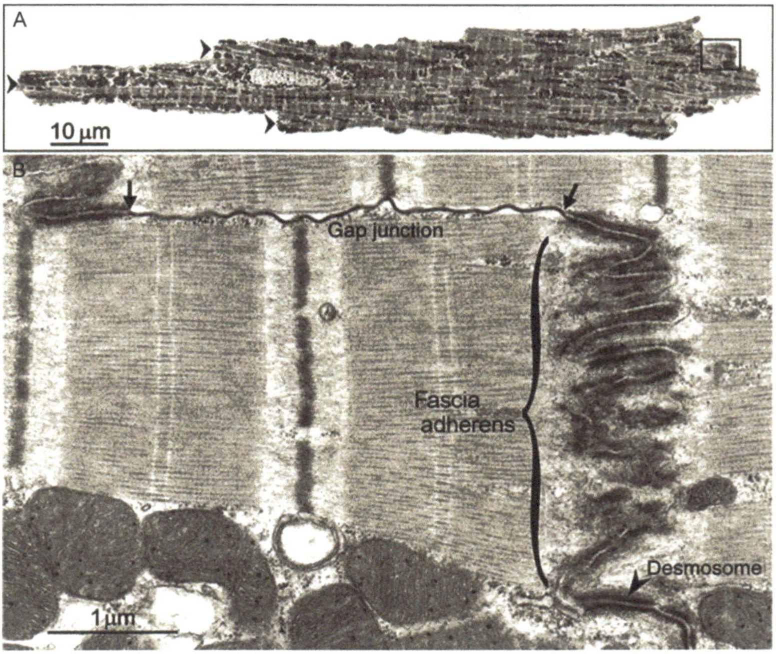

Figure 4.2 Intercalated disc junction. (A) In single myocyte disc is shown by arrowheads on left. Box on right is magnified in B. (B) In this image, electron-dense vertical plicate zones of the disc (fasciae adherentes) are shown, and gap junctions and desmosomes are located primarily at the lateral-facing zones. Here, the distal region of the gap junction is adjacent to the *fascia adherens* (arrows). *Reprinted from Severs, Bruce, Dupont, and Rothery (2008) with permission from Oxford Press.*

1.1. Intercalated disc anchoring junctions

Anchoring cell–cell junctions maintain the mechanical integrity of cardiac myocytes. Anchoring junctions connect adjacent cells through direct association with cytoskeletal components from both cells and include *fascia adherens* junctions and desmosomes.

1.1.1 Fascia adherens *junctions*

Cardiac myocyte adherens junctions are composed of *fascia adherens* junctions primarily generated from membrane-spanning cadherins that link actin microfilaments in the cytoplasm with cadherins from neighboring cells. The molecular components of *fascia adherens* junctions are (1) N-cadherin and other transmembrane and catenin-binding proteins (coxsackievirus and

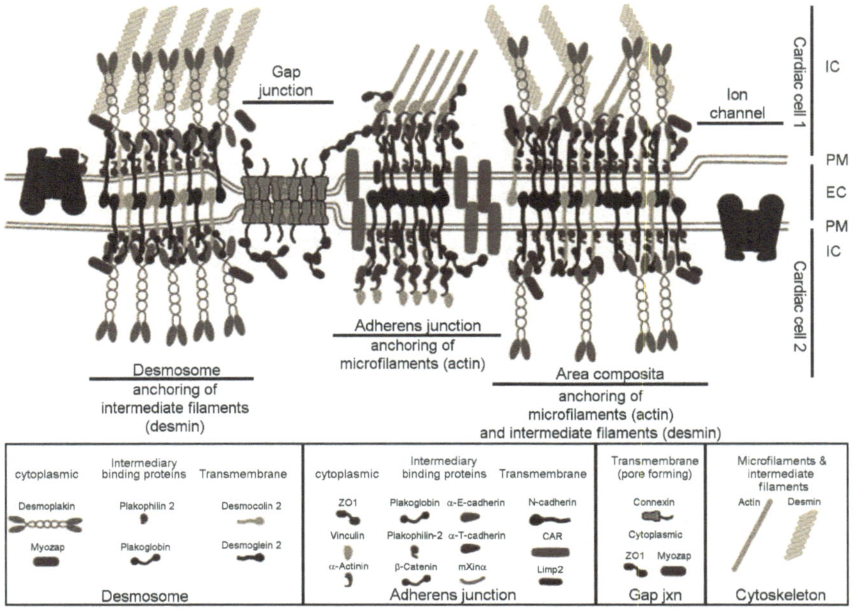

Figure 4.3 Molecular composition of cardiac cell–cell junctions. EC, extracellular; IC, intracellular; PM, plasma membrane. *Reprinted from Mezzano and Sheikh (2012) with permission from Elsevier.* (See Color Plate.)

adenovirus receptor (CAR) and lysosomal integral membrane protein 2 (LIMP-2)); (2) the catenins/armadillo proteins; and (3) cytoskeletal actin-binding proteins, including vinculin/metavinculin, zonula occludens-1 (ZO-1), Xin repeat-containing protein (mXinα), and α-actinin (Fig. 4.3).

1.1.1.1 N-cadherin

N-cadherin (~88 kD) is a transmembrane single-pass glycoprotein that mediates calcium-dependent cell–cell adhesion via homophilic interactions. Structurally, N-cadherin has one transmembrane domain, a cytoplasmic domain, and five extracellular domains. The recognition sites for cadherin molecules consist of an HAV motif localized on the adhesion dimer interface. Ca^{2+} is involved in the association of the five subdomains to create the rodlike morphology of cadherins. The extracellular domain dimerizes, where two monomers are arranged in parallel to the sarcolemma. The dimers are arranged zipper-like in the intercellular space, where adjacent cells become connected in a calcium-dependent manner. N-cadherin provides strong cell–cell adhesion through this cadherin–catenin linkage to the actin cytoskeleton and also serves as the site of attachment of myofibrils, enabling transmission of contractile forces across the sarcolemma. Because actin filaments pull against adherens junctions, cadherins facilitate bidirectional transmission of mechanical tension between cells. In essence, cadherins mediate the transduction of cytoskeletal tension between cells, responding to intercellular mechanical load by remodeling the cytoskeleton to reinforce the adherens junction.

The involvement of N-cadherin in morphoregulation and cardiac differentiation and formation and function is well established. To elucidate the function of cardiac N-cadherins, a number of studies have been performed in animal models in which N-cadherin has been deleted, either entirely or partially. Structural disruption of the intercalated disc follows downregulation of N-cadherin in a hereditary hamster model of dilated cardiomyopathy (Fujio et al., 1995). Complete knockout of N-cadherin results in embryonic lethality shortly after implantation (E10). Myocardial tissue had been formed at this stage, but myocytes dissociated and the heart failed to form properly, primarily due to adhesion dysfunction (Radice et al., 1997). Interestingly, isolated myocytes from these embryos could aggregate and weakly contract, suggesting that N-cadherin is not exclusively required for cell adhesion and electric coupling at this stage in development (Radice et al., 1997). Conversely, mice engineered to overexpress N-cadherin in the adult myocardium demonstrate dilated cardiomyopathy resulting from cadherin-mediated modulation of intercalated disc function (Ferreira-Cornwell et al., 2002). To reveal the function of N-cadherin in the adult heart, a conditional knockout mouse possessing a cardiac-specific, inducible *Cre* transgene to specifically delete N-cadherin in the adult myocardium has been established (Li et al., 2005). These mice display loss of the intercalated disc structure and loss of components within the *fascia adherens* junction and desmosomes, resulting in modest dilated cardiomyopathy, impaired left ventricular function, ventricular tachyarrhythmias, and abnormal conduction (Li et al., 2005). Mechanistically, depletion of N-cadherin resulted in a reduction of connexin 43 (Cx43) expression, correlating with a decrease in conduction velocity, inducible and spontaneous arrhythmias, and sudden cardiac death within 2 months after deletion of the N-cadherin gene (Li et al., 2005).

1.1.1.2 Catenins

α-Catenins ($\sim$100–102 kD) are cytoplasmic molecules that link the cytoplasmic domain of N-cadherin to the actin cytoskeleton. Molecularly, α-catenin interacts with β- or γ-catenin via its N-terminal domain and, with actin, directly through its C-terminus. α-Catenin also indirectly interacts with other actin-binding proteins, including vinculin and α-actinin. There are three α-catenin subtypes, with α-E-catenin the most widely studied and highly expressed in the adherens junction of the intercalated disc. Perturbations in α-E-catenin expression are associated with dilated cardiomyopathy, with cardiac myocyte-specific deletion of the murine α-E-cadherin gene resulting in progressive dilated cardiomyopathy, unique defects in the right ventricle, and complete loss of vinculin from the intercalated disc (Sheikh et al., 2006). These mice were also predisposed to ventricular free wall rupture following myocardial infarction (Sheikh et al., 2006). Reduced expression of α-E-catenin has also been associated with postmyocardial infarction ventricular rupture in humans (van den Borne et al., 2008). Interestingly, patients with reduced α-E-catenin expression demonstrated normal expression and distribution of other *fascia adherens* components (β-catenin, γ-catenin, and N-cadherin).

β-Catenin (~88 kD) is a multifunctional protein with capabilities that vary depending on cellular localization. In the cardiac myocyte, β-catenin is localized to the *fascia adherens* junction, participating in the N-cadherin/actin complex. The role of β-catenin in adult myocytes is elusive based on evidence suggesting that upregulation of plakoglobin compensated for the loss of β-catenin in the adult myocardium of various cardiac-specific β-catenin-deficient mouse models (Chen et al., 2006). Plakoglobin is functionally and structurally similar to β-catenin and has been demonstrated to interfere with β-catenin signaling. Interestingly, both increased and decreased expressions of β-catenin are observed in human cardiomyopathy (Masuelli et al., 2003). β-Catenin expression increases in the intercalated discs of patients with hypertrophic cardiomyopathy (Masuelli et al., 2003). Conversely, β-catenin expression is reduced in end-stage heart failure, where expression of β-catenin and one of its binding partners, estrogen receptor-α, is lost from the intercalated disc. These studies suggest that some intercalated disc proteins may have unique functions in different types of myopathies or in different stages of cardiac disease.

1.1.1.3 Vinculin

Vinculin (~117 kD) and its splice variant, metavinculin (~124 kD), are membrane-associated proteins involved in linking cell–matrix adhesions, cell–cell adhesions (*fascia adherens* junctions), and costameres (subsarcolemmal adhesion plaques) to the actin cytoskeleton. Heterozygous vinculin-deficient mice display cardiac dysfunction and increased mortality after acute hemodynamic stress imposed by transverse aortic constriction (TAC; Zemljic-Harpf et al., 2004). Mechanistically, TAC induces ultrastructural defects in the intercalated discs (Zemljic-Harpf et al., 2004). Cardiac myocyte-specific knockout of vinculin results in adherens junction/intercalated disc abnormalities with reduced expression of N-cadherin and β1D integrin, mislocalization of Cx43, and sudden cardiac death before 3 months of age (Zemljic-Harpf et al., 2007). These studies demonstrate the importance of vinculin in the preservation of intercalated disc ultrastructure, in addition to cardiac contractile and electric function.

Mutations in vinculin and metavinculin have also been associated with both dilated cardiomyopathy and hypertrophic cardiomyopathy. Metavinculin deficiency has been demonstrated due to a defect in mRNA alternative splicing in a patient with dilated cardiomyopathy (Maeda, Holder, Lowes, Valent, & Bies, 1997). Additionally, a comprehensive evaluation of 350 unrelated patients with dilated cardiomyopathy revealed three metavinculin gene variants (R975W, L954del, and A934V). *In vitro* work determined that these variants resulted in defective actin filament organization, suggesting that the interaction between metavinculin and actin may alter force transmission within the myocyte and between adjacent myocytes. A single gene variant in vinculin (resulting in R975W) has been identified in patients that exhibit either dilated cardiomyopathy or hypertrophic cardiomyopathy, suggesting a role for modifier genes and/or environmental components in the phenotypic presentation of vinculin/metavinculin mutations (Vasile, Ommen, Edwards, & Ackerman, 2006). More recent evidence has identified additional vinculin and metavinculin mutations in patients with obstructive hypertrophic cardiomyopathy (Vasile et al., 2006). It is proposed that these gene variants (resulting in A934V, P943A, and L277M) may cause disruptions in the secondary structure of the protein or may lead to a reduction in the stability/expression of vinculin/metavinculin at the intercalated disc.

1.1.1.4 Novel *fascia adherens* junction proteins

Recently, a number of novel *fascia adherens* junction proteins have been identified as important players in cardiac dysfunction and arrhythmias. These novel proteins are *fascia adherens* components via their interactions with N-cadherin and β-catenin: muscle-specific mouse Xin α (mXinα), LIMP-2, and the CAR.

Muscle-specific mouse Xin α (~155 kD) is a striated muscle-specific protein. mXinα colocalizes with N-cadherin and β-catenin, beginning in embryogenesis and continuing through adulthood. It is proposed that mXinα plays a role in cell–cell adhesion and modulating the Wnt/β-catenin/N-cadherin-mediated signaling pathway, as well as in organizing actin filament assembly. Deletion of mXinα results in defects in intercalated disc ultrastructure, myofilament assembly, and abnormal expression of *fascia adherens* and desmosomal proteins (β-catenin, N-cadherin, desmoplakin, and Cx43). These animals exhibit cardiac hypertrophy, cardiomyopathy, and conduction defects. The role of Xin in humans has yet to be determined with regard to cardiomyopathies and arrhythmias.

LIMP-2 (~54 kD) was originally identified as a component of lysosomal membranes. However, recent studies in humans and mice demonstrate localization of LIMP-2 to the cardiac intercalated disc, based on its association with N-cadherin and the ability to regulate interactions between phosphorylated β-catenin and N-cadherin (Schroen et al., 2007). LIMP-2 knockout mice are unable to produce a hypertrophic response after TAC and display intercalated disc abnormalities, N-cadherin mislocalization, and the development of dilated cardiomyopathy following angiotensin II treatment (Schroen et al., 2007). Interestingly, LIMP-2 expression is increased in the hearts of patients with severe pressure overload, suggesting an important role in cardiac hemodynamics (Schroen et al., 2007).

CAR (~40 kD) is a transmembrane protein with functions as an adhesion molecule and as a common receptor for the coxsackievirus and adenovirus (Lim et al., 2008). Its identification as a novel *fascia adherens* junction component is based on its myocyte subcellular localization and its interactions with β-catenin and gap junction proteins, Cx45 and ZO-1 (Lim et al., 2008). CAR knockout mice exhibit AV block associated with loss of Cx45, β-catenin, and ZO-1 localization at the intercalated disc and intercalated disc structural abnormalities (Lim et al., 2008). As a result, these mice develop dilated cardiomyopathy. Further investigation also revealed sinus node dysfunction, defects in AV morphology, and abnormal Cx43 expression. CAR has also been implicated in human myocarditis and dilated cardiomyopathy. Based on these studies, a role for CAR in the development of cardiac disease and arrhythmias has been demonstrated.

1.1.2 Intercalated disc desmosomal junctions

The desmosome was first observed in the spinous layer of the epidermis by the Italian physiologist Giulio Bizzozero. His observations led him to the interpretation that the structures were sites of cell–cell adhesive contact. The term desmosome was applied in 1920, derived from the Greek words "desmo," meaning bond or fastening, and "soma," meaning body. The advent of electron microscopy revealed desmosome organization at the ultrastructural level. From these studies, the desmosome was further divided into three morphological zones: the outer dense plaque (ODP), the inner dense plaque (IDP), and the extracellular core region (desmoglea) (Kowalczyk et al., 1994). While *fascia adherens* junctions link the actin cytoskeleton of adjacent myocytes, desmosomes provide continuity in the intermediate filament network (primarily desmin). In the intercellular space, desmosomal cadherins (desmocollins and desmogleins) associate tightly with each other. In the intracellular space, the intermediate filaments bind to desmoplakin. The interaction between desmoplakin and the desmosomal cadherin is mediated by plakophilin and plakoglobin (γ-catenin).

Human genetic studies have identified mutations in all known components of the desmosomes in individuals harboring arrhythmogenic right ventricular cardiomyopathy/dysplasia (ARVC/D). ARVC/D is considered a desmosomal cardiomyopathy (Awad, Calkins, & Judge, 2008). ARVC/D is an autosomal-dominant disease presenting with ventricular arrhythmia, progressive myocardial atrophy with fibroadipocytic replacement of myocytes, and sudden death in youth (Ellinor, MacRae, & Thierfelder, 2010). Recent studies in a murine model of ARVC/D (Yang et al., 2006) and in humans with ARVC/D (Syrris et al., 2007) indicate that ARVC/D is not exclusively localized to the right ventricle and that the left ventricle is significantly affected. Nearly 50% of patients with ARVC/D harbor a mutation in one of the various components of the desmosome (Hermida et al., 1997). Several human studies have revealed heterozygous mutations of the desmosomal cadherins desmocollin-2 (Syrris et al., 2006) and desmoglein-2 (Awad et al., 2006) in cases of ARVC/D. Mutations in plakophilin-2, plakoglobin, and desmoplakin are also common in ARVC/D (Tsatsopoulou, Protonotarios, & McKenna, 2006). Functionally, mutations in desmosomal components destabilize the desmosomal complex and alter the integrity of cell–cell junctions. This destabilization is sufficient to promote cardiac myocyte cell death under conditions of mechanical stress. The resulting tissue repair via fibrofatty replacement/scar formation is the structural basis for reentrant arrhythmias and heart failure. Additionally, the concurrent remodeling of the gap junction may enhance structural defects in ARVC to provide a substrate for malignant arrhythmias.

1.1.2.1 Desmocollin-2/desmoglein-2

Desmocollins and desmogleins, members of the cadherin superfamily, mediate adhesion at desmosomal junctions. The extracellular domains of the desmocollins and desmogleins mediate cell adhesion, whereas the cytoplasmic tails associate with the desmosomal plaque proteins. In humans, desmoglein-2 (~122 kD) and desmocollin-2 (~100 kD) are the primary isoforms identified at the intercalated disc. Desmosomes demonstrate Ca^{2+}-dependent adhesion, though the precise mechanism of desmosomal cadherin adhesion and specificity are not understood (Sheikh, Ross, & Chen, 2009). Specifically, the intracellular tails of desmosomal cadherins associate with plakoglobin and plakophilin, while their intercellular portions interact with the intercellular portion of desmosomal components from the adjacent cell.

Mutations in desmoglein-2 may result in ARVC/D, most likely reflecting the strict requirement of desmoglein-2 in resisting mechanical stresses associated with cardiac contraction. Interestingly, mutations in desmoglein-2 have variable penetrance and heterogeneity (Pilichou et al., 2006). Mutations resulting in premature truncation or mislocalization of desmocollin-2 have been identified in autosomal-dominant ARVC/D (Heuser et al., 2006; Syrris et al., 2006). An additional truncation mutant in the *desmocollin-2* gene has been associated with recessive ARVC/D with mild palmoplantar keratoderma and woolly hair.

1.1.2.2 Plakophilin-2

Plakophilin-2 (~97 kD) is a cytoplasmic cadherin-binding desmosomal component involved in regulating cadherin adhesive activity and signaling. Plakophilin-2 mutations have been found to be the most common genetic cause of ARVC/D, with about 70% of familial cases of ARVC/D resulting from a pathogenic plakophilin-2 mutation (van Tintelen et al., 2006). Molecular studies in an ARVC/D patient with a plakophilin-2 mutation revealed a reduction of cardiac plakophilin-2 and a reduction in plakoglobin, desmoplakin, and Cx43, while N-cadherin remained unchanged. Endomyocardial biopsies from ARVC/D patients with plakophilin mutations demonstrated decreased Cx43 expression regardless of the plakophilin mutation. This change in Cx43 expression in ARVC/D patients correlates with plakophilin siRNA experiments performed in a mouse cardiomyocyte cell line.

A canine model of ARVC/D shows striking similarities to humans harboring ARVC/D gene variants. Specifically, these animals demonstrate histological abnormalities, such as right ventricular loss of cardiac myocytes and fibrofatty replacement. Immunofluorescence microscopy revealed reduced localization of N-cadherin and plakophilin-2, with larger reductions in desmoplakin, plakoglobin, and Cx43. In the majority of dogs, changes in intercalated disc structure could be directly linked to the induction of ventricular arrhythmias, implying that the abnormalities produced an arrhythmogenic substrate.

1.1.2.3 Desmoplakin

Desmoplakin is one of the major components of the desmosome and likely the most abundant (Mueller & Franke, 1983). It is a member of a large family of proteins termed the plakin family of cytolinkers. These large molecules connect cytoskeletal networks to the plasma membrane, integrating actin, microtubules, and intermediate filaments. It is a large protein (~210–250 kD depending on isoform) and is proposed to specifically link the desmocollins and desmogleins to the intermediate filament network (desmin) through interaction with plakoglobin and plakophilin.

Several studies have demonstrated the importance of desmoplakin in desmosomal structure and function. Mice deficient in desmoplakin die around the time of implantation (E6.5) and show fewer desmosomes (Gallicano et al., 1998). Additionally, the desmosome seen in desmoplakin-null embryos does not attach to the intermediate filaments (Gallicano et al., 1998). These observations strengthen *in vitro* studies and clarify the importance of desmoplakin in the linkage of intermediate filaments to the sarcolemma. A role for desmoplakin in ARVC/D has also been established in mouse models, where deletion of the desmoplakin gene produced embryonic lethality (E10–E12) (Garcia-Gras et al., 2006). In these embryos, there were poorly formed hearts with unorganized myocytes and absent chamber specification (Garcia-Gras et al., 2006). Desmoplakin heterozygous mice are viable and display some ARVC/D symptoms. While the exact mechanism has not been established, work with desmoplakin homozygous and heterozygous mice suggests that desmoplakin deficiency produces mislocalization of plakoglobin from the intercalated disc with decreased canonical Wnt/β-catenin signaling. As a result, there was a promotion in adipogenic/fibrinogenic gene expression (fibrofatty replacement).

A number of human desmoplakin mutations have been linked with ARVC/D. R2834H was found to disrupt binding of desmoplakin to intermediate filaments, plakoglobin, desmin, β-catenin, and plakophilin-2 (Yang et al., 2006). Moreover, myocyte-specific knockouts of desmoplakin reiterate these findings and exhibit features of ARVC/D, clearly revealing an important role for desmoplakin in this disease. Other desmoplakin mutations have been shown to demonstrate a variety of ARVC/D phenotypes (Rampazzo et al., 2002). Two recessive mutations, G2375R and 7901delG, both disrupt desmoplakin/intermediate filament interactions but present with different phenotypes. G2375R patients present with the typical ARVC/D symptoms (Alcalai, Metzger, Rosenheck, Meiner, & Chajek-Shaul, 2003), whereas 7901Gdel patients' presentation predominantly involves the left ventricle and has been termed Carvajal syndrome (Norgett et al., 2000). Carvajal syndrome, that is, a cardiocutaneous syndrome characterized by palmoplantar keratoderma and woolly hair, primarily presents as dilated cardiomyopathy rather than ARVC/D (Protonotarios & Tsatsopoulou, 2004). Specifically, studies have shown that these individuals demonstrate ventricular hypertrophy and dilatation, focal ventricular aneurysms, and distinct desmosome ultrastructural abnormalities (Kaplan, Gard, Carvajal-Huerta, et al., 2004). Unlike ARVC/D, no fibrofatty infiltration is present in Carvajal syndrome (Kaplan, Gard, Carvajal-Huerta, et al., 2004). Molecularly, there is no appreciable expression of desmoplakin, with significant reductions in plakoglobin, desmin, and Cx43 (Kaplan, Gard, Carvajal-Huerta, et al., 2004). It is believed that mutation in desmoplakin may interfere with the interactions among desmoplakin, plakoglobin, and the desmosomal cadherins. These interruptions are proposed to cause the contractile and electric dysfunction present in Carvajal syndrome.

Desmoplakin is located in the ODP of the desmosome. It is suggested that mutations in the ODP proteins that disrupt association with desmin may produce the ARVC/D phenotype with left ventricular involvement. In contrast, mutations in proteins located in the IDP, which disrupt binding between cells, may explain a more right ventricular phenotype (Tsatsopoulou et al., 2006).

1.1.2.4 Plakoglobin

Plakoglobin (γ-catenin) is an armadillo protein that is present in the intracellular domain of desmosomes and *fascia adherens* junctions. Structurally, it is ~82 kD and consists of 12 armadillo repeats flanked by unique C-terminal and N-terminal domains. The central armadillo domain of plakoglobin also binds to desmoplakin that tethers intermediate filaments to the desmosomal plaque.

The first gene variation discovered that was linked to ARVC/D was a homozygous two-base pair deletion (2057del2) in the plakoglobin gene, resulting in a premature truncation of the plakoglobin protein. The disease was originally identified in individuals on the Greek island of Naxos and has since been referred to as Naxos disease (Protonotarios, Tsatsopoulou, & Gatzoulis, 2002). Naxos disease is an autosomal-recessive cardiocutaneous subtype of ARVC/D (Protonotarios et al., 1986). Naxos patients exhibit palmoplantar keratoderma and woolly hair (Protonotarios et al., 1986). When the truncated form of plakoglobin is transfected into HEK293 cells, the cells display mislocalization of plakoglobin, leading to decreased cell–cell adhesion (Huang, Asimaki, Lo, McKenna, & Saffitz, 2008). Follow-up studies in humans demonstrated that the truncated form of plakoglobin was expressed in the cardiac myocyte but failed to localize at the desmosomal junctions, while N-cadherin, desmocollin-2, desmoplakin, α- and β-catenins, and plakophilin-2 were localized normally in the cell junctions (Kaplan, Gard, Protonotarios, et al., 2004). Cx43 was reduced in the left and right ventricles of Naxos patients, with significant reduction in the right ventricle (Kaplan, Gard, Protonotarios, et al., 2004). The resulting inference is that defective mechanical coupling results in defective electric coupling, slowing impulse conduction and underlying the etiology of ARVC/D (Kaplan, Gard, Protonotarios, et al., 2004).

After its initial discovery, it was recognized that plakoglobin mutations may play a causal role in both dominant and recessive forms of ARVC/D. The dominant plakoglobin mutation, S39_K40insS (which inserts an extra serine residue at position 39), causes a decrease in plakoglobin at the intercalated disc (Asimaki et al., 2007). Transfection of this mutant into HEK293 cells demonstrates mislocalization of the plakoglobin protein away from the plasma membrane to the cytoplasm, perhaps due to altered turnover kinetics (Asimaki et al., 2007). A dominant mutation in desmoplakin that disrupts association with plakoglobin (S299R) also causes a dominant form of ARVC/D.

The use of mouse models has given great insight into the role of plakoglobin in the etiology of ARVC/D. Plakoglobin-null mice die due to fragility of the myocardium but demonstrate acantholysis, which is indicative of compromised desmosome function. Heterozygous-deficient mice show increased right ventricular volume, reduced right ventricular function, and spontaneous ventricular ectopic activity. These mice also demonstrate normal left ventricular size and function and normal desmosome structure. Furthermore, it was observed that endurance training of these mice accelerated the development of cardiac dysfunction and increased arrhythmia vulnerability (Kirchhof et al., 2006). Using a conditional knockout with depletion of plakoglobin in the cardiac myocyte, researchers revealed altered desmosomal structure and decreased cardiac function as determined by echocardiography (Li et al., 2011). The cardiac pathology mimics many features of human ARVC/D, including progressive loss of myocytes, extensive inflammatory infiltration, and replacement fibrosis (Li et al., 2011).

1.1.2.5 Ankyrin-G

Ankyrins are a family of polypeptides implicated in the targeting, stability, and regulation of select ion channels, transporters, pumps, and cytosolic proteins (Bennett & Baines, 2001; Bennett & Healy, 2008, 2009; Mohler, Gramolini, & Bennett, 2002a). Initial work in the brain established a necessary role for ankyrin-G, encoded by *ANK3*, for the targeting of the voltage-gated Na^+ channels to specific neuronal membrane domains (Jenkins & Bennett, 2001; Zhou et al., 1998). Specifically, loss of ankyrin-G results in a loss of $Na_v1.6$ at the axon initial segments of Purkinje neurons. Related to this chapter, ankyrin-G is concentrated at the cardiac intercalated disc, where it regulates $Na_v1.5$ localization (Lowe et al., 2008; Mohler, Rivolta, et al., 2004). $Na_v1.5$ in the heart co-immunoprecipitates and colocalizes with ankyrin-G (Lowe et al., 2008; Mohler, Rivolta, et al., 2004). Moreover, a human mutation (E1053K) in the DII/DIII loop of $Na_v1.5$ that blocks ankyrin-G binding is associated with a potentially fatal arrhythmia disease termed Brugada syndrome (Mohler, Rivolta, et al., 2004). Consistent with the Brugada syndrome phenotype of reduced inward cardiac I_{Na}, the $Na_v1.5$ E1053K mutation prevents delivery of $Na_v1.5$ to the intercalated disc plasma membrane (Mohler, Rivolta, et al., 2004).

Ankyrin-G forms a complex and co-immunoprecipitates with E-cadherin in human bronchial epithelial cells. Specifically, E-cadherin binds directly to ankyrin-G and this interaction is necessary for proper E-cadherin localization to specialized membrane domains (Kizhatil et al., 2007). Moreover, Bennett and colleagues further reported an analogous interaction between ankyrin-G and cardiac N-cadherin, suggesting that ankyrin-G may play a similar role in regulating N-cadherin expression and localization in the cardiomyocyte, as well as morphoregulation, cardiac differentiation, and intercalated disc formation (Kizhatil et al., 2007). Further work will be necessary to establish such a role for ankyrin-G in the cardiac myocyte.

Finally, ankyrin-G has been proposed as a key functional component at the intersection of three independent complexes: the voltage-gated sodium channel, gap junctions, and the cardiac desmosome. Specifically, the use ofsiRNA to reduce ankyrin-G expression in cultured neonatal cardiomyocytes demonstrated significant changes in the distribution and/or abundance of plakophilin-2 and connexin 43, in addition to decreased intercellular adhesion strength and electric coupling (Sato et al., 2011). While these studies were performed in isolated cell systems, these data do provide support to the hypothesis that desmosomes, gap junctions, and the voltage-gated sodium channel complex may form a higher-order integrated network whereby alterations in the composition of one may affect the function of the others. Understanding this regulation *in vivo* will be critical in aiding our understanding of rare desmosomal diseases and acquired cardiomyopathies that largely affect the integrity of the intercalated disc.

1.2. Communicating junctions

Communicating junctions form cell–cell pathways for the propagation of intercellular communication. Intercellular communication plays a key role in tissue homeostasis and growth, development, and differentiation. In the heart, intercellular communication is crucial for the electric coupling between cardiac myocytes, allowing for the transmission of electric waves necessary for synchronous contraction. Gap junctions provide this function, forming conduits for the orchestrated patterns of current flow that regulate the regular rhythm of the heart.

1.2.1 Gap junctions

Gap junctions are clusters of transmembrane-spanning channels that directly link the cytoplasmic compartments of neighboring cells. Normal heart rhythm depends on the coupling of cardiac myocytes by gap junctions. Cardiac gap junctions are comprised of a pair of adjoining connexons (hemichannels), one contributed by each of the adjacent sarcolemmal membranes. Each connexon is comprised of six connexin molecules and completely spans the membrane. There are twenty different connexin types that have been identified in humans. The specificity of connexin type and their arrangement confers functional properties of the channel.

1.2.1.1 Connexins

Connexins are tetraspan membrane proteins that form interlocking hexamers at each cell membrane to create a pore between two adjacent cells.

Specifically, a gap junction channel is composed of 12 connexin proteins (six contributed by each cell). The six connexin hexamer is a hemichannel, also referred to as a connexon. Three primary connexins are expressed in the cardiac myocyte: connexin 43 (Cx43), connexin 40 (Cx40), and connexin 45 (Cx45). These three connexins are coexpressed in unique and relative combinations and in a chamber-specific and myocyte type-specific manner. Ventricular myocytes mainly express Cx43-formed gap junctions, whereas the gap junctions of atrial myocytes contain Cx43 and Cx40 (Severs et al., 2001). Cx45 is expressed in a lower quantity than Cx43 and Cx40 but has higher expression in the atria compared to the ventricles. Effective gap junction coupling is determined by a number of factors: amount and types of the connexin molecules expressed, the size and distribution of the gap junction complexes, the proportion of each connexin within the junction, and the gating and arrangement of individual gap junction channels (Severs et al., 2008). Connexin localization occurs via a microtubule-based targeted delivery system. Specifically, work from Shaw and colleagues demonstrated that EB1 and p150^{GLUED}, microtubule plus-end-binding proteins, cooperate in the targeting of Cx43 vesicles to specific subdomains using heterologous cells (Shaw et al., 2007). While work in primary cardiomyocytes will be critical to confirm the role of the microtubule system in connexin localization in gap junction plaque formation, this study demonstrates the importance of the microtubules in the delivery of gap junction proteins to the cell–cell border. In addition, it has been observed that EB1 is lost from microtubules in stressed mouse and human myocardium, limiting the delivery of connexins to the plasma membrane.

In a mouse model, heterozygous expression of Cx43 resulted in reduced expression of Cx43 along with a decrease in the number and size of gap junctions (Saffitz, Green, Kraft, Schechtman, & Yamada, 2000). Moreover, the amount of N-cadherin was unchanged. Additional work demonstrated that *fascia adherens* junction proteins and associated catenins are independently regulated from the gap junction (Gutstein, Liu, Meyers, Choo, & Fishman, 2003). Cx43 knockout mice exhibit normal *fascia adherens* junctions and desmosomes in the absence of Cx43, with normal localization of corresponding component proteins (Gutstein et al., 2003). From this study, it can be concluded that Cx43 is not necessary for the organization of mechanical cell junctions at the intercalated disc.

In humans, a number of Cx43 mutations have been linked to oculodentodigital dysplasia (ODDD). In some cases, patients present with similar hair and skin abnormalities present in a number of desmosomal diseases (Kelly et al., 2006). Additionally, some mutations have been associated with cardiac developmental defects and cardiac rhythm disturbances (Paznekas et al., 2003). In an ODDD mouse model, abnormal cardiac myocyte Cx43 trafficking was observed, resulting in a decrease in Cx43-based gap junctions; however, the reduction was not beyond levels that would affect normal cardiac conduction. It would appear, therefore, that dysfunctional Cx43 is not a primary cause of mechanical or electric coupling defects associated with ventricular arrhythmia (Manias et al., 2008). Mutations of the Cx43 affecting phosphorylation sites in the Cx43 carboxy domain are associated with complex cardiac malformations and visceroatrial heterotaxia or hypoplastic left heart syndrome. However, follow-up studies have yet to confirm a direct link with Cx43 mutation and phenotype. Mutation of the Cx40 gene has been reported in a few cases of congenital heart disease with anomalies of the aortic arch. Heterozygous somatic missense mutations and polymorphisms have also been identified with the Cx40 regulatory region and have been linked with atrial fibrillation.

1.2.1.2 Zona occludens-1

ZO-1 is a large scaffolding protein (~220 kD) that is a member of the membrane-associated guanylate kinase (MAGUK) protein family. ZO-1 interacts with C-terminus of Cx43 via the ZO-1 PDZ domain (Toyofuku et al., 1998). The interaction between ZO-1 and Cx43 has been demonstrated in primary cells, where the linkage between ZO-1 and Cx43 is necessary for the localization of Cx43 at the intercalated disc (Toyofuku et al., 1998). While ZO-1 is found in normal heart; interestingly, ZO-1 is almost completely absent in the failing heart. In addition to Cx43, ZO-1 can associate with α-catenin. Microscopic evaluation of rat ventricle has demonstrated colocalization of ZO-1, α-catenin, and N-cadherin (Barker, Price, & Gourdie, 2002). More specifically, there is a high level of colocalization of ZO-1 and N-cadherin, in contrast to low colocalization of ZO-1 and Cx43. Interestingly, the colocalization levels of Cx43 and ZO-1 increased following the induction of gap junction endocytosis (Barker et al., 2002), indicating that ZO-1 participates in gap junction turnover during cardiac remodeling.

1.3. Area composita

Recent studies have revealed that the localization of desmosomal and *fascia adherens* junction components is not as distinct as previously assumed. In fact, these components overlap in a structure called the *area composita*. Using comprehensive immunoelectron microscopy, it has been observed that desmoplakin is located at desmosomal and *fascia adherens* junctions (Franke, Borrmann, Grund, & Pieperhoff, 2006). Using a number of antibodies to other desmosomal components, it was also determined that the localization of plakoglobin, plakophilin, desmocollin, and desmoglein was not restricted to the desmosomes but could also be detected in *fascia adherens* junctions as well (Franke et al., 2006). Further studies have determined that the typical components of the *fascia adherens* (N-cadherin, α-catenin, and β-catenin) colocalize with desmosomal proteins in the *area composita* junctions (Borrmann et al., 2006). These studies suggest that the *area composita* is a complex hybrid with intimately associated components (Borrmann et al., 2006). Interestingly, it is believed that the formation of the *area composita* junction occurs late in development and is primarily a postnatal process in mammalian heart development. Conversely, nonmammalian species have separate *fascia adherens* junctions and desmosomes, suggesting that the formation of the *area composita* is a relatively late process in vertebrate evolution.

An impressive level of cross talk has been shown to exist among the various junctional proteins within the intercalated disc. This occurs due to the close spatial proximity of the junctional complexes and shared junctional components. Deficiencies in plakoglobin, for example, result in cardiac rupture in the embryonic heart, while heterozygous loss/mutations in the adult myocardium are associated with ARVC/D (Kirchhof et al., 2006). Mice with deficiencies in desmosomal components also demonstrate this phenotype, suggesting that defects in plakoglobin cause issues related to the desmosomal as opposed to the *fascia adherens* complex. Recent studies with reduced N-cadherin expression have also suggested a hierarchal relationship among junctional complex components (Li, Patel, & Radice, 2006). These mice exhibit a loss in all components of the *fascia adherens* junction and the desmosomes, implying that desmosomal stability relies on N-cadherin, a traditional *fascia adherens* junctional protein (Li et al., 2006, 2008). While desmosomal and *fascia adherens* proteins are essential for mechanical integrity, it is observed that their loss results in destabilization/loss gap junctions. On the other hand, conditional loss of Cx43 has no effect on desmosomes or *fascia adherens* junctions (Gutstein et al., 2003), suggesting that gap junctions are not necessary for the establishment of these complexes at the intercalated disc. These findings are strengthened by studies that the mechanical junctions form prior to gap junctions in the developing cardiac myocyte.

2. TRANSVERSE TUBULES

Transverse tubules (T-tubules) are complex and tightly regulated invaginations of the cell membrane enriched in a host of cardiac ion channels, transporters, pumps, receptors, and signaling molecules critical for excitation–contraction coupling in myocytes. In the ventricle, this network is essential for the spatial and temporal synchronous release and spread of Ca^{2+} throughout the myocyte. Changes in T-tubule structure and composition occur not only during development but also in disease during the transition from compensated cardiac hypertrophy to heart failure.

2.1. T-tubule structural components

T-tubules are primarily mammalian membrane structures, present in the cardiac tissue of all species of mammals investigated (mice, rats, guinea pigs, rabbits, dogs, pigs, and humans), but are absent in avian, reptile, and amphibian heart. T-tubules are predominantly in ventricular myocytes, with no or significantly less developed presence in atrial, conducting, and pacemaking tissues (Ayettey & Navaratnam, 1978). T-tubules are found at regular intervals along the long axis of the cardiac myocyte; roughly at every Z-line. Along the Z-line, T-tubules run deep into the cell, where they also branch into longitudinal extensions (Fig. 4.4; Forbes, Hawkey, & Sperelakis, 1984). The T-tubule system represents the network of membrane within the cell that conducts the action potential into the center of the cardiac myocyte.

There is a labile nature to T-tubules: they are absent in neonatal cells (Haddock et al., 1999) and decrease when cells are kept in culture (Mitcheson, Hancox, & Levi, 1996); however, little is understood regarding the underlying mechanisms of T-tubule expression and maintenance. Nonetheless, T-tubules appear to be regulated by biochemical and biophysical factors. The ability of the T-tubule system to maintain its structure while under the constant forces of contraction may be due to the presence of "scaffold" molecules, membrane-associated proteins, and basal lamina proteins (Kostin et al., 1998). In Chinese hamster ovary (CHO) cells, expression of *amphiphysin-2* (also known as *bridging integrator 1, BIN1*), a linker between the sarcolemma and the submembranous cytosolic scaffolds, generates narrow tubules that are continuous with the sarcolemma (Lee et al., 2002). When expressed in nonmuscle cells, amphiphysin-2 results in tubular formation and is highly concentrated at the sites of developing membrane striations in the muscle. Knockdown of amphiphysin-2 is perinatal lethal,

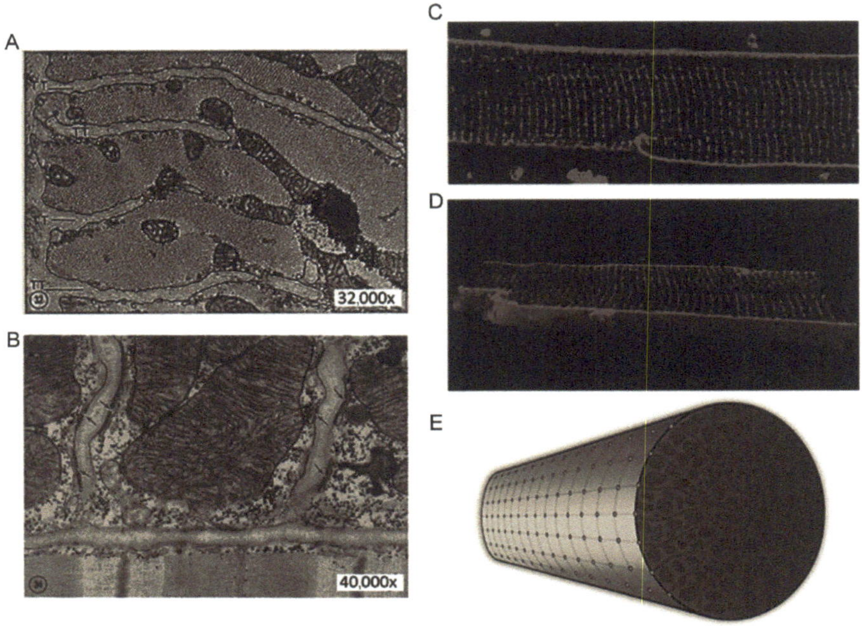

Figure 4.4 T-tubule structure in vertebrate ventricle. (A) Image of four T-tubules in cat cardiac muscle (32,000 ×). (B) Image of two T-tubules in cat papillary muscle. Note coupling between SR and T-tubules (arrows, 40,000 ×). (C and D) Image and 3D constructions from Song and colleagues of T-tubule network in ventricular myocyte stained with Di-8 ANEPPS. (E) Transverse diagram of ventricular cardiomyocyte T-tubule network. *Reprinted from Guo, Zhang, Wei, Chen, and Song (2013) with permission from Oxford Press.* (See Color Plate.)

with a dilated cardiomyopathy phenotype. Moreover, in nonmyocyte cell lines, amphiphysin-2/BIN1 is responsible for trafficking L-type Ca^{2+} channels (LTCCs) (Hong et al., 2010). The confirmation of these findings *in vivo* or in primary cardiomyocytes will be critical to confirm the role of amphiphysin-2/BIN1 in LTCC targeting to the T-tubule. Detubulation that occurs in diseased cardiac myocytes results in the loss or uncoupling of LTCCs, suggesting that amphiphysin-2/BIN1 expression may play a role in T-tubule structure and protein localization (Hong et al., 2010).

Junctophilin-2 also plays a role in promoting junction formation between the sarcoplasmic membrane and the T-tubule (Takeshima, Komazaki, Nishi, Iino, & Kangawa, 2000), a site also referred to as the cardiac "dyad." Junctophilin-2 is downregulated during the progression from hypertrophy to heart failure, possibly mediating the uncoupling observed between the

T-tubule and the sarcoplasmic reticulum (SR; Song et al., 2006). Knockdown of junctophilin-2 suggests that its deficiency is sufficient to disrupt dyadic structure and alter calcium-induced calcium release, which is associated with dilated cardiomyopathy and increased mortality (van Oort et al., 2011). These changes may be due to the function of junctophilin in membrane binding or perhaps due to a possible direct association with the cardiac ryanodine receptor (RyR_2), a calcium-release channel necessary for mediating calcium-release events, termed "sparks," from the SR.

Telethonin (*Tcap*) is a load-dependent regulator of the T-tubule network (Hayashi et al., 2004; Knoll et al., 2002). Loss of Tcap in a zebrafish model produces a form of muscular dystrophy, suggesting a role in force production (Zhang, Yang, Zhu, & Xu, 2009). Interestingly, this protein is stretch-sensitive and its expression is increased by stretch (Zhang et al., 2009). Additionally, Tcap is thought to promote T-tubule formation in response to stretch (Zhang et al., 2009). Tcap mutations can result in dilated cardiomyopathy (by reducing stretch sensitivity) or hypertrophic cardiomyopathy (by increasing stretch sensitivity) (Hayashi et al., 2004). In patients, it was observed that dilated cardiomyopathy-associated Tcap mutations disrupted the association of Tcap with its binding partners in the stretch-signaling complex (Hayashi et al., 2004), while hypertrophic cardiomyopathy-associated Tcap mutations promoted interactions with its binding partners (Hayashi et al., 2004).

Caveolae are vesicular invaginations that participate in signal transduction and vesicular transport. Caveolae, unlike most phospholipid-containing plasma membrane regions, are mainly composed of cholesterol–sphingolipid-rich domains (Smart et al., 1999). The primary protein component of caveolae is the cholesterol-binding caveolin (Smart et al., 1999). *Caveolin-3* is found primarily in striated muscle, where it localizes to the sarcolemma in a complex with dystrophin and associated glycoproteins and the T-tubules (Parton, Way, Zorzi, & Stang, 1997). Several muscle disorders are associated with mutations in caveolin-3. Specifically, mutations resulting in a decreased expression of caveolin-3 are linked with limb-girdle muscular dystrophy, an autosomal-dominant disease characterized by mild to moderate proximal muscle weakness (Minetti et al., 1998). Additional human variants have been linked with rippling muscle disease, familial hypertrophic cardiomyopathy (Hayashi et al., 2004), and long QT syndrome 9 (Vatta et al., 2006). Mouse models of caveolin-3 deficiency have revealed a mild myopathic phenotype similar to human pathologies (Galbiati et al., 2001). Reduced caveolin-3 expression in the T-tubule leads to T-tubule disorganization with dilation and loss of transverse orientation, suggesting a pivotal role for caveolin-3 in the biogenesis of T-tubules (Galbiati et al., 2001). Notably, caveolin-3 is present not only in cardiac T-tubules but also within the peripheral myocyte sarcolemmal membrane structure.

In addition to amphiphysin-2, junctophilin-2, caveolin, and Tcap, there are a number of other proteins associated with the biogenesis and regulation of the T-tubule network. *Myotubularin* is a phosphoinositide phosphatase involved in the biogenesis and maintenance of muscle structure and membrane homeostasis, where mutation results in the rare congenital disease, X-linked myotubular myopathy (Al-Qusairi et al., 2013). Additionally, overexpression of myotubularin results in the accumulation of membrane saccules. Myotubularin may also play a role in modulating LTCC and ryanodine receptor expression and function (Al-Qusairi et al., 2009). *Tropomyosin* is an actin-binding protein that plays a necessary role in the stabilization of muscle fibers and linking myofibrillar networks to the membrane systems. Ablation of cytoskeletal tropomyosin (Tm5NM1) disrupts T-tubules, suggesting that myofilaments may contribute to the maintenance of the T-tubule system. *Mitsugumin 29* (MG29), a member of the synaptophysin family, is located between the cell surface T-tubule and the SR where it is proposed to play a role in the proper formation and refinement of junctional complexes between the sarcolemma and intracellular membranes (Takeshima et al., 1998). MG29 knockout mice exhibit swollen T-tubules, irregular sarcoplasmic reticular structures, and abnormal formation of the triad muscle, resulting in a limited myopathy. *Dysferlin* is a transmembrane protein involved in calcium binding and calcium-dependent membrane fusion and repair. Loss-of-function human variants in dysferlin may be associated with limb-girdle muscular dystrophy type 2B, Miyoshi myopathy, and distal anterior compartment myopathy. While dysferlin has a sarcolemmal localization in the skeletal muscle, during myotube differentiation, dysferlin is associated to the T-tubule system and can migrate to the sarcolemma in response to myofiber injury. In fact, regeneration studies in rat have demonstrated that dysferlin is localized to the T-tubules during the early stage of myofiber regeneration but translocates to the sarcolemma in the later stages. Mice deficient in dysferlin exhibit alterations in T-tubule structure, with dilated and longitudinally oriented tubules. While the exact role of dysferlin in T-tubule biogenesis is not completely understood, it is proposed that dysferlin assists in the fusion of caveolae with the T-tubules. This hypothesis is based on evidence demonstrating that dysferlin associates with the LTCC and caveolin-3, that there is evidence of colocalization of caveolin-3 and dysferlin during early myogenesis, and on the observed accumulation of subsarcolemmal vacuoles connecting with the T-tubule network in patients with dysferlinopathies.

2.2. T-tubule function

T-tubules are a specialized membrane domain that facilitates transmembrane ion flux, providing an electric pathway from the surface of the cardiac myocyte to the cell interior. For many years, however, it was assumed that T-tubules were simply invaginations of the sarcolemma that allowed for the propagation of the action potential to the cell interior. Now, it appears that T-tubules contain many of the key proteins involved in excitation–contraction coupling, specifically specializing in Ca^{2+} handling. Detubulation protocols have allowed quantification of the distribution of membrane currents between the T-tubules and the surface membrane (Despa, Brette, Orchard, & Bers, 2003). Currents involved in Ca^{2+} handling are predominantly located within the T-tubule network. This is an important finding because it places Ca^{2+} handling within a domain that has restricted diffusion access to the bulk of extracellular space. It appears that an intracellular-restricted diffusion domain exists under the sarcolemma (Lederer, Niggli, & Hadley, 1990), allowing for a difference in the concentrations between the bulk intracellular space and the intracellular face of ion flux pathways. Computer modeling suggests that this may allow for Ca^{2+} depletion during activity, limiting Ca^{2+} influx.

2.3. Transverse-tubule components

The function of the T-tubules is dependent not only on their structure but also on the proteins within and adjacent to the T-tubule membrane. Many important proteins are present in the T-tubule membrane, including Ca^{2+} channels, Na^{+} channels, K^{+} channels, and anion channels (Fig. 4.5). Additionally, key proteins involved in the regulation of these channels have also been identified, including various receptors, second messenger-related proteins, and scaffolding complexes.

2.3.1 Ca^{2+}-handling proteins

Cardiac Ca^{2+}-handling proteins in the sarcolemma are important because they play a vital role in excitation–contraction coupling and reside close to Ca^{2+} release channels in the SR (specifically, the ryanodine receptor), forming the cardiac dyad (Carl et al., 1995). Ca^{2+} influx via the *LTCC* triggers Ca^{2+} release from the SR. This released Ca^{2+} predominantly activates

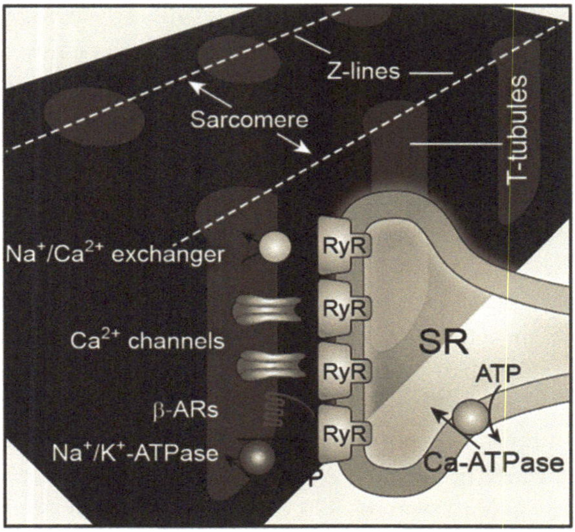

Figure 4.5 Model of molecular membrane components of T-tubule/SR junction in ventricular cardiomyocytes. Note close spatial organization/orientation of T-tubule L-type calcium channels with SR ryanodine receptors. *Reprinted from Guo et al. (2013) with permission from Oxford Press.* (See Color Plate.)

the contractile proteins and produces contraction in the ventricular myocyte. The Ca^{2+} ions entering the cardiac myocytes via the LTCC $Ca_v1.2$ also shape the plateau phase of the action potential. In both rabbit and rat heart, it was demonstrated that the T-tubule system displayed the greatest density of LTCCs versus the rest of the sarcolemma. As such, the LTCC current, I_{Ca}, is also found to be highly concentrated in the T-tubules (Carl et al., 1995).

In the mid-1990s, a severe form of cardiac arrhythmia termed long QT syndrome was identified in young children with a specific phenotype, syndactyly. After a decade of research, the mutations were linked to disruption of the $Ca_v1.2$ channel (*CACNA1C*) and referred to as Timothy syndrome. Timothy syndrome is a rare autosomal-dominant inherited form of long QT syndrome (LQT-8) with less than twenty patients identified worldwide to date (Bidaud & Lory, 2011). It is a multiorgan disorder where the average survival is 2–3 years (Bidaud & Lory, 2011). While patients exhibit a number of symptoms (dysmorphic facial features, syndactyly, developmental delays, and immune deficiencies), cardiac defects in Timothy syndrome are severe and include ventricular fibrillation and hypertrophic or dilated cardiomyopathy (Splawski et al., 2004).

Likely due to its phenotypic severity, only a limited number of Timothy syndrome-associated missense mutations in exons 8 and 8A of *CACNA1C* have been described (Splawski et al., 2004). $Ca_v1.2$ G406R has been found in exon 8 and 8A, while G402S has only been observed in exon 8. Mutations in exon 8 are proposed to lead to a more severe form of the disease (Timothy syndrome 2) than mutations in exon 8A (Timothy syndrome 1) due to the broad expression of the gene splice variant (Splawski et al., 2004). Mechanistically, these gene variants produce a gain of function with an impaired voltage-dependent inactivation of the channel. Ultimately, this phenotype produces a sustained plateau phase of the action potential and prolonged repolarization rates. Although Timothy syndrome results from *de novo* mutations, germline mosaicism has also been observed (Splawski et al., 2004).

While Timothy syndrome mutations can present as gain-of-function mutations of $Ca_v1.2$, $Ca_v1.2$ loss-of-function variants have been linked to the Brugada syndrome (syndrome typically associated with voltage-gated Na+ channel dysfunction). Brugada syndrome is an inherited (autosomal-dominant) cardiac arrhythmia associated with a high risk of ventricular fibrillation and sudden death (Antzelevitch et al., 2005). It is proposed that dysfunction results from an increased dispersion of the repolarization enabling local reexcitation and/or right ventricular conduction delay in the outflow tract. Two missense mutations (A39V and G490R) in *CACNA1C* have been identified in patients with an atypical Brugada syndrome (Brugada syndrome 3) (Antzelevitch et al., 2007). In addition to the ST-segment elevations in the right precordial leads, these patients also demonstrated a shorter than normal QTc, corresponding to a shortened action potential (a subgroup of the short QT syndrome, SQT4) (Antzelevitch et al., 2007). This suggests that loss-of-function mutations in *CACNA1C* lead to a phenotype composed of Brugada syndrome and short QT syndrome.

Following myocyte contraction, cytosolic calcium is rapidly sequestered through the dual activities of the SR Ca^{+2}-ATPase (SERCA2) and the T-tubule *Na^+/Ca^{2+} exchanger* (NCX1). Genetic knockout or overexpression of NCX1 in mouse models has allowed for a more precise examination of the role of NCX1 in cardiac function and arrhythmogenesis (Pott et al., 2012). Ventricular loss of NCX1 results in dramatic adaptations in the myocyte. Specifically, Ca^{2+} influx is decreased by 80% due to downregulation of the LTCC current and shortening of the action potential (Pott, Philipson, & Goldhaber, 2005). In the absence of NCX1, Ca^{2+} likely accumulates in the dyadic cleft and directly inactivates LTCCs, creating a feedback mechanism to limit the degree of Ca^{2+} influx (Pott, Henderson, Goldhaber, & Philipson, 2007). Single-cell studies using cardiac myocytes in NCX1-overexpressing mice revealed an increased vulnerability to delayed afterdepolarizations and succeeding ventricular tachycardia (Pott et al., 2012). Early afterdepolarizations and delayed afterdepolarizations can be elicited by spontaneous Ca^{2+} release for the SR in the presence of calcium overload (Pott et al., 2012).

While no human disease variants in the NCX1 gene have been identified, variants in genes that encode auxiliary proteins responsible for NCX1 regulation have been identified that lead to abnormal NCX1 activity. For example, ankyrin-B, a member of the larger ankyrin family of adapter proteins, is necessary for NCX1 T-tubule membrane targeting in the heart (Cunha, Bhasin, & Mohler, 2007; Mohler, 2006). Ventricular cardiomyocytes from ankyrin-B-deficient mice display significant loss of T-tubule NCX1 and Na/K ATPase (also a resident T-tubule protein). Moreover, ankyrin-$B^{+/-}$ myocytes exhibit increased SR Ca^{2+} load and Ca^{2+} transient amplitudes. While stable at rest, these myocytes display afterdepolarizations in response to catecholaminergic stimulation (Le Scouarnec et al., 2008), with ankyrin-$B^{+/-}$ mice displaying polymorphic arrhythmia and sudden death in response to catecholaminergic stimulation. A correlative phenotype identified in humans is described later in this chapter.

2.3.2 Na^+ regulatory proteins

In addition to the NCX1, other T-tubule proteins allow for the flux of Na^+ across the sarcolemma. The Na^+/H^+ exchanger that regulates intracellular pH by extruding H^+ from the myocyte is concentrated at the T-tubules and intercalated discs. The Na^+/K^+ ATPase (NKA) as noted earlier has also been shown to localize at the T-tubules, where it functions to maintain the electrochemical membrane gradient.

2.3.3 K^+ channels

A number of published works have evaluated the distribution of myocyte K^+ channels. Most results have been limited by low resolution; however, $K_v4.2$, the channeling underlying the transient outward current (I_{to}), has been shown to be localized predominantly to the T-tubule system, and, more recently, Kir6.2 has been linked with T-tubule localization (Alekseev et al., 2010; Flagg et al., 2008; Morrissey et al., 2005). Likewise, TASK-1, which is thought to mediate the steady-state outward current (I_{ss}), and Kir2.1, which underlies the inward rectifier current (I_{K1}), have also been observed to be localized to the T-tubule network.

2.3.4 Ankyrin-B

Ankyrin-B, encoded by *ANK2*, is localized to both the M-line and Z-line in adult ventricular cardiomyocytes (Abdi, Mohler, Davis, & Bennett, 2006; Mohler, Gramolini, & Bennett, 2002b). The importance of ankyrin-B in heart physiology is highlighted by a number of cardiac phenotypes associated with ankyrin-B dysfunction. Type 4 long QT syndrome (LQT4) was originally described in a large French kindred who displayed prolonged QTc. However, individuals harboring the human ANK2 loss of function variants display a number of cardiac phenotypes including severe sinus node dysfunction, atrial fibrillation, notched biphasic T-wave morphology, and syncope (Mohler et al., 2003; Schott et al., 1995) leading to a terminology shift to "ankyrin-B syndrome." (Cunha et al., 2011; Le Scouarnec et al., 2008; Mohler et al., 2007; Mohler, Splawski, et al., 2004) As noted earlier, at the molecular level, loss of ankyrin-B function results in a loss of Na/Ca exchanger and Na/K ATPase expression and localization (Cunha et al., 2007; Mohler, Yoon, & Bennett, 2004). These molecular phenotypes result in elevated cytosolic Na^{+}, mimicking the action of digitalis, and ultimately lead to Ca^{2+} overload and increased likelihood for catecholamine-induced afterdepolarizations. At the population level, *ANK2* gene variants have been linked with alterations in cardiac electric activity (Sedlacek et al., 2008).

Studies of ankyrin-B function in the myocyte have revealed the role for a new family of trafficking proteins in the human heart. Members of the family of EH (Eps15 homology) domain-containing (EHD) proteins regulate endosomal anterograde and retrograde trafficking and membrane protein recycling and lipid homeostasis (Caplan et al., 2002; Daumke et al., 2007; Rapaport et al., 2006). EHD1–4 proteins directly associate with ankyrin in the perinuclear region with increased EHD expression in ankyrin-B-deficient hearts (Gudmundsson et al., 2010). Additionally, myocytes deficient in EHD3 have a significant loss of NCX membrane trafficking and function (Gudmundsson et al., 2010). Moreover, the link between heart disease and EHD has revealed that EHD3 levels are consistently elevated in heart failure, along with a concomitant increase in Na/Ca exchanger expression (Gudmundsson et al., 2012). These findings not only identified a new class of cardiac membrane trafficking proteins but also identified EHD3 as a component of the cardiac remodeling pathway.

2.3.5 Second messenger pathways

The β-adrenergic system modulates the function of a number of key Ca^{2+} channels (via phospholamban) and contractile proteins (via troponin I). As such, a number of key proteins in this signaling pathway are localized at the T-tubule (stimulatory G-protein, G_s, and A-kinase anchoring protein, AKAP) (Laflamme, Domingue, Guillemette, & Guillemette, 2002). Studies have suggested that the β_2-adrenergic receptor is tightly linked with the modulation of T-tubule LTCCs, since β-adrenergic stimulation results in a greater increase in Ca^{2+} current in normal cells versus detubulated cells. There is evidence that β_2-adrenergic stimulation can produce an increase in I_{Ca} without phosphorylation of proteins, such as phospholamban, suggesting that there is a close association of the β_2-adrenergic signaling pathway and the LTCC. Studies have shown that this β_2-adrenergic receptor/G-protein macromolecular complex may also play a structural role. A subpopulation of LTCCs resides in the caveolae associated with the T-tubule network (Balijepalli, Foell, Hall, Hell, & Kamp, 2006; Davare et al., 2001). It has been suggested that these complexes are able to process small changes in Ca^{2+} signals that affect hypertrophic responses and contractility. Unfortunately, there are conflicting reports as to the effect of caveolae disruption on β_2-adrenergic-mediated LTCC stimulation. While some studies suggest that caveolae are necessary (Balijepalli et al., 2006), there are other studies that suggest they are not. Nonetheless, it remains clear that caveolae-based signaling mechanisms modulate Ca^{2+} signaling in cardiac myocytes.

2.4. T-tubules, physiology, and disease

T-tubule structure and function is severely compromised in human heart failure and in small and large animal models of cardiovascular disease (Balijepalli et al., 2003; Louch et al., 2004). Animal models of heart failure exhibit T-tubule disorganization that appears to result from various myocardial insults, including sustained tachycardia (Balijepalli et al., 2003), spontaneous hypertension (Gomez et al., 1997), and myocardial infarction (Heinzel et al., 2008). The application of laser damage along with fluorescent lipophilic membrane markers in a canine model of tachycardia-induced heart failure revealed a marked loss of T-tubules in ventricular myocytes, although the overall organized pattern of the T-tubule network was unaffected (Balijepalli et al., 2003). In a spontaneously hypertensive rat model, distinct changes in the T-tubule system were reported, specifically a dramatic reorganization of the T-tubule system (Song et al., 2006). The

T-tubule system in this model was characterized by a loss of transverse elements but a gain in longitudinal elements, producing a T-tubule network with a chaotic appearance (Song et al., 2006). Notably, in this same study, the organization of the SR as noted by RyR_2 distribution was unaffected. Disruption or loss of the T-tubule network is also observed in human cardiac myocytes from patients with heart failure (Lyon et al., 2009). Patients with hypertrophic cardiomyopathy, dilated cardiomyopathy, and ischemic heart disease all exhibit profound T-tubule abnormalities, including reduction in T-tubule density and cell surface regularity (Lyon et al., 2009). Additionally, human failing myocytes have T-tubules that lie on the longitudinal axis and are more often dilated and bifurcated. As a whole, these studies have demonstrated that T-tubule alteration is a common process in failing myocytes, including different animal heart failure model of different species and in human heart failure patients with different etiologies. Interestingly, in all models, T-tubule changes occur early in the transition from hypertrophy to heart failure, suggesting they are an early occurrence in the pathogenic process.

Surprisingly, the majority of reports demonstrate no change in the density of I_{Ca} during hypertrophy and failure (Benitah et al., 2002). This can be due to a number of factors, including upregulation of the remaining channels, consistent with studies showing an increase in the activity of single channels from failing myocytes. This upregulation may result from channel phosphorylation, which may partly explain the blunted response of failing hearts to β-adrenergic stimulation. The ability of I_{Ca} to trigger CICR from the SR is reduced in myocytes from failing rat hearts (Gomez et al., 1997). Possible changes in the colocalization of LTCC with RyR_2, increased physical separation of the T-tubule from the SR, and T-tubule remodeling processes have been suggested as underlying causes for the lack of LTCC/RyR2 integration. A porcine postischemic cardiomyopathy model suggests that T-tubule dysfunction in the failing heart is related to contractile abnormalities (Heinzel et al., 2008). In this model, impaired contractility was associated with reduced Ca^{2+} release synchronicity, a longer time to reach peak Ca^{2+} release, and a lower peak concentration of Ca^{2+} (Heinzel et al., 2008). Furthermore, the authors observed a significant reduction in T-tubule density, while the LTCC current and SR Ca^{2+} content remained unaffected (Heinzel et al., 2008). From their data, the authors proposed that the dysfunction was due to a gain in the CICR process, implicating T-tubule structural disorganization in the pathophysiology of the failing myocyte. Follow-up studies also demonstrated that failing myocytes have increased Ca^{2+} spark frequency, consistent with an uncoupling of the Ca^{2+} release machinery and disruption of the T-tubules. It has also been argued that such a mechanism would produce only transient/short-lived changes in the size of the Ca^{2+} transient and that it is more likely that a decrease in SR Ca^{2+} content underlies the decrease in Ca^{2+} transients. However, a reduction in T-tubule density could desynchronize Ca^{2+} following electric stimulation, reducing peak Ca^{2+} and slowing the time course. Although the loss of T-tubules cannot be completely responsible for the changes in Ca^{2+} handling observed in heart failure, it is important to note that many of the functional changes exhibited in heart failure are observed in detubulation experiments, suggesting that the loss of T-tubules in heart failure may contribute to or exacerbate the phenotype observed in models of heart failure.

3. CONCLUDING REMARKS

Intercalated disc and T-tubule proteins have evolved to tightly synchronize the activities of excitation–contraction coupling and conduction in the vertebrate cardiomyocyte. In fact, as noted in this chapter, defects in either membrane domain are now clearly implicated in aberrant vertebrate cardiovascular physiology and potentially fatal acquired and congenital human disease. However, while the past decade has witnessed major breakthroughs in understanding the cell biology of these membrane domains, major questions still remain due to the difficulty of studying these structures in their physiologic milieu. For example, experiments in heterologous cells or even neonatal myocytes have been unsuccessful in generating great insight into mature intercalated disc or T-tubule function in myocytes—in part because many of the proteins and lipids clearly essential for disc or T-tubule biogenesis and maintenance simply are not present in these cell types. Clearly new cell models, potentially differentiated human iPS cells may offer new cell biological tools to understand these amazingly complex and dynamic structures.

ACKNOWLEDGMENTS

Supported by the NIH [HL084583, HL083422]; Saving Tiny Hearts Society, Fondation LeDucq; and American Heart Association.

REFERENCES

Abdi, K. M., Mohler, P. J., Davis, J. Q., & Bennett, V. (2006). Isoform Specificity of Ankyrin-B: A site in the divergent C-terminal domain is required for intramolecular association. *The Journal of Biological Chemistry, 281*(9), 5741–5749.

Alcalai, R., Metzger, S., Rosenheck, S., Meiner, V., & Chajek-Shaul, T. (2003). A recessive mutation in desmoplakin causes arrhythmogenic right ventricular dysplasia, skin disorder, and woolly hair. *Journal of the American College of Cardiology, 42*(2), 319–327.

Alekseev, A. E., Reyes, S., Yamada, S., Hodgson-Zingman, D. M., Sattiraju, S., Zhu, Z., et al. (2010). Sarcolemmal ATP-sensitive K(+) channels control energy expenditure determining body weight. *Cell Metabolism, 11*(1), 58–69.

Al-Qusairi, L., Prokic, I., Amoasii, L., Kretz, C., Messaddeq, N., Mandel, J. L., et al. (2013). Lack of myotubularin (MTM1) leads to muscle hypotrophy through unbalanced regulation of the autophagy and ubiquitin-proteasome pathways. *The FASEB Journal, 27*(8), 3384–3394.

Al-Qusairi, L., Weiss, N., Toussaint, A., Berbey, C., Messaddeq, N., Kretz, C., et al. (2009). T-tubule disorganization and defective excitation-contraction coupling in muscle fibers lacking myotubularin lipid phosphatase. *Proceedings of the National Academy of Sciences of the United States of America, 106*(44), 18763–18768.

Antzelevitch, C., Brugada, P., Borggrefe, M., Brugada, J., Brugada, R., Corrado, D., et al. (2005). Brugada syndrome: Report of the second consensus conference: Endorsed by the Heart Rhythm Society and the European Heart Rhythm Association. *Circulation, 111*(5), 659–670.

Antzelevitch, C., Pollevick, G. D., Cordeiro, J. M., Casis, O., Sanguinetti, M. C., Aizawa, Y., et al. (2007). Loss-of-function mutations in the cardiac calcium channel underlie a new clinical entity characterized by ST-segment elevation, short QT intervals, and sudden cardiac death. *Circulation, 115*(4), 442–449.

Asimaki, A., Syrris, P., Wichter, T., Matthias, P., Saffitz, J. E., & McKenna, W. J. (2007). A novel dominant mutation in plakoglobin causes arrhythmogenic right ventricular cardiomyopathy. *American Journal of Human Genetics, 81*(5), 964–973.

Awad, M. M., Calkins, H., & Judge, D. P. (2008). Mechanisms of disease: Molecular genetics of arrhythmogenic right ventricular dysplasia/cardiomyopathy. *Nature Clinical Practice, 5*(5), 258–267.

Awad, M. M., Dalal, D., Cho, E., Amat-Alarcon, N., James, C., Tichnell, C., et al. (2006). DSG2 mutations contribute to arrhythmogenic right ventricular dysplasia/cardiomyopathy. *American Journal of Human Genetics, 79*(1), 136–142.

Ayettey, A. S., & Navaratnam, V. (1978). The T-tubule system in the specialized and general myocardium of the rat. *Journal of Anatomy, 127*(Pt 1), 125–140.

Balijepalli, R. C., Foell, J. D., Hall, D. D., Hell, J. W., & Kamp, T. J. (2006). Localization of cardiac L-type Ca(2+) channels to a caveolar macromolecular signaling complex is required for beta(2)-adrenergic regulation. *Proceedings of the National Academy of Sciences of the United States of America, 103*(19), 7500–7505.

Balijepalli, R. C., Lokuta, A. J., Maertz, N. A., Buck, J. M., Haworth, R. A., Valdivia, H. H., et al. (2003). Depletion of T-tubules and specific subcellular changes in sarcolemmal proteins in tachycardia-induced heart failure. *Cardiovascular Research, 59*(1), 67–77.

Barker, R. J., Price, R. L., & Gourdie, R. G. (2002). Increased association of ZO-1 with connexin43 during remodeling of cardiac gap junctions. *Circulation Research, 90*(3), 317–324.

Benitah, J. P., Gomez, A. M., Fauconnier, J., Kerfant, B. G., Perrier, E., Vassort, G., et al. (2002). Voltage-gated Ca2+ currents in the human pathophysiologic heart: A review. *Basic Research in Cardiology, 97*(Suppl. 1), I11–I18.

Bennett, V., & Baines, A. J. (2001). Spectrin and ankyrin-based pathways: Metazoan inventions for integrating cells into tissues. *Physiological Reviews, 81*(3), 1353–1392.

Bennett, V., & Healy, J. (2008). Organizing the fluid membrane bilayer: Diseases linked to spectrin and ankyrin. *Trends in Molecular Medicine, 14*(1), 28–36.

Bennett, V., & Healy, J. (2009). Membrane domains based on ankyrin and spectrin associated with cell-cell interactions. *Cold Spring Harbor Perspectives in Biology, 1*(6), a003012.

Bidaud, I., & Lory, P. (2011). Hallmarks of the channelopathies associated with L-type calcium channels: A focus on the Timothy mutations in Ca(v)1.2 channels. *Biochimie, 93*(12), 2080–2086.

Borrmann, C. M., Grund, C., Kuhn, C., Hofmann, I., Pieperhoff, S., & Franke, W. W. (2006). The area composita of adhering junctions connecting heart muscle cells of vertebrates. II. Colocalizations of desmosomal and fascia adhaerens molecules in the intercalated disk. *European Journal of Cell Biology, 85*(6), 469–485.

Caplan, S., Naslavsky, N., Hartnell, L. M., Lodge, R., Polishchuk, R. S., Donaldson, J. G., et al. (2002). A tubular EHD1-containing compartment involved in the recycling of major histocompatibility complex class I molecules to the plasma membrane. *The EMBO Journal, 21*(11), 2557–2567.

Carl, S. L., Felix, K., Caswell, A. H., Brandt, N. R., Ball, W. J., Jr., Vaghy, P. L., et al. (1995). Immunolocalization of sarcolemmal dihydropyridine receptor and sarcoplasmic reticular triadin and ryanodine receptor in rabbit ventricle and atrium. *The Journal of Cell Biology, 129*(3), 672–682.

Chen, X., Shevtsov, S. P., Hsich, E., Cui, L., Haq, S., Aronovitz, M., et al. (2006). The beta-catenin/T-cell factor/lymphocyte enhancer factor signaling pathway is required for normal and stress-induced cardiac hypertrophy. *Molecular and Cellular Biology, 26*(12), 4462–4473.

Cunha, S. R., Bhasin, N., & Mohler, P. J. (2007). Targeting and stability of Na/Ca exchanger 1 in cardiomyocytes requires direct interaction with the membrane adaptor ankyrin-B. *The Journal of Biological Chemistry, 282*(7), 4875–4883.

Cunha, S. R., Hund, T. J., Hashemi, S., Voigt, N., Li, N., Wright, P., et al. (2011). Defects in ankyrin-based membrane protein targeting pathways underlie atrial fibrillation. *Circulation, 124*(11), 1212–1222.

Daumke, O., Lundmark, R., Vallis, Y., Martens, S., Butler, P. J., & McMahon, H. T. (2007). Architectural and mechanistic insights into an EHD ATPase involved in membrane remodelling. *Nature, 449*(7164), 923–927.

Davare, M. A., Avdonin, V., Hall, D. D., Peden, E. M., Burette, A., Weinberg, R. J., et al. (2001). A beta2 adrenergic receptor signaling complex assembled with the Ca^{2+} channel Cav1.2. *Science, 293*(5527), 98–101.

Despa, S., Brette, F., Orchard, C. H., & Bers, D. M. (2003). Na/Ca exchange and Na/K-ATPase function are equally concentrated in transverse tubules of rat ventricular myocytes. *Biophysical Journal, 85*(5), 3388–3396.

Ellinor, P. T., MacRae, C. A., & Thierfelder, L. (2010). Arrhythmogenic right ventricular cardiomyopathy. *Heart Failure Clinics, 6*(2), 161–177.

Ferreira-Cornwell, M. C., Luo, Y., Narula, N., Lenox, J. M., Lieberman, M., & Radice, G. L. (2002). Remodeling the intercalated disc leads to cardiomyopathy in mice misexpressing cadherins in the heart. *Journal of Cell Science, 115*(Pt 8), 1623–1634.

Flagg, T. P., Kurata, H. T., Masia, R., Caputa, G., Magnuson, M. A., Lefer, D. J., et al. (2008). Differential structure of atrial and ventricular KATP: Atrial KATP channels require SUR1. *Circulation Research, 103*(12), 1458–1465.

Forbes, M. S., Hawkey, L. A., & Sperelakis, N. (1984). The transverse-axial tubular system (TATS) of mouse myocardium: Its morphology in the developing and adult animal. *The American Journal of Anatomy, 170*(2), 143–162.

Franke, W. W., Borrmann, C. M., Grund, C., & Pieperhoff, S. (2006). The area composita of adhering junctions connecting heart muscle cells of vertebrates. I. Molecular definition in intercalated disks of cardiomyocytes by immunoelectron microscopy of desmosomal proteins. *European Journal of Cell Biology, 85*(2), 69–82.

Fujio, Y., Yamada-Honda, F., Sato, N., Funai, H., Wada, A., Awata, N., et al. (1995). Disruption of cell-cell adhesion in an inbred strain of hereditary cardiomyopathic hamster (Bio 14.6). *Cardiovascular Research, 30*(6), 899–904.

Galbiati, F., Engelman, J. A., Volonte, D., Zhang, X. L., Minetti, C., Li, M., et al. (2001). Caveolin-3 null mice show a loss of caveolae, changes in the microdomain distribution of the dystrophin-glycoprotein complex, and t-tubule abnormalities. *The Journal of Biological Chemistry, 276*(24), 21425–21433.
Gallicano, G. I., Kouklis, P., Bauer, C., Yin, M., Vasioukhin, V., Degenstein, L., et al. (1998). Desmoplakin is required early in development for assembly of desmosomes and cytoskeletal linkage. *The Journal of Cell Biology, 143*(7), 2009–2022.
Garcia-Gras, E., Lombardi, R., Giocondo, M. J., Willerson, J. T., Schneider, M. D., Khoury, D. S., et al. (2006). Suppression of canonical Wnt/beta-catenin signaling by nuclear plakoglobin recapitulates phenotype of arrhythmogenic right ventricular cardiomyopathy. *The Journal of Clinical Investigation, 116*(7), 2012–2021.
Gomez, A. M., Valdivia, H. H., Cheng, H., Lederer, M. R., Santana, L. F., Cannell, M. B., et al. (1997). Defective excitation-contraction coupling in experimental cardiac hypertrophy and heart failure. *Science, 276*(5313), 800–806.
Gudmundsson, H., Curran, J., Kashef, F., Snyder, J. S., Smith, S. A., Vargas-Pinto, P., et al. (2012). Differential regulation of EHD3 in human and mammalian heart failure. *Journal of Molecular and Cellular Cardiology, 52*(5), 1183–1190.
Gudmundsson, H., Hund, T. J., Wright, P. J., Kline, C. F., Snyder, J. S., Qian, L., et al. (2010). EH domain proteins regulate cardiac membrane protein targeting. *Circulation Research, 107*(1), 84–95.
Guo, A., Zhang, C., Wei, S., Chen, B., & Song, L. S. (2013). Emerging mechanisms of T-tubule remodelling in heart failure. *Cardiovascular Research, 98*(2), 204–215.
Gutstein, D. E., Liu, F. Y., Meyers, M. B., Choo, A., & Fishman, G. I. (2003). The organization of adherens junctions and desmosomes at the cardiac intercalated disc is independent of gap junctions. *Journal of Cell Science, 116*(Pt 5), 875–885.
Haddock, P. S., Coetzee, W. A., Cho, E., Porter, L., Katoh, H., Bers, D. M., et al. (1999). Subcellular $[Ca2+]i$ gradients during excitation-contraction coupling in newborn rabbit ventricular myocytes. *Circulation Research, 85*(5), 415–427.
Hayashi, T., Arimura, T., Itoh-Satoh, M., Ueda, K., Hohda, S., Inagaki, N., et al. (2004). Tcap gene mutations in hypertrophic cardiomyopathy and dilated cardiomyopathy. *Journal of the American College of Cardiology, 44*(11), 2192–2201.
Heinzel, F. R., Bito, V., Biesmans, L., Wu, M., Detre, E., von Wegner, F., et al. (2008). Remodeling of T-tubules and reduced synchrony of $Ca2+$ release in myocytes from chronically ischemic myocardium. *Circulation Research, 102*(3), 338–346.
Hermida, J. S., Minassian, A., Jarry, G., Delonca, J., Rey, J. L., Quiret, J. C., et al. (1997). Familial incidence of late ventricular potentials and electrocardiographic abnormalities in arrhythmogenic right ventricular dysplasia. *The American Journal of Cardiology, 79*(10), 1375–1380.
Heuser, A., Plovie, E. R., Ellinor, P. T., Grossmann, K. S., Shin, J. T., Wichter, T., et al. (2006). Mutant desmocollin-2 causes arrhythmogenic right ventricular cardiomyopathy. *American Journal of Human Genetics, 79*(6), 1081–1088.
Hong, T. T., Smyth, J. W., Gao, D., Chu, K. Y., Vogan, J. M., Fong, T. S., et al. (2010). BIN1 localizes the L-type calcium channel to cardiac T-tubules. *PLoS Biology, 8*(2), e1000312.
Huang, H., Asimaki, A., Lo, D., McKenna, W., & Saffitz, J. (2008). Disparate effects of different mutations in plakoglobin on cell mechanical behavior. *Cell Motility and the Cytoskeleton, 65*(12), 964–978.
Jenkins, S. M., & Bennett, V. (2001). Ankyrin-G coordinates assembly of the spectrin-based membrane skeleton, voltage-gated sodium channels, and L1 CAMs at Purkinje neuron initial segments. *The Journal of Cell Biology, 155*(5), 739–746.
Kaplan, S. R., Gard, J. J., Carvajal-Huerta, L., Ruiz-Cabezas, J. C., Thiene, G., & Saffitz, J. E. (2004). Structural and molecular pathology of the heart in Carvajal syndrome. *Cardiovascular Pathology, 13*(1), 26–32.

Kaplan, S. R., Gard, J. J., Protonotarios, N., Tsatsopoulou, A., Spiliopoulou, C., Anastasakis, A., et al. (2004). Remodeling of myocyte gap junctions in arrhythmogenic right ventricular cardiomyopathy due to a deletion in plakoglobin (Naxos disease). *Heart Rhythm*, *1*(1), 3–11.

Kelly, S. C., Ratajczak, P., Keller, M., Purcell, S. M., Griffin, T., & Richard, G. (2006). A novel GJA 1 mutation in oculo-dento-digital dysplasia with curly hair and hyperkeratosis. *European Journal of Dermatology*, *16*(3), 241–245.

Kirchhof, P., Fabritz, L., Zwiener, M., Witt, H., Schafers, M., Zellerhoff, S., et al. (2006). Age- and training-dependent development of arrhythmogenic right ventricular cardiomyopathy in heterozygous plakoglobin-deficient mice. *Circulation*, *114*(17), 1799–1806.

Kizhatil, K., Davis, J. Q., Davis, L., Hoffman, J., Hogan, B. L., & Bennett, V. (2007). Ankyrin-G is a molecular partner of E-cadherin in epithelial cells and early embryos. *The Journal of Biological Chemistry*, *282*(36), 26552–26561.

Knoll, R., Hoshijima, M., Hoffman, H. M., Person, V., Lorenzen-Schmidt, I., Bang, M. L., et al. (2002). The cardiac mechanical stretch sensor machinery involves a Z disc complex that is defective in a subset of human dilated cardiomyopathy. *Cell*, *111*(7), 943–955.

Kostin, S., Scholz, D., Shimada, T., Maeno, Y., Mollnau, H., Hein, S., et al. (1998). The internal and external protein scaffold of the T-tubular system in cardiomyocytes. *Cell and Tissue Research*, *294*(3), 449–460.

Kowalczyk, A. P., Stappenbeck, T. S., Parry, D. A., Palka, H. L., Virata, M. L., Bornslaeger, E. A., et al. (1994). Structure and function of desmosomal transmembrane core and plaque molecules. *Biophysical Chemistry*, *50*(1–2), 97–112.

Laflamme, K., Domingue, O., Guillemette, B. I., & Guillemette, G. (2002). Immunohistochemical localization of type 2 inositol 1,4,5-trisphosphate receptor to the nucleus of different mammalian cells. *Journal of Cellular Biochemistry*, *85*(1), 219–228.

Lederer, W. J., Niggli, E., & Hadley, R. W. (1990). Sodium-calcium exchange in excitable cells: Fuzzy space. *Science*, *248*(4953), 283.

Lee, E., Marcucci, M., Daniell, L., Pypaert, M., Weisz, O. A., Ochoa, G. C., et al. (2002). Amphiphysin 2 (Bin1) and T-tubule biogenesis in muscle. *Science*, *297*(5584), 1193–1196.

Le Scouarnec, S., Bhasin, N., Vieyres, C., Hund, T. J., Cunha, S. R., Koval, O., et al. (2008). Dysfunction in ankyrin-B-dependent ion channel and transporter targeting causes human sinus node disease. *Proceedings of the National Academy of Sciences of the United States of America*, *105*(40), 15617–15622.

Li, J., Levin, M. D., Xiong, Y., Petrenko, N., Patel, V. V., & Radice, G. L. (2008). N-cadherin haploinsufficiency affects cardiac gap junctions and arrhythmic susceptibility. *Journal of Molecular and Cellular Cardiology*, *44*(3), 597–606.

Li, J., Patel, V. V., Kostetskii, I., Xiong, Y., Chu, A. F., Jacobson, J. T., et al. (2005). Cardiac-specific loss of N-cadherin leads to alteration in connexins with conduction slowing and arrhythmogenesis. *Circulation Research*, *97*(5), 474–481.

Li, J., Patel, V. V., & Radice, G. L. (2006). Dysregulation of cell adhesion proteins and cardiac arrhythmogenesis. *Clinical Medicine & Research*, *4*(1), 42–52.

Li, J., Swope, D., Raess, N., Cheng, L., Muller, E. J., & Radice, G. L. (2011). Cardiac tissue-restricted deletion of plakoglobin results in progressive cardiomyopathy and activation of {beta}-catenin signaling. *Molecular and Cellular Biology*, *31*(6), 1134–1144.

Lim, B. K., Xiong, D., Dorner, A., Youn, T. J., Yung, A., Liu, T. I., et al. (2008). Coxsackievirus and adenovirus receptor (CAR) mediates atrioventricular-node function and connexin 45 localization in the murine heart. *The Journal of Clinical Investigation*, *118*(8), 2758–2770.

Louch, W. E., Bito, V., Heinzel, F. R., Macianskiene, R., Vanhaecke, J., Flameng, W., et al. (2004). Reduced synchrony of Ca2+ release with loss of T-tubules-a comparison to Ca2+ release in human failing cardiomyocytes. *Cardiovascular Research*, *62*(1), 63–73.

Lowe, J. S., Palygin, O., Bhasin, N., Hund, T. J., Boyden, P. A., Shibata, E., et al. (2008). Voltage-gated Nav channel targeting in the heart requires an ankyrin-G dependent cellular pathway. *The Journal of Cell Biology*, *180*(1), 173–186.
Lyon, A. R., MacLeod, K. T., Zhang, Y., Garcia, E., Kanda, G. K., Lab, M. J., et al. (2009). Loss of T-tubules and other changes to surface topography in ventricular myocytes from failing human and rat heart. *Proceedings of the National Academy of Sciences of the United States of America*, *106*(16), 6854–6859.
Maeda, M., Holder, E., Lowes, B., Valent, S., & Bies, R. D. (1997). Dilated cardiomyopathy associated with deficiency of the cytoskeletal protein metavinculin. *Circulation*, *95*(1), 17–20.
Manias, J. L., Plante, I., Gong, X. Q., Shao, Q., Churko, J., Bai, D., et al. (2008). Fate of connexin43 in cardiac tissue harbouring a disease-linked connexin43 mutant. *Cardiovascular Research*, *80*(3), 385–395.
Masuelli, L., Bei, R., Sacchetti, P., Scappaticci, I., Francalanci, P., Albonici, L., et al. (2003). Beta-catenin accumulates in intercalated disks of hypertrophic cardiomyopathic hearts. *Cardiovascular Research*, *60*(2), 376–387.
Mezzano, V., & Sheikh, F. (2012). Cell-cell junction remodeling in the heart: Possible role in cardiac conduction system function and arrhythmias? *Life Sciences*, *90*(9–10), 313–321.
Minetti, C., Sotgia, F., Bruno, C., Scartezzini, P., Broda, P., Bado, M., et al. (1998). Mutations in the caveolin-3 gene cause autosomal dominant limb-girdle muscular dystrophy. *Nature Genetics*, *18*(4), 365–368.
Mitcheson, J. S., Hancox, J. C., & Levi, A. J. (1996). Action potentials, ion channel currents and transverse tubule density in adult rabbit ventricular myocytes maintained for 6 days in cell culture. *Pflügers Archiv*, *431*(6), 814–827.
Mohler, P. J. (2006). Ankyrins and human disease: What the electrophysiologist should know. *Journal of Cardiovascular Electrophysiology*, *17*(10), 1153–1159.
Mohler, P. J., Gramolini, A. O., & Bennett, V. (2002a). Ankyrins. *Journal of Cell Science*, *115*(Pt 8), 1565–1566.
Mohler, P. J., Gramolini, A. O., & Bennett, V. (2002b). The ankyrin-B C-terminal domain determines activity of ankyrin-B/G chimeras in rescue of abnormal inositol 1,4,5-trisphosphate and ryanodine receptor distribution in ankyrin-B (-/-) neonatal cardiomyocytes. *The Journal of Biological Chemistry*, *277*(12), 10599–10607.
Mohler, P. J., Le Scouarnec, S., Denjoy, I., Lowe, J. S., Guicheney, P., Caron, L., et al. (2007). Defining the cellular phenotype of "ankyrin-B syndrome" variants: Human ANK2 variants associated with clinical phenotypes display a spectrum of activities in cardiomyocytes. *Circulation*, *115*(4), 432–441.
Mohler, P. J., Rivolta, I., Napolitano, C., LeMaillet, G., Lambert, S., Priori, S. G., et al. (2004). Nav1.5 E1053K mutation causing Brugada syndrome blocks binding to ankyrin-G and expression of Nav1.5 on the surface of cardiomyocytes. *Proceedings of the National Academy of Sciences of the United States of America*, *101*(50), 17533–17538.
Mohler, P. J., Schott, J. J., Gramolini, A. O., Dilly, K. W., Guatimosim, S., duBell, W. H., et al. (2003). Ankyrin-B mutation causes type 4 long-QT cardiac arrhythmia and sudden cardiac death. *Nature*, *421*(6923), 634–639.
Mohler, P. J., Splawski, I., Napolitano, C., Bottelli, G., Sharpe, L., Timothy, K., et al. (2004). A cardiac arrhythmia syndrome caused by loss of ankyrin-B function. *Proceedings of the National Academy of Sciences of the United States of America*, *101*(24), 9137–9142.
Mohler, P. J., Yoon, W., & Bennett, V. (2004). Ankyrin-B targets beta2-spectrin to an intracellular compartment in neonatal cardiomyocytes. *The Journal of Biological Chemistry*, *279*(38), 40185–40193.
Morrissey, A., Rosner, E., Lanning, J., Parachuru, L., Dhar Chowdhury, P., Han, S., et al. (2005). Immunolocalization of KATP channel subunits in mouse and rat cardiac myocytes and the coronary vasculature. *BMC Physiology*, *5*(1), 1.

Mueller, H., & Franke, W. W. (1983). Biochemical and immunological characterization of desmoplakins I and II, the major polypeptides of the desmosomal plaque. *Journal of Molecular Biology, 163*(4), 647–671.

Norgett, E. E., Hatsell, S. J., Carvajal-Huerta, L., Cabezas, J. C., Common, J., Purkis, P. E., et al. (2000). Recessive mutation in desmoplakin disrupts desmoplakin-intermediate filament interactions and causes dilated cardiomyopathy, woolly hair and keratoderma. *Human Molecular Genetics, 9*(18), 2761–2766.

Parton, R. G., Way, M., Zorzi, N., & Stang, E. (1997). Caveolin-3 associates with developing T-tubules during muscle differentiation. *The Journal of Cell Biology, 136*(1), 137–154.

Paznekas, W. A., Boyadjiev, S. A., Shapiro, R. E., Daniels, O., Wollnik, B., Keegan, C. E., et al. (2003). Connexin 43 (GJA1) mutations cause the pleiotropic phenotype of oculodentodigital dysplasia. *American Journal of Human Genetics, 72*(2), 408–418.

Pilichou, K., Nava, A., Basso, C., Beffagna, G., Bauce, B., Lorenzon, A., et al. (2006). Mutations in desmoglein-2 gene are associated with arrhythmogenic right ventricular cardiomyopathy. *Circulation, 113*(9), 1171–1179.

Pott, C., Henderson, S. A., Goldhaber, J. I., & Philipson, K. D. (2007). Na+/Ca2+ exchanger knockout mice: Plasticity of cardiac excitation-contraction coupling. *Annals of the New York Academy of Sciences, 1099*, 270–275.

Pott, C., Muszynski, A., Ruhe, M., Bogeholz, N., Schulte, J. S., Milberg, P., et al. (2012). Proarrhythmia in a non-failing murine model of cardiac-specific Na+/Ca 2+ exchanger overexpression: Whole heart and cellular mechanisms. *Basic Research in Cardiology, 107*(2), 247.

Pott, C., Philipson, K. D., & Goldhaber, J. I. (2005). Excitation-contraction coupling in Na^+-Ca^{2+} exchanger knockout mice: Reduced transsarcolemmal Ca^{2+} flux. *Circulation Research, 97*(12), 1288–1295.

Protonotarios, N., & Tsatsopoulou, A. (2004). Naxos disease and Carvajal syndrome: cardiocutaneous disorders that highlight the pathogenesis and broaden the spectrum of arrhythmogenic right ventricular cardiomyopathy. *Cardiovascular Pathology, 13*(4), 185–194.

Protonotarios, N. I., Tsatsopoulou, A. A., & Gatzoulis, K. A. (2002). Arrhythmogenic right ventricular cardiomyopathy caused by a deletion in plakoglobin (Naxos disease). *Cardiac Electrophysiology Review, 6*(1–2), 72–80.

Protonotarios, N., Tsatsopoulou, A., Patsourakos, P., Alexopoulos, D., Gezerlis, P., Simitsis, S., et al. (1986). Cardiac abnormalities in familial palmoplantar keratosis. *British Heart Journal, 56*(4), 321–326.

Radice, G. L., Rayburn, H., Matsunami, H., Knudsen, K. A., Takeichi, M., & Hynes, R. O. (1997). Developmental defects in mouse embryos lacking N-cadherin. *Developmental Biology, 181*(1), 64–78.

Rampazzo, A., Nava, A., Malacrida, S., Beffagna, G., Bauce, B., Rossi, V., et al. (2002). Mutation in human desmoplakin domain binding to plakoglobin causes a dominant form of arrhythmogenic right ventricular cardiomyopathy. *American Journal of Human Genetics, 71*(5), 1200–1206.

Rapaport, D., Auerbach, W., Naslavsky, N., Pasmanik-Chor, M., Galperin, E., Fein, A., et al. (2006). Recycling to the plasma membrane is delayed in EHD1 knockout mice. *Traffic, 7*(1), 52–60.

Saffitz, J. E., Green, K. G., Kraft, W. J., Schechtman, K. B., & Yamada, K. A. (2000). Effects of diminished expression of connexin43 on gap junction number and size in ventricular myocardium. *American Journal of Physiology Heart and Circulatory Physiology, 278*(5), H1662–H1670.

Sato, P. Y., Coombs, W., Lin, X., Nekrasova, O., Green, K. J., Isom, L. L., et al. (2011). Interactions between ankyrin-G, Plakophilin-2, and Connexin43 at the cardiac intercalated disc. *Circulation Research, 109*(2), 193–201.

Schott, J. J., Charpentier, F., Peltier, S., Foley, P., Drouin, E., Bouhour, J. B., et al. (1995). Mapping of a gene for long QT syndrome to chromosome 4q25-27. *American Journal of Human Genetics*, *57*(5), 1114–1122.

Schroen, B., Leenders, J. J., van Erk, A., Bertrand, A. T., van Loon, M., van Leeuwen, R. E., et al. (2007). Lysosomal integral membrane protein 2 is a novel component of the cardiac intercalated disc and vital for load-induced cardiac myocyte hypertrophy. *The Journal of Experimental Medicine*, *204*(5), 1227–1235.

Sedlacek, K., Stark, K., Cunha, S. R., Pfeufer, A., Weber, S., Berger, I., et al. (2008). Common genetic variants in ANK2 modulate QT interval: Results from the KORA study. *Circulation Cardiovascular Genetics*, *1*(2), 93–99.

Severs, N. J., Bruce, A. F., Dupont, E., & Rothery, S. (2008). Remodelling of gap junctions and connexin expression in diseased myocardium. *Cardiovascular Research*, *80*(1), 9–19.

Severs, N. J., Rothery, S., Dupont, E., Coppen, S. R., Yeh, H. I., Ko, Y. S., et al. (2001). Immunocytochemical analysis of connexin expression in the healthy and diseased cardiovascular system. *Microscopy Research and Technique*, *52*(3), 301–322.

Shaw, R. M., Fay, A. J., Puthenveedu, M. A., von Zastrow, M., Jan, Y. N., & Jan, L. Y. (2007). Microtubule plus-end-tracking proteins target gap junctions directly from the cell interior to adherens junctions. *Cell*, *128*(3), 547–560.

Sheikh, F., Chen, Y., Liang, X., Hirschy, A., Stenbit, A. E., Gu, Y., et al. (2006). alpha-E-catenin inactivation disrupts the cardiomyocyte adherens junction, resulting in cardiomyopathy and susceptibility to wall rupture. *Circulation*, *114*(10), 1046–1055.

Sheikh, F., Ross, R. S., & Chen, J. (2009). Cell-cell connection to cardiac disease. *Trends in Cardiovascular Medicine*, *19*(6), 182–190.

Shimada, T., Kawazato, H., Yasuda, A., Ono, N., & Sueda, K. (2004). Cytoarchitecture and intercalated disks of the working myocardium and the conduction system in the mammalian heart. *The Anatomical Record Part A, Discoveries in Molecular, Cellular, and Evolutionary Biology*, *280*(2), 940–951.

Smart, E. J., Graf, G. A., McNiven, M. A., Sessa, W. C., Engelman, J. A., Scherer, P. E., et al. (1999). Caveolins, liquid-ordered domains, and signal transduction. *Molecular and Cellular Biology*, *19*(11), 7289–7304.

Song, L. S., Sobie, E. A., McCulle, S., Lederer, W. J., Balke, C. W., & Cheng, H. (2006). Orphaned ryanodine receptors in the failing heart. *Proceedings of the National Academy of Sciences of the United States of America*, *103*(11), 4305–4310.

Splawski, I., Timothy, K. W., Sharpe, L. M., Decher, N., Kumar, P., Bloise, R., et al. (2004). Ca(V)1.2 calcium channel dysfunction causes a multisystem disorder including arrhythmia and autism. *Cell*, *119*(1), 19–31.

Syrris, P., Ward, D., Asimaki, A., Evans, A., Sen-Chowdhry, S., Hughes, S. E., et al. (2007). Desmoglein-2 mutations in arrhythmogenic right ventricular cardiomyopathy: A genotype-phenotype characterization of familial disease. *European Heart Journal*, *28*(5), 581–588.

Syrris, P., Ward, D., Evans, A., Asimaki, A., Gandjbakhch, E., Sen-Chowdhry, S., et al. (2006). Arrhythmogenic right ventricular dysplasia/cardiomyopathy associated with mutations in the desmosomal gene desmocollin-2. *American Journal of Human Genetics*, *79*(5), 978–984.

Takeshima, H., Komazaki, S., Nishi, M., Iino, M., & Kangawa, K. (2000). Junctophilins: A novel family of junctional membrane complex proteins. *Molecular Cell*, *6*(1), 11–22.

Takeshima, H., Shimuta, M., Komazaki, S., Ohmi, K., Nishi, M., Iino, M., et al. (1998). Mitsugumin29, a novel synaptophysin family member from the triad junction in skeletal muscle. *The Biochemical Journal*, *331*(Pt 1), 317–322.

Toyofuku, T., Yabuki, M., Otsu, K., Kuzuya, T., Hori, M., & Tada, M. (1998). Direct association of the gap junction protein connexin-43 with ZO-1 in cardiac myocytes. *The Journal of Biological Chemistry*, *273*(21), 12725–12731.

Tsatsopoulou, A. A., Protonotarios, N. I., & McKenna, W. J. (2006). Arrhythmogenic right ventricular dysplasia, a cell adhesion cardiomyopathy: Insights into disease pathogenesis from preliminary genotype–phenotype assessment. *Heart (British Cardiac Society), 92*(12), 1720–1723.

van den Borne, S. W., Narula, J., Voncken, J. W., Lijnen, P. M., Vervoort-Peters, H. T., Dahlmans, V. E., et al. (2008). Defective intercellular adhesion complex in myocardium predisposes to infarct rupture in humans. *Journal of the American College of Cardiology, 51*(22), 2184–2192.

van Oort, R. J., Garbino, A., Wang, W., Dixit, S. S., Landstrom, A. P., Gaur, N., et al. (2011). Disrupted junctional membrane complexes and hyperactive ryanodine receptors after acute junctophilin knockdown in mice. *Circulation, 123*(9), 979–988.

van Tintelen, J. P., Entius, M. M., Bhuiyan, Z. A., Jongbloed, R., Wiesfeld, A. C., Wilde, A. A., et al. (2006). Plakophilin-2 mutations are the major determinant of familial arrhythmogenic right ventricular dysplasia/cardiomyopathy. *Circulation, 113*(13), 1650–1658.

Vasile, V. C., Ommen, S. R., Edwards, W. D., & Ackerman, M. J. (2006). A missense mutation in a ubiquitously expressed protein, vinculin, confers susceptibility to hypertrophic cardiomyopathy. *Biochemical and Biophysical Research Communications, 345*(3), 998–1003.

Vatta, M., Ackerman, M. J., Ye, B., Makielski, J. C., Ughanze, E. E., Taylor, E. W., et al. (2006). Mutant caveolin-3 induces persistent late sodium current and is associated with long-QT syndrome. *Circulation, 114*(20), 2104–2112.

Yang, Z., Bowles, N. E., Scherer, S. E., Taylor, M. D., Kearney, D. L., Ge, S., et al. (2006). Desmosomal dysfunction due to mutations in desmoplakin causes arrhythmogenic right ventricular dysplasia/cardiomyopathy. *Circulation Research, 99*(6), 646–655.

Zemljic-Harpf, A. E., Miller, J. C., Henderson, S. A., Wright, A. T., Manso, A. M., Elsherif, L., et al. (2007). Cardiac-myocyte-specific excision of the vinculin gene disrupts cellular junctions, causing sudden death or dilated cardiomyopathy. *Molecular and Cellular Biology, 27*(21), 7522–7537.

Zemljic-Harpf, A. E., Ponrartana, S., Avalos, R. T., Jordan, M. C., Roos, K. P., Dalton, N. D., et al. (2004). Heterozygous inactivation of the vinculin gene predisposes to stress-induced cardiomyopathy. *The American Journal of Pathology, 165*(3), 1033–1044.

Zhang, R., Yang, J., Zhu, J., & Xu, X. (2009). Depletion of zebrafish Tcap leads to muscular dystrophy via disrupting sarcomere-membrane interaction, not sarcomere assembly. *Human Molecular Genetics, 18*(21), 4130–4140.

Zhou, D., Lambert, S., Malen, P. L., Carpenter, S., Boland, L. M., & Bennett, V. (1998). AnkyrinG is required for clustering of voltage-gated Na channels at axon initial segments and for normal action potential firing. *The Journal of Cell Biology, 143*(5), 1295–1304.

CHAPTER FIVE

Excitable Domains of Myelinated Nerves: Axon Initial Segments and Nodes of Ranvier

Kae-Jiun Chang[*], **Matthew N. Rasband**[*,†,1]
[*]Program in Developmental Biology, Baylor College of Medicine, Houston, Texas, USA
[†]Department of Neuroscience, Baylor College of Medicine, Houston, Texas, USA
[1]Corresponding author: e-mail address: rasband@bcm.edu

Contents

Abstract

Neurons are highly polarized cells. They can be subdivided into at least two structurally and functionally distinct domains: somatodendritic and axonal domains. The somatodendritic domain receives and integrates upstream input signals, and the axonal domain generates and relays outputs in the form of action potentials to the downstream target. Demand for quick response to the harsh surroundings prompted evolution to equip vertebrates' neurons with a remarkable glia-derived structure called myelin. Not only Insulating the axon, myelinating glia also rearrange the axonal components and elaborate functional subdomains along the axon. Proper functioning of all theses domains and subdomains is vital for a normal, efficient nervous system.

Current Topics in Membranes, Volume 72
ISSN 1063-5823
http://dx.doi.org/10.1016/B978-0-12-417027-8.00005-2

1. INTRODUCTION

In the nervous system, neurons conduct electrical signals, called action potentials, that enable rapid perception of sensory modalities, conduction of behavior, and orchestration of the vital activities of various tissues and organs. To efficiently fulfill these functions, neurons are anatomically, functionally, and molecularly highly polarized (Figs. 5.1 and 5.2). The upstream synaptic inputs are mostly received at dendrites and the cell body (the somatodendritic domain). The decision of a neuron to fire an action potential is made at the axon initial segment (AIS), where the somatodendritic inputs are integrated. Once an action potential is generated at the AIS, it

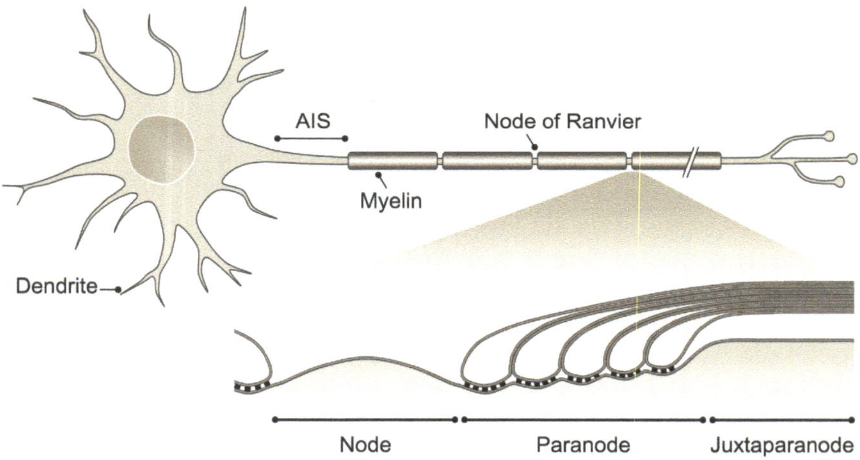

Figure 5.1 A schema of the neuron and myelinated axon. The neuron is polarized into the somatodendritic domain (with dendrites extending from the cell body) and the axonal domain. Axons can be wrapped and myelinated by Schwann cells in the PNS or oligodendrocytes in the CNS. Once the summed synaptic input in the somatodendritic domain exceeds the firing threshold, action potentials are generated at the axon initial segment (AIS). Action potentials travel along the axon and are regenerated at nodes of Ranvier until reaching the axon terminals. Nodes are the gaps between myelin sheaths. Myelin sheaths end at the paranodal domain with cytoplasm-containing glial loops closely attached to the axon; these contact sites form the paranodal junctions (five loops are shown for simplicity). Adjacent to the paranode, a short region underneath the myelin is highly enriched with Kv1 channels and is called the juxtaparanode. The internodal region comprising the majority of the myelinated axon is located between two juxtaparanodes beneath the myelin sheath and is not shown in this close-up view. (For color version of this figure, the reader is referred to the online version of this chapter.)

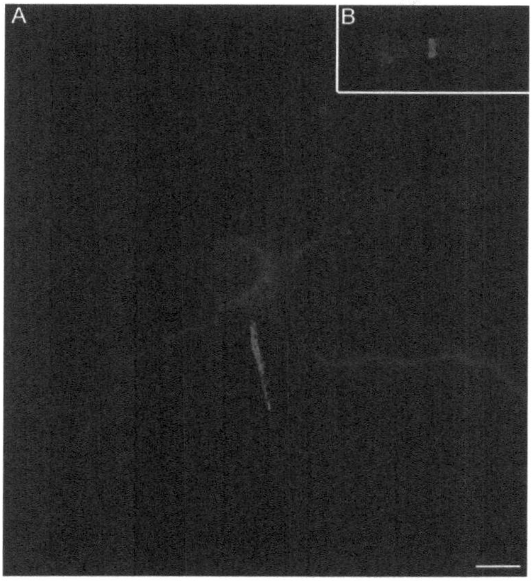

Figure 5.2 Immunostaining of different neuronal domains. (A) A cultured mouse hippocampal neuron was stained to visualize the somatodendritic domain (Map2, microtubule-associated protein 2, in red) and the AIS (Nav channels in green). Nav channels are highly enriched at the AIS, whereas Map2 is excluded from the axon. (B) A longitudinal section of a mouse sciatic nerve shows the nodal (βIV spectrin in green), paranodal (Caspr in blue), and juxtaparanodal (Kv1.2 channels in red) domains. Scale bar = 10 μm for (A) and 5 μm for (B). (See Color Plate.)

is propagated along the axon to the downstream neuron or target organ. The voltage-gated sodium (Nav) channels, required for action potential generation, are highly enriched at the AIS (Kole et al., 2008; Wollner & Catterall, 1986).

The AIS also functions as a diffusion barrier to separate somatodendritic components from axonal ones, maintaining axonal identity and therefore neuronal polarity. Studies tracing phospholipids, membrane proteins, dextrans, and transport vesicles collectively showed the AIS acts as a membrane and cytoplasmic barrier to selectively restrict mobility of these molecules and organelles through the AIS (Kobayashi, Storrie, Simons, & Dotti, 1992; Nakada et al., 2003; Song et al., 2009; Winckler, Forscher, & Mellman, 1999).

To further expedite action potential propagation, axons can be wrapped by glial membranes, which form an elaborate structure called myelin. Myelin is an evolutionary innovation extensively utilized by gnathostomata (jawed vertebrates) (Zalc, Goujet, & Colman, 2008). It consists of multilamellar sheets of glial membranes, most of which are compacted and almost devoid of cytoplasmic and extracellular space. The resultant lipid-rich sheath functions to reduce the passive attenuation of the axonal current propagation by increasing the resistance and decreasing the capacitance across the axonal membrane. Myelination enables more rapid communication, in a smaller space and at lower metabolic expense, allowing the nervous systems of jawed vertebrates to be much more complex than those of other species. To illustrate how myelination facilitates the increased complexity of the nervous system, the space occupied by a squid giant axon can accommodate more than 1000 vertebrate myelinated axons that are able to conduct action potentials at the same speed or even faster than the giant axon.

In myelinated axons, the action potential generated at the AIS travels along the axon and is repeatedly regenerated at regularly spaced gaps between myelin sheaths, called nodes of Ranvier. In contrast with the continuous regeneration of action potentials along unmyelinated axons, the propagation by regenerating action potentials at nodes is called saltatory conduction. As another excitable membrane domain, the node is also highly enriched in Nav channels (Ellisman & Levinson, 1982) and, not surprisingly, has a similar molecular organization with the AIS (see below).

As domains crucial for neuronal function, the AIS and node are composed of highly clustered macromolecular complexes. In this chapter, we discuss the current knowledge about the molecular compositions of the AIS and node underlying their functional importance. We also discuss how these macromolecular membrane domains are formed during development, how they are maintained afterward, and the consequences of their disruption in disease. We will not focus on the physiology of the axon or the domains discussed here and instead refer interested readers to the several excellent reviews recently published on axon physiology (Debanne, Campanac, Bialowas, Carlier, & Alcaraz, 2011; Kole & Stuart, 2012).

2. MOLECULAR COMPOSITION OF AISs AND NODES OF RANVIER

2.1. Axon initial segments

Ultrastructurally, the AIS is characterized by fasciculated microtubules, an electron-dense axolemmal undercoating, and a high density of freeze-fracture intramembranous particles (Matsumoto & Rosenbluth, 1985; Palay, Sotelo, Peters, & Orkand, 1968; Peters, Proskauer, & Kaiserman-Abramof, 1968). Demarcating the somatodendritic and axonal domains and firing action potentials, the AIS is composed of and can be molecularly identified by high-density clusters of several molecules including ion channels, cell adhesion molecules (CAMs), cytoskeletal scaffolds, and extracellular matrix (ECM) components.

In addition to the Nav channels Nav1.1, Nav1.2, and Nav1.6, voltage-gated potassium (Kv) channels Kv1.1, Kv1.2, Kv1.4, Kv2.1, Kv2.2, Kv7.2 (KCNQ2), and Kv7.3 (KCNQ3); voltage-gated calcium (Cav) channels Cav2.1 and Cav2.2; ion channel auxiliary subunits Navβ1, Navβ2, Navβ4, and Kvβ2 (associated with Kv1); and Fgf13 and Fgf14 (Nav channel binding and modulating intracellular homologous factors of fibroblast growth factors) have all been reported to be enriched at the AISs of various types of neurons (Brackenbury et al., 2010; Buffington & Rasband, 2013; Cooper, 2011; Debanne et al., 2011; Ogawa et al., 2008; Sarmiere, Weigle, & Tamkun, 2008; Zhang, Bao, Yang, Wu, & Li, 2012). Robust activity of T-type and R-type Cav channels was also detected at the AIS and shown to affect action potential firing (Bender & Trussell, 2009). However, the enrichment of T-type and R-type Cav channels at the AIS has not been established by immunohistochemistry.

As the site of action potential initiation, not surprisingly, the AIS shapes and regulates neuronal firing properties. Consistent with the large diversity of neuronal cell types and their variety of functions in the nervous system, cell type-specific physiology and firing frequencies of action potentials have been described and certainly depend on the heterogeneity of channels, channel subtypes, and channel subunits enriched at the AISs of these different neurons (Bean, 2007; Debanne et al., 2011; Lorincz & Nusser, 2008). For example, Purkinje neurons fire action potentials at a very high frequency, and the main Nav subtype and auxiliary subunit enriched at their AISs are Nav1.6 (with a lower firing threshold) and Navβ4 (for generating resurgent sodium currents), respectively (Buffington & Rasband, 2013; Khaliq, Gouwens, & Raman, 2003; Lorincz & Nusser, 2008). Besides the molecular differences among AISs, it is noteworthy that even in a single AIS, some channel subtypes tend to localize at the proximal part of the AIS and others at the distal half (Lorincz & Nusser, 2008; Van Wart, Trimmer, & Matthews, 2007). Although the molecular mechanisms regulating the proximodistal gradient are not known, the distal Nav1.6 has been proposed to initiate action potentials in the distal AIS, whereas the proximal Nav1.2 is thought to be important for action potential backpropagation (Hu et al., 2009).

In addition to the differential compositions of channels that shape many aspects of action potentials, neurotransmitters and neuromodulators may also play a role in AIS excitability. For example, although most synaptic inputs are received at the dendrites and soma, some synapses target the AIS. Basket cells and chandelier cells can form GABAergic synapses at the AISs of several neurons at the same time (Somogyi, Freund, & Cowey, 1982; Somogyi, Nunzi, Gorio, & Smith, 1983; Somogyi, Smith, et al., 1983) and may contribute to synchronization of neuronal activity (Cobb, Buhl, Halasy, Paulsen, & Somogyi, 1995). The neuromodulator dopamine was also recently shown to reduce neuronal activity by downregulating T-type Cav channel activity at the AISs of cartwheel interneurons (Bender, Ford, & Trussell, 2010).

In addition to ion channels, the AIS is highly enriched with CAMs including NF186 (neurofascin 186 kDa isoform), NrCAM (neuron–glia cell adhesion molecule-related cell adhesion molecule), Caspr2 (contactin-associated protein-like 2), Tag1 (transient axonal glycoprotein-1), and Adam22 (a disintegrin and metalloprotease 22). The AIS also contains CK2 (casein kinase 2), Schip1 (schwannomin-interacting protein 1), and cytoskeleton-associated proteins and scaffolds including ankyrinG (AnkG), βIV spectrin, EB1 and EB3 (microtubule plus end-binding proteins), and Psd93 (postsynaptic density 93) (Fig. 5.3; Bennett & Chen, 2001; Inda, DeFelipe, & Muñoz, 2006; Leterrier et al., 2011; Martin et al., 2008; Ogawa et al., 2008, 2010; Rasband, 2010; Vacher et al., 2011). Another signaling molecule phosphorylated IκBα (inhibitor of NF-κB-alpha) was reported to be concentrated at the AIS and has been widely used as an AIS marker (Schultz et al., 2006). However, the staining has recently been shown to be nonspecific (Buffington, Sobotzik, Schultz, & Rasband, 2012).

Ankyrins interact with membrane proteins and spectrins, and the latter link the ankyrin–membrane protein complexes to the actin cytoskeleton (Bennett & Baines, 2001). Since a short actin filament can interact with multiple spectrins, ankyrin/spectrin complexes are capable of linking large networks of membrane protein complexes to the cytoskeleton. This scheme is reiteratively utilized in the structures of the AIS, node of Ranvier, and node-flanking domains. For example, AnkG interacts with Nav and KCNQ channels, NF186, and NrCAM through its membrane-binding domain and βIV spectrin through its spectrin-binding domain. βIV spectrin then links the whole membrane protein complex to the actin cytoskeleton (Bennett & Baines, 2001; Rasband, 2010). The interaction with the cytoskeleton is so tight that it renders the AIS protein complex detergent-resistant and even caused the patch clamp recordings of the AIS to underestimate the AIS Nav

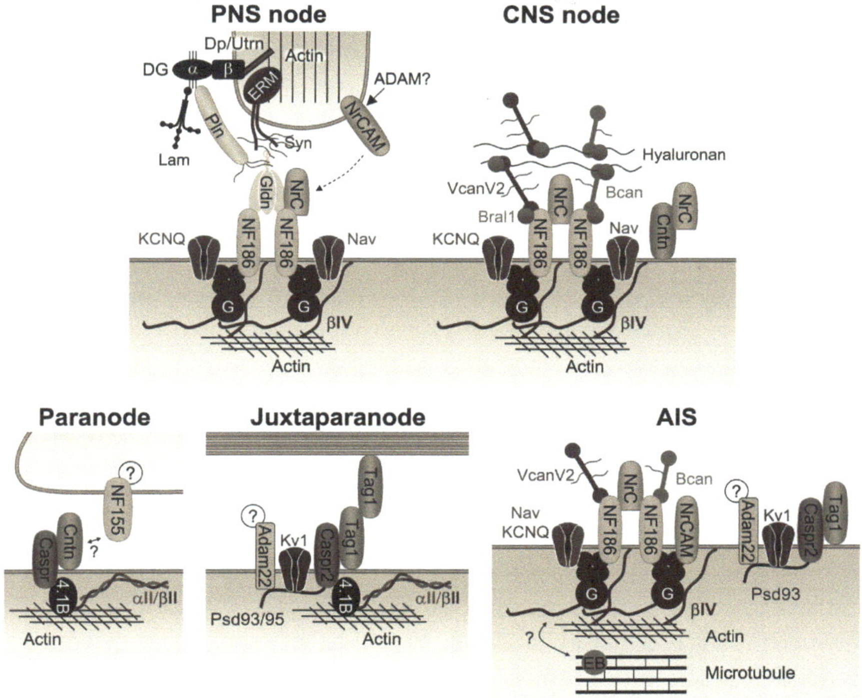

Figure 5.3 Molecular organizations of various axonal subdomains. Simplified cartoons display some of the molecules and interactions involved in the nodal, paranodal, juxtaparanodal, and AIS domains. The PNS nodes are contacted by microvilli of Schwann cells. Dystroglycan (DG) is autocleaved into α- and β-chains, which remain associated (Akhavan, Crivelli, Singh, Lingappa, & Muschler, 2008; Barresi & Campbell, 2006). β-DG interacts with dystrophin (Dp) and utrophin (Utrn). NrCAM is cleaved by furin and the two resultant fragments remain associated as a transmembrane form. Once NrCAM is cleaved again by as yet unidentified metalloproteases (probably by a disintegrin and metalloproteases (ADAMs)), its extracellular domain (NrC) is released from the plasma membrane and incorporated into the ECM (Davis, Lambert, & Bennett, 1996; Kayyem, Roman, de la Rosa, Schwarz, & Dreyer, 1992; Susuki et al., 2013). The transmembrane form of NrCAM is present at the microvilli (Feinberg et al., 2010). Interaction of laminins (Lam) and perlecan (Pln) with α-DG requires proper glycosylation of α-DG (thin black lines) (Barresi & Campbell, 2006). Pln and syndecans 3/4 (Syn) are modified by heparan sulfate side chains (green curves). The furin-shed gliomedin (Gldn) trimerizes and is associated with heparan sulfate through its N-terminal region and collagen-like domain and interacts with NrC and NF186 through its olfactomedin domain (Eshed, Feinberg, Carey, & Peles, 2007). G and βIV represent AnkG and βIV spectrin, respectively. In the CNS, the nodal ECM is also enriched with NrC (Susuki et al., 2013), but its cellular source is unknown. Vcan V2 and Bcan interact with Bral1 and hyaluronan through their G1 globular domains and with NF186 through the G3 domains (Susuki et al., 2013). The intervening regions of Vcan V2 and Bcan are modified by chondroitin sulfate side chains

channel density (Garrido et al., 2003; Kole et al., 2008; Winckler et al., 1999). Intriguingly, instead of a polygonal meshwork as found in erythrocytes, a recent superresolution study suggests a cylindrical rebar cage-like framework of the actin/spectrin-based cytoskeleton in the axon: actin filaments form regularly spaced rings, connected by spectrin tetramer beams with ankyrins located around the middle (Xu, Zhong, & Zhuang, 2013).

A specialized ECM composed of tenascin-R (Tn-R) and the chondroitin sulfate proteoglycans (CSPGs) aggrecan, brevican, neurocan, and versican is observed at the AIS both *in vitro* and *in vivo*, either as an extension of perineuronal nets or as an extracellular milieu established by interactions between NF186, the CSPGs brevican (Bcan) and versican (Vcan), and Tn-R independent of hyaluronan (Brückner, Szeöke, Pavlica, Grosche, & Kacza, 2006; Frischknecht et al., 2009; Hedstrom et al., 2007; John et al., 2006; Susuki et al., 2013; Volkmer, Zacharias, Nörenberg, & Rathjen, 1998). This AIS ECM may constitute the molecular basis for the extracellular electron-dense material surrounding the AIS (Peters et al., 1968). The specialized ECM at the AIS may function in ionic buffering and/or stabilization of axo-axonic synapses (Celio, Spreafico, De Biasi, & Vitellaro-Zuccarello, 1998; Frischknecht et al., 2009; Hedstrom et al., 2007).

2.2. Nodes of Ranvier

Both the AIS and nodes are responsible for action potential generation and have very similar molecular compositions (Fig. 5.3). Assembly of the AIS is a cell-autonomous event in vertebrate neurons no matter whether their axons are myelinated or not. On the other hand, nodes are the products of neuron–glia interactions that assemble both nodes and additional membrane domains underneath the flanking myelin sheaths. Since myelinating glia in the peripheral nervous system (PNS) and central nervous system (CNS) are

Figure 5.3—cont'd (light blue curves). Contactin (Cntn) was found at CNS nodes but only weakly at a few PNS nodes (Kazarinova-Noyes et al., 2001). The mode of the interaction between Caspr/Cntn and NF155 is obscure. The cytoplasmic partners of NF155 at the paranodal junction and ligands of Adam22 at the juxtaparanode and AIS are currently unknown. The AIS domain of a CNS neuron is a fusion of the CNS nodal and juxtaparanodal domains but without Cntn, 4.1B, αII/βII spectrins (αII/βII), and Psd95. Furthermore, the AIS also has fasciculated microtubules, EB1 and EB3, and the transmembrane form of NrCAM. Whether there are NrC and Bral1 at the AIS is unclear. The domain of AnkG that interacts with EB1 and EB3 is currently unknown. (See Color Plate.)

different, both the nodal molecular compositions and mechanisms of assembly have some unique differences.

2.2.1 Nodes proper

Nodes of Ranvier are ultrastructurally very similar to the AIS since they have an electron-dense axolemmal undercoating, extracellular material, and a high density of freeze-fracture intramembranous particles; however, the nodal microtubules are not fasciculated (Elfvin, 1961; Peters, 1966; Peters et al., 1968; Rosenbluth, 1976; Sloper & Powell, 1973).

To accomplish saltatory conduction, nodes may be highly enriched with one or more types of ion channels and accessory subunits including Nav1.1, Nav1.2, Nav1.6 (the major Nav subtype at nodes during adulthood), Nav1.8, Nav1.9, Kv3.1b, KCNQ2, and KCNQ3 channels and the auxiliary subunits Navβ1, Navβ2, and Navβ4 (and presumably Navβ3) (Buffington & Rasband, 2013; Cooper, 2011; Poliak & Peles, 2003; Salzer, 2003). Nodes and the AIS also share NF186, NrCAM, AnkG, βIV spectrin, CK2, Fgf13 (only at nodes of peripheral sensory neurons), and Schip1 (Bréchet et al., 2008; Martin et al., 2008; Poliak & Peles, 2003; Salzer, 2003; Wittmack et al., 2004), but not Caspr2, Tag1, Adam22, Psd93, Kv1 channels, and Kvβ2, which instead accumulate at juxtaparanodes (see below).

The nodes in the PNS are surrounded and contacted by Schwann cell microvilli, which are enriched with the microvillar components ezrin, radixin, moesin (collectively ERM proteins) and EBP50 (ezrin-binding protein 50), dystrophin isoform 116, and utrophin (Occhi et al., 2005; Poliak & Peles, 2003; Salzer, 2003). Schwann cell microvilli build up a complex extracellular microenvironment surrounding PNS nodes that is composed of the transmembrane glycoprotein dystroglycan, the transmembrane heparan sulfate proteoglycans (HSPGs) syndecan3 and syndecan4, and ECM components including laminins α2β1γ1 and α5β1γ1, the shed extracellular domains of gliomedin and NrCAM, the HSPG perlecan, the CSPG Vcan V1 isoform, the shed CSPG NG2, and collagen V (Colognato & Tzvetanova, 2011; Feinberg et al., 2010; Martin, Levine, Chen, Ughrin, & Levine, 2001; Melendez-Vasquez et al., 2005; Occhi et al., 2005). Ezrin may interact with syndecans3/4 and link them to the microvillar actin bundles (Tkachenko, Rhodes, & Simons, 2005). The microvilli and PNS nodal ECM are involved in node of Ranvier formation and stabilization (see below).

Oligodendrocytes do not have microvillar structures, but there is also a specialized ECM surrounding the nodes consisting of Tn-R, Bral1 (brain link protein 1), the shed extracellular domain of NrCAM, and the CSPGs phosphacan, neurocan, Bcan, and Vcan V2 isoform (Bekku & Oohashi, 2010; Hedstrom et al., 2007; Susuki et al., 2013; Zimmermann & Dours-Zimmermann, 2008). The CNS nodal ECM participates in node formation and is also proposed to function as a cationic pool facilitating node function through the negative charges of chondroitin sulfate chains (Bekku et al., 2010; Susuki et al., 2013).

2.2.2 Paranodes

The nodal axolemma is exposed to the external environment without being covered by myelin. Immediately flanking the node are paranodes, where the compact myelin opens to form cytoplasm-containing loops (Fig. 5.1). The glial loop membranes indent and closely attach to the axolemma with a distance of only 2.5–3 nm (Elfvin, 1961; Peters, 1966). The loops are actually a single cytoplasmic collar that forms a spiral wrap around the axon (Arroyo & Scherer, 2000). The intimate neuron–glia interactions at this spiral wrap form a septate-like paranodal junction, which is characterized by electron-dense transverse bands between glial and axonal membranes. The paranodal junction is not a complete seal like a tight junction, but instead forms a narrow, spiral, long passage that makes diffusion of ions and other substances much slower between the nodal extracellular environment and the space between myelin and internodal axolemma (Rosenbluth, 2009). Therefore, paranodal junctions provide high resistance indispensable for the insulating function of myelin.

Freeze-fracture studies of myelinated axons showed that the E face of the axolemma has a high density of intramembranous particles at the node and juxtaparanode that are separated by undulating indentations (paranodes) containing much fewer particles. Thus, the paranodal junction has been proposed to function as a membrane protein diffusion barrier that restricts the position of the nodal particles (presumably Nav channel complexes) between myelin sheaths and separates the juxtaparanodal particles (presumably Kv1 channel complexes, see below) from the nodal ones (Rosenbluth, 1999). Paranodal junctions have characteristic parallel rows of particles in the axolemmal P face and glial P and E faces, which may be related to the transverse bands (Schnapp & Mugnaini, 1978; Wiley & Ellisman, 1980).

The paranodal axoglial junction is formed mainly by three CAMs: glial NF155 (neurofascin 155 kDa isoform) and an axonal complex consisting of

Caspr (contactin-associated protein) and contactin (Fig. 5.3; Einheber et al., 1997; Menegoz et al., 1997; Rios et al., 2000; Tait et al., 2000). However, whether the Caspr/contactin complex and NF155 interact directly or indirectly remains controversial (Charles et al., 2002; Gollan, Salomon, Salzer, & Peles, 2003). Knocking out any one of these CAMs disrupts paranodal junctions and delays nerve conduction, but causes no gross defect in myelination (Bhat et al., 2001; Boyle et al., 2001; Pillai et al., 2009; Rosenbluth, Petzold, & Peles, 2012). These mutant mice exhibit severe ataxia, hypomobility, trembling, and limb weakness, similar to myelin mutants. Some mutant strains lacking these proteins die between 2 and 5 weeks of age. In these mutants, the paranodal enrichment of the other two CAMs is impaired, the transverse bands are absent, the axolemmal indentations are absent or reduced, and the distance between the glial loops and axolemma becomes wider. In addition, the nodal length increases, the density of the nodal intramembranous particles decreases, and the juxtaparanodal molecular complexes translocate to paranodes, consistent with the idea that an intact paranodal junction functions as a membrane protein diffusion barrier. In the PNS, the alignment of glial loops toward the axon is relatively preserved, but in the CNS, paranodal loops are often everted and face away from the axon.

The cytoskeletal scaffolds 4.1B, αII spectrin and βII spectrin are present at the cytoplasmic side of the paranodal axolemma (Fig. 5.3; Einheber et al., 2013; Ogawa et al., 2006; Ohara, Yamakawa, Nakayama, & Ohara, 2000). αII spectrin and βII spectrin form a functional tetramer (Bennett & Baines, 2001). 4.1B interacts with Caspr and βII spectrin, and 4.1B and αII/βII spectrins may stabilize each other's association with the actin cytoskeleton (Bennett & Baines, 2001; Cifuentes-Diaz et al., 2011; Denisenko-Nehrbass et al., 2003; Einheber et al., 2013; Ogawa et al., 2006). The junctional CAM complex may in turn be stabilized through 4.1B and βII spectrin's interaction with the actin cytoskeleton. This view of the role of these cytoskeletal adaptors is consistent with the observation that in 4.1B knockouts (KOs), some paranodal Caspr immunoreactivity appears attenuated and fragmented. In some instances, proteins normally restricted to juxtaparanodes can be observed in paranodal regions (Buttermore et al., 2011; Cifuentes-Diaz et al., 2011; Duflocq, Chareyre, Giovannini, Couraud, & Davenne, 2011; Einheber et al., 2013). Furthermore, without its 4.1B-binding site, paranodal Caspr clusters are less stable and allow invasion of juxtaparanodal proteins (Horresh, Bar, Kissil, & Peles, 2010). These results suggest that the interaction of Caspr with paranodal cytoskeletal scaffolds may be important for the stabilization of the paranodal domain and its function as a membrane protein diffusion barrier. It will be very interesting to directly test these ideas by analyzing loss-of-function phenotypes in mice lacking axonal αII/βII spectrins.

In addition to the aforementioned paranodal cytoskeletal proteins, AnkB and AnkG have also been described at paranodes (Ogawa et al., 2006; Rasband, Peles, et al., 1999). However, their cellular locations and functions remain poorly understood.

Besides the protein components described earlier, a specialized lipid environment also exists at paranodes. Along with the assembly of Caspr/contactin/NF155 clusters, the paranodal membrane domain acquires a specialized detergent-insoluble property (Schafer, Bansal, Hedstrom, Pfeiffer, & Rasband, 2004). Galactocerebroside and its sulfated derivative sulfatide are highly enriched in myelin preparations of the similar detergent-insoluble property (Lee, 2001). Interestingly, mice lacking CGT (UDP–galactose–ceramide galactosyltransferase, the enzyme that catalyzes production of these two galactolipids) exhibit paranodal abnormalities including the disappearance of transverse bands, disrupted paranodal CAM clusters, and translocation of the juxtaparanodal Kv1 channels into paranodes (Dupree, Girault, & Popko, 1999; Dupree & Popko, 1999; Poliak et al., 2001). Loss of CST (3′-phosphoadenylylsulfate–galactosylceramide 3′-sulfotransferase, the enzyme that catalyzes production of sulfatide) results in similar, but milder, paranodal defects (Honke et al., 2002; Hoshi et al., 2007; Ishibashi et al., 2002). In paranodal mutants with disrupted axon–glia interactions (e.g., Caspr KOs), partitioning of NF155 into the insoluble fractions is attenuated (Ogawa & Rasband, 2009). Altogether, these observations suggest that assembly of the paranodal junctional complex may involve a mutual cooperative recruitment and stabilization of CAMs and special lipid components.

During development, the nodal Nav channel subtypes switch from Nav1.2 to Nav1.6 in the CNS and from Nav1.2/Nav1.6 to Nav1.6 in the PNS (Boiko et al., 2001; Kaplan et al., 2001; Schafer, Custer, Shrager, & Rasband, 2006). However, this switch is impaired in paranodal mutants (Rios et al., 2003; Suzuki et al., 2004), indicating that paranodal neuron–glia interactions play key roles in regulating the expression of distinct Nav channel subtypes at nodes. However, the molecular basis of this regulation remains unknown.

Several other molecules are enriched in the noncompact myelin and therefore also in the paranodal loops. These include CNP (2′,3′-cyclic nucleotide 3′ phosphodiesterase), MAG (myelin-associated glycoprotein), 4.1G, and Necl4 (nectin-like CAM 4) (Ohno et al., 2006; Spiegel et al.,

2007; Trapp, Bernier, Andrews, & Colman, 1988; Trapp & Quarles, 1982). However, loss of these proteins causes only mild paranodal phenotypes (Ivanovic et al., 2012; Rasband et al., 2005). Finally, the paranodal loops are interconnected by autotypic tight, gap, and adherens junctions, and the components of these junctions can also be found at paranodal loops (Poliak, Matlis, Ullmer, Scherer, & Peles, 2002; Salzer, 2003).

2.2.3 Juxtaparanodes

The regions underneath the compact myelin and immediately abutting paranodes are juxtaparanodes (Figs. 5.1 and 5.2). At the glial side, juxtaparanodes are enriched with Tag1; in axons, juxtaparanodes are enriched with Kv1 channels, Kvβ2, Caspr2, Tag1, Adam22, Psd93, and Psd95 (Fig. 5.3; Horresh et al., 2008; Ogawa et al., 2010; Poliak et al., 2003; Salzer, 2003). Axonal Caspr2 forms a heterodimer with axonal Tag1; glial Tag1 and axonal Tag1 interact *in trans* (Poliak et al., 2003; Traka et al., 2003). Kv1 channels, Caspr2, and Adam22 have PDZ (Psd95/discs large/zona occludens-1) domain-binding motifs in their cytoplasmic domains that can interact with Psd93 and Psd95. Hence, Caspr2/Tag1, Kv1 channels/Kvβ2, Adam22, Psd93, and Psd95 may form a macromolecular protein complex (Ogawa et al., 2010; Poliak et al., 1999; Rasband et al., 2002). 4.1B, αII spectrin and βII spectrin are also present in the juxtaparanodal domain in the axon (Fig. 5.3; Denisenko-Nehrbass et al., 2003; Ogawa et al., 2006; Poliak et al., 2001). Caspr2 has a 4.1B-binding site and may link the whole complex to the actin cytoskeleton through 4.1B and αII/βII spectrins.

Caspr2, Tag1, and 4.1B are essential components to assemble the juxtaparanodal complex. Caspr2 and Tag1 are interdependent for their juxtaparanodal targeting, and lacking any one of them disrupts juxtaparanodal clusters of Kv1 channels (Horresh et al., 2008; Ogawa et al., 2010; Poliak et al., 2003; Traka et al., 2003). The PDZ domain-containing scaffolds Psd95 and Psd93 may bridge Kv1 channels and Caspr2; surprisingly, however, Psd95, Psd93, and even the PDZ-binding motif of Caspr2 are not required for Caspr2 to recruit Kv1 channels (Horresh et al., 2008). The 4.1B-binding site is essential for the association of Caspr2 with Kv1 channels, and lacking either 4.1B or the 4.1B-binding site of Caspr2 disrupts the establishment of the juxtaparanodal complex (Horresh et al., 2008, 2010). However, the exact molecular link between Caspr2/4.1B and Kv1 channels is still obscure. Interestingly, since 4.1B is also distributed along the internode (Cifuentes-Diaz et al., 2011; Horresh et al., 2010), how the juxtaparanodal components are selectively clustered at juxtaparanodes remains an open question. An unknown mechanism might push the juxtaparanodal complex toward nodes, which is repulsed by the barrier of paranodal junctions; these two forces may then reach equilibrium at juxtaparanodes.

Psd95 and Psd93 are not necessary for Adam22 localization; instead, Adam22 recruits both scaffolds to the juxtaparanodal complex (Ogawa et al., 2010). Despite these interactions, the functions of Adam22 at juxtaparanodes are currently unknown because Adam22 is not required for clustering Kv1 channels and Caspr2/Tag1 at juxtaparanodes (Ogawa et al., 2010). In addition, how Caspr2/Tag1 recruits Adam22 to juxtaparanodes in the absence of Psd95 and Psd93 (i.e., the molecular link between Caspr2/Tag1 and Adam22 other than Psd95/93) is also unclear.

Without juxtaparanodal clustering of Kv1 channels in the absence of Caspr2 or Tag1, there is little or no defect in conduction and excitability of myelinated axons (Poliak et al., 2003; Traka et al., 2003). This is different from Kv1.1-deficient mice, which show neuromyotonia and nerve hyperexcitability at lower temperatures (Chiu, Zhou, Zhang, & Messing, 1999). Computer simulations suggest that the high-density clusters of Kv1 channels at juxtaparanodes may be required to protect myelinated axons from reentrant excitation when nerve conduction is delayed by 40% with compromised myelin or disrupted paranodal junctions (Poliak et al., 2003). Consistent with this idea, peripheral nerve hyperexcitability and disrupted juxtaparanodal clustering of Kv1 channels have been described in db/db mice and type 2 diabetes mellitus patients with delayed nerve conduction (Zenker et al., 2012).

3. ASSEMBLY AND MAINTENANCE OF AISs AND NODES OF RANVIER

3.1. Axon initial segments

AnkG is the master scaffold responsible for building the AIS molecular domain. AnkG recruits βIV spectrin to the AIS and these two scaffolds accumulate at the AIS earlier than Nav1.6 or NF186 (Jenkins & Bennett, 2001; Yang, Ogawa, Hedstrom, & Rasband, 2007). *In vitro* and *in vivo* studies have shown AnkG is essential for the initial AIS clustering of all other AIS proteins tested so far (Hedstrom et al., 2007; Jenkins & Bennett, 2001; Zhou et al., 1998); thus, AnkG defines the AIS. Without AnkG, the electron-dense undercoating and fasciculated microtubules normally found at the AIS also fail to form (Sobotzik et al., 2009). Removal of AnkG from the assembled AIS disrupts its molecular organization, indicating that AnkG is also required for maintenance of the AIS (Hedstrom, Ogawa, & Rasband, 2008). Studies have now shown that the AIS is assembled after the axon is specified and AnkG is not essential for axon specification; however, the axon acquires dendritic features without AnkG and the associated AIS, indicating that AnkG and the AIS are necessary to maintain neuronal polarity (Boiko et al., 2007; Galiano et al., 2012; Hedstrom et al., 2008; Sobotzik et al., 2009).

What upstream mechanism accounts for the localization of this master scaffold to the AIS? Recent studies showed that a complementary scaffold consisting of AnkB/αII/βII spectrins is enriched at the axon distal to the AIS. Furthermore, loss of αII/βII spectrins disrupted proper clustering of AnkG at the AIS both *in vitro* and *in vivo*, and overexpression of AnkB shortens the length of AIS AnkG clusters (Galiano et al., 2012). Based on these observations, a model was proposed where the distal axonal cytoskeletal scaffolds AnkB/αII/βII spectrins are carried to the distal axon by the motor complex Kif3/Kap3 (kinesin superfamily protein 3/kinesin superfamily-associated protein 3) to assemble a distal axonal cytoskeleton. Subsequently, AnkG is expressed and fills the proximal axon demarcated by AnkB/αII/βII spectrins in a mutually repulsive manner (Galiano et al., 2012; Fig. 5.4). Still, the molecular mechanisms linking axon specification and its downstream event—AIS domain assembly—remain unknown and are an active area of investigation.

Although the AIS CAM NF186 is recruited by AnkG and not required for AIS domain assembly, it does play essential roles in establishing the AIS ECM microenvironment and maintenance of the AIS. By conditionally knocking out NF186 in mature Purkinje neurons, the AIS disassembles and the spontaneous firing of Purkinje neurons is abolished (Zonta et al., 2011). Therefore, it is conceivable that the AIS domain is maintained in a somewhat different way from how it is established immediately after axon formation. This may be similar to maintenance of the nodal domain and will be discussed later.

βIV spectrin is also not required for AIS assembly *in vitro* (Hedstrom et al., 2007). Instead, its *in vivo* function at the AIS has been proposed to be stabilization of the AIS domain through its actin-binding and pleckstrin homology (PH) domains (Komada & Soriano, 2002; Uemoto et al., 2007; Yang et al., 2007). EB1 and EB3 have recently been reported to be enriched at the AIS compared to their localization in the cell body and dendrites,

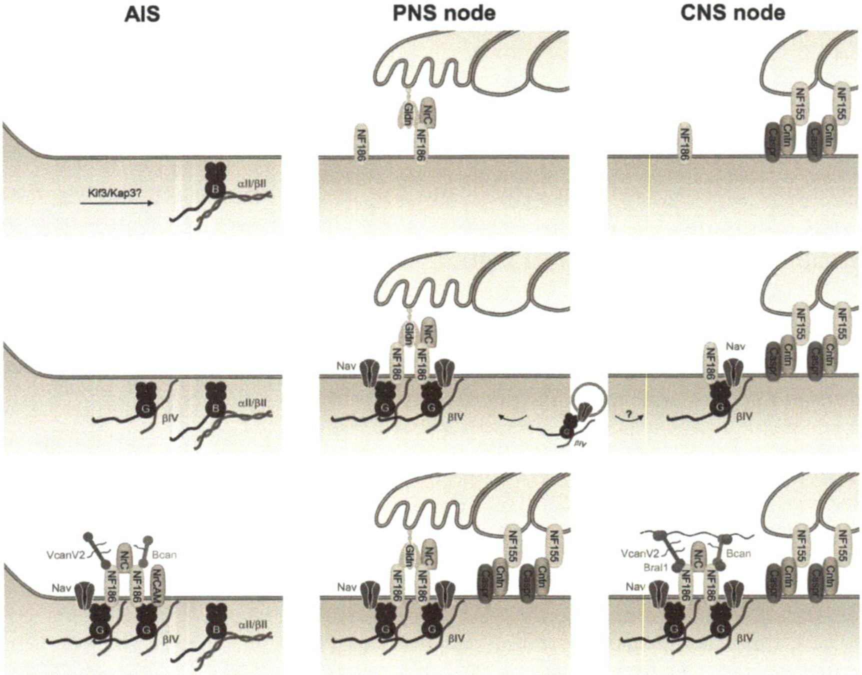

Figure 5.4 Models of axonal subdomain assembly. AIS assembly follows an intrinsic mechanism without the need of glial cells. After the axon is specified, AnkB/αII/βII spectrins establish the distal axonal cytoskeleton, probably through transport by the Kif3/Kap3 motor complex. Then, AnkG fills the proximal axon and recruits other AIS components as a master scaffold. In contrast, node assembly is mediated by neuron–glia interactions. In the PNS, assembly starts with interactions between the ECM components Gldn and NrC at Schwann cell microvilli and the axolemmal CAM NF186. These interactions cluster NF186, which in turn recruits AnkG, Nav channels, and βIV spectrin from the transport vesicles. The nascent nodal complex is further restricted by paranodal junctions. In the CNS, paranodal junctions form early and collect nodal components adjacent to myelin sheaths. The nascent nodal clusters are stabilized by the link to the actin cytoskeleton (not shown) through βIV spectrin and by interactions between the ECM components and NF186 and are further restricted by paranodal junctions. Whether AnkG, Nav channels, and βIV spectrin are recruited to CNS nodes from a pool of transport vesicles is unclear. The illustrations and designations of various molecules are the same as those in Fig. 5.3. (See Color Plate.)

through their binding to microtubules and AnkG (Leterrier et al., 2011; Vacher et al., 2011). Knockdown of EB1 or EB3 decreases AnkG and Nav channel clustering at the AISs of cultured neurons, suggesting they may play a role in stabilizing the AIS complex by linking it to the underlying fasciculated microtubule network.

Nav and KCNQ2/3 channels have similar AnkG-binding motifs and are recruited to the AIS via direct binding to AnkG (Garrido et al., 2003; Lemaillet, Walker, & Lambert, 2003; Pan et al., 2006). Phosphorylation of the AnkG-binding motif in Nav channels by CK2 dramatically increases the binding affinity *in vitro* (Bréchet et al., 2008). Recent experiments showed that phosphorylation-incompetent mutant channels are targeted to the AIS less efficiently than wild-type channels *in vivo* (Gasser et al., 2012). At least one putative CK2 phosphorylation site is conserved in KCNQ2/3 channels. Whether its phosphorylation regulates AIS targeting of KCNQ2/3 remains to be tested. Intriguingly, inhibition and knockdown of CK2 impair AIS localization of not only Nav channels but also AnkG (Bréchet et al., 2008; Sánchez-Ponce, Muñoz, & Garrido, 2011), suggesting CK2 may also act in an unknown upstream phosphorylation event in AIS assembly or mutual stabilization of the AIS protein complex. Meanwhile, this observation further emphasizes the lack of knowledge about the roles of posttranslational modifications in AIS assembly and stabilization (Buffington et al., 2012). Recently, He et al. identified palmitoylation of AnkG, which is required for AnkG to be clustered at and assemble the AIS *in vitro* (He, Jenkins, & Bennett, 2012), suggesting AnkG may need to be deployed at submembranous positions in order for proper functioning.

Although a "juxtaparanode-like" complex exists at the AIS, the proteins that comprise this complex have no known AnkG-interacting sites. Thus, how Kv1 channels are recruited to the AIS is still unclear although they still depend on AnkG (Sánchez-Ponce, DeFelipe, Garrido, & Muñoz, 2012). In contrast to the juxtaparanode, neither 4.1B nor Psd95 are enriched at the AIS (Duflocq et al., 2011; Ogawa et al., 2008). However, Psd93 is required for clustering of Kv1 channels at the AISs of cultured neurons but not neurons *in vivo* (Ogawa et al., 2008, 2010). In further contrast to the juxtaparanode, Kv1 channel clustering at the AIS is independent of Caspr2 and Tag1 (Duflocq et al., 2011; Ogawa et al., 2008, 2010). One clue to the mechanism of assembly of this complex has come from the study of interactions between Kvβ2 and EB1. It was found that phosphorylation of Kvβ2 by Cdk2 and Cdk5 (cyclin-dependent kinases 2 and 5) inhibits Kvβ2 interaction with EB1 and inhibition of Kvβ2 phosphorylation by Cdks enhances AIS targeting of Kv1 channels/Kvβ2 and EB1 (Vacher et al., 2011). However, only the intracellular pool of Kv1 channels/Kvβ2 at the AIS increases, not the surface pool, suggesting that EB1 mediates Kv1 channel/Kvβ2 transport along microtubules to the AIS, but continuous interaction with EB1 prevents Kv1 channels from being inserted into the AIS plasma membrane.

Thus, it is still unclear what critical molecular interactions regulate anchoring of AIS Kv1 channels *in vivo*.

3.2. Nodes of Ranvier

Despite their similar molecular compositions, in contrast to the AIS, node assembly requires the interaction between the axon and myelinating glia, which makes the involved molecular mechanisms different and more complicated. In the PNS and CNS, myelin sheaths are made by different cell types: Schwann cells and oligodendrocytes, respectively. Although similar, the PNS and CNS myelin compositions and fine structures are not identical (Arroyo & Scherer, 2000). Therefore, it is not surprising that the molecular mechanisms of CNS and PNS node formation are not identical (Fig. 5.4).

3.2.1 PNS nodes

The analysis of different developmental stages of node formation in the PNS showed nodal clustering of NF186 and NrCAM precedes that of AnkG and Nav channels. This, in turn, is followed by paranodal Caspr accumulation (Custer et al., 2003; Eshed et al., 2005; Lambert, Davis, & Bennett, 1997; Melendez-Vasquez et al., 2001; Schafer et al., 2006). The initial clustering of axonal NF186 and NrCAM is induced by their binding to the PNS nodal ECM components gliomedin and NrCAM, which are deployed to microvilli and secreted by Schwann cells (Eshed et al., 2005, 2007; Feinberg et al., 2010). In contrast to the AIS, in myelinated axons, NF186 interacts with and recruits AnkG; Nav channels and βIV spectrin are in turn recruited by AnkG (Bennett & Baines, 2001; Dzhashiashvili et al., 2007; Eshed et al., 2005; Yang et al., 2007). Paranodal junctions then restrict the nascent nodal complex between myelin internodes (Feinberg et al., 2010). The nodal complex initially assembles at the edge of a forming myelin sheath as a heminode; two heminodes then fuse to form a full node when two myelin segments elongate and meet each other. After Caspr starts to accumulate at paranodes, the Caspr2/Tag1/Kv1 channel complex appears transiently at nodes and paranodes and then translocates to juxtaparanodes when paranodal junctions mature (Rasband, 2004).

For PNS node formation, Schwann cell microvilli and nodal ECM components play an important role in clustering NF186 in the axolemma and nucleating node assembly. Glial gliomedin and NrCAM interact with each other and are interdependent for clustering at the nodal ECM. In the absence of gliomedin or NrCAM, NF186 and the whole nodal complex fail to be clustered at heminodes, but the assembly at full nodes is rescued by the diffusion barrier function of paranodal junctions (Custer et al., 2003; Feinberg et al., 2010). Similar heminodal defects were observed in the absence of NF186 but not axonal NrCAM (Feinberg et al., 2010), indicating NF186 is the main ligand of glial signals for nucleating node assembly. Mice lacking both Caspr and gliomedin or Caspr and NrCAM exhibit more severe defects in node formation than mice lacking either Caspr, NrCAM, or gliomedin alone. Furthermore, mice lacking both NF186 and NF155 lack nodes, indicating the nodal ECM and paranodal junctions function together and can compensate for the loss of each other (Feinberg et al., 2010; Zonta et al., 2008).

Deposition of gliomedin in the PNS nodal ECM requires its binding to heparan sulfate (Eshed et al., 2007). Hypomorphs of the HSPG perlecan show slightly longer nodal gaps (Bangratz et al., 2012). Mutants of laminins α2 and γ1 and dystroglycan (the receptor of perlecan and laminins) have abnormal microvillar structure and exhibit reduced nodal Nav channel immunoreactivity (Occhi et al., 2005; Saito et al., 2003). Because dystroglycan links the nodal ECM molecules laminins and perlecan to the microvillar intracellular scaffolds dystrophin 116 and utrophin, which interact with actin (Masaki & Matsumura, 2010), this complex may organize the microvilli and ECM heparan sulfate microenvironment and promote the function of gliomedin.

During the initial clustering of nodal components, diffusely distributed NF186 in the axolemma is gathered to the edges of nascent myelin sheaths through an extracellular interaction with glial gliomedin and NrCAM. In contrast, the nodal clusters of AnkG, βIV spectrin, and Nav and KCNQ channels are thought to be delivered through transport vesicles from neuronal cell bodies, with the AnkG-binding region of NF186's intracellular domain functioning as a potential docking signal (Zhang, Bekku, et al., 2012). At mature nodes, only the intracellular domain of NF186 is required for newly made NF186 to be targeted to nodes. The docking signal for NF186 at mature nodes may be the interaction with preexistent nodal AnkG and vice versa (Zhang, Bekku, et al., 2012). This finding of the node maintenance mechanism may explain the requirement of NF186 for AIS maintenance if the newly made Nav channels and AnkG together need the intracellular domain of NF186 for efficient docking.

3.2.2 CNS nodes

In the CNS, the analysis of different developmental stages of node formation showed a sequence of events different from that in the PNS. In particular, the paranodal Caspr clusters assemble first and are then followed by nodal clustering of βIV spectrin, NF186, and Nav channels (Jenkins & Bennett, 2002; Rasband, Peles, et al., 1999; Susuki et al., 2013). AnkG clusters early as well, but the interpretation is complicated by the fact that it is present at both nodes and paranodes in the CNS, and early clusters could be long regions of paranodal AnkG or paranodal+nodal AnkG (Jenkins & Bennett, 2002; Rasband, Peles, et al., 1999). In contrast to PNS nodes, the key ECM molecule gliomedin is not detected at CNS nodes (Eshed et al., 2005). Instead, CNS nodal ECM components include Bcan, Vcan V2, and Bral1 and accumulate at nodes relatively late in development (Susuki et al., 2013). Another difference between the CNS and PNS is the developmental clustering of Caspr2/Tag1/Kv1 channels in myelinated axons. In the CNS, this complex appears first at juxtaparanodes (Rasband, Trimmer, Peles, Levinson, & Shrager, 1999), while in the PNS, it appears first at the nodes and paranodes and finally juxtaparanodes (Vabnick et al., 1999). This difference may reflect the early formation of paranodal junctions in the CNS and differences in the timing of Caspr2/Tag1/Kv1 channel protein expression.

Although paranodal junction formation is the earliest event in CNS node formation, node assembly is still robust in the CNS of Caspr-deficient mice. However, lower densities of nodal Nav channels and wider nodal gaps were observed in the absence of paranodal junctions (Bhat et al., 2001; Rios et al., 2003), indicating a decreased efficiency in node formation or maintenance.

Unidentified oligodendrocyte-secreted molecules were previously proposed to instruct Nav channel clustering at CNS nodes (Kaplan et al., 1997), as putative CNS equivalents of gliomedin. By analyzing the CNS nodal ECM, we have shown that the core ECM molecules Bcan, Vcan V2, Bral1 and NrCAM interact with NF186, and such interactions are sufficient to induce node-like clusters along the axons of cultured neurons in the absence of glia (Susuki et al., 2013). However, node formation is not or only very mildly affected in the single and double KOs of these ECM molecules.

The *qv3J* mutant allele of βIV spectrin removes its C-terminal PH domain and part of the variable region that is less conserved among β spectrins. In these mutants, CNS nodes exhibit age-dependent elongation and membrane protrusion, but nodes can still be formed albeit with a reduced number (Susuki et al., 2013; Yang, Lacas-Gervais, Morest, Solimena, & Rasband, 2004).

Analysis of these single-mechanism mutants revealed only mildly affected node assembly. When these mutations were combined to make mutants with two mechanisms disrupted simultaneously, additive and even synergistic defects in node assembly were observed, supporting the notion that the ECM, paranodal junction, and cytoskeletal scaffold βIV spectrin function together in node formation during early development. Thus, a mechanism of assembly emerges where paranodal diffusion barriers collect NF186 and AnkG to the edges of nascent myelin sheaths and the immature nodal clusters are stabilized by interactions between NF186 and the ECM and between AnkG-recruited βIV spectrin and actin (Susuki et al., 2013; Fig. 5.4). The mild defects observed in single-mechanism mutants result from compensation or partial redundancy by the other two mechanisms.

The Nav channel auxiliary β subunits can also interact with AnkG, NF186, NrCAM, contactin, and Tn-R, which are all found at CNS nodes (the former three at PNS nodes as well), and thus have been proposed to be involved in nodal targeting of Nav channels in addition to their role in efficient surface expression (Patino & Isom, 2010). However, the analysis of a CD4 chimera containing Nav1.2's AnkG-binding motif showed the interaction between Nav1.2 and AnkG is sufficient for nodal enrichment. Additional experiments analyzing full-length Nav1.6 with a point mutation in the AnkG-binding motif suggested the interaction between Nav1.6 and AnkG is essential for its nodal targeting (Gasser et al., 2012). Furthermore, recent experiments have now shown that nodal targeting of Navβ2 and Navβ4 requires their linkage to Nav channels (Buffington & Rasband, 2013; Chen et al., 2012). Taken together, these observations indicate that the direct interaction between Nav channels and AnkG plays a crucial role in recruiting Nav channels to nodes. Nevertheless, future experiments using AnkG-deficient mice will be necessary to finally prove the essential role this scaffold has in node of Ranvier formation.

3.3. Assembly of AISs and nodes of Ranvier during evolution

When did neurons start to concentrate ion channels at the AIS and nodes of Ranvier? As mentioned earlier, their high-density clustering at the nodes and AIS depends on their binding to AnkG. Phylogenetic analysis of Nav channels' AnkG-binding motifs suggested their advent when chordates diverged from invertebrates and before the evolution of myelin. Thus, the nodal molecular clustering is an adoption and adaptation of AIS clustering (Hill et al., 2008). Immunostaining of Nav channels confirmed the enrichment and detergent resistance of Nav channels at the morphological AISs of lampreys, which lack myelin, suggesting lampreys have functional AIS domains.

Interestingly, KCNQ2 and KCNQ3 share conserved AnkG-binding motifs with Nav channels, and they acquired these motifs when jawed vertebrates diverged from jawless vertebrates, close to the time when myelin is thought to have evolved (Hill et al., 2008). This observation suggests the AIS domain has evolved from a primitive form to a more advanced state and that nodes were built by adapting the blueprint of the advanced AIS. Nevertheless, several key questions remain. For example, it is not known when Kv1 channels began to accumulate at the AIS and what key evolutionary event separated them from the AIS-like node. Interestingly, amphibians have retained nodal Kv1 channels (Rasband, 2004), suggesting that mammals have further modified the nodal organization to meet their unique needs. To begin to understand how Kv1 channels came to be enriched at the AIS and juxtaparanodes will require identification of the molecular links between Kv1 channels and AnkG and between Kv1 channels and Caspr2, respectively.

4. DISRUPTION OF AISs AND NODES OF RANVIER IN DISEASE AND INJURY

Recent studies have shown that both the AIS and nodes are involved in the pathophysiology of many diseases and injuries. Indeed, mutations in nodal and AIS proteins; changes in ion channel expression levels, function, and location; and even frank dismantling of the AIS and nodes have all been reported to cause or result from disease or injury. Although the contributions of the AIS to injury and disease have been reviewed recently (Buffington & Rasband, 2011; Wimmer, Reid, So, Berkovic, & Petrou, 2010), we wish to highlight several examples that illustrate how disease can alter AIS and node structure and function.

As described earlier, AnkG is required to maintain both the molecular organization of the AIS and neuronal polarity. Thus, diseases or injuries that disrupt AnkG might be expected to dramatically alter nervous system function through the loss of channel clustering and neuronal polarity. Schafer et al. (Schafer et al., 2009) examined this possibility in a stroke model. We found that stroke results in the rapid activation of the Ca^{2+}-dependent protease calpain. Furthermore, we found that both AnkG and βIV spectrin are potent substrates for calpain and are rapidly proteolyzed after injury.

The proteolysis of these AIS scaffolds causes loss of clustering of AIS proteins and neuronal polarity without accompanying cell death. Surprisingly, loss of the AIS was found to be an irreversible event: many neurons that survived the initial hypoxic insult remained viable, but without an AIS and without a molecularly defined axon (i.e., neuronal polarity was disrupted). However, the addition of calpain inhibitors effectively blocked proteolysis and dismantling of the AIS and preserved polarity. In contrast, neuroprotection was not sufficient to inhibit proteolysis of AnkG and βIV spectrin. These results are important since they illustrate that therapeutic strategies aimed at neuroprotection must be accompanied by treatments that preserve the AIS cytoskeleton in order to maintain neuronal polarity and AIS ion channel clustering.

Not all injuries or diseases result in dismantling of the AIS. Some injuries lead to structural alterations in the AIS that are thought to be homeostatic responses to altered neuronal activity. For example, in one model of injury where neurons were deprived of synaptic input, the length of the AIS increased to facilitate neuronal activity (Kuba, Oichi, & Ohmori, 2010). In contrast, in a model of blast-induced mild traumatic brain injury, the length of the AIS decreased, likely in response to hyperexcitability of injured neurons (Baalman, Cotton, Rasband, & Rasband, 2013). Similarly, in a photothrombotic model of focal cortical stroke, AIS length decreased, likely in response to the excitotoxic environment surrounding the lesion (Hinman, Rasband, & Carmichael, 2013). Taken together, these observations illustrate how the AIS can respond to a variety of CNS injuries and emphasize the central role this structure plays in normal and diseased nervous system function.

As described earlier, node of Ranvier assembly and maintenance depends on neuron–glia interactions. Therefore, it is not surprising that demyelination, dysmyelination, or hypomyelination lead to altered node structure and function. For example, studies of human tissues from patients with multiple sclerosis, as well as a variety of animal models of demyelination, have shown that demyelinated axons have diffuse immunoreactivity for nodal, paranodal, and juxtaparanodal proteins (Coman et al., 2006; Craner et al., 2004; Dugandžija-Novaković, Koszowski, Levinson, & Shrager, 1995). Furthermore, animal models of Charcot–Marie–Tooth disease also show aberrant channel clustering, expression, and localization (Devaux & Scherer, 2005). In fact, demyelination and hypomyelination have even been shown to cause changes in the kinds of Nav channels that are expressed in axons (Noebels, Marcom, & Jalilian-Tehrani, 1991; Rasband, Kagawa,

Park, Ikenaka, & Trimmer, 2003). These observations have led some investigators to propose that demyelinating diseases should be considered "transcriptional channelopathies" (Waxman, 2001).

Although not comprehensive, the few examples cited earlier illustrate how disruption of ion channel clustering and the normal polarization of the axonal membrane is a common and previously unappreciated consequence of nervous system injuries and diseases. Efforts to develop therapeutic strategies aimed at nervous system preservation, repair, and regeneration will require a detailed understanding of the mechanisms leading to disruption of these important functional domains. Fortunately, the recent advances described in this chapter, and the accelerating pace of discoveries to understand the molecular composition of the AIS, nodes, paranodes, and juxtaparanodes, and the developmental mechanisms responsible for their assembly, bring us that much closer to realization of this important goal.

ACKNOWLEDGMENTS

We wish to thank Pei-Jung Lee for the graphic design of the figures. This work was supported by NIH grant R01 NS063688. We respectfully acknowledge and apologize to our colleagues whose work we have not discussed due to space limitations.

REFERENCES

Akhavan, A., Crivelli, S. N., Singh, M., Lingappa, V. R., & Muschler, J. L. (2008). SEA domain proteolysis determines the functional composition of dystroglycan. *The FASEB Journal*, *22*, 612–621.

Arroyo, E. J., & Scherer, S. S. (2000). On the molecular architecture of myelinated fibers. *Histochemistry and Cell Biology*, *113*, 1–18.

Baalman, K. L., Cotton, R. J., Rasband, S. N., & Rasband, M. N. (2013). Blast wave exposure impairs memory and decreases axon initial segment length. *Journal of Neurotrauma*, *30*, 741–751.

Bangratz, M., Sarrazin, N., Devaux, J., Zambroni, D., Echaniz-Laguna, A., René, F., et al. (2012). A mouse model of Schwartz-Jampel syndrome reveals myelinating Schwann cell dysfunction with persistent axonal depolarization in vitro and distal peripheral nerve hyperexcitability when perlecan is lacking. *The American Journal of Pathology*, *180*, 2040–2055.

Barresi, R., & Campbell, K. P. (2006). Dystroglycan: From biosynthesis to pathogenesis of human disease. *Journal of Cell Science*, *119*, 199–207.

Bean, B. P. (2007). The action potential in mammalian central neurons. *Nature Reviews Neuroscience*, *8*, 451–465.

Bekku, Y., & Oohashi, T. (2010). Neurocan contributes to the molecular heterogeneity of the perinodal ECM. *Archives of Histology and Cytology*, *73*, 95–102.

Bekku, Y., Vargová, L., Goto, Y., Vorísek, I., Dmytrenko, L., Narasaki, M., et al. (2010). Bral1: Its role in diffusion barrier formation and conduction velocity in the CNS. *The Journal of Neuroscience*, *30*, 3113–3123.

Bender, K. J., Ford, C. P., & Trussell, L. O. (2010). Dopaminergic modulation of axon initial segment calcium channels regulates action potential initiation. *Neuron*, *68*, 500–511.

Bender, K. J., & Trussell, L. O. (2009). Axon initial segment Ca^{2+} channels influence action potential generation and timing. *Neuron, 61*, 259–271.

Bennett, V., & Baines, A. J. (2001). Spectrin and ankyrin-based pathways: Metazoan inventions for integrating cells into tissues. *Physiological Reviews, 81*, 1353–1392.

Bennett, V., & Chen, L. (2001). Ankyrins and cellular targeting of diverse membrane proteins to physiological sites. *Current Opinion in Cell Biology, 13*, 61–67.

Bhat, M. A., Rios, J. C., Lu, Y., Garcia-Fresco, G. P., Ching, W., St Martin, M., et al. (2001). Axon-glia interactions and the domain organization of myelinated axons requires neurexin IV/Caspr/Paranodin. *Neuron, 30*, 369–383.

Boiko, T., Rasband, M. N., Levinson, S. R., Caldwell, J. H., Mandel, G., Trimmer, J. S., et al. (2001). Compact myelin dictates the differential targeting of two sodium channel isoforms in the same axon. *Neuron, 30*, 91–104.

Boiko, T., Vakulenko, M., Ewers, H., Yap, C. C., Norden, C., & Winckler, B. (2007). Ankyrin-dependent and -independent mechanisms orchestrate axonal compartmentalization of L1 family members neurofascin and L1/neuron-glia cell adhesion molecule. *The Journal of Neuroscience, 27*, 590–603.

Boyle, M. E., Berglund, E. O., Murai, K. K., Weber, L., Peles, E., & Ranscht, B. (2001). Contactin orchestrates assembly of the septate-like junctions at the paranode in myelinated peripheral nerve. *Neuron, 30*, 385–397.

Brackenbury, W. J., Calhoun, J. D., Chen, C., Miyazaki, H., Nukina, N., Oyama, F., et al. (2010). Functional reciprocity between Na^+ channel $Na_v1.6$ and β1 subunits in the coordinated regulation of excitability and neurite outgrowth. *Proceedings of the National Academy of Sciences of the United States of America, 107*, 2283–2288.

Bréchet, A., Fache, M. P., Brachet, A., Ferracci, G., Baude, A., Irondelle, M., et al. (2008). Protein kinase CK2 contributes to the organization of sodium channels in axonal membranes by regulating their interactions with ankyrin G. *The Journal of Cell Biology, 183*, 1101–1114.

Brückner, G., Szeöke, S., Pavlica, S., Grosche, J., & Kacza, J. (2006). Axon initial segment ensheathed by extracellular matrix in perineuronal nets. *Neuroscience, 138*, 365–375.

Buffington, S. A., & Rasband, M. N. (2011). The axon initial segment in nervous system disease and injury. *The European Journal of Neuroscience, 34*, 1609–1619.

Buffington, S. A., & Rasband, M. N. (2013). Na^+ channel-dependent recruitment of $Na_v\beta4$ to axon initial segments and nodes of Ranvier. *The Journal of Neuroscience, 33*, 6191–6202.

Buffington, S. A., Sobotzik, J. M., Schultz, C., & Rasband, M. N. (2012). IκBα is not required for axon initial segment assembly. *Molecular and Cellular Neurosciences, 50*, 1–9.

Buttermore, E. D., Dupree, J. L., Cheng, J., An, X., Tessarollo, L., & Bhat, M. A. (2011). The cytoskeletal adaptor protein band 4.1B is required for the maintenance of paranodal axoglial septate junctions in myelinated axons. *The Journal of Neuroscience, 31*, 8013–8024.

Celio, M. R., Spreafico, R., De Biasi, S., & Vitellaro-Zuccarello, L. (1998). Perineuronal nets: Past and present. *Trends in Neurosciences, 21*, 510–515.

Charles, P., Tait, S., Faivre-Sarrailh, C., Barbin, G., Gunn-Moore, F., Denisenko-Nehrbass, N., et al. (2002). Neurofascin is a glial receptor for the paranodin/Caspr-contactin axonal complex at the axoglial junction. *Current Biology, 12*, 217–220.

Chen, C., Calhoun, J. D., Zhang, Y., Lopez-Santiago, L., Zhou, N., Davis, T. H., et al. (2012). Identification of the cysteine residue responsible for disulfide linkage of Na^+ channel α and β2 subunits. *The Journal of Biological Chemistry, 287*, 39061–39069.

Chiu, S. Y., Zhou, L., Zhang, C. L., & Messing, A. (1999). Analysis of potassium channel functions in mammalian axons by gene knockouts. *Journal of Neurocytology, 28*, 349–364.

Cifuentes-Diaz, C., Chareyre, F., Garcia, M., Devaux, J., Carnaud, M., Levasseur, G., et al. (2011). Protein 4.1B contributes to the organization of peripheral myelinated axons. *PLoS One, 6*, e25043.

Cobb, S. R., Buhl, E. H., Halasy, K., Paulsen, O., & Somogyi, P. (1995). Synchronization of neuronal activity in hippocampus by individual GABAergic interneurons. *Nature*, *378*, 75–78.

Colognato, H., & Tzvetanova, I. D. (2011). Glia unglued: How signals from the extracellular matrix regulate the development of myelinating glia. *Developmental Neurobiology*, *71*, 924–955.

Coman, I., Aigrot, M. S., Seilhean, D., Reynolds, R., Girault, J. A., Zalc, B., et al. (2006). Nodal, paranodal and juxtaparanodal axonal proteins during demyelination and remyelination in multiple sclerosis. *Brain*, *129*, 3186–3195.

Cooper, E. C. (2011). Made for "anchorin": Kv7.2/7.3 (KCNQ2/KCNQ3) channels and the modulation of neuronal excitability in vertebrate axons. *Seminars in Cell & Developmental Biology*, *22*, 185–192.

Craner, M. J., Newcombe, J., Black, J. A., Hartle, C., Cuzner, M. L., & Waxman, S. G. (2004). Molecular changes in neurons in multiple sclerosis: Altered axonal expression of $Na_v1.2$ and $Na_v1.6$ sodium channels and Na^+/Ca^{2+} exchanger. *Proceedings of the National Academy of Sciences of the United States of America*, *101*, 8168–8173.

Custer, A. W., Kazarinova-Noyes, K., Sakurai, T., Xu, X., Simon, W., Grumet, M., et al. (2003). The role of the ankyrin-binding protein NrCAM in node of Ranvier formation. *The Journal of Neuroscience*, *23*, 10032–10039.

Davis, J. Q., Lambert, S., & Bennett, V. (1996). Molecular composition of the node of Ranvier: Identification of ankyrin-binding cell adhesion molecules neurofascin (mucin+/third FNIII domain-) and NrCAM at nodal axon segments. *The Journal of Cell Biology*, *135*, 1355–1367.

Debanne, D., Campanac, E., Bialowas, A., Carlier, E., & Alcaraz, G. (2011). Axon physiology. *Physiological Reviews*, *91*, 555–602.

Denisenko-Nehrbass, N., Oguievetskaia, K., Goutebroze, L., Galvez, T., Yamakawa, H., Ohara, O., et al. (2003). Protein 4.1B associates with both Caspr/paranodin and Caspr2 at paranodes and juxtaparanodes of myelinated fibres. *The European Journal of Neuroscience*, *17*, 411–416.

Devaux, J. J., & Scherer, S. S. (2005). Altered ion channels in an animal model of Charcot-Marie-Tooth disease type IA. *The Journal of Neuroscience*, *25*, 1470–1480.

Duflocq, A., Chareyre, F., Giovannini, M., Couraud, F., & Davenne, M. (2011). Characterization of the axon initial segment (AIS) of motor neurons and identification of a para-AIS and a juxtapara-AIS, organized by protein 4.1B. *BMC Biology*, *9*, 66.

Dugandžija-Novaković, S., Koszowski, A. G., Levinson, S. R., & Shrager, P. (1995). Clustering of Na^+ channels and node of Ranvier formation in remyelinating axons. *The Journal of Neuroscience*, *15*, 492–503.

Dupree, J. L., Girault, J. A., & Popko, B. (1999). Axo-glial interactions regulate the localization of axonal paranodal proteins. *The Journal of Cell Biology*, *147*, 1145–1152.

Dupree, J. L., & Popko, B. (1999). Genetic dissection of myelin galactolipid function. *Journal of Neurocytology*, *28*, 271–279.

Dzhashiashvili, Y., Zhang, Y., Galinska, J., Lam, I., Grumet, M., & Salzer, J. L. (2007). Nodes of Ranvier and axon initial segments are ankyrin G-dependent domains that assemble by distinct mechanisms. *The Journal of Cell Biology*, *177*, 857–870.

Einheber, S., Meng, X., Rubin, M., Lam, I., Mohandas, N., An, X., et al. (2013). The 4.1B cytoskeletal protein regulates the domain organization and sheath thickness of myelinated axons. *Glia*, *61*, 240–253.

Einheber, S., Zanazzi, G., Ching, W., Scherer, S., Milner, T. A., Peles, E., et al. (1997). The axonal membrane protein Caspr, a homologue of neurexin IV, is a component of the septate-like paranodal junctions that assemble during myelination. *The Journal of Cell Biology*, *139*, 1495–1506.

Elfvin, L. G. (1961). The ultrastructure of the nodes of Ranvier in cat sympathetic nerve fibers. *Journal of Ultrastructure Research*, *5*, 374–387.

Ellisman, M. H., & Levinson, S. R. (1982). Immunocytochemical localization of sodium channel distributions in the excitable membranes of *Electrophorus electricus*. *Proceedings of the National Academy of Sciences of the United States of America*, *79*, 6707–6711.

Eshed, Y., Feinberg, K., Carey, D. J., & Peles, E. (2007). Secreted gliomedin is a perinodal matrix component of peripheral nerves. *The Journal of Cell Biology*, *177*, 551–562.

Eshed, Y., Feinberg, K., Poliak, S., Sabanay, H., Sarig-Nadir, O., Spiegel, I., et al. (2005). Gliomedin mediates Schwann cell-axon interaction and the molecular assembly of the nodes of Ranvier. *Neuron*, *47*, 215–229.

Feinberg, K., Eshed-Eisenbach, Y., Frechter, S., Amor, V., Salomon, D., Sabanay, H., et al. (2010). A glial signal consisting of gliomedin and NrCAM clusters axonal Na^+ channels during the formation of nodes of Ranvier. *Neuron*, *65*, 490–502.

Frischknecht, R., Heine, M., Perrais, D., Seidenbecher, C. I., Choquet, D., & Gundelfinger, E. D. (2009). Brain extracellular matrix affects AMPA receptor lateral mobility and short-term synaptic plasticity. *Nature Neuroscience*, *12*, 897–904.

Galiano, M. R., Jha, S., Ho, T. S., Zhang, C., Ogawa, Y., Chang, K. J., et al. (2012). A distal axonal cytoskeleton forms an intra-axonal boundary that controls axon initial segment assembly. *Cell*, *149*, 1125–1139.

Garrido, J. J., Giraud, P., Carlier, E., Fernandes, F., Moussif, A., Fache, M. P., et al. (2003). A targeting motif involved in sodium channel clustering at the axonal initial segment. *Science*, *300*, 2091–2094.

Gasser, A., Ho, T. S., Cheng, X., Chang, K. J., Waxman, S. G., Rasband, M. N., et al. (2012). An ankyrinG-binding motif is necessary and sufficient for targeting $Na_v1.6$ sodium channels to axon initial segments and nodes of Ranvier. *The Journal of Neuroscience*, *32*, 7232–7243.

Gollan, L., Salomon, D., Salzer, J. L., & Peles, E. (2003). Caspr regulates the processing of contactin and inhibits its binding to neurofascin. *The Journal of Cell Biology*, *163*, 1213–1218.

He, M., Jenkins, P., & Bennett, V. (2012). Cysteine 70 of ankyrin-G is S-palmitoylated and is required for function of ankyrin-G in membrane domain assembly. *The Journal of Biological Chemistry*, *287*, 43995–44005.

Hedstrom, K. L., Ogawa, Y., & Rasband, M. N. (2008). AnkyrinG is required for maintenance of the axon initial segment and neuronal polarity. *The Journal of Cell Biology*, *183*, 635–640.

Hedstrom, K. L., Xu, X., Ogawa, Y., Frischknecht, R., Seidenbecher, C. I., Shrager, P., et al. (2007). Neurofascin assembles a specialized extracellular matrix at the axon initial segment. *The Journal of Cell Biology*, *178*, 875–886.

Hill, A. S., Nishino, A., Nakajo, K., Zhang, G., Fineman, J. R., Selzer, M. E., et al. (2008). Ion channel clustering at the axon initial segment and node of Ranvier evolved sequentially in early chordates. *PLoS Genetics*, *4*, e1000317.

Hinman, J. D., Rasband, M. N., & Carmichael, S. T. (2013). Remodeling of the axon initial segment after focal cortical and white matter stroke. *Stroke*, *44*, 182–189.

Honke, K., Hirahara, Y., Dupree, J., Suzuki, K., Popko, B., Fukushima, K., et al. (2002). Paranodal junction formation and spermatogenesis require sulfoglycolipids. *Proceedings of the National Academy of Sciences of the United States of America*, *99*, 4227–4232.

Horresh, I., Bar, V., Kissil, J. L., & Peles, E. (2010). Organization of myelinated axons by Caspr and Caspr2 requires the cytoskeletal adapter protein 4.1B. *The Journal of Neuroscience*, *30*, 2480–2489.

Horresh, I., Poliak, S., Grant, S., Bredt, D., Rasband, M. N., & Peles, E. (2008). Multiple molecular interactions determine the clustering of Caspr2 and Kv1 channels in myelinated axons. *The Journal of Neuroscience*, *28*, 14213–14222.

Hoshi, T., Suzuki, A., Hayashi, S., Tohyama, K., Hayashi, A., Yamaguchi, Y., et al. (2007). Nodal protrusions, increased Schmidt-Lanterman incisures, and paranodal disorganization are characteristic features of sulfatide-deficient peripheral nerves. *Glia, 55*, 584–594.

Hu, W., Tian, C., Li, T., Yang, M., Hou, H., & Shu, Y. (2009). Distinct contributions of $Na_v1.6$ and $Na_v1.2$ in action potential initiation and backpropagation. *Nature Neuroscience, 12*, 996–1002.

Inda, M. C., DeFelipe, J., & Muñoz, A. (2006). Voltage-gated ion channels in the axon initial segment of human cortical pyramidal cells and their relationship with chandelier cells. *Proceedings of the National Academy of Sciences of the United States of America, 103*, 2920–2925.

Ishibashi, T., Dupree, J. L., Ikenaka, K., Hirahara, Y., Honke, K., Peles, E., et al. (2002). A myelin galactolipid, sulfatide, is essential for maintenance of ion channels on myelinated axon but not essential for initial cluster formation. *The Journal of Neuroscience, 22*, 6507–6514.

Ivanovic, A., Horresh, I., Golan, N., Spiegel, I., Sabanay, H., Frechter, S., et al. (2012). The cytoskeletal adapter protein 4.1G organizes the internodes in peripheral myelinated nerves. *The Journal of Cell Biology, 196*, 337–344.

Jenkins, S. M., & Bennett, V. (2001). Ankyrin-G coordinates assembly of the spectrin-based membrane skeleton, voltage-gated sodium channels, and L1 CAMs at Purkinje neuron initial segments. *The Journal of Cell Biology, 155*, 739–746.

Jenkins, S. M., & Bennett, V. (2002). Developing nodes of Ranvier are defined by ankyrin-G clustering and are independent of paranodal axoglial adhesion. *Proceedings of the National Academy of Sciences of the United States of America, 99*, 2303–2308.

John, N., Krügel, H., Frischknecht, R., Smalla, K. H., Schultz, C., Kreutz, M. R., et al. (2006). Brevican-containing perineuronal nets of extracellular matrix in dissociated hippocampal primary cultures. *Molecular and Cellular Neurosciences, 31*, 774–784.

Kaplan, M. R., Cho, M. H., Ullian, E. M., Isom, L. L., Levinson, S. R., & Barres, B. A. (2001). Differential control of clustering of the sodium channels $Na_v1.2$ and $Na_v1.6$ at developing CNS nodes of Ranvier. *Neuron, 30*, 105–119.

Kaplan, M. R., Meyer-Franke, A., Lambert, S., Bennett, V., Duncan, I. D., Levinson, S. R., et al. (1997). Induction of sodium channel clustering by oligodendrocytes. *Nature, 386*, 724–728.

Kayyem, J. F., Roman, J. M., de la Rosa, E. J., Schwarz, U., & Dreyer, W. J. (1992). Bravo/Nr-CAM is closely related to the cell adhesion molecules L1 and Ng-CAM and has a similar heterodimer structure. *The Journal of Cell Biology, 118*, 1259–1270.

Kazarinova-Noyes, K., Malhotra, J. D., McEwen, D. P., Mattei, L. N., Berglund, E. O., Ranscht, B., et al. (2001). Contactin associates with Na^+ channels and increases their functional expression. *The Journal of Neuroscience, 21*, 7517–7525.

Khaliq, Z. M., Gouwens, N. W., & Raman, I. M. (2003). The contribution of resurgent sodium current to high-frequency firing in Purkinje neurons: An experimental and modeling study. *The Journal of Neuroscience, 23*, 4899–4912.

Kobayashi, T., Storrie, B., Simons, K., & Dotti, C. G. (1992). A functional barrier to movement of lipids in polarized neurons. *Nature, 359*, 647–650.

Kole, M. H., Ilschner, S. U., Kampa, B. M., Williams, S. R., Ruben, P. C., & Stuart, G. J. (2008). Action potential generation requires a high sodium channel density in the axon initial segment. *Nature Neuroscience, 11*, 178–186.

Kole, M. H., & Stuart, G. J. (2012). Signal processing in the axon initial segment. *Neuron, 73*, 235–247.

Komada, M., & Soriano, P. (2002). βIV-spectrin regulates sodium channel clustering through ankyrin-G at axon initial segments and nodes of Ranvier. *The Journal of Cell Biology, 156*, 337–348.

Kuba, H., Oichi, Y., & Ohmori, H. (2010). Presynaptic activity regulates Na^+ channel distribution at the axon initial segment. *Nature, 465*, 1075–1078.

Lambert, S., Davis, J. Q., & Bennett, V. (1997). Morphogenesis of the node of Ranvier: Co-clusters of ankyrin and ankyrin-binding integral proteins define early developmental intermediates. *The Journal of Neuroscience, 17*, 7025–7036.

Lee, A. G. (2001). Myelin: Delivery by raft. *Current Biology, 11*, R60–R62.

Lemaillet, G., Walker, B., & Lambert, S. (2003). Identification of a conserved ankyrin-binding motif in the family of sodium channel alpha subunits. *The Journal of Biological Chemistry, 278*, 27333–27339.

Leterrier, C., Vacher, H., Fache, M. P., d'Ortoli, S. A., Castets, F., Autillo-Touati, A., et al. (2011). End-binding proteins EB3 and EB1 link microtubules to ankyrin G in the axon initial segment. *Proceedings of the National Academy of Sciences of the United States of America, 108*, 8826–8831.

Lorincz, A., & Nusser, Z. (2008). Cell-type-dependent molecular composition of the axon initial segment. *The Journal of Neuroscience, 28*, 14329–14340.

Martin, P. M., Carnaud, M., Garcia del Caño, G., Irondelle, M., Irinopoulou, T., Girault, J. A., et al. (2008). Schwannomin-interacting protein-1 isoform IQCJ-SCHIP-1 is a late component of nodes of Ranvier and axon initial segments. *The Journal of Neuroscience, 28*, 6111–6117.

Martin, S., Levine, A. K., Chen, Z. J., Ughrin, Y., & Levine, J. M. (2001). Deposition of the NG2 proteoglycan at nodes of Ranvier in the peripheral nervous system. *The Journal of Neuroscience, 21*, 8119–8128.

Masaki, T., & Matsumura, K. (2010). Biological role of dystroglycan in Schwann cell function and its implications in peripheral nervous system diseases. *Journal of Biomedicine & Biotechnology, 2010*, 740403.

Matsumoto, E., & Rosenbluth, J. (1985). Plasma membrane structure at the axon hillock, initial segment and cell body of frog dorsal root ganglion cells. *Journal of Neurocytology, 14*, 731–747.

Melendez-Vasquez, C., Carey, D. J., Zanazzi, G., Reizes, O., Maurel, P., & Salzer, J. L. (2005). Differential expression of proteoglycans at central and peripheral nodes of Ranvier. *Glia, 52*, 301–308.

Melendez-Vasquez, C. V., Rios, J. C., Zanazzi, G., Lambert, S., Bretscher, A., & Salzer, J. L. (2001). Nodes of Ranvier form in association with ezrin-radixin-moesin (ERM)-positive Schwann cell processes. *Proceedings of the National Academy of Sciences of the United States of America, 98*, 1235–1240.

Menegoz, M., Gaspar, P., Le Bert, M., Galvez, T., Burgaya, F., Palfrey, C., et al. (1997). Paranodin, a glycoprotein of neuronal paranodal membranes. *Neuron, 19*, 319–331.

Nakada, C., Ritchie, K., Oba, Y., Nakamura, M., Hotta, Y., Iino, R., et al. (2003). Accumulation of anchored proteins forms membrane diffusion barriers during neuronal polarization. *Nature Cell Biology, 5*, 626–632.

Noebels, J. L., Marcom, P. K., & Jalilian-Tehrani, M. H. (1991). Sodium channel density in hypomyelinated brain increased by myelin basic protein gene deletion. *Nature, 352*, 431–434.

Occhi, S., Zambroni, D., Del Carro, U., Amadio, S., Sirkowski, E. E., Scherer, S. S., et al. (2005). Both laminin and Schwann cell dystroglycan are necessary for proper clustering of sodium channels at nodes of Ranvier. *The Journal of Neuroscience, 25*, 9418–9427.

Ogawa, Y., Horresh, I., Trimmer, J. S., Bredt, D. S., Peles, E., & Rasband, M. N. (2008). Postsynaptic density-93 clusters Kv1 channels at axon initial segments independently of Caspr2. *The Journal of Neuroscience, 28*, 5731–5739.

Ogawa, Y., Oses-Prieto, J., Kim, M. Y., Horresh, I., Peles, E., Burlingame, A. L., et al. (2010). ADAM22, a Kv1 channel-interacting protein, recruits membrane-associated

guanylate kinases to juxtaparanodes of myelinated axons. *The Journal of Neuroscience*, *30*, 1038–1048.
Ogawa, Y., & Rasband, M. N. (2009). Proteomic analysis of optic nerve lipid rafts reveals new paranodal proteins. *Journal of Neuroscience Research*, *87*, 3502–3510.
Ogawa, Y., Schafer, D. P., Horresh, I., Bar, V., Hales, K., Yang, Y., et al. (2006). Spectrins and ankyrinB constitute a specialized paranodal cytoskeleton. *The Journal of Neuroscience*, *26*, 5230–5239.
Ohara, R., Yamakawa, H., Nakayama, M., & Ohara, O. (2000). Type II brain 4.1 (4.1B/KIAA0987), a member of the protein 4.1 family, is localized to neuronal paranodes. *Brain Research Molecular Brain Research*, *85*, 41–52.
Ohno, N., Terada, N., Yamakawa, H., Komada, M., Ohara, O., Trapp, B. D., et al. (2006). Expression of protein 4.1G in Schwann cells of the peripheral nervous system. *Journal of Neuroscience Research*, *84*, 568–577.
Palay, S. L., Sotelo, C., Peters, A., & Orkand, P. M. (1968). The axon hillock and the initial segment. *The Journal of Cell Biology*, *38*, 193–201.
Pan, Z., Kao, T., Horvath, Z., Lemos, J., Sul, J. Y., Cranstoun, S. D., et al. (2006). A common ankyrin-G-based mechanism retains KCNQ and NaV channels at electrically active domains of the axon. *The Journal of Neuroscience*, *26*, 2599–2613.
Patino, G. A., & Isom, L. L. (2010). Electrophysiology and beyond: Multiple roles of Na^+ channel β subunits in development and disease. *Neuroscience Letters*, *486*, 53–59.
Peters, A. (1966). The node of Ranvier in the central nervous system. *Quarterly Journal of Experimental Physiology and Cognate Medical Sciences*, *51*, 229–236.
Peters, A., Proskauer, C. C., & Kaiserman-Abramof, I. R. (1968). The small pyramidal neuron of the rat cerebral cortex. The axon hillock and initial segment. *The Journal of Cell Biology*, *39*, 604–619.
Pillai, A. M., Thaxton, C., Pribisko, A. L., Cheng, J. G., Dupree, J. L., & Bhat, M. A. (2009). Spatiotemporal ablation of myelinating glia-specific *Neurofascin* ($Nfasc^{NF155}$) in mice reveals gradual loss of paranodal axoglial junctions and concomitant disorganization of axonal domains. *Journal of Neuroscience Research*, *87*, 1773–1793.
Poliak, S., Gollan, L., Martinez, R., Custer, A., Einheber, S., Salzer, J. L., et al. (1999). Caspr2, a new member of the neurexin superfamily, is localized at the juxtaparanodes of myelinated axons and associates with K^+ channels. *Neuron*, *24*, 1037–1047.
Poliak, S., Gollan, L., Salomon, D., Berglund, E. O., Ohara, R., Ranscht, B., et al. (2001). Localization of Caspr2 in myelinated nerves depends on axon-glia interactions and the generation of barriers along the axon. *The Journal of Neuroscience*, *21*, 7568–7575.
Poliak, S., Matlis, S., Ullmer, C., Scherer, S. S., & Peles, E. (2002). Distinct claudins and associated PDZ proteins form different autotypic tight junctions in myelinating Schwann cells. *The Journal of Cell Biology*, *159*, 361–372.
Poliak, S., & Peles, E. (2003). The local differentiation of myelinated axons at nodes of Ranvier. *Nature Reviews Neuroscience*, *4*, 968–980.
Poliak, S., Salomon, D., Elhanany, H., Sabanay, H., Kiernan, B., Pevny, L., et al. (2003). Juxtaparanodal clustering of *Shaker*-like K^+ channels in myelinated axons depends on Caspr2 and TAG-1. *The Journal of Cell Biology*, *162*, 1149–1160.
Rasband, M. N. (2004). It's "juxta" potassium channel!. *Journal of Neuroscience Research*, *76*, 749–757.
Rasband, M. N. (2010). The axon initial segment and the maintenance of neuronal polarity. *Nature Reviews Neuroscience*, *11*, 552–562.
Rasband, M. N., Kagawa, T., Park, E. W., Ikenaka, K., & Trimmer, J. S. (2003). Dysregulation of axonal sodium channel isoforms after adult-onset chronic demyelination. *Journal of Neuroscience Research*, *73*, 465–470.

Rasband, M. N., Park, E. W., Zhen, D., Arbuckle, M. I., Poliak, S., Peles, E., et al. (2002). Clustering of neuronal potassium channels is independent of their interaction with PSD-95. *The Journal of Cell Biology, 159*, 663–672.

Rasband, M. N., Peles, E., Trimmer, J. S., Levinson, S. R., Lux, S. E., & Shrager, P. (1999). Dependence of nodal sodium channel clustering on paranodal axoglial contact in the developing CNS. *The Journal of Neuroscience, 19*, 7516–7528.

Rasband, M. N., Tayler, J., Kaga, Y., Yang, Y., Lappe-Siefke, C., Nave, K. A., et al. (2005). CNP is required for maintenance of axon-glia interactions at nodes of Ranvier in the CNS. *Glia, 50*, 86–90.

Rasband, M. N., Trimmer, J. S., Peles, E., Levinson, S. R., & Shrager, P. (1999). K^+ channel distribution and clustering in developing and hypomyelinated axons of the optic nerve. *Journal of Neurocytology, 28*, 319–331.

Rios, J. C., Melendez-Vasquez, C. V., Einheber, S., Lustig, M., Grumet, M., Hemperly, J., et al. (2000). Contactin-associated protein (Caspr) and contactin form a complex that is targeted to the paranodal junctions during myelination. *The Journal of Neuroscience, 20*, 8354–8364.

Rios, J. C., Rubin, M., St Martin, M., Downey, R. T., Einheber, S., Rosenbluth, J., et al. (2003). Paranodal interactions regulate expression of sodium channel subtypes and provide a diffusion barrier for the node of Ranvier. *The Journal of Neuroscience, 23*, 7001–7011.

Rosenbluth, J. (1976). Intramembranous particle distribution at the node of Ranvier and adjacent axolemma in myelinated axons of the frog brain. *Journal of Neurocytology, 5*, 731–745.

Rosenbluth, J. (1999). A brief history of myelinated nerve fibers: One hundred and fifty years of controversy. *Journal of Neurocytology, 28*, 251–262.

Rosenbluth, J. (2009). Multiple functions of the paranodal junction of myelinated nerve fibers. *Journal of Neuroscience Research, 87*, 3250–3258.

Rosenbluth, J., Petzold, C., & Peles, E. (2012). Dependence of paranodal junctional gap width on transverse bands. *The Journal of Comparative Neurology, 520*, 2774–2784.

Saito, F., Moore, S. A., Barresi, R., Henry, M. D., Messing, A., Ross-Barta, S. E., et al. (2003). Unique role of dystroglycan in peripheral nerve myelination, nodal structure, and sodium channel stabilization. *Neuron, 38*, 747–758.

Salzer, J. L. (2003). Polarized domains of myelinated axons. *Neuron, 40*, 297–318.

Sánchez-Ponce, D., DeFelipe, J., Garrido, J. J., & Muñoz, A. (2012). Developmental expression of Kv potassium channels at the axon initial segment of cultured hippocampal neurons. *PLoS One, 7*, e48557.

Sánchez-Ponce, D., Muñoz, A., & Garrido, J. J. (2011). Casein kinase 2 and microtubules control axon initial segment formation. *Molecular and Cellular Neurosciences, 46*, 222–234.

Sarmiere, P. D., Weigle, C. M., & Tamkun, M. M. (2008). The Kv2.1 K^+ channel targets to the axon initial segment of hippocampal and cortical neurons in culture and *in situ*. *BMC Neuroscience, 9*, 112.

Schafer, D. P., Bansal, R., Hedstrom, K. L., Pfeiffer, S. E., & Rasband, M. N. (2004). Does paranode formation and maintenance require partitioning of neurofascin 155 into lipid rafts? *The Journal of Neuroscience, 24*, 3176–3185.

Schafer, D. P., Custer, A. W., Shrager, P., & Rasband, M. N. (2006). Early events in node of Ranvier formation during myelination and remyelination in the PNS. *Neuron Glia Biology, 2*, 69–79.

Schafer, D. P., Jha, S., Liu, F., Akella, T., McCullough, L. D., & Rasband, M. N. (2009). Disruption of the axon initial segment cytoskeleton is a new mechanism for neuronal injury. *The Journal of Neuroscience, 29*, 13242–13254.

Schnapp, B., & Mugnaini, E. (1978). Membrane architecture of myelinated fibers as seen by freeze-fracture. In S. G. Waxman (Ed.), *Physiology and pathobiology of axons* (pp. 83–123). New York: Raven Press.

Schultz, C., König, H. G., Del Turco, D., Politi, C., Eckert, G. P., Ghebremedhin, E., et al. (2006). Coincident enrichment of phosphorylated IκBα, activated IKK, and phosphorylated p65 in the axon initial segment of neurons. *Molecular and Cellular Neurosciences*, *33*, 68–80.

Sloper, J. J., & Powell, T. P. (1973). Observations on the axon initial segment and other structures in the neocortex using conventional staining and ethanolic phosphotungstic acid. *Brain Research*, *50*, 163–169.

Sobotzik, J. M., Sie, J. M., Politi, C., Del Turco, D., Bennett, V., Deller, T., et al. (2009). AnkyrinG is required to maintain axo-dendritic polarity in vivo. *Proceedings of the National Academy of Sciences of the United States of America*, *106*, 17564–17569.

Somogyi, P., Freund, T. F., & Cowey, A. (1982). The axo-axonic interneuron in the cerebral cortex of the rat, cat and monkey. *Neuroscience*, *7*, 2577–2607.

Somogyi, P., Nunzi, M. G., Gorio, A., & Smith, A. D. (1983). A new type of specific interneuron in the monkey hippocampus forming synapses exclusively with the axon initial segments of pyramidal cells. *Brain Research*, *259*, 137–142.

Somogyi, P., Smith, A. D., Nunzi, M. G., Gorio, A., Takagi, H., & Wu, J. Y. (1983). Glutamate decarboxylase immunoreactivity in the hippocampus of the cat: Distribution of immunoreactive synaptic terminals with special reference to the axon initial segment of pyramidal neurons. *The Journal of Neuroscience*, *3*, 1450–1468.

Song, A.-H., Wang, D., Chen, G., Li, Y., Luo, J., Duan, S., et al. (2009). A selective filter for cytoplasmic transport at the axon initial segment. *Cell*, *136*, 1148–1160.

Spiegel, I., Adamsky, K., Eshed, Y., Milo, R., Sabanay, H., Sarig-Nadir, O., et al. (2007). A central role for Necl4 (SynCAM4) in Schwann cell-axon interaction and myelination. *Nature Neuroscience*, *10*, 861–869.

Susuki, K., Chang, K. J., Zollinger, D. R., Liu, Y., Ogawa, Y., Eshed-Eisenbach, Y., et al. (2013). Three mechanisms assemble central nervous system nodes of Ranvier. *Neuron*, *78*, 469–482.

Suzuki, A., Hoshi, T., Ishibashi, T., Hayashi, A., Yamaguchi, Y., & Baba, H. (2004). Paranodal axoglial junction is required for the maintenance of the Nav1.6-type sodium channel in the node of Ranvier in the optic nerves but not in peripheral nerve fibers in the sulfatide-deficient mice. *Glia*, *46*, 274–283.

Tait, S., Gunn-Moore, F., Collinson, J. M., Huang, J., Lubetzki, C., Pedraza, L., et al. (2000). An oligodendrocyte cell adhesion molecule at the site of assembly of the paranodal axo-glial junction. *The Journal of Cell Biology*, *150*, 657–666.

Tkachenko, E., Rhodes, J. M., & Simons, M. (2005). Syndecans: New kids on the signaling block. *Circulation Research*, *96*, 488–500.

Traka, M., Goutebroze, L., Denisenko, N., Bessa, M., Nifli, A., Havaki, S., et al. (2003). Association of TAG-1 with Caspr2 is essential for the molecular organization of juxtaparanodal regions of myelinated fibers. *The Journal of Cell Biology*, *162*, 1161–1172.

Trapp, B. D., Bernier, L., Andrews, S. B., & Colman, D. R. (1988). Cellular and subcellular distribution of 2',3'-cyclic nucleotide 3'-phosphodiesterase and its mRNA in the rat central nervous system. *Journal of Neurochemistry*, *51*, 859–868.

Trapp, B. D., & Quarles, R. H. (1982). Presence of the myelin-associated glycoprotein correlates with alterations in the periodicity of peripheral myelin. *The Journal of Cell Biology*, *92*, 877–882.

Uemoto, Y., Suzuki, S., Terada, N., Ohno, N., Ohno, S., Yamanaka, S., et al. (2007). Specific role of the truncated βIV-spectrin Σ6 in sodium channel clustering at axon initial segments and nodes of Ranvier. *The Journal of Biological Chemistry*, *282*, 6548–6555.

Vabnick, I., Trimmer, J. S., Schwarz, T. L., Levinson, S. R., Risal, D., & Shrager, P. (1999). Dynamic potassium channel distributions during axonal development prevent aberrant firing patterns. *The Journal of Neuroscience*, *19*, 747–758.

Vacher, H., Yang, J. W., Cerda, O., Autillo-Touati, A., Dargent, B., & Trimmer, J. S. (2011). Cdk-mediated phosphorylation of the Kvβ2 auxiliary subunit regulates Kv1 channel axonal targeting. *The Journal of Cell Biology*, *192*, 813–824.

Van Wart, A., Trimmer, J. S., & Matthews, G. (2007). Polarized distribution of ion channels within microdomains of the axon initial segment. *The Journal of Comparative Neurology*, *500*, 339–352.

Volkmer, H., Zacharias, U., Nörenberg, U., & Rathjen, F. G. (1998). Dissection of complex molecular interactions of neurofascin with axonin-1, F11, and tenascin-R, which promote attachment and neurite formation of tectal cells. *The Journal of Cell Biology*, *142*, 1083–1093.

Waxman, S. G. (2001). Transcriptional channelopathies: An emerging class of disorders. *Nature Reviews Neuroscience*, *2*, 652–659.

Wiley, C. A., & Ellisman, M. H. (1980). Rows of dimeric-particles within the axolemma and juxtaposed particles within glia, incorporated into a new model for the paranodal glial-axonal junction at the node of Ranvier. *The Journal of Cell Biology*, *84*, 261–280.

Wimmer, V. C., Reid, C. A., So, E. Y., Berkovic, S. F., & Petrou, S. (2010). Axon initial segment dysfunction in epilepsy. *The Journal of Physiology*, *588*, 1829–1840.

Winckler, B., Forscher, P., & Mellman, I. (1999). A diffusion barrier maintains distribution of membrane proteins in polarized neurons. *Nature*, *397*, 698–701.

Wittmack, E. K., Rush, A. M., Craner, M. J., Goldfarb, M., Waxman, S. G., & Dib-Hajj, S. D. (2004). Fibroblast growth factor homologous factor 2B: Association with $Na_v1.6$ and selective colocalization at nodes of Ranvier of dorsal root axons. *The Journal of Neuroscience*, *24*, 6765–6775.

Wollner, D. A., & Catterall, W. A. (1986). Localization of sodium channels in axon hillocks and initial segments of retinal ganglion cells. *Proceedings of the National Academy of Sciences of the United States of America*, *83*, 8424–8428.

Xu, K., Zhong, G., & Zhuang, X. (2013). Actin, spectrin, and associated proteins form a periodic cytoskeletal structure in axons. *Science*, *339*, 452–456.

Yang, Y., Lacas-Gervais, S., Morest, D. K., Solimena, M., & Rasband, M. N. (2004). βIV spectrins are essential for membrane stability and the molecular organization of nodes of Ranvier. *The Journal of Neuroscience*, *24*, 7230–7240.

Yang, Y., Ogawa, Y., Hedstrom, K. L., & Rasband, M. N. (2007). βIV spectrin is recruited to axon initial segments and nodes of Ranvier by ankyrinG. *The Journal of Cell Biology*, *176*, 509–519.

Zalc, B., Goujet, D., & Colman, D. (2008). The origin of the myelination program in vertebrates. *Current Biology*, *18*, R511–R512.

Zenker, J., Poirot, O., de Preux Charles, A. S., Arnaud, E., Médard, J. J., Lacroix, C., et al. (2012). Altered distribution of juxtaparanodal $K_v1.2$ subunits mediates peripheral nerve hyperexcitability in type 2 diabetes mellitus. *The Journal of Neuroscience*, *32*, 7493–7498.

Zhang, X., Bao, L., Yang, L., Wu, Q., & Li, S. (2012). Roles of intracellular fibroblast growth factors in neural development and functions. *Science China Life Sciences*, *55*, 1038–1044.

Zhang, Y., Bekku, Y., Dzhashiashvili, Y., Armenti, S., Meng, X., Sasaki, Y., et al. (2012). Assembly and maintenance of nodes of Ranvier rely on distinct sources of proteins and targeting mechanisms. *Neuron*, *73*, 92–107.

Zhou, D., Lambert, S., Malen, P. L., Carpenter, S., Boland, L. M., & Bennett, V. (1998). AnkyrinG is required for clustering of voltage-gated Na channels at axon initial segments and for normal action potential firing. *The Journal of Cell Biology*, *143*, 1295–1304.

Zimmermann, D. R., & Dours-Zimmermann, M. T. (2008). Extracellular matrix of the central nervous system: From neglect to challenge. *Histochemistry and Cell Biology*, *130*, 635–653.
Zonta, B., Desmazieres, A., Rinaldi, A., Tait, S., Sherman, D. L., Nolan, M. F., et al. (2011). A critical role for Neurofascin in regulating action potential initiation through maintenance of the axon initial segment. *Neuron*, *69*, 945–956.
Zonta, B., Tait, S., Melrose, S., Anderson, H., Harroch, S., Higginson, J., et al. (2008). Glial and neuronal isoforms of Neurofascin have distinct roles in the assembly of nodes of Ranvier in the central nervous system. *The Journal of Cell Biology*, *181*, 1169–1177.

CHAPTER SIX

Microdomains of SNARE Proteins in the Plasma Membrane

Geert van den Bogaart*, Thorsten Lang†, Reinhard Jahn‡,1
*Department of Tumor Immunology, Nijmegen Centre for Molecular Life Sciences, Radboud University Nijmegen Medical Centre, Nijmegen, The Netherlands
†Department of Membrane Biochemistry, Life & Medical Sciences (LIMES) Institute, University of Bonn, Bonn, Germany
‡Department of Neurobiology, Max Planck Institute for Biophysical Chemistry, Göttingen, Germany
[1]Corresponding author: e-mail address: r.jahn@gwdg.de

Contents

Current Topics in Membranes, Volume 72
ISSN 1063-5823
http://dx.doi.org/10.1016/B978-0-12-417027-8.00006-4

Abstract

Exocytosis is catalyzed by the engagement of SNARE proteins embedded in the plasma membrane with complementary SNAREs in the membrane of trafficking vesicles undergoing exocytosis. In most cells studied so far, SNAREs are not randomly distributed across the plasma membrane but are clustered and segregated in discrete membrane domains of defined size, composition, and stability. SNARE clusters have been intensively studied for more than a decade. Different mechanisms have been proposed to be responsible for SNARE clustering such as partitioning into cholesterol-enriched lipid rafts, hydrophobic mismatch, posttranslational modifications of the SNAREs including phosphorylation and palmitoylation, electrostatic protein–protein and protein–lipid interactions, homotypic and heterotypic protein interactions, and anchoring to the cortical cytoskeleton. Although several of these proposed mechanisms are still controversially discussed, it is becoming apparent that independent physicochemical principles must cooperate in a synergistic manner to yield SNARE microdomains. Here, we discuss the architecture and function of SNARE domains. We also discuss the various factors influencing SNARE clustering, resulting in a model that we believe may be of general use to explain domain formation of proteins in the plasma membrane.

1. SYNOPSIS

Exocytosis, that is, the fusion of a trafficking vesicle with the plasma membrane, is a fundamental feature of all eukaryotic cells. Exocytosis is not only needed for membrane addition during cell growth or for the insertion of proteins and lipids into the plasma membrane, but it is also the final step for releasing intracellularly synthesized proteins, peptides, and small molecules such as neurotransmitters into the extracellular environment. The proteins responsible for the exocytotic fusion between the vesicle and the plasma membrane have been identified. They include members of conserved protein families including the SNAREs, the SM proteins, and scaffolding proteins of the CATCHR family. These proteins form dynamic supramolecular complexes that assemble at the site of fusion and interact with a range of additional proteins that regulate their function. The molecular mechanisms are best understood for exocytosis of synaptic vesicles in neurons (see Jahn & Fasshauer, 2012; Ramakrishnan, Drescher, & Drescher, 2012 for reviews).

In recent years, it has become apparent that the components of the secretory apparatus are specifically localized to distinct sites in the cell. In neurons, such spatial organization is required for releasing neurotransmitters exclusively at synapses. Similarly, in adrenal chromaffin cells, the components of the exocytotic machinery are organized in such a way that catecholamines are only released at discrete areas of the plasma membrane (the so-called exocytotic "hot spots") (Oheim & Stuhmer, 2000; Robinson, Finnegan, Monck, Wightman, & Fernandez, 1995). Restriction of exocytosis to distinct sites is also observed in many other cell types such as in polarized cells where specific vesicles are selectively directed to exocytotic sites at the apical or basal surfaces, respectively. How is such a spatial organization of the exocytotic machinery at the plasma membrane achieved? In this chapter, we try to address this question. We specifically focus on the spatial organization of the engines of membrane fusion, namely, the SNARE proteins. SNAREs are concentrated in discrete membrane areas of the membranes (membrane domains or clusters), and this regulates both intracellular trafficking and local SNARE function. SNARE clustering has been studied for more than a decade, with the key principles becoming increasingly clear. As we will explain in this review, a picture is emerging according to which multiple forces contribute to SNARE clustering, shedding light on the principles behind membrane domain formation that may be of general significance.

2. MEMBRANE PARTITIONING OF SNARE PROTEINS

2.1. Function and structure of SNARE proteins

SNAREs are an evolutionarily conserved superfamily of small proteins that have a conserved motif of about 60–70 amino acids (SNARE motif) in their cytoplasmic domain. SNARE proteins catalyze intracellular membrane fusion in eukaryotes and are essential for a wide range of cellular processes, including cell growth, cytokinesis, and synaptic transmission (reviewed in Jahn & Scheller, 2006; Wickner & Schekman, 2008). Many SNAREs are tail-anchored proteins that contain a single C-terminal transmembrane domain. In addition, some of them are peripheral membrane proteins that are membrane-anchored via palmitoyl chains. Isolated SNARE motifs can be unfolded and have no secondary structure. Complementary sets of SNARE proteins that are embedded in adjacent membranes can form a tight coiled-coil alpha-helical complex (core complex) bridging the membrane (called *trans*-SNAREs), consisting of a total of four SNARE motifs with one of each subtypes (referred to as Qa-, Qb-, Qc-, and R-SNAREs). Formation of this core complex pulls the membranes together and overcomes the energy barrier for membrane fusion (Fig. 6.1). Various membrane fusion steps are catalyzed by distinct sets of SNARE proteins. For instance, Ca^{2+}-regulated neurotransmitter release from neurons and neuroendocrine cells is

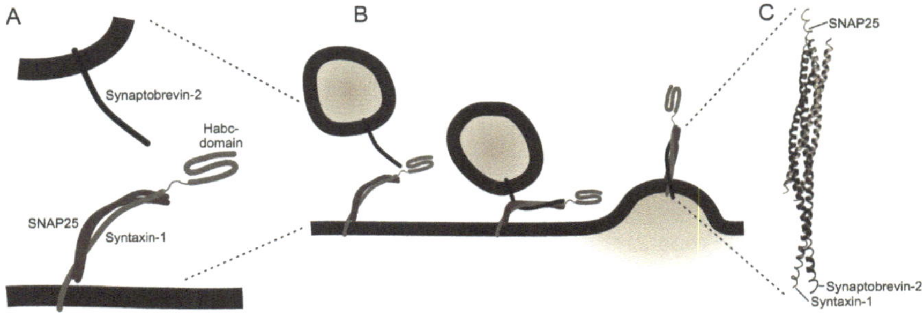

Figure 6.1 SNARE proteins catalyze membrane fusion. (A) Domain topology of the neuronal SNAREs. Vesicular SNARE, synaptobrevin-2; acceptor plasma membrane SNAREs, SNAP25 and syntaxin-1. The N-terminal regulatory Habc domain of syntaxin-1 is indicated. (B) Scheme of exocytosis where a SNARE protein in the vesicular membrane engages with a complementary acceptor SNARE complex in the plasma membrane. This results in the formation of a tight alpha-helical coiled-coil SNARE domain, which overcomes the energy barrier for membrane fusion. (C) Crystal structure of the neuronal SNARE complex (Protein Data Bank accession code: 3HD7; Stein, Weber, Wahl, & Jahn, 2009). The Habc domain of syntaxin-1 and the palmitoylated linker region that connects the two SNARE domains of SNAP25 are not present in the structure.

mediated by the (*vesicular*) *v*-SNARE synaptobrevin-2 (also called VAMP2, the R-SNARE) in the vesicle membrane and the (*target*) *t*-SNAREs syntaxin-1 (Qa) and SNAP25 (Qb and Qc) in the plasma membrane (Jahn & Scheller, 2006), henceforth referred to as neuronal SNARE complex (note that, like all eukaryotic cells, neurons also contain many additional SNAREs). The assignment of the SNAREs functioning in constitutive exocytosis (i.e., via the Golgi apparatus) is more ambiguous but involves the *v*-SNARE cellubrevin (or VAMP3; R-type) and the *t*-SNAREs syntaxin-2 or syntaxin-4 (both Qa; Fig. 6.2) and SNAP23 (Qb and Qc) (Lang, 2007; Predescu, Predescu, Shimizu, Klein, & Malik, 2005).

2.2. Domain partitioning of SNARE proteins

It is well established that plasma membrane SNAREs are organized in membrane domains in all cell types investigated (see below) instead of being evenly distributed across the plasma membrane. Membrane clustering is best characterized for the neuronal SNAREs, because these are highly abundant proteins in neurons and neuroendocrine cells (in contrast to most other SNAREs). For (pheochromocytoma) PC12 cells that are derived from rat chromaffin cells and release catecholamines (e.g., adrenalin) from dense-core vesicles by calcium-dependent exocytosis, estimates of the average surface density of syntaxin-1 range from 540 molecules per μm^2 membrane area

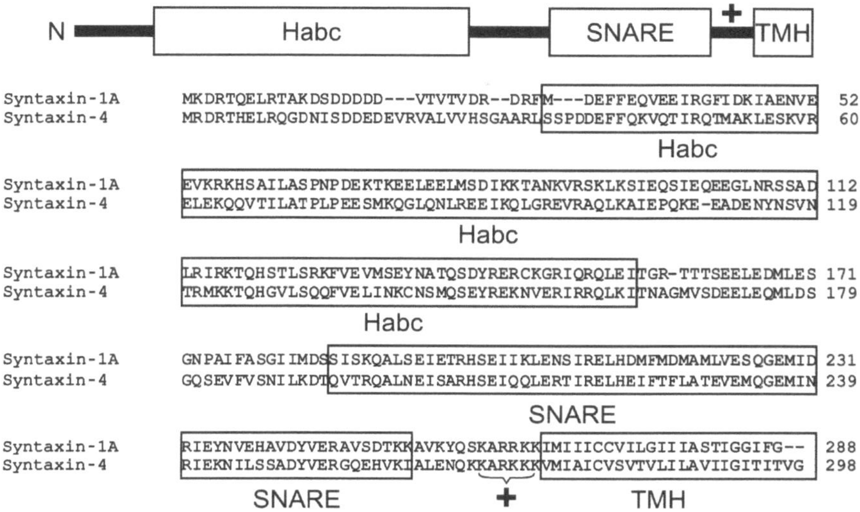

Figure 6.2 Protein sequence alignment of syntaxin-1 and syntaxin-4 (43% sequence identity). The positions of the N-terminal regulatory Habc domains, the SNARE motifs, the polybasic juxtamembrane regions (indicated with a plus symbol), and the C-terminal transmembrane helices (TMH) are indicated.

measured by capacitance (Knowles et al., 2010) to about 2000 molecules per μm^2 imaged membrane (James, Kowalchyk, Daily, Petrie, & Martin, 2009; Sieber et al., 2007). Here, it should be noted that this latter value is an upper estimate of the molecular density, since the 2D projection from potentially curved membranes of microscope images may lead to an underestimation of the membrane surface area. The densities for SNAP25 were estimated to be about 7500 molecules per μm^2 of plasma membrane in PC12 cells (Knowles et al., 2010). Finally, synaptic vesicles contain even higher SNARE densities amounting to approximately 12,000 copies of synaptobrevin-2 per μm^2 (70 copies per synaptic vesicle of 42 nm diameter; Takamori et al., 2006). These SNARE densities are average values that do not take known clustering effects into account, which result in very high local surface densities within the membrane domains (see Section 3).

Evidence for the widespread presence of SNARE domains in the plasma membrane is derived primarily from microscopy experiments where the SNAREs are visualized by two approaches: (i) labeling of the SNAREs with antibodies and (ii) overexpression of SNARE proteins fused to GFP or other tags.

(i) Using antibody labeling for the SNAREs syntaxin-1 and syntaxin-4 and for SNAP25, it was shown in many studies that SNAREs are clustered in the plasma membrane of neuroendocrine PC12 cells (Aikawa, Xia, & Martin, 2006; Aoyagi et al., 2005; Lang et al., 2001; Rickman et al., 2010; Sieber, Willig, Heintzmann, Hell, & Lang, 2006; Taverna et al., 2007). In general, immunolabeling experiments have to be interpreted with caution because the use of antibodies and detergents to permeabilize the cells may lead to clustering artifacts as recently discussed (Lang & Rizzoli, 2010). Importantly, chemical fixation of the sample prior to antibody labeling may not completely prevent such clustering artifacts, because at least some integral membrane proteins and in particular lipid-anchored proteins remain to some extent mobile, even when very rigid fixing procedures are applied (Tanaka et al., 2010).

(ii) Clustering of syntaxin-1, syntaxin-4, and SNAP25 in the plasma membrane of PC12 cells was also demonstrated by overexpression of constructs containing GFP or other tags (monomeric GFP, GFP analogs, HA-tag, and myc-tag) (Aikawa et al., 2006; Barg, Knowles, Chen, Midorikawa, & Almers, 2010; Knowles et al., 2010; Lang et al., 2001; Rickman et al., 2010; Sieber et al., 2006). Of course, the overexpression of fusion proteins from strong viral promoters can also lead to artifacts and fusion proteins may not behave equal to endogenous proteins, especially considering the relative size of the GFP tag (27 kDa) that is comparable to or even larger than most SNARE proteins.

Despite such inherent problems, the data from these two approaches agree well with each other and suggest clustering of SNAREs in small membrane domains. Further support is derived from a third labeling approach that does not rely on antibodies or fusion proteins (Bar-On et al., 2008; Lang, Margittai, Hölzler, & Jahn, 2002; Zilly et al., 2011). In these studies, a fluorescently labeled and soluble fragment of the SNARE synaptobrevin-2 was added to membrane sheets from PC12 cells. This fragment readily forms SNARE complexes with endogenous syntaxin 1 and SNAP-25, which again resulted in staining of small membrane domains containing these SNAREs.

In addition to PC12 cells, SNARE clustering was demonstrated in many other cell types by means of immunolabeling and overexpression of tagged proteins including the plasmalemmal SNAREs syntaxin-1, syntaxin-2, syntaxin-3, and syntaxin-4, SNAP23, and SNAP25: primary bovine or mouse chromaffin cells (López et al., 2009; Nagy et al., 2005; Rickman, Meunier, Binz, & Davletov, 2004; Villanueva et al., 2010; Zilly,

Sørensen, Jahn, & Lang, 2006), neurons (Foletti, Lin, Finley, & Scheller, 2000; Khuong et al., 2013), blood vessel endothelial cells (Predescu et al., 2005), the kidney epithelial cell line MDCK (Bulbarelli, Sprocati, Barberi, Pedrazzini, & Borgese, 2002; Low et al., 2006; Torres, Funk, Zegers, & ter Beest, 2011), the helper T-cell line Jurkat (Low et al., 2006), the pancreatic beta cell lines MIN6 and INS-1 (Ohara-Imaizumi et al., 2004; Yang, Xu, Xiao, Xiong, & Xu, 2006), CHO cells (Yang et al., 2006), and 3T3 fibroblasts (Chamberlain & Gould, 2002; Low et al., 2006). Together, the combined evidence from all these approaches firmly establishes that SNARE proteins cluster in domains in the plasma membrane of a large variety of cell types, which thus probably constitutes an inherent feature of all eukaryotic cells.

Other than this microscopy-based evidence, attempts were made to isolate SNARE clusters using biochemical techniques. For instance, SNARE-enriched fractions were separated from other membrane constituents following solubilization by certain detergents such as Triton X-100 (referred to as detergent-resistant membranes or DRMs, reviewed in Lang, 2007). However, due to methodical problems, DRM fractions are no longer considered to be identical to membrane domains (Lingwood & Simons, 2010) and we will primarily focus on microscopy-based evidence in this review (but see Section 4.1 for further discussion).

3. INNER ARCHITECTURE OF SNARE CLUSTERS

3.1. Size and shape of SNARE clusters

Since SNARE clusters are present at high surface densities and are smaller than the diffraction limit of conventional light microscopy, superresolution microscopy techniques such as stimulated emission depletion (STED) microscopy, stochastic optical reconstruction microscopy (STORM), and photoactivated localization microscopy (PALM) have been instrumental for their analysis. It was shown that membrane domains containing syntaxin-1 and SNAP25 in the plasma membrane of PC12 cells are significantly smaller than 100 nm (Bar-On et al., 2012; Rickman et al., 2010; Sieber et al., 2007; Willig, Keller, Bossi, & Hell, 2006). STED microscopy studies indicated that SNAP25 and syntaxin-1 domains have a circular shape (Sieber et al., 2007; Willig et al., 2006). In contrast, in a recent study employing STORM microscopy (Bar-On et al., 2012), SNARE domains were found to be more elliptical in shape. These authors suggested that small clusters of SNAP25 or syntaxin-1 can combine to form larger membrane domains whereby the individual clusters within these domains are still recognizable. In this study (Bar-On et al., 2012), the highest densities of SNAP25 and syntaxin-1 were observed in the centers of the individual clusters and less tightly clustered molecules seemed concentrated in areas adjacent to these clusters. Because this study employed labeling with antibodies directed against the cytoplasmic domains of these SNAREs, it still remains to be established whether the transmembrane regions of the SNAREs are also more clustered in the centers of the domains or they are more uniformly distributed across the clusters or even tightly clustered in 20 nm sized bundles (as discussed in detail in Sieber et al., 2007).

Interestingly, despite considerable size variability between different reports, the sizes of SNARE domains within a particular cell type seem rather uniform. In PC12 cells, the size of syntaxin-1 clusters was found to be apparently independent of its expression levels, with higher expression levels resulting in more clusters rather than an increase in size of individual clusters (Sieber et al., 2006). Similarly, no change in cluster size was observed upon overexpression of syntaxin-3 in the plasma membrane of MDCK cells although, in this study, no superresolution microscopy technique was used (Low et al., 2006).

In addition to the micropatterning described earlier, SNAREs can be restricted to certain membrane areas as in polar differentiated cells. For instance, in the plasma membranes of the kidney epithelial cell line MDCK, certain SNAREs are specifically enriched in the basolateral (syntaxin-4) and apical (syntaxin-3) membranes (Bulbarelli et al., 2002; Low et al., 1996). Other examples include pancreatic acinar cells where syntaxin-2 and syntaxin-4 are predominantly localized to the apical and basolateral plasma membrane, respectively (Gaisano et al., 1996), and migrating NIH 3T3 cells where both syntaxin-3 and syntaxin-4 are enriched at the leading edge (Low et al., 2006). Due to limitations of the microscopy techniques employed in these studies, it remains open whether in these cell types the SNAREs are also organized into smaller (submicrometer) domains.

3.2. Number of SNARE molecules per domain

For the neuronal SNAREs, the copy number of SNARE molecules per cluster has been determined by three different approaches in the membrane of PC12 cells. First, an estimate of about 75 molecules per domain was obtained by determining the average total number of syntaxin-1 molecules per PC12 cell using quantitative Western blotting, divided by the number of syntaxin-1 clusters per PC12 cell as determined by microscopy and corrected for the fraction of syntaxin-1 that is not located in the clusters (Sieber et al., 2007). Second, using overexpression of syntaxin-1 tagged with a photoactivatable RFP analog, a copy number of 30–40 syntaxin-1 molecules concentrated in 50 nm diameter membrane areas was obtained using PALM (Rickman et al., 2010). Because of the limitations of PALM and because the presence of endogenous unlabeled syntaxin-1 in the membrane domains was not accounted for, this is a lower estimate of the number of molecules per cluster. Finally, using overexpression of GFP-tagged syntaxin-1 and SNAP25, TIRF microscopy was employed for quantitative analysis (Knowles et al., 2010). Here, the signal arising from membrane clusters was divided by the signal arising from single molecules. In this study, the numbers were corrected for endogenous SNAREs, resulting in an estimate of about 50–70 molecules per cluster for both syntaxin-1 and SNAP25 (Knowles et al., 2010). Taken together, these studies are in excellent agreement with each other showing that a 50-nm sized membrane domain contains about 70 syntaxin-1 molecules.

3.3. Dynamics of SNARE domains

Membrane domains enriched in SNAREs are relatively immobile as demonstrated for syntaxin-1 clusters in live MIN6 and chromaffin cells (Barg et al., 2010; Ohara-Imaizumi et al., 2004) and plasma membrane sheets of PC12 cells (Sieber et al., 2007). In bovine chromaffin cells, the lateral mobility of SNAP25 clusters was determined to be 20–30 nm s^{-1} (López et al., 2009; Torregrosa-Hetland et al., 2013). Intriguingly, in these studies, SNARE domains did not only move in the plane of the membrane (x, y) but also move perpendicular to the membrane (z). This velocity would correspond to a lateral diffusion coefficient of about 4.6×10^{-5} $\mu m^2\ s^{-1}$, which is three to four orders of magnitude lower than that of free syntaxin-1 ($\sim$0.1 $\mu m^2\ s^{-1}$; Knowles et al., 2010) and phosphatidylcholine ($\sim$0.5 $\mu m^2\ s^{-1}$; Honigmann et al., 2013) in the plasma membrane.

Albeit membrane domains containing syntaxin-1 and SNAP25 appear rather static or only move with a low speed, syntaxin-1 and SNAP25 in the plasma membrane of PC12 cells show alternating periods of slow and rapid diffusion at the level of single molecules (Barg et al., 2010; Knowles et al., 2010), which is in line with these SNAREs being transiently captured and released from membrane domains. Such dynamic behavior has been postulated earlier by fitting experimental fluorescence recovery after photobleaching (FRAP) curves of GFP-tagged syntaxin-1 in PC12 cells to a Brownian dynamics simulation (Sieber et al., 2007). From these FRAP experiments, it was estimated that about 84% of syntaxin-1 located in the membrane clusters and the remaining syntaxin-1 molecules diffused between the clusters. By single-particle tracking, about a third of GFP-tagged syntaxin-1 molecules were essentially immobile in PC12 cells in a time window of 200 ms, suggesting that this fraction of syntaxin-1 was sequestered in membrane domains (Knowles et al., 2010).

3.4. Overlap of SNARE domains

The molecular composition of SNARE domains may be rather heterogeneous. For instance, it is clear that not all syntaxin-1 domains contain SNAP25, although the precise degree of overlap of SNAP25 and syntaxin-1 clusters is still controversial. By immunostaining of endogenous proteins, several studies indicate that only a fraction (about 25%) of the syntaxin-1 clusters contain SNAP25 in the plasma membrane of PC12 cells and mouse chromaffin cells (Nagy et al., 2005; Taverna et al., 2007). Such a low degree of overlap between SNARE domains has also been observed in the plasma membrane of blood vessel endothelial cells, where only some syntaxin-4 clusters were found to contain SNAP23 (Predescu et al., 2005). In contrast, other studies show much higher fractions of syntaxin-1 clusters that contain SNAP25 or even near-complete overlap, for example, in the plasma membrane of PC12 cells, bovine chromaffin cells, neuroblastoma 2a cells, and the pancreatic beta cell line MIN6 (Lang et al., 2001; Lopez et al., 2007; López et al., 2009; Ohara-Imaizumi et al., 2004; Rickman et al., 2004, 2010). These discrepancies may be due, at least in part, to technical limitations. First, as discussed in Section 3.1, SNARE domains are considerably smaller than the diffraction-limited resolution of optical microscopy. This may lead to overestimation of the extent of overlap between SNAP25 and syntaxin-1, because overlapping domains cannot be discerned from domains that are in close proximity but do not overlap. Second, fluorescence cross talk or bleed-through between the different spectral channels of the microscope may lead to overestimation of the extent of overlap. Finally, molecular crowding of SNAREs in the membrane domains (see Section 3.2), conformational dynamics, and interactions of the SNARE proteins with each other or with other proteins may render epitopes inaccessible to external probes. As a consequence, not every protein may bind to an antibody.

For syntaxin-1, epitope masking is not expected to be a major problem, because staining PC12 cells overexpressing GFP-tagged syntaxin-1 with an antibody directed against syntaxin-1 (i.e., the N-terminal regulatory Habc domain) revealed identical images in both the GFP and the antibody channels (Zilly et al., 2011), indicating that all membrane domains were labeled with at least a single antibody. However, similar experiments for SNAP25 showed that a pool of GFP-tagged SNAP25 was not recognized by the antibody (Zilly et al., 2011). Thus, for SNAP25, protein–protein interactions, molecular crowding effects, and/or the conformation of SNAP25 can result in masking of the binding sites for (at least some) antibodies.

Work from our own laboratories strongly favors the view that the degree of overlap between SNAP25 and syntaxin-1 is rather limited. First, in a study designed to overcome the limitations of antibody accessibility for SNAP25 and of diffraction-limited microscopy, PC12 cells overexpressing GFP-tagged SNAP25 were immunostained for both GFP and syntaxin-1 and imaged by superresolution STED microscopy. Here, although membrane domains enriched in SNAP25 and syntaxin-1 located in close proximity to each other, they did not show concentric overlap (Halemani, Bethani, Rizzoli, & Lang, 2010). Second, the study by Lang et al. (2002) provides further evidence for a low degree of colocalization between SNAP25 and syntaxin-1, because here, endogenous syntaxin-1 did not compete with exogenous syntaxin-1 for binding to SNAP25. Finally, botulinum neurotoxin C1 cleavage of endogenous syntaxin-1 did not increase the number of binding sites for exogenous syntaxin-1 (Lang et al., 2002). Syntaxin-1 clusters that overlap with SNAP25 clusters may be heterogeneous, as demonstrated in a FRET study where YFP-tagged SNAP25 bound to cerulean-tagged syntaxin-1 in two conformations and this was suggested to reflect the engagement of either single or both of the Qb- and Qc-SNARE motifs of SNAP25 in the plasma membrane of PC12 cells (Rickman et al., 2010).

Taken together, despite lingering controversies, there is growing consensus that the SNAREs segregate at least partially into different domains. This is of rather fundamental importance and has implications for SNARE function as these protein pools may signify functional differentiation. Only syntaxin-1 and SNAP-25 together can mediate neurotransmitter release, and thus, it is likely that such differentiation into different domains plays an important, albeit still incompletely understood, role in tuning vesicle docking and release (see Section 5). In addition, the complete segregation of the closely related syntaxin-1 and syntaxin-4 into different domains (Sieber et al., 2006) correlates with their functional differentiation: Syntaxin-1 is involved in regulated exocytosis, whereas syntaxin-4 is involved in constitutive exocytosis (43% sequence identity; Fig. 6.2). Segregation in nonoverlapping clusters has also been observed for syntaxin-3 and syntaxin-4 (40% sequence identity; both thought to be involved in constitutive exocytosis) in the plasma membrane of the helper T-cell line Jurkat and for syntaxin-2, syntaxin-3, and syntaxin-4 in the kidney epithelial MDCK cell line (Low et al., 2006). This differential sorting of highly homologous proteins into different clusters may be a means to make sure that SNAREs are segregated for different fusion jobs. Moreover, the diversity of SNARE domains cannot be explained by a uniform clustering mechanism or by simple two-phase partitioning as has been suggested for lipid rafts (see Section 4.1). Rather, the combined action of multiple clustering factors is required, which will be discussed in the following sections.

4. MECHANISMS OF SNARE PARTITIONING

Due to their high abundance and the availability of sensitive probes, SNAREs have served as models for understanding protein partitioning in membranes. Due to technical hurdles, this field is still in its infancy, and albeit likely, it is presently not clear to which extent the mechanisms that have been shown to contribute to SNARE clustering are of general significance for other membrane proteins. In principle, SNARE clustering may involve both the membrane anchoring domains (mostly single transmembrane domains) and the cytoplasmic domains, the latter consisting of the SNARE motifs involved in SNARE complex formation and adjacent domains such as the N-terminal regulatory Habc domains of syntaxin-1 and syntaxin-4 and/or the linker regions that connect these domains (Fig. 6.2). Indeed, there is evidence for all segments of the SNAREs contributing to clustering.

Generally, SNARE clustering has been attributed to a variety of factors (Fig. 6.3). First, clustering may be promoted by lipid segregation and/or interactions between SNAREs and membrane lipids. Factors presently discussed include partitioning of SNAREs into cholesterol-enriched membrane rafts (Chamberlain, Burgoyne, & Gould, 2001; Chamberlain & Gould, 2002; Lang, 2007; Predescu et al., 2005; Salaün, Gould, & Chamberlain, 2005a, 2005b; Salaün, James, & Chamberlain, 2004; Taverna et al., 2004, 2007), cholesterol competition with SNAREs for solvation by bulk lipids ("salting out") (Murray & Tamm, 2009, 2011), and clustering of SNAREs by electrostatic protein–lipid interactions (Khuong

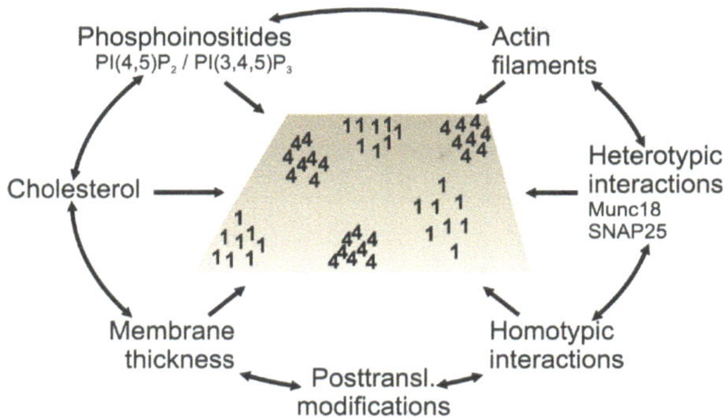

Figure 6.3 SNARE partitioning in the plasma membrane is multifactorial. The scheme depicts the tight interplay of factors affecting membrane partitioning of the SNAREs syntaxin-1 and syntaxin-4. See text for details.

et al., 2013; Van den Bogaart et al., 2011). Second, clustering may be promoted by homo- or heterophilic protein–protein interactions such as electrostatic protein–protein interactions (Zilly et al., 2011), homotypic interactions between the SNARE motifs (Sieber et al., 2006), heterotypic protein–protein interactions (Yang et al., 2006), anchoring to dense-core vesicles that are in the proximity of the plasma membrane (Barg et al., 2010; Knowles et al., 2010), and anchoring to the cortical cytoskeleton (Low et al., 2006; Torregrosa-Hetland et al., 2011). We will now discuss the evidence for these various clustering mechanisms. In Section 5, we will then discuss how the interplay between these mechanisms may regulate cluster dynamics and functions.

4.1. Cholesterol-dependent SNARE partitioning

Given the high 30–40% cholesterol content of the plasma membrane (Daum, 1985) and the membrane condensing effect of cholesterol (Lingwood & Simons, 2010), it seems not surprising that SNARE clustering depends on cholesterol. Nevertheless, the mechanism and extent of cholesterol-mediated clustering of SNAREs are still unclear. For instance, the reported effects of cholesterol extraction from cells by methyl-β-cyclodextrin range from partial to complete disintegration of plasma membrane clusters enriched in syntaxin-1 in PC12 cells (Lang et al., 2001), syntaxin-1 and SNAP25 in chromaffin cells (Lopez et al., 2007), syntaxin-1 in MIN6 pancreatic beta cells (Ohara-Imaizumi et al., 2004), and SNAP23 clusters in 3T3 fibroblasts (Chamberlain & Gould, 2002), whereas similar treatment has none or only moderate effects on clustering of syntaxin-4 and SNAP23 in microvascular endothelial cells (Predescu et al., 2005), syntaxin-1 and SNAP25 in neuroblastoma 2a cells (Rickman et al., 2010), and SNAP25 in PC12 cells (Taverna et al., 2007). We believe that at least some of these discrepancies may be due to variable experimental conditions such as the concentration of methyl-β-cyclodextrin, incubation times, temperature, and composition of the growth medium (other hydrophobic molecules may compete with cholesterol for binding to methyl-β-cyclodextrin). Interestingly, the extent of cholesterol-mediated clustering differs among SNAREs, because in kidney epithelial MDCK cells, cholesterol extraction completely disrupted syntaxin-3 clusters, but clustering of syntaxin-4 was not affected (Low et al., 2006).

SNAREs were initially believed to localize to the so-called lipid rafts. Lipid rafts were originally defined as a separate phase in the membrane that is enriched in cholesterol, saturated (sphingo)lipids, and specific sets of membrane-anchored proteins. Lipid rafts are in a condensed and ordered state (liquid-ordered) compared to the rest of the membrane, which is in the liquid-disordered phase (or fluid phase), and such phases can be reconstituted in artificial membranes (reviewed in Lang, 2007; Lingwood & Simons, 2010; Salaün et al., 2004). A problem with the original concept of lipid rafts has been that many conclusions were based on the isolation of detergent-resistant membranes (DRMs; see Section 2.2), which in the meantime has been recognized as being unreliable, prone to artifacts, and very sensitive to the precise experimental conditions (detergent/protein ratio and detergent/lipid ratio). Thus, it is not surprising that in the early years, there were conflicting reports as to whether SNAREs are associated with lipid rafts or not (reviewed in Lang, 2007). Presently, the partitioning of membrane proteins in DRMs is no longer being accepted as criterion for localization in lipid rafts. Furthermore, SNARE association with DRMs has been challenged since syntaxin-1 can be separated from raft markers such as flotillin-1 (MIN6 cells; Ohara-Imaizumi et al., 2004) and Thy-1 (PC12 cells; Lang et al., 2001) with independent approaches (fractionation after detergent solubilization, copatching on intact membranes using antibodies; see Lang et al., 2001). More recent data suggest that rafts are small (2–100 nm) and very short-lived (below a millisecond), unless they are somehow stabilized (e.g., in caveolae) (Lingwood & Simons, 2010), which contrasts with the emerging properties of the SNARE domains.

Experiments with SNAREs reconstituted in artificial membrane systems have shed new light on the physicochemical parameters involved in cholesterol-dependent clustering. First, syntaxin-1 and synaptobrevin-2 were found to clearly partition in the liquid-disordered domains in artificial membranes (Bacia, Schuette, Kahya, Jahn, & Schwille, 2004; Saslowsky, Lawrence, Henderson, & Edwardson, 2003). These studies provide further support for clustering of SNAREs away from classical lipid rafts, although it may be argued that in these studies, the simplified lipid composition (an equimolar ratio of unsaturated phosphatidylcholine, saturated sphingomyelin, and cholesterol) does not adequately mimic the plasma membrane. Murray and Tamm proposed an alternative "ice-breaker" model based on Förster resonance energy transfer (FRET) experiments on small liposomes, where SNARE clustering is driven by a competition with cholesterol for solvation by (bulk) phospholipids such as phosphatidylcholine (Murray & Tamm, 2009, 2011). Recently, these domains were also reconstituted in giant unilamellar vesicles and imaged by microscopy (Van den Bogaart et al., 2011). As we will explain in the following Sections 4.2–4.8, cholesterol-dependent clustering likely cannot be described by a simple two-phase model but rather relies on an intricate interplay between various additional factors such as local recruitment of phosphoinositides, hydrophobic mismatch, homophilic and heterophilic interactions between proteins, and the presence of hydrophobic posttranslational modifications such as palmitoylation.

4.2. Palmitoylation of SNAREs

S-palmitoylation involves a labile posttranslational thioester linkage between (generally) a palmitate (C16:0) and a cysteine residue. Palmitoylation modulates protein activity, stability, and subcellular localization (reviewed in Greaves, Prescott, Gorleku, & Chamberlain, 2009; Levental, Grzybek, & Simons, 2010) and results in membrane anchoring of otherwise soluble proteins (such as SNAP23 and SNAP25). SNAP23 and SNAP25 are synthesized as soluble proteins and get palmitoylated on a region located between the two SNARE domains. Newly synthesized SNAP25 is predominantly palmitoylated at the Golgi (Gonzalo & Linder, 1998). Palmitoylation is dynamic, but for SNAP25, turnover of palmitate is slow and occurs with a lifetime of hours. SNAP25 is expressed in two different splicing isoforms called SNAP25a and SNAP25b (Bark & Wilson, 1994). SNAP25b is the major isoform in the brain and the minimal region required and sufficient for membrane targeting is residues 85–120. This region contains four cysteines at positions 85, 88, 90, and 92 (rat sequence; Gonzalo, Greentree, & Linder, 1999), and by metabolic labeling with [^{3}H] palmitate in CV-1 kidney cells, it was estimated that all four cysteines are palmitoylated (Veit, Söllner, & Rothman, 1996). Interestingly, three of the nine amino acids that differ between SNAP25a and SNAP25b occur within the palmitoylated cysteine-rich domain, suggesting that functional differences of these two isoforms may be related to their palmitoylated sites (Prescott, Gorleku, Greaves, & Chamberlain, 2009). Palmitoylation is important for intracellular sorting of SNAP25 and regulates recycling of SNAP25 (Greaves & Chamberlain, 2011).

SNAP23 is 62% identical to SNAP25b at the amino acid level (rat sequences) and contains 5 cysteines in the palmitoylated region (positions 79, 80, 83, 85, and 87). Palmitoylation of SNAP23 and SNAP25 proteins may promote their association to cholesterol-rich raft domains (see Section 4.1) (Prescott et al., 2009; Salaün et al., 2005a,2005b). Importantly, different cellular sorting and membrane partitioning of SNAP23 compared to SNAP25 may be due to the additional palmitoylation site, as suggested by the finding that introduction of an additional cysteine into the SNAP25-palmitoylated region (to mimic SNAP23) enhanced DRM association and mutation of a single cysteine in SNAP23 (to mimic SNAP25) and reduced DRM association to similar levels as observed for wild-type SNAP23 and SNAP25, respectively (Salaün et al., 2005a).

Not only SNAP23 and SNAP25 but also many other synaptic proteins are palmitoylated as well (reviewed in Fukata & Fukata, 2010; Prescott et al., 2009). Recently, the proteome of palmitoylated proteins was determined of rat brain synaptosomal fractions and cultured cortical neurons (Kang et al., 2008). In this study, many putative palmitoylated proteins were identified, including syntaxin-1 and synaptobrevin-2. Synaptobrevin-2 was already known to be palmitoylated by labeling with [^{3}H] palmitic acid (Veit, Becher, & Ahnert-Hilger, 2000). Palmitoylation of both isoforms of syntaxin-1 (a and b) was also verified by incorporation of [^{3}H] palmitic acid (Kang et al., 2008). Interestingly, the palmitoylated sites of both synaptobrevin-2 (at cysteine 103; rat sequence) and syntaxin-1 (cysteines 271 and/or 272; rat syntaxin-1a) are located within their transmembrane helices, and (other than for SNAP23 and SNAP25) palmitoylation is clearly not required for membrane association of these SNAREs. Palmitoylation of syntaxin-1 and synaptobrevin-2 is proposed to promote localization to cholesterol-enriched raft domains in a similar manner as for SNAP23 and SNAP25 (Fukata & Fukata, 2010; Prescott et al., 2009; Salaün et al., 2004; Salaün et al., 2005a, 2005b). In fact, palmitoylation may be a general mechanism for cells to target peripheral and integral membrane proteins to raft-like membrane domains (Greaves et al., 2009; Levental, Grzybek, et al., 2010), as recently suggested in a study employing giant plasma membrane vesicles (Levental, Lingwood, Grzybek, Coskun, & Simons, 2010).

4.3. Hydrophobic mismatch

A recent bioinformatics study showed that fungal SNAREs involved in ER/Golgi trafficking generally have about 4 residues shorter transmembrane helices than SNAREs that are located in the plasma membrane (Sharpe, Stevens, & Munro, 2010). By a solution X-ray scattering technique, the apical plasma membrane was estimated to be about 3 Å thicker than Golgi membranes in rat hepatocytes (Mitra, Ubarretxena-Belandia, Taguchi, Warren, & Engelman, 2004), which would correspond to about 2 residues in an alpha helix. Since the plasma membrane is thicker than ER/Golgi membrane, it was proposed that hydrophobic matching between the length of the transmembrane helices and the thicknesses of the various organellar membranes could act as an intracellular protein-sorting mechanism (Bulbarelli et al., 2002; De Planque & Killian, 2003; Greaves et al., 2009; Kaiser et al., 2011). Here, proteins with similar lengths of their transmembrane helices could cluster together in the membranes of the ER or Golgi apparatus, which would facilitate their sorting into specific intracellular trafficking compartments. Similarly, hydrophobic mismatch may also mediate the domain organization of proteins in the plasma membrane. Direct evidence for this comes from a study by Bulbarelli et al. (2002), where GFP-tagged transmembrane domains of syntaxin-3 and syntaxin-4 were sufficient for localization to the apical and basolateral membranes of MDCK cells, respectively, and this was attributed to the (1 residue) longer transmembrane helix of syntaxin-3.

Hydrophobic mismatch between transmembrane helices and lipid bilayers could also trigger clustering of SNAREs in small membrane domains, as is well described for synthetic transmembrane peptides (De Planque & Killian, 2003; Kaiser et al., 2011). Especially when the transmembrane helix is shorter than the thickness of the surrounding lipid membrane (called negative hydrophobic mismatch), the resulting line tension can drive clustering of membrane proteins into domains. Since the recent crystal structure of the neuronal coiled-coil SNARE complex (Fig. 6.1C) showed that the (alpha-helical) transmembrane helices of syntaxin-1 and synaptobrevin-2 may not be long enough to span the entire plasma membrane (Stein et al., 2009), it seems possible that hydrophobic mismatch contributes to membrane clustering of these SNAREs. In fact, it may well be that the clustering of syntaxin-1 by a competition with cholesterol for solvation by bulk lipids (Section 4.1; Murray & Tamm, 2009, 2011; Van den Bogaart et al., 2011) can be explained by hydrophobic mismatch.

4.4. Phosphorylation of SNAREs

In addition to palmitoylation (Section 4.2), phosphorylation is another posttranslational modification that affects membrane partitioning of SNAREs. Many SNAREs are phosphorylated, including syntaxin-1, syntaxin-4, SNAP25, and synaptobrevin-2 by a variety of kinases (Foletti et al., 2000; Rickman & Duncan, 2010; Risinger & Bennett, 1999). For most of these phosphorylation modifications, their functional significances, their occurrences, and the stimuli that control phosphorylation remain largely unknown. The best-studied effect of phosphorylation on membrane partitioning of SNAREs is the phosphorylation of syntaxin-1 at serine 14 by casein kinase II. In PC12 cells, overexpression of a phosphomimetic mutant of syntaxin1 (S14E) significantly reduced exocytosis of dense-core vesicles (Rickman & Duncan, 2010). In rat brain, up to 40% of syntaxin-1 is phosphorylated at serine 14 (for both rat syntaxin-1a and syntaxin-1b) (Foletti et al., 2000). In this study, it was shown by a phosphorylation-specific antibody that phosphorylated syntaxin-1 was clustered in discrete domains in the axonal membrane at positions that did not colocalize with active zones or synaptic vesicles. How syntaxin-1 phosphorylated at serine 14 was segregated from unmodified syntaxin-1 is unclear, but this may involve interactions with other proteins and/or lipids. For instance, serine 14-phosphorylated syntaxin-1 preferentially associates with SNAP25 (Foletti et al., 2000), whereas phosphorylation results in decreased affinities of syntaxin-1 for synaptotagmin-1 (Risinger & Bennett, 1999) and Munc18a (Rickman & Duncan, 2010) (see Section 4.8), and these altered protein interactions could affect membrane partitioning. Alternatively, interactions with lipids could be affected by phosphorylation, as suggested by recent molecular dynamics simulations where the interaction of the N-terminus of syntaxin-1 with phosphatidylinositol (4,5)-bisphosphate (PI(4,5)P_2) was disrupted by phosphorylation of serine 14 (Khelashvili, Galli, & Weinstein, 2012) (see Section 4.5).

4.5. Ca^{2+} and phosphoinositide interactions

Although only a minor component of the plasma membrane (~1% of total lipids), phosphatidylinositol (4,5)-bisphosphate (PI(4,5)P_2) is essential for endo- and exocytosis (reviewed in Cremona & De Camilli, 2001; Di Paolo & De Camilli, 2006; Koch & Holt, 2012; Martin, 2012; Wen, Osborne, & Meunier, 2011). Neuronal exocytosis requires the presence of PI(4,5)P_2 at the plasma membrane (Milosevic et al., 2005). PI(4,5)P_2 is not uniformly distributed over the plasma membrane but is accumulated in at least a fraction of the syntaxin-1- and SNAP25-enriched membrane domains (Aikawa et al., 2006; Aoyagi et al., 2005) where it is the dominant inner-leaflet lipid by far (Van den Bogaart et al., 2011). Many proteins involved in membrane trafficking bind to PI(4,5)P_2, including rabphilin, CAPS, synaptotagmin-1, SCAMP2, and Mint proteins. As reviewed elsewhere (Cremona & De Camilli, 2001; Di Paolo & De Camilli, 2006; Koch & Holt, 2012; Martin, 2012; Wen et al., 2011), PI(4,5)P_2 is predominantly believed to act as a factor in docking or priming of exocytotic vesicles upstream of the actual membrane fusion. Indeed, we recently showed targeting of synaptotagmin-1 (a calcium sensor essential for vesicular docking; De Wit et al., 2009) to syntaxin-1-enriched membrane domains containing PI(4,5)P_2 (Honigmann et al., 2013).

Various SNAREs including syntaxin-1 and synaptobrevin-2 directly interact with PI(4,5)P_2 via well-conserved polybasic juxtamembrane domains that connect the SNARE motifs to the transmembrane helices (Fig. 6.2) (Lam, Tryoen-Toth, Tsai, Vitale, & Stuenkel, 2008; Murray & Tamm, 2009, 2011; Van den Bogaart et al., 2011; Williams, Vicôgne, Zaitseva, McLaughlin, & Pessin, 2009). In addition, various other cationic residues located in the cytoplasmic domain of syntaxin-1 are predicted to interact with PI(4,5)P_2 as well (Khelashvili et al., 2012). PI(4,5)P_2 interactions are required for domain partitioning of syntaxin-1, because removal of cationic residues by mutations in the polybasic motif or depletion of plasma membrane PI(4,5)P_2 by overexpression of a membrane-targeted region of the phosphatase domain of synaptojanin-1 leads to reduced clustering of syntaxin-1 in the plasma membrane of PC12 cells (Van den Bogaart et al., 2011). In fact, clustering of syntaxin-1 can be directly mediated by electrostatic interactions with PI(4,5)P_2, as evidenced by the domain formation when PI(4,5)P_2 and syntaxin-1 are reconstituted at approximately equimolar levels in giant unilamellar vesicles. Here, the polyanionic PI(4,5)P_2 is suggested to act as "charge bridge" and thereby clusters (the polybasic) syntaxin-1 molecules together (Van den Bogaart et al., 2011).

A recent study demonstrates a role for phosphatidylinositol (3,4,5)-trisphosphate ($PI(3,4,5)P_3$) instead of $PI(4,5)P_2$ in neurotransmitter release (Khuong et al., 2013). In this study, it was demonstrated that $PI(3,4,5)P_3$ was localized to syntaxin-1-enriched membrane clusters, both at *Drosophila melanogaster* neuromuscular junction boutons and at the plasma membrane of PC12 cells. Reducing $PI(3,4,5)P_3$ (but not $PI(4,5)P_2$) levels impaired neurotransmitter release and led to temperature-sensitive paralysis of *Drosophila*. Importantly, and similar to $PI(4,5)P_2$ (Van den Bogaart et al., 2011), reduction of cellular levels of $PI(3,4,5)P_3$ or mutating cationic residues from the juxtamembrane linker of syntaxin-1 led to a dispersal of syntaxin-1 membrane clusters (Khuong et al., 2013). These findings raise the question whether reported effects of $PI(4,5)P_2$ in exocytosis may in fact be attributable to $PI(3,4,5)P_3$. We consider this probable because many of the perturbation experiments used in the literature (including in our own study; Van den Bogaart et al., 2011) cannot adequately distinguish between $PI(4,5)P_2$ and $PI(3,4,5)P_3$. For instance, the synaptojanin-1 construct used to deplete plasma membrane $PI(4,5)P_2$ (Milosevic et al., 2005; Van den Bogaart et al., 2011) will also deplete $PI(3,4,5)P_3$. Moreover, syntaxin-1-enriched clusters overlap more with $PI(3,4,5)P_3$ than with $PI(4,5)P_2$ in the plasma membrane of PC12 cells (compare Khuong et al., 2013 with Van den Bogaart et al., 2011). On the other hand, membrane domains enriched in these phosphoinositides do not overlap in the plasma membrane of PC12 cells (Wang & Richards, 2012), rendering it unlikely that individual syntaxin-1 clusters contain both $PI(4,5)P_2$ and $PI(3,4,5)P_3$.

Finally, physiological concentrations of Ca^{2+} (μM range) specifically promote the clustering of SNAREs and other membrane proteins in the plasma membrane of PC12 and primary chromaffin cells (Zilly et al., 2011). The extent of Ca^{2+}-mediated clustering correlated with the net negative charge of the proteins, suggesting that this effect is driven by electrostatic protein–protein interactions. Interestingly, Ca^{2+} modulation of SNARE domains may have functional implications for (Ca^{2+}-triggered) neurotransmitter release, since the SNAREs that catalyze neurotransmitter release have a higher negative charge (rat syntaxin-1a, -12; SNAP25, -14) than the SNAREs specific for constitutive release (syntaxin-4, -5; SNAP23, -9; charges calculated by subtracting the total number of negatively charged residues (Asp + Glu) from the number of positively charged residues (Arg + Lys)). Indeed, Ca^{2+} directly modulates the activity of SNAP25 and syntaxin-1, since

Ca^{2+}-promoted SNARE clustering inhibited the binding of exogenously added SNARE motif of synaptobrevin-2 to these SNAREs (Zilly et al., 2011). In addition to these protein–protein interactions, Ca^{2+} could also promote SNARE clustering via $PI(4,5)P_2$. Ca^{2+} is well known to specifically induce the formation of $PI(4,5)P_2$ membrane domains in artificial membranes (Carvalho, Ramos, Roy, & Picart, 2008; Levental et al., 2009; Wang et al., 2012) and may thereby rearrange syntaxin-1–$PI(4,5)P_2$ membrane domains.

4.6. Homotypic protein–protein interactions

Homo-oligomerization of syntaxin-1 has been implicated in SNARE cluster formation. Sieber et al. (2006) showed that although the transmembrane helix (with the polybasic juxtamembrane domain) was sufficient for membrane clustering of truncated GFP-tagged syntaxin-1 in the plasma membrane of PC12 cells (probably by thermodynamic properties of the membrane; see Sections 4.1, 4.3, and 4.5), the presence of the SNARE domain was required for correct colocalization with myc-tagged full-length syntaxin-1. Since the SNARE domain of syntaxin-1 can self-interact, as shown for recombinant fragments of syntaxin-1 (Lerman, Robblee, Fairman, & Hughson, 2000), such homotypic interactions of the SNARE domain were suggested to mediate SNARE membrane partitioning (Sieber et al., 2006). Importantly, in this study, deletion of the N-terminal regulatory Habc domain did not seem to affect partitioning in the plasma membrane of PC12 cells. Additional evidence for a role of the SNARE region in SNARE clustering comes from FRAP measurements in PC12 cells (Sieber et al., 2007) and rat spinal cord neurons (Ribrault et al., 2011). Here, deletion of the SNARE motif (but not of deletion of the N-terminal Habc-region) increased the mobility of syntaxin-1. However, these studies and the clustering mechanisms described in Section 4.1–4.5 seem to contradict findings in pancreatic INS-1 beta cells where deletion of the regulatory Habc domain of syntaxin-1 resulted in dispersal of the membrane clusters (Yang et al., 2006) and we currently have no explanation for this discrepancy. Finally, although the transmembrane helices of both synaptobrevin-2 and syntaxin-1 form stable dimers (Kroch & Fleming, 2006; Laage, Rohde, Brosig, & Langosch, 2000), mutagenesis of the interacting residues demonstrated that these interactions do not play a role in syntaxin-1 cluster formation, at least in the plasma membrane of PC12 cells (Sieber et al., 2006; Van den Bogaart et al., 2011).

In addition to the homotypic protein–protein interactions discussed earlier, more specific SNARE sorting mechanisms may exist, because overexpression of both syntaxin-1 and syntaxin-4 weakens the segregation of domains enriched in these SNAREs in the plasma membrane of PC12 cells (Sieber et al., 2006). Specifically the N-terminal regulatory Habc domain might influence partitioning of syntaxin-1 in (at least a population of) membrane domains. Evidence for such a role of the Habc domain comes from findings that the Habc domain is essential for specific clustering of syntaxin-1 below docked vesicles, although truncation mutants of syntaxin-1 lacking the N-terminal Habc domain still clustered in membrane domains (Barg et al., 2010).

4.7. Cytoskeleton anchoring

The cortical cytoskeleton has a well-studied role in vesicle docking, vesicle priming, exocytosis, and endocytosis (see Gutiérrez, 2012; Torregrosa-Hetland et al., 2010; Villanueva et al., 2012 for recent reviews). Overall, the resting actin cytoskeleton forms a complex and dynamic network that serves as a barrier and limits access of exocytotic granules to the plasma membrane, although this is controversial and some studies show no or only a limited effect of actin on exocytosis (particularly in neurons). The cortical dense actin mesh contains open polygonal spaces through which vesicles can access the membrane. Syntaxin-1 and SNAP25 membrane domains are located at the edges of these actin-devoid regions in chromaffin cells and exocytosis takes place here (Torregrosa-Hetland et al., 2010, 2011; Villanueva et al., 2010; Wang & Richards, 2011).

Some SNARE proteins associate with the cortical cytoskeleton, for instance syntaxin-1 and syntaxin-4. In some cases, SNARE interaction with F-actin is not direct but requires other proteins such as the motor protein myosin-Va for syntaxin-1 (Watanabe et al., 2005) or the actin adapter protein α-fodrin for syntaxin-1, syntaxin-3, and syntaxin-4 (Nakano, Nogami, Sato, Terano, & Shirataki, 2001). In other cases, SNAREs bind directly, such as syntaxin-4 that interacts with F-actin, via the N-terminal regulatory Habc-region (Jewell, Luo, Oh, Wang, & Thurmond, 2008; Woronowicz et al., 2010). Recent data have also suggested that SNAP25 can interact with actin fibers. In this study, FRET has been observed between a DsRed-tagged overexpressed version of SNAP25 and GFP-tagged LifeAct, an actin-binding probe, in chromaffin cells (Torregrosa-Hetland et al., 2013). Interaction of SNAREs with the cortical cytoskeleton may well be involved in the organization of the secretory machinery at the plasma membrane. Interestingly, depolymerization of actin fibers in kidney renal epithelial MDCK cells by latrunculin B resulted in disruption of the syntaxin-4 (but not syntaxin-3) membrane clusters, indicating a role of F-actin in clustering of syntaxin-4 membrane domains (Low et al., 2006). In contrast, syntaxin-3 clustering was mediated by the microtubular network, as evidenced by nocodazole disruption of microtubules that resulted in the opposite effect (disruption of syntaxin-3 but not syntaxin-4 clusters) (Low et al., 2006). These results indicate that different interactions of SNAREs with cortical actin or microtubules may explain their segregation in distinct compartments. F-actin may also influence the location of other proteins involved in exocytosis, such as calcium channels that interact not only with SNAREs (see Section 5.3) but also with F-actin structures (Gutiérrez, 2012; Lopez et al., 2007; Torregrosa-Hetland et al., 2011).

4.8. Heterotypic protein–protein interactions

The partitioning of syntaxin-1 and syntaxin-4 may require interactions with "accessory" proteins. For instance, in lung endothelial cells, SNAP23 and syntaxin-4 colocalize with one of the structural proteins that stabilizes caveolae: caveolin-1 (Predescu et al., 2005). However, a general role of caveolin in SNARE clustering is unlikely because the expression levels of caveolin are too low to account for SNARE clustering in PC12 cells (Bilderback, Gazula, Lisanti, & Dobrowsky, 1999; Chamberlain et al., 2001; Schmidt, Hannah, & Huttner, 1997; note, however, that some expression of caveolin in these cells was observed in another study: Galbiati et al., 1998).

Other candidate proteins that may be involved in organization of syntaxin-1 and syntaxin-4 include the Munc18 isoforms. Munc18 is a member of the SM (Sec1/Munc18) protein family, a group of proteins that are essential for exocytosis. Munc18a (or Munc18-1) is expressed and functions in exocytosis of neurons and neuroendocrine cells (reviewed in Gutiérrez, 2012; Jahn & Fasshauer, 2012; Ramakrishnan et al., 2012). Two other isoforms of the Munc18 protein family, Munc18b and Munc18c (or Munc18-2 and Munc18-3, respectively), are ubiquitously expressed. In mice, Munc18a deletion leads to a complete block of neurotransmitter release (Verhage et al., 2000) and Munc18a is a well-established factor for docking of synaptic vesicles to the plasma membrane (Gutiérrez, 2012; Ramakrishnan et al., 2012; Toonen et al., 2006). Indeed, Munc18a bound to syntaxin-1-enriched domains in the plasma membrane of PC12 cells (Zilly et al., 2006) and the presence of Munc18a was essential for tethering of dense-core vesicles to syntaxin-1 domains in chromaffin cells (Voets et al., 2001). In addition to a function in vesicular docking or exocytosis, Munc18a is also involved in syntaxin-1 sorting from the Golgi network to the plasma membrane, although the precise extent and role of Munc18a in SNARE trafficking are still controversial (see for discussion Arunachalam et al., 2008; Kurps & de Wit, 2012; Voets et al., 2001; Yang et al., 2006).

The association of Munc18a with syntaxin-1 does not only play a role in the morphological docking of dense-core vesicles but also influence the structure of the F-actin cortical cytoskeleton (Kurps & de Wit, 2012). Interestingly, the cortical actin layer is markedly (about 30%) increased in chromaffin cells isolated from Munc18a knockout mice compared to wild-type cells (Kurps & de Wit, 2012; Toonen et al., 2006). Very similar to these observations, actin levels are substantially increased in kidney epithelial MDCK cells upon Munc18c knockdown by shRNA. In this case, the increase of actin by Munc18c depletion can be attributed to decreased levels of syntaxin-4. Expression levels of Munc18c and syntaxin-4 are tightly related and knockdown of one of these proteins leads to reduction of the other protein in MDCK cells. However, rescue of Munc18c levels in syntaxin-4-depleted MDCK cells (by Munc18c expression from a strong viral promoter) was not able to restore actin levels to wild-type, indicating that increased levels of actin upon Munc18c knockdown are the result of reduced syntaxin-4 levels (Martin ter Beest, personal communication). Whether a similar mechanism explains the increased F-actin layer in Munc18a-null chromaffin cells remains elusive, but at least syntaxin-1 and Munc18a levels are correlated in chromaffin and PC12 cells (Arunachalam et al., 2008; Voets et al., 2001).

5. THE BIOLOGICAL ROLE OF SNARE CLUSTERING

The clustering of SNAREs in distinct membrane regions offers at least two potential advantages for the cell. The first potential advantage is called a "local concentration effect" or reduction of dimensionality, and this is an effective concentration that could be three orders of magnitude higher than if the SNAREs were randomly dispersed and is caused by the alignment in dense membrane clusters (see McLaughlin, Wang, Gambhir, & Murray, 2002 for discussion). However, at very high SNARE densities, molecular crowding effects may result in shielding of binding sites on the SNARE proteins and thereby actually lower the effective concentration and decrease the accessibility and activity of SNARE proteins (see Section 5.2). The second potential advantage is that domain partitioning of SNAREs allows localized regulation of the cellular functions of these proteins, for instance by defining sites for vesicle docking and/or fusion or by constituting reserve pools of active SNAREs.

5.1. SNARE domains as sites for vesicle docking and fusion

Because SNAREs such as syntaxin-1 (Kurps & de Wit, 2012; Toonen et al., 2006) are involved in morphological vesicle docking, membrane domains of SNAREs may be preferential docking sites that would be ideal to act as molecular targeting signals ("beacons") for exocytotic vesicles. Indeed, there is evidence that membrane domains enriched in SNAREs form functional platforms for vesicle docking and/or exocytosis. First, several studies demonstrate that in PC12 cells, dense-core vesicles colocalize with syntaxin-1 clusters in the plasma membrane (Aoyagi et al., 2005; Lang et al., 2001; Zilly et al., 2006), albeit the likelihood of fusion is independent of cluster proximity (Lang et al., 2001). Second, insulin-containing vesicles colocalize with syntaxin-1- and SNAP25-enriched membrane domains in the MIN6 and INS-1 pancreatic beta cell lines (Ohara-Imaizumi et al., 2004; Somanath, Barg, Marshall, Silwood, & Turner, 2009). Third, in chromaffin cells, SNAP25- and syntaxin-1-enriched membrane domains nearly completely colocalize with single-vesicle secretory spots as shown by overlap with a granule content marker (Lopez et al., 2007). Fourth, SNAP25 microdomains in the plasma membrane of chromaffin cells preexist and the motion of these SNARE domains correlates with vesicle motion (López et al., 2009). Fifth, caveolae were found to colocalize to SNAP23- and syntaxin-4-enriched domains and fusion occurred at these domains in blood vessel endothelial cells (Predescu et al., 2005). Finally, synaptotagmin-1, a calcium sensor essential for vesicular docking in chromaffin cells (De Wit et al., 2009), specifically binds to syntaxin-1-enriched membrane domains in the plasma membrane of PC12 cells (Honigmann et al., 2013). All these studies are in line with the view that (preexisting) SNARE domains operate as functional platforms for vesicular docking and/or fusion. Whether SNARE domains remain clustered throughout the complete exocytotic cycle (Lopez et al., 2007) or disassemble upon granule fusion (Barg et al., 2010) is still ambiguous.

Despite the evidence listed earlier, several recent studies challenge the view that vesicle docking and/or fusion takes place at SNARE membrane clusters. Rather, they suggest that this may actually take place away from the clusters. Two live-cell imaging studies demonstrated that vesicles do not dock to preexisting SNARE-enriched membrane domains, but instead to syntaxin-1 clusters that are dynamically assembled beneath docked vesicles (Barg et al., 2010; Knowles et al., 2010). These studies suggest that SNARE domains may not play a role in vesicular docking/tethering to the plasma membrane, although they could still be involved in the final exocytosis step (i.e., membrane fusion). Such a function of SNARE clusters in exocytosis was questioned in a more recent PALM study, where vesicles were found to dock and remain docked at areas of low SNARE density in the plasma membrane of PC12 cells (Yang et al., 2012). Actually, the local accumulation of many SNARE proteins does not seem a requirement for exocytosis per se, since recent findings indicate that only a very low number of one to three SNARE complexes are required for membrane fusion (see Van den Bogaart & Jahn, 2011 for review). In fact, recent evidence suggest that SNAREs in membrane domains may actually be less active than SNAREs that locate away from the clusters, as we will discuss in the next section.

5.2. Membrane domains to modulate SNARE activity

It is increasingly clear that membrane domains may modulate SNARE activity by preventing the formation of unproductive SNARE complexes, such as ternary SNARE complexes (i.e., SNAP25 and syntaxin-1 with synaptobrevin-2 already bound in *cis*) in the plasma membrane that can no longer engage with synaptobrevin-2 in the synaptic vesicle. Another example of an unproductive SNARE complex is the so-called 2:1 complex, consisting of two copies of syntaxin-1 and a single copy of SNAP25. At least *in vitro*, such a 2:1 complex, where the second copy of syntaxin-1 takes the place of synaptobrevin-2, is well known to form a kinetic dead end that inhibits binding of synaptobrevin-2 (Fasshauer & Margittai, 2004). By preventing the formation of such unproductive SNARE complexes, membrane domains could assure the presence of sufficient free acceptor SNAREs in the plasma membrane even during periods of sustained exocytotic activity.

The most direct evidence for a role of SNARE domains in modulating SNARE activity comes from a study where exogenously added fluorescently labeled SNARE motifs of synaptobrevin-2 were suggested to more rapidly engage with syntaxin-1 located outside compared to within membrane domains in the plasma membrane of PC12 cells (Bar-On et al., 2009). There are several potential mechanisms how clustering in membrane domains may inhibit such binding of syntaxin-1 to synaptobrevin-2 (and/or SNAP25; see Section 3.4). First, steric hindrance caused by the high molecular crowding in the membrane domains might limit the accessibility of syntaxin-1 for its cognate SNARE proteins (see Section 3). Second, engagement of the SNARE motifs in homotypic interactions (see Section 4.6) might prevent the binding of cognate SNARE proteins to syntaxin-1 domains. Third, syntaxin-1 located in membrane domains might be kept in a so-called closed conformation where the N-terminal Habc domain is folded back on the SNARE motif. In this closed conformation, syntaxin-1 is unable to bind to synaptobrevin-2 and SNAP25. The regulatory protein Munc18a (see Section 4.8) regulates activity of syntaxin-1 by stabilization of this closed conformation and thereby can inhibit binding to its cognate SNAREs (reviewed in Gutiérrez, 2012; Jahn & Fasshauer, 2012; Ramakrishnan et al., 2012). Nevertheless, we consider stabilization of a closed conformation of a large fraction of syntaxin-1 by Munc18a unlikely, because syntaxin-1 is present in PC12 cells at a ~20-fold molecular excess over Munc18a as determined by quantitative Western blotting (Schütz, Zilly, Lang, Jahn, & Bruns, 2005). In addition, even although Munc18a is bound to syntaxin-1 domains in PC12 membrane sheets, it can be readily dissociated from syntaxin-1 by addition of the cytoplasmic domain of synaptobrevin-2 (Zilly et al., 2006). This synaptobrevin-2-induced dissociation of Munc18a required the presence of endogenous SNAP25, indicating that most syntaxin-1 in the domains was accessible to their cognate SNAREs and suggesting that only a small fraction (if any) of syntaxin-1 located in the domains is in a fully closed conformation.

Membrane domains of syntaxin-1 and SNAP25 may also prevent the formation of 2:1-complexes (see above). Indeed, in a recent study (Halemani et al., 2010) employing truncation mutants of SNAP25 in PC12 cells, it was demonstrated that the N-terminal Qb- (but not the C-terminal Qc-) SNARE motif of SNAP25 was both required and sufficient for binding to syntaxin-1, indicating that 2:1 complexes (which require both the Qb- and Qc-SNARE motifs) were not present in these membranes. This absence of 2:1-complexes in the plasma membrane of PC12 cells is further supported by the lack of FRET from cerulean- to YFP-tagged syntaxin-1 (Rickman et al., 2010).

Finally, clustering of SNAREs in membrane domains may facilitate the disassembly of inactive *cis*-SNARE complexes (i.e., syntaxin-1–SNAP25–synaptobrevin-2 complexes). Following membrane fusion, the SNARE proteins are recycled by disassembly of these *cis*-SNARE complexes by the ATPase NSF in conjunction with α-SNAP (reviewed in Jahn & Fasshauer, 2012). In fact, by fitting the dissociation of prebound SNARE motif of synaptobrevin-2 from PC12 membrane sheets with a kinetic model, NSF/α-SNAP-mediated disassembly of SNARE complexes was estimated about 30-fold faster in the domains compared to outside (Bar-On et al., 2009). This indicates that indeed membrane domains might facilitate the disassembly of SNARE complexes, probably via a local increase of the concentration of substrate for NSF/α-SNAP in the SNARE domains (see Section 5).

5.3. Other functions of SNARE domains

In addition to the putative roles in vesicular docking and membrane fusion and/or functioning as a reservoir described earlier, SNARE domains may have other functions as well. For instance, it is possible that domain organization prevents endocytosis of plasma membrane SNAREs and/or promotes the recycling of vesicular SNAREs and other vesicular components. Since synaptic vesicles contain an average of about 70 copies of synaptobrevin-2 (Takamori et al., 2006), the clustering of these SNAREs would dramatically facilitate vesicular recycling (see Geumann et al., 2010 for discussion). Domain partitioning would also allow localized regulation of the cellular functions of the SNAREs and this function is supported by various findings. First, as we described in Section 3.4, SNAREs that catalyze different exocytotic trafficking routes are completely spatially segregated, which would allow to independently regulate their functions. Second, the colocalization of SNAREs with regulatory proteins in membrane domains provides further evidence that SNARE clusters play a role in modulation of SNARE function and this is best described for calcium channels. Many types of calcium channels directly bind to SNAREs, including L- (Cav1.2), N- (Cav2.2), P/Q- (Cav2.1), and T-type (Cav3.2) calcium channels (Davies, Jarvis, & Zamponi, 2011; Lopez et al., 2007; Taverna et al., 2004; Weiss et al., 2012) and high levels of colocalization between calcium channels and SNARE domains have been observed in adrenal slices (Lopez et al., 2007; Torregrosa-Hetland et al., 2011). The close proximity of calcium channels to sites of exocytosis would facilitate exocytosis by leading to faster increases and higher final local calcium concentrations upon calcium influx, and this effect is well understood in the literature (e.g., Sheng, Westenbroek, & Catterall, 1998; Torregrosa-Hetland et al., 2010).

6. CONCLUSIONS AND OUTLOOK

The domain organization of SNAREs in the plasma membrane has been extensively studied for more than a decade. Although, as we have reviewed in the earlier sections, some questions and controversies remain regarding the compositions, functional significances, and mechanisms of SNARE clustering in the plasma membrane, we believe a clear picture is emerging of how SNAREs are organized in membrane domains.

6.1. Towards a synergistic model of SNARE partitioning

At a first glance, it may seem surprising that so many different clustering mechanisms (Section 4) seem to influence the partitioning of SNAREs in the plasma membrane. However, all these clustering mechanisms are somehow interrelated (Fig. 6.3). For instance, cholesterol is essential for PI(4,5)P_2 membrane clustering (Milosevic et al., 2005); cholesterol increases the interaction between syntaxin-1 and PI(4,5)P_2 (Murray & Tamm, 2009); PI(4,5)P_2 may affect phosphorylation of syntaxin-1 (Khelashvili et al., 2012); phosphorylation of syntaxin-1 affects interactions with SNAP25, Munc18a, and synaptotagmin-1 (Foletti et al., 2000; Rickman & Duncan, 2010; Risinger & Bennett, 1999); Munc18 affects the cortical actin network (Kurps & de Wit, 2012); and PI(4,5)P_2 regulates the organization and dynamics of the actin cytoskeleton (reviewed in Saarikangas, Zhao, & Lappalainen, 2010; Zhang, Mao, Janmey, & Yin, 2012). This tight interrelationship between these clustering mechanisms has two implications for SNARE partitioning: First, the possibility that SNARE partitioning is multifactorial has to be taken into account in future studies addressing SNARE partitioning. It will be very difficult (if not impossible) to design (or interpret) experiments that selectively target only a single of these clustering mechanisms without affecting others. Second, and more importantly, the complex network of membrane clustering mechanisms offers an elaborate regulatory mechanism whereby cells can precisely modulate the activity and localization of SNAREs in the plasma membrane. We propose this regulatory mechanism consists of three layers of organization:

The first layer of organization relates to the intrinsic property of SNAREs to cluster in biological membranes. Truncation mutants of syntaxin isoforms still cluster in the plasma membrane, even when their entire cytoplasmic domains are deleted (Sieber et al., 2006; Van den Bogaart et al., 2011). As demonstrated by perturbation experiments that affected the lipid composition of the plasma membrane (e.g., cholesterol depletion, Section 4.1; PI(4,5)P_2 depletion, Section 4.5), this clustering is driven by hydrophobic and electrostatic properties of the membrane and is clearly required for the partitioning of SNARE proteins in membrane domains. However, these membrane interactions do not seem to be sufficiently specific to explain the segregation of highly homologous SNARE proteins in different membrane domains, such as for syntaxin-1 and syntaxin-4 in neuroendocrine cells (Section 5.3). Membrane interactions also cannot explain the structural diversity of SNARE domains in the plasma membrane, such as subpopulations of syntaxin-1 domains with or without SNAP25 (Section 3.4). This clearly requires a second layer of organization by more specific protein–protein interactions, which appears to be based upon homotypic interactions of the SNARE motifs (Section 4.6) and the binding of (multiple) SNAREs to exocytotic vesicles in the proximity of the plasma membrane (Section 4.8). Finally, a third layer of organization regulates SNARE sorting from ER/Golgi and other intracellular compartments to the plasma membrane. Here, interactions with the cortical cytoskeleton (Section 4.7) or with Munc18 isoforms (Section 4.8) directly influence overall densities and polar distributions of SNAREs in the plasma membrane (e.g., syntaxin-3 and syntaxin-4 over the apical and basal membranes of epithelial cells; Section 3.1). Together, we believe such a synergistic and hierarchical model explains how SNAREs (and other proteins; see below) are organized in the plasma membrane.

6.2. Beyond SNARE partitioning in the plasma membrane

In this review, we have limited our discussion to SNARE cluster formation in the plasma membrane. However, SNAREs are present in all organelles of the secretory pathway where they display highly distinct localizations (see Jahn & Scheller, 2006 for review). The underlying sorting mechanisms are only poorly understood. For instance, the highly homologous SNAREs syntaxin-12 (also called syntaxin-13) and syntaxin-7 (50% sequence identity for rat proteins) are sorted to early and late endosomal compartments, respectively. The transition from syntaxin-12 to syntaxin-7 upon endosome maturations occurs relatively fast, as shown, for instance, during the transport of a phagocytotic vesicle to the lysosome in RAW 264.7 macrophages

(Collins, Schreiber, Grinstein, & Trimble, 2002). How these SNAREs are sorted apart during the transition from an early to a late endosome is still completely unclear, but this sorting likely involves membrane clustering of SNAREs in endosomal membranes (Gruenberg, 2001). Indeed, by electron and STED microscopy, it was demonstrated that endosomal SNAREs (VAMP4, syntaxin-12, and syntaxin-16) and syntaxin-1, synaptobrevin-2, and SNAP25 are clustered in early endosomes of PC12 cells (Geumann, Schäfer, Riedel, Jahn, & Rizzoli, 2010), and it is therefore conceivable that some of the mechanisms responsible for SNARE clustering in the plasma membrane may also be involved in the sorting of SNAREs over intracellular membrane compartments.

Finally, there is growing evidence that the majority (if not all) of integral and peripheral membrane proteins are not uniformly distributed across the plane of the membrane but rather are partitioned in subdomains and clusters. Thus, many of the mechanisms responsible for SNARE clustering may also contribute to membrane partitioning of other proteins, thus being of general significance for the organization of eukaryotic plasma membranes. The structural simplicity of SNARE proteins constitutes an experimental advantage, thus providing an excellent model system for understanding the general principles of membrane partitioning.

ACKNOWLEDGMENTS

We thank Martin ter Beest (Nijmegen Centre for Molecular Life Sciences, the Netherlands), Rory Duncan (Heriot-Watt University, United Kingdom), Dragomir Milovanovic (Max Planck Institute for Biophysical Chemistry, Germany), Colin Rickman (Heriot-Watt University, United Kingdom), Lukas Tamm (University of Virginia School of Medicine, United States), and Heidi de Wit (Center for Neurogenomics and Cognitive Research, the Netherlands) for suggestions and comments.

REFERENCES

Aikawa, Y., Xia, X., & Martin, T. F. J. (2006). SNAP25, but not syntaxin 1A, recycles via an ARF6-regulated pathway in neuroendocrine cells. *Molecular Biology of the Cell*, *17*, 711–722.

Aoyagi, K., Sugaya, T., Umeda, M., Yamamoto, S., Terakawa, S., & Takahashi, M. (2005). The activation of exocytotic sites by the formation of phosphatidylinositol 4,5-bisphosphate microdomains at syntaxin clusters. *The Journal of Biological Chemistry*, *280*, 17346–17352.

Arunachalam, L., Han, L., Tassew, N. G., He, Y., Wang, L., Xie, L., et al. (2008). Munc18-1 is critical for plasma membrane localization of syntaxin1 but not of SNAP-25 in PC12 cells. *Molecular Biology of the Cell*, *19*, 722–734.

Bacia, K., Schuette, C. G., Kahya, N., Jahn, R., & Schwille, P. (2004). SNAREs prefer liquid-disordered over "raft" (liquid-ordered) domains when reconstituted into giant unilamellar vesicles. *The Journal of Biological Chemistry*, *279*, 37951–37955.

Barg, S., Knowles, M. K., Chen, X., Midorikawa, M., & Almers, W. (2010). Syntaxin clusters assemble reversibly at sites of secretory granules in live cells. *Proceedings of the National Academy of Sciences of the United States of America*, *107*, 20804–20809.
Bark, I. C., & Wilson, M. C. (1994). Human cDNA clones encoding two different isoforms of the nerve terminal protein SNAP-25. *Gene*, *139*, 291–292.
Bar-On, D., Gutman, M., Mezer, A., Ashery, U., Lang, T., & Nachliel, E. (2009). Evaluation of the heterogeneous reactivity of the syntaxin molecules on the inner leaflet of the plasma membrane. *The Journal of Neuroscience*, *29*, 12292–12301.
Bar-On, D., Winter, U., Nachliel, E., Gutman, M., Fasshauer, D., Lang, T., et al. (2008). Imaging the assembly and disassembly kinetics of cis-SNARE complexes on native plasma membranes. *FEBS Letters*, *582*, 3563–3568.
Bar-On, D., Wolter, S., van de Linde, S., Heilemann, M., Nudelman, G., Nachliel, E., et al. (2012). Super-resolution imaging reveals the internal architecture of nano-sized syntaxin clusters. *The Journal of Biological Chemistry*, *287*, 27158–27167.
Bilderback, T. R., Gazula, V. R., Lisanti, M. P., & Dobrowsky, R. T. (1999). Caveolin interacts with Trk A and p75(NTR) and regulates neurotrophin signaling pathways. *The Journal of Biological Chemistry*, *274*, 257–263.
Bulbarelli, A., Sprocati, T., Barberi, M., Pedrazzini, E., & Borgese, N. (2002). Trafficking of tail-anchored proteins: Transport from the endoplasmic reticulum to the plasma membrane and sorting between surface domains in polarised epithelial cells. *Journal of Cell Science*, *115*, 1689–1702.
Carvalho, K., Ramos, L., Roy, C., & Picart, C. (2008). Giant unilamellar vesicles containing phosphatidylinositol(4,5)bisphosphate: Characterization and functionality. *Biophysical Journal*, *95*, 4348–4360.
Chamberlain, L. H., Burgoyne, R. D., & Gould, G. W. (2001). SNARE proteins are highly enriched in lipid rafts in PC12 cells: Implications for the spatial control of exocytosis. *Proceedings of the National Academy of Sciences of the United States of America*, *98*, 5619–5624.
Chamberlain, L. H., & Gould, G. W. (2002). The vesicle- and target-SNARE proteins that mediate Glut4 vesicle fusion are localized in detergent-insoluble lipid rafts present on distinct intracellular membranes. *The Journal of Biological Chemistry*, *277*, 49750–49754.
Collins, R. F., Schreiber, A. D., Grinstein, S., & Trimble, W. S. (2002). Syntaxins 13 and 7 function at distinct steps during phagocytosis. *Journal of Immunology*, *169*, 3250–3256.
Cremona, O., & De Camilli, P. (2001). Phosphoinositides in membrane traffic at the synapse. *Journal of Cell Science*, *114*, 1041–1052.
Daum, G. (1985). Lipids of mitochondria. *Biochimica et Biophysica Acta*, *822*, 1–42.
Davies, J. N., Jarvis, S. E., & Zamponi, G. W. (2011). Bipartite syntaxin 1A interactions mediate CaV2.2 calcium channel regulation. *Biochemical and Biophysical Research Communications*, *411*, 562–568.
De Planque, M. R., & Killian, J. A. (2003). Protein-lipid interactions studied with designed transmembrane peptides: Role of hydrophobic matching and interfacial anchoring. *Molecular and Membrane Biology*, *20*, 271–284.
De Wit, H., Walter, A. M., Milosevic, I., Gulyás-Kovács, A., Riedel, D., Sørensen, J. B., et al. (2009). Synaptotagmin-1 docks secretory vesicles to syntaxin-1/SNAP-25 acceptor complexes. *Cell*, *138*, 935–946.
Di Paolo, G., & De Camilli, P. (2006). Phosphoinositides in cell regulation and membrane dynamics. *Nature*, *443*, 651–657.
Fasshauer, D., & Margittai, M. (2004). A Transient N-terminal interaction of SNAP-25 and syntaxin nucleates SNARE assembly. *The Journal of Biological Chemistry*, *279*, 7613–7621.
Foletti, D. L., Lin, R., Finley, M. A. F., & Scheller, R. H. (2000). Phosphorylated syntaxin 1 is localized to discrete domains along a subset of axons. *The Journal of Neuroscience*, *20*, 4535–4544.

Fukata, Y., & Fukata, M. (2010). Protein palmitoylation in neuronal development and synaptic plasticity. *Nature Reviews Neuroscience*, *11*, 161–175.
Gaisano, H. Y., Ghai, M., Malkus, P. N., Sheu, L., Bouquillon, A., Bennett, M. K., et al. (1996). Distinct cellular locations and protein protein interactions of the syntaxin family of proteins in rat pancreatic acinar cells. *Molecular Biology of the Cell*, 7, 2019–2027.
Galbiati, F., Volonte, D., Gil, O., Zanazzi, G., Salzer, J. L., Sargiacomo, M., et al. (1998). Expression of caveolin-1 and -2 in differentiating PC12 cells and dorsal root ganglion neurons: Caveolin-2 is up-regulated in response to cell injury. *Proceedings of the National Academy of Sciences of the United States of America*, *95*, 10257–10262.
Geumann, U., Schäfer, C., Riedel, D., Jahn, R., & Rizzoli, S. O. (2010). Synaptic membrane proteins form stable microdomains in early endosomes. *Microscopy Research and Technique*, *73*, 606–617.
Gonzalo, S., Greentree, W. K., & Linder, M. E. (1999). SNAP-25 is targeted to the plasma membrane through a novel membrane-binding domain. *The Journal of Biological Chemistry*, *274*, 21313–21318.
Gonzalo, S., & Linder, M. E. (1998). SNAP-25 palmitoylation and plasma membrane targeting require a functional secretory pathway. *Molecular Biology of the Cell*, *9*, 585–597.
Greaves, J., & Chamberlain, L. H. (2011). Differential palmitoylation regulates intracellular patterning of SNAP25. *Journal of Cell Science*, *124*, 1351–1360.
Greaves, J., Prescott, G. R., Gorleku, O. A., & Chamberlain, L. H. (2009). The fat controller: Roles of palmitoylation in intracellular protein trafficking and targeting to membrane microdomains. *Molecular and Membrane Biology*, *26*, 67–79.
Gruenberg, J. (2001). The endocytic pathway: A mosaic of domains. *Nature Reviews Molecular Cell Biology*, *2*, 721–730.
Gutiérrez, L. M. (2012). New insights into the role of the cortical cytoskeleton in exocytosis from neuroendocrine cells. *International Review of Cell and Molecular Biology*, *295*, 109–137.
Halemani, N. D., Bethani, I., Rizzoli, S. O., & Lang, T. (2010). Structure and dynamics of a two-helix SNARE complex in live cells. *Traffic*, *11*, 394–404.
Honigmann, A., van den Bogaart, G., Iraheta, E., Risselada, H. J., Milovanovic, D., Mueller, V., et al. (2013). Phosphatidylinositol 4,5-bisphosphate clusters act as molecular beacons for vesicle recruitment. *Nature Structural & Molecular Biology*, *20*, 679–686.
Jahn, R., & Fasshauer, D. (2012). Molecular machines governing exocytosis of synaptic vesicles. *Nature*, *490*, 201–207.
Jahn, R., & Scheller, R. H. (2006). SNAREs-engines for membrane fusion. *Nature Reviews Molecular Cell Biology*, *7*, 631–643.
James, D. J., Kowalchyk, J., Daily, N., Petrie, M., & Martin, T. F. J. (2009). CAPS drives trans-SNARE complex formation and membrane fusion through syntaxin interactions. *Proceedings of the National Academy of Sciences of the United States of America*, *106*, 17308–17313.
Jewell, J. L., Luo, W., Oh, E., Wang, Z., & Thurmond, D. C. (2008). Filamentous actin regulates insulin exocytosis through direct interaction with syntaxin 4. *The Journal of Biological Chemistry*, *283*, 10716–10726.
Kaiser, H. J., Orłowski, A., Róg, T., Nyholm, T. K., Chai, W., Feizi, T., et al. (2011). Lateral sorting in model membranes by cholesterol-mediated hydrophobic matching. *Proceedings of the National Academy of Sciences of the United States of America*, *108*, 16628–16633.
Kang, R., Wan, J., Arstikaitis, P., Takahashi, H., Huang, K., Bailey, A. O., et al. (2008). Neural palmitoyl-proteomics reveals dynamic synaptic palmitoylation. *Nature*, *456*, 904–909.
Khelashvili, G., Galli, A., & Weinstein, H. (2012). Phosphatidylinositol 4,5-biphosphate (PIP(2)) lipids regulate the phosphorylation of syntaxin N-terminus by modulating both its position and local structure. *Biochemistry*, *51*, 7685–7698.

Khuong, T. M., Habets, R. L., Kuenen, S., Witkowska, A., Kasprowicz, J., Swerts, J., et al. (2013). Synaptic PI(3,4,5)P_3 is required for syntaxin1A clustering and neurotransmitter release. *Neuron, 77*, 1097–1108.
Knowles, M. K., Barg, S., Wan, L., Midorikawa, M., Chen, X., & Almers, W. (2010). Single secretory granules of live cells recruit syntaxin-1 and synaptosomal associated protein 25 (SNAP-25) in large copy numbers. *Proceedings of the National Academy of Sciences of the United States of America, 107*, 20810–20815.
Koch, M., & Holt, M. (2012). Coupling exo- and endocytosis: An essential role for PIP_2 at the synapse. *Biochimica et Biophysica Acta, 1821*, 1114–1132.
Kroch, A. E., & Fleming, K. G. (2006). Alternate interfaces may mediate homomeric and heteromeric assembly in the transmembrane domains of SNARE proteins. *Journal of Molecular Biology, 357*, 184–194.
Kurps, J., & de Wit, H. (2012). The role of Munc18-1 and its orthologs in modulation of cortical F-actin in chromaffin cells. *Journal of Molecular Neuroscience, 48*, 339–346.
Laage, R., Rohde, J., Brosig, B., & Langosch, D. (2000). A conserved membrane-spanning amino acid motif drives homomeric and supports heteromeric assembly of presynaptic SNARE proteins. *The Journal of Biological Chemistry, 275*, 17481–17487.
Lam, A. D., Tryoen-Toth, P., Tsai, B., Vitale, N., & Stuenkel, E. L. (2008). SNARE-catalyzed fusion events are regulated by Syntaxin1A-lipid interactions. *Molecular Biology of the Cell, 19*, 485–497.
Lang, T. (2007). SNARE proteins and 'membrane rafts'. *The Journal of Physiology, 585*, 693–698.
Lang, T., Bruns, D., Wenzel, D., Riedel, D., Holroyd, P., Thiele, C., et al. (2001). SNAREs are concentrated in cholesterol-dependent clusters that define docking and fusion sites for exocytosis. *The EMBO Journal, 20*, 2202–2213.
Lang, T., Margittai, M., Hölzler, H., & Jahn, R. (2002). SNAREs in native plasma membranes are active and readily form core complexes with endogenous and exogenous SNAREs. *The Journal of Cell Biology, 158*, 751–760.
Lang, T., & Rizzoli, S. O. (2010). Membrane protein clusters at nanoscale resolution: More than pretty pictures. *Physiology (Bethesda, MD), 25*, 116–124.
Lerman, J. C., Robblee, J., Fairman, R., & Hughson, F. M. (2000). Structural analysis of the neuronal SNARE protein syntaxin-1A. *Biochemistry, 39*, 8470–8479.
Levental, I., Christian, D. A., Wang, Y. H., Madara, J. J., Discher, D. E., & Janmey, P. A. (2009). Calcium-dependent lateral organization in phosphatidylinositol 4,5-bisphosphate (PIP2)- and cholesterol-containing monolayers. *Biochemistry, 48*, 8241–8248.
Levental, I., Grzybek, M., & Simons, K. (2010). Greasing their way: Lipid modifications determine protein association with membrane rafts. *Biochemistry, 49*, 6305–6316.
Levental, I., Lingwood, D., Grzybek, M., Coskun, U., & Simons, K. (2010). Palmitoylation regulates raft affinity for the majority of integral raft proteins. *Proceedings of the National Academy of Sciences of the United States of America, 107*, 22050–22054.
Lingwood, D., & Simons, K. (2010). Lipid rafts as a membrane-organizing principle. *Science, 327*, 46–50.
Lopez, I., Giner, D., Ruiz-Nuno, A., Fuentealba, J., Viniegra, S., Garcia, A. G., et al. (2007). Tight coupling of the t-SNARE and calcium channel microdomains in adrenomedullary slices and not in cultured chromaffin cells. *Cell Calcium, 41*, 547–558.
López, I., Ortiz, J. A., Villanueva, J., Torres, V., Torregrosa-Hetland, C. J., del Mar Francés, M., et al. (2009). Vesicle motion and fusion are altered in chromaffin cells with increased SNARE cluster dynamics. *Traffic, 10*, 172–185.
Low, S. H., Chapin, S. J., Weimbs, T., Kömüves, L. G., Bennett, M. K., & Mostov, K. E. (1996). Differential localization of syntaxin isoforms in polarized MDCK cells. *Molecular Biology of the Cell, 7*, 2007–2018.

Low, S. H., Vasanji, A., Nanduri, J., He, M., Sharma, N., Koo, M., et al. (2006). Syntaxins 3 and 4 are concentrated in separate clusters on the plasma membrane before the establishment of cell polarity. *Molecular Biology of the Cell, 17*, 977–989.

Martin, T. F. J. (2012). Role of PI(4,5)P(2) in vesicle exocytosis and membrane fusion. *Subcellular Biochemistry, 59*, 111–130.

McLaughlin, S., Wang, J., Gambhir, A., & Murray, D. (2002). PIP(2) and proteins: Interactions, organization, and information flow. *Annual Review of Biophysics and Biomolecular Structure, 31*, 151–175.

Milosevic, I., Sørensen, J. B., Lang, T., Krauss, M., Nagy, G., Haucke, V., et al. (2005). Plasmalemmal phosphatidylinositol-4,5-bisphosphate level regulates the releasable vesicle pool size in chromaffin cells. *The Journal of Neuroscience, 25*, 2557–2565.

Mitra, K., Ubarretxena-Belandia, I., Taguchi, T., Warren, G., & Engelman, D. M. (2004). Modulation of the bilayer thickness of exocytic pathway membranes by membrane proteins rather than cholesterol. *Proceedings of the National Academy of Sciences of the United States of America, 101*, 4083–4088.

Murray, D. H., & Tamm, L. K. (2009). Clustering of syntaxin-1A in model membranes is modulated by phosphatidylinositol 4,5-bisphosphate and cholesterol. *Biochemistry, 48*, 4617–4625.

Murray, D. H., & Tamm, L. K. (2011). Molecular mechanism of cholesterol- and polyphosphoinositide-mediated syntaxin clustering. *Biochemistry, 50*, 9014–9022.

Nagy, G., Milosevic, I., Fasshauer, D., Müller, E. M., de Groot, B. L., Lang, T., et al. (2005). Alternative splicing of SNAP-25 regulates secretion through nonconservative substitutions in the SNARE domain. *Molecular Biology of the Cell, 16*, 5675–5685.

Nakano, M., Nogami, S., Sato, S., Terano, A., & Shirataki, H. (2001). Interaction of syntaxin with alpha-fodrin, a major component of the submembranous cytoskeleton. *Biochemical and Biophysical Research Communications, 288*, 468–475.

Ohara-Imaizumi, M., Nishiwaki, C., Kikuta, T., Kumakura, K., Nakamichi, Y., & Nagamatsu, S. (2004). Site of docking and fusion of insulin secretory granules in live MIN6 beta cells analyzed by TAT-conjugated anti-syntaxin 1 antibody and total internal reflection fluorescence microscopy. *The Journal of Biological Chemistry, 279*, 8403–8408.

Oheim, M., & Stuhmer, W. (2000). Interaction of secretory organelles with the membrane. *The Journal of Membrane Biology, 178*, 163–173.

Predescu, S. A., Predescu, D. N., Shimizu, K., Klein, I. K., & Malik, A. B. (2005). Cholesterol-dependent syntaxin-4 and SNAP-23 clustering regulates caveolar fusion with the endothelial plasma membrane. *The Journal of Biological Chemistry, 280*, 37130–37138.

Prescott, G. R., Gorleku, O. A., Greaves, J., & Chamberlain, L. H. (2009). Palmitoylation of the synaptic vesicle fusion machinery. *Journal of Neurochemistry, 110*, 1135–1149.

Ramakrishnan, N. A., Drescher, M. J., & Drescher, D. G. (2012). The SNARE complex in neuronal and sensory cells. *Molecular and Cellular Neurosciences, 50*, 58–69.

Ribrault, C., Reingruber, J., Petković, M., Galli, T., Ziv, N. E., Holcman, D., et al. (2011). Syntaxin1A lateral diffusion reveals transient and local SNARE interactions. *The Journal of Neuroscience, 31*, 17590–17602.

Rickman, C., & Duncan, R. R. (2010). Munc18/Syntaxin interaction kinetics control secretory vesicle dynamics. *The Journal of Biological Chemistry, 285*, 3965–3972.

Rickman, C., Medine, C. N., Dun, A. R., Moulton, D. J., Mandula, O., Halemani, N. D., et al. (2010). t-SNARE protein conformations patterned by the lipid microenvironment. *The Journal of Biological Chemistry, 285*, 13535–13541.

Rickman, C., Meunier, F. A., Binz, T., & Davletov, B. (2004). High affinity interaction of syntaxin and SNAP-25 on the plasma membrane is abolished by botulinum toxin E. *The Journal of Biological Chemistry, 279*, 644–651.

Risinger, C., & Bennett, M. K. (1999). Differential phosphorylation of syntaxin and synaptosome-associated protein of 25 kDa (SNAP-25) isoforms. *Journal of Neurochemistry*, *72*, 614–624.
Robinson, I. M., Finnegan, J. M., Monck, J. R., Wightman, R. M., & Fernandez, J. M. (1995). Colocalization of calcium entry and exocytotic release sites in adrenal chromaffin cells. *Proceedings of the National Academy of Sciences of the United States of America*, *92*, 2474–2478.
Saarikangas, J., Zhao, H., & Lappalainen, P. (2010). Regulation of the actin cytoskeleton-plasma membrane interplay by phosphoinositides. *Physiological Reviews*, *90*, 259–289.
Salaün, C., Gould, G. W., & Chamberlain, L. H. (2005a). The SNARE proteins SNAP-25 and SNAP-23 display different affinities for lipid rafts in PC12 cells. Regulation by distinct cysteine-rich domains. *The Journal of Biological Chemistry*, *280*, 1236–1240.
Salaün, C., Gould, G. W., & Chamberlain, L. H. (2005b). Lipid raft association of SNARE proteins regulates exocytosis in PC12 cells. *The Journal of Biological Chemistry*, *280*, 19449–19453.
Salaün, C., James, D. J., & Chamberlain, L. H. (2004). Lipid rafts and the regulation of exocytosis. *Traffic*, *5*, 255–264.
Saslowsky, D. E., Lawrence, J. C., Henderson, R. M., & Edwardson, J. M. (2003). Syntaxin is efficiently excluded from sphingomyelin-enriched domains in supported lipid bilayers containing cholesterol. *The Journal of Membrane Biology*, *194*, 153–164.
Schmidt, A., Hannah, M. J., & Huttner, W. B. (1997). Synaptic-like microvesicles of neuroendocrine cells originate from a novel compartment that is continuous with the plasma membrane and devoid of transferrin receptor. *The Journal of Cell Biology*, *137*, 445–458.
Schütz, D., Zilly, F., Lang, T., Jahn, R., & Bruns, D. (2005). A dual function for Munc-18 in exocytosis of PC12 cells. *The European Journal of Neuroscience*, *21*, 2419–2432.
Sharpe, H. J., Stevens, T. J., & Munro, S. (2010). A comprehensive comparison of transmembrane domains reveals organelle-specific properties. *Cell*, *142*, 158–169.
Sheng, Z. H., Westenbroek, R. E., & Catterall, W. A. (1998). Physical link and functional coupling of presynaptic calcium channels and the synaptic vesicle docking/fusion machinery. *Journal of Bioenergetics and Biomembranes*, *30*, 335–345.
Sieber, J. J., Willig, K. I., Heintzmann, R., Hell, S. W., & Lang, T. (2006). The SNARE motif is essential for the formation of syntaxin clusters in the plasma membrane. *Biophysical Journal*, *90*, 2843–2851.
Sieber, J. J., Willig, K. I., Kutzner, C., Gerding-Reimers, C., Harke, B., Donnert, G., et al. (2007). Anatomy and dynamics of a supramolecular membrane protein cluster. *Science*, *317*, 1072–1076.
Somanath, S., Barg, S., Marshall, C., Silwood, C. J., & Turner, M. D. (2009). High extracellular glucose inhibits exocytosis through disruption of syntaxin 1A-containing lipid rafts. *Biochemical and Biophysical Research Communications*, *389*, 241–246.
Stein, A., Weber, G., Wahl, M. C., & Jahn, R. (2009). Helical extension of the neuronal SNARE complex into the membrane. *Nature*, *460*, 525–528.
Takamori, S., Holt, M., Stenius, K., Lemke, E. A., Grønborg, M., Riedel, D., et al. (2006). Molecular anatomy of a trafficking organelle. *Cell*, *127*, 831–846.
Tanaka, K. A., Suzuki, K. G., Shirai, Y. M., Shibutani, S. T., Miyahara, M. S., Tsuboi, H., et al. (2010). Membrane molecules mobile even after chemical fixation. *Nature Methods*, *7*, 865–866.
Taverna, E., Saba, E., Linetti, A., Longhi, R., Jeromin, A., Righi, M., et al. (2007). Localization of synaptic proteins involved in neurosecretion in different membrane microdomains. *Journal of Neurochemistry*, *100*, 664–677.
Taverna, E., Saba, E., Rowe, J., Francolini, M., Clementi, F., & Rosa, P. (2004). Role of lipid microdomains in P/Q-type calcium channel (Cav2.1) clustering and function in presynaptic membranes. *The Journal of Biological Chemistry*, *279*, 5127–5134.

Toonen, R. F., Kochubey, O., de Wit, H., Gulyas-Kovacs, A., Konijnenburg, B., Sørensen, J. B., et al. (2006). Dissecting docking and tethering of secretory vesicles at the target membrane. *The EMBO Journal, 25*, 3725–3737.
Torregrosa-Hetland, C. J., Villanueva, J., Garcia-Martínez, V., Expósito-Romero, G., Francés Mdel, M., & Gutiérrez, L. M. (2013). Cortical F-actin affects the localization and dynamics of SNAP-25 membrane clusters in chromaffin cells. *The International Journal of Biochemistry & Cell Biology, 45*, 583–592.
Torregrosa-Hetland, C. J., Villanueva, J., Giner, D., Lopez-Font, I., Nadal, A., Quesada, I., et al. (2011). The F-actin cortical network is a major factor influencing the organization of the secretory machinery in chromaffin cells. *Journal of Cell Science, 124*, 727–734.
Torregrosa-Hetland, C. J., Villanueva, J., López-Font, I., Garcia-Martinez, V., Gil, A., Gonzalez-Vélez, V., et al. (2010). Association of SNAREs and calcium channels with the borders of cytoskeletal cages organizes the secretory machinery in chromaffin cells. *Cellular and Molecular Neurobiology, 30*, 1315–1319.
Torres, J., Funk, H. M., Zegers, M. M., & ter Beest, M. B. (2011). The syntaxin 4 N terminus regulates its basolateral targeting by munc18c-dependent and -independent mechanisms. *The Journal of Biological Chemistry, 286*, 10834–10846.
Van den Bogaart, G., & Jahn, R. (2011). Counting the SNAREs needed for membrane fusion. *Journal of Molecular Cell Biology, 3*, 204–205.
Van den Bogaart, G., Meyenberg, K., Risselada, H. J., Amin, H., Willig, K. I., Hubrich, B. E., et al. (2011). Membrane protein sequestering by ionic protein-lipid interactions. *Nature, 479*, 552–555.
Veit, M., Becher, A., & Ahnert-Hilger, G. (2000). Synaptobrevin 2 is palmitoylated in synaptic vesicles prepared from adult, but not from embryonic brain. *Molecular and Cellular Neurosciences, 15*, 408–416.
Veit, M., Söllner, T. H., & Rothman, J. E. (1996). Multiple palmitoylation of synaptotagmin and the t-SNARE SNAP-25. *FEBS Letters, 385*, 119–123.
Verhage, M., Maia, A. S., Plomp, J. J., Brussaard, A. B., Heeroma, J. H., Vermeer, H., et al. (2000). Synaptic assembly of the brain in the absence of neurotransmitter secretion. *Science, 287*, 864–869.
Villanueva, J., Torregrosa-Hetland, C. J., García-Martínez, V., del Mar Francés, M., Viniegra, S., & Gutiérrez, L. M. (2012). The F-actin cortex in chromaffin granule dynamics and fusion: A minireview. *Journal of Molecular Neuroscience, 48*, 323–327.
Villanueva, J., Torregrosa-Hetland, C. J., Gil, A., González-Vélez, V., Segura, J., Viniegra, S., et al. (2010). The organization of the secretory machinery in chromaffin cells as a major factors in modelling exocytosis. *Human Frontier Science Program, 4*, 85–92.
Voets, T., Toonen, R. F., Brian, E. C., de Wit, H., Moser, T., Rettig, J., et al. (2001). Munc18-1 promotes large dense-core vesicle docking. *Neuron, 31*, 581–591.
Wang, Y. H., Collins, A., Guo, L., Smith-Dupont, K. B., Gai, F., Svitkina, T., et al. (2012). Divalent cation-induced cluster formation by polyphosphoinositides in model membranes. *Journal of the American Chemical Society, 134*, 3387–3395.
Wang, J., & Richards, D. A. (2011). Spatial regulation of exocytic site and vesicle mobilization by the actin cytoskeleton. *PLoS One, 6*, e29162.
Wang, J., & Richards, D. A. (2012). Segregation of PIP2 and PIP3 into distinct nanoscale regions within the plasma membrane. *Biology Open, 1*, 857–862.
Watanabe, M., Nomura, K., Ohyama, A., Ishikawa, R., Komiya, Y., Hosaka, K., et al. (2005). Myosin-Va regulates exocytosis through the submicromolar Ca^{2+}-dependent binding of syntaxin-1A. *Molecular Biology of the Cell, 16*, 4519–4530.
Weiss, N., Hameed, S., Fernández-Fernández, J. M., Fablet, K., Karmazinova, M., Poillot, C., et al. (2012). A Ca(v)3.2/syntaxin-1A signaling complex controls T-type channel activity and low-threshold exocytosis. *The Journal of Biological Chemistry, 287*, 2810–2818.

Wen, P. J., Osborne, S. L., & Meunier, F. A. (2011). Dynamic control of neuroexocytosis by phosphoinositides in health and disease. *Progress in Lipid Research*, *50*, 52–61.

Wickner, W., & Schekman, R. (2008). Membrane fusion. *Nature Structural & Molecular Biology*, *15*, 658–664.

Williams, D., Vicôgne, J., Zaitseva, I., McLaughlin, S., & Pessin, J. E. (2009). Evidence that electrostatic interactions between vesicle-associated membrane protein 2 and acidic phospholipids may modulate the fusion of transport vesicles with the plasma membrane. *Molecular Biology of the Cell*, *20*, 4910–4919.

Willig, K. I., Keller, J., Bossi, M., & Hell, S. W. (2006). STED microscopy resolves nanoparticle assemblies. *New Journal of Physics*, *8*, 106.

Woronowicz, K., Dilks, J. R., Rozenvayn, N., Dowal, L., Blair, P. S., Peters, C. G., et al. (2010). The platelet actin cytoskeleton associates with SNAREs and participates in alpha-granule secretion. *Biochemistry*, *49*, 4533–4542.

Yang, L., Dun, A. R., Martin, K. J., Qiu, Z., Dunn, A., Lord, G. J., et al. (2012). Secretory vesicles are preferentially targeted to areas of low molecular SNARE density. *PLoS One*, *7*, e49514.

Yang, X., Xu, P., Xiao, Y., Xiong, X., & Xu, T. (2006). Domain requirement for the membrane trafficking and targeting of syntaxin 1A. *The Journal of Biological Chemistry*, *281*, 15457–15463.

Zhang, L., Mao, Y. S., Janmey, P. A., & Yin, H. L. (2012). Phosphatidylinositol 4, 5 bisphosphate and the actin cytoskeleton. *Subcellular Biochemistry*, *59*, 177–215.

Zilly, F. E., Halemani, N. D., Walrafen, D., Spitta, L., Schreiber, A., Jahn, R., et al. (2011). Ca^{2+} induces clustering of membrane proteins in the plasma membrane via electrostatic interactions. *The EMBO Journal*, *30*, 1209–1220.

Zilly, F. E., Sørensen, J. B., Jahn, R., & Lang, T. (2006). Munc18-bound syntaxin readily forms SNARE complexes with synaptobrevin in native plasma membranes. *PLoS Biology*, *4*, e330.

CHAPTER SEVEN

Photoreceptor Inner and Outer Segments

Sheila A. Baker[1], Vasily Kerov
Department of Biochemistry, Carver College of Medicine, University of Iowa, Iowa City, Iowa, USA
[1]Corresponding author: e-mail address: sheila-baker@uiowa.edu

Contents

Abstract

Photoreceptors are exquisitely adapted to transform light stimuli into electrical signals that modulate neurotransmitter release. These cells are organized into several compartments including the unique outer segment (OS). Its whole function is to absorb light and transduce this signal into a change of membrane potential. Another compartment is the inner segment where much of metabolism and regulation of membrane potential takes place and that connects the OS and synapse. The synapse is the compartment where changes in membrane potentials are relayed to other neurons in the retina via release of neurotransmitter. The composition of the plasma membrane surrounding these compartments varies to accommodate their specific functions. In this chapter, we discuss the organization of the plasma membrane emphasizing the protein composition of each region as it relates to visual signaling. We also point out examples where mutations in these proteins cause visual impairment.

Current Topics in Membranes, Volume 72
ISSN 1063-5823
http://dx.doi.org/10.1016/B978-0-12-417027-8.00007-6

1. INTRODUCTION

The first link in the chain of visual perception is the photoreceptor cell, which functions as a photon detector. Photoreceptors are of two types, rods or cones. Rods are used in very dim lighting conditions as they are so sensitive that they can respond to the absorption of a single photon. Cones operate in the ranges of mid to bright light intensity that we most frequently use to light our homes and work places. Cones are responsible for color vision as subtypes of cones are maximally sensitive to different wavelengths of light. It is often said that rods are used for night vision, while cones are for daytime vision. This is a helpful referent, but even when stargazing, we are using both systems as evident from our ability to detect the color of the red star, Betelgeuse.

Rods and cones operate on the same molecular principles. In the dark, or resting state, a flow of ions across the plasma membrane of the photoreceptor generates a circulating current that keeps the cell depolarized. Absorption of photons triggers a signaling cascade that results in the closure of channels found exclusively in the outer segment (OS) of the cell that breaks the circulating current and drives the cell to hyperpolarized potentials. This in turn affects the voltage-gated Ca^{2+} channels at the synapse that control neurotransmitter release.

Many inherited visual disorders are a consequence of mutations in photoreceptor-specific genes that lead to dysfunction and frequently the death of photoreceptors (https://sph.uth.edu/retnet/). Some of these genes are necessary for the development of photoreceptors or the transduction of the light signal. Yet others play roles in establishing and maintaining the compartmentalization of the cell. In this chapter, we will discuss the organization and function of the major photoreceptor compartments with a particular emphasis on the plasma membrane. For simplicity, we will focus on rods.

2. THE ANATOMY OF VERTEBRATE PHOTORECEPTORS

The anatomy of the vertebrate retina and photoreceptors was described by Ramon Y Cajal at the beginning of the twentieth century. The seminal work of Cajal and his contemporaries created a solid framework on which the field has built a deeper understanding of the organization and physiology of the retina. Anatomically, photoreceptors are composed of five compartments: OS, inner segment (IS), soma, axon, and synaptic terminal (ST; Fig. 7.1). This organization is maintained in all vertebrates although there are species-specific differences in the size or morphology of these compartments.

The OS is an elaborated primary cilium where photon capture takes place. The OS is joined to the remainder of the cell by a thin connecting cilium that also serves as the boundary between the differing plasma membrane compositions of the outer and ISs, reviewed in Breslow and Nachury

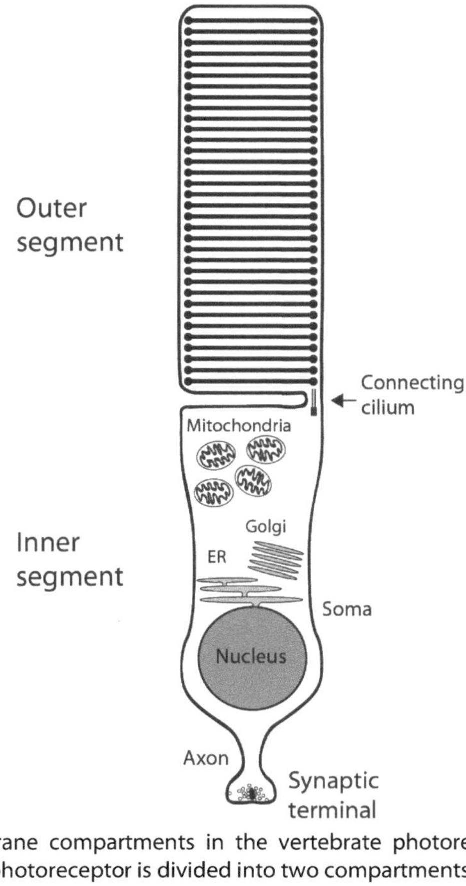

Figure 7.1 Membrane compartments in the vertebrate photoreceptor. The plasma membrane of the photoreceptor is divided into two compartments: the outer and inner segments separated by the connecting cilium. The outer segment is filled with disk membranes. The inner segment encompassing the soma, nucleus, and axon is filled with organelles such as the ER, Golgi, and mitochondria. The synaptic terminal is a specialization of the inner segment plasma membrane. (For color version of this figure, the reader is referred to the online version of this chapter.)

(2011). The IS is where most of the housekeeping functions such as protein and lipid synthesis occur; it contains the Golgi, ER, and mitochondria. The diameter of the IS in many species, such as the human and mouse, constricts just below an aligned band of adherens junctions between photoreceptors and Muller glia cells. This junction forms the morphological marker between the IS layer and the outer nuclear layer (containing the soma) seen histologically. In mammals, the soma is a thin cytoplasmic process that widens to accommodate the nucleus and then transitions to a thin, unmyelinated axon. In species such as the frog where the packing density of photoreceptors is lower, the diameter of the cell remains similar through the IS and soma. In either case, there is no barrier between the cytoplasm of the IS, soma, and axon. There also seems to be no barrier to the distribution of plasma membrane-resident proteins between these compartments. Therefore, for the purposes of this chapter, we will not make a distinction between the plasma membranes surrounding these areas and will generally refer to it as the IS plasma membrane. The axon ends with a pre-ST that has a specialized membrane organization associated with the regulated synaptic vesicle cycle. All together, there are three distinct plasma membrane compartments (OS, IS, and ST).

3. PHOTORECEPTOR SIGNALING

Before delving into the unique features of the photoreceptor plasma membrane compartments, it is helpful to briefly consider how they work together in light detection. A current generated by the activity of multiple ion channels and exchangers, including potassium channels and sodium/potassium ATPase (NKA) in the IS and cyclic nucleotide-gated (CNG) channels in the OS, keeps the membrane potential at approximately -40 mV in the dark (Baylor, Matthews, & Nunn, 1984). This current is often referred to as the dark current or photocurrent. With the cell depolarized, voltage-gated Ca^{2+} channels in the synapse are open, and the resulting influx of calcium at the synapse triggers fusion of synaptic vesicles with the plasma membrane and release of the neurotransmitter, glutamate (Fig. 7.2).

Photon absorption in the OS initiates a signaling cascade that causes a drop in intracellular cGMP. This results in the closure of CNG channels, without affecting the other participants in the dark current, and the cell hyperpolarizes to approximately -75 mV. Several reviews are recommended for an in-depth discussion of phototransduction (Arshavsky & Burns, 2012; Burns & Baylor, 2001; Fain, Matthews, Cornwall, & Koutalos, 2001; Kefalov, 2012; Palczewski, 2012; Pugh & Lamb, 1993; Yau & Hardie,

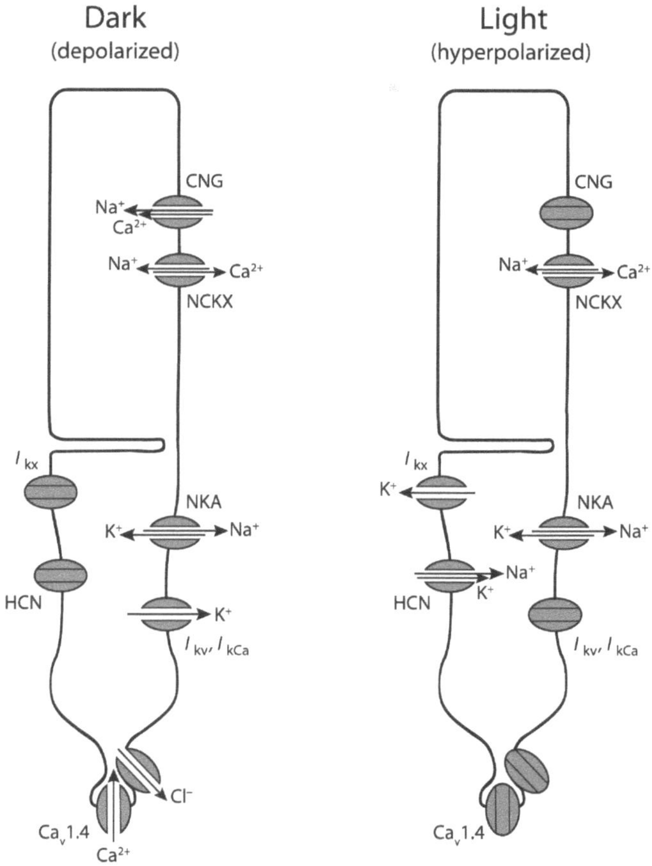

Figure 7.2 Light-regulated changes in ion flow across the plasma membrane. In the dark, sodium enters the outer segment principally through the cyclic nucleotide-gated (CNG) channels and is extruded by sodium/potassium ATPase (NKA) in the inner segment. Calcium entering the outer segment through CNG channels is balanced by the activity of the cation exchanger (NCKX). Potassium efflux is mediated by two classes of channels, I_{kv} and I_{kCa}. Under these conditions, the cell is depolarized and the voltage-gated calcium channel ($Ca_v1.4$) in the synapse carries an inward current. Calcium-regulated chloride efflux occurs in the synapse. In the presence of light, CNG channels close. This hyperpolarizes the cell, which decreases the activity of the potassium channels and $Ca_v1.4$. Secondarily, hyperpolarization activates HCN and I_{kx} channels in the inner segment that shape the photoresponse. (See Color Plate.)

2009). Hyperpolarization causes Ca^{2+} channels in the synapse to close and neurotransmitter release to decrease. Altogether, signaling by photoreceptors requires one electrically continuous plasma membrane, and each of the three principle plasma membrane compartments (OS, IS, and ST) makes a unique contribution.

4. ORGANIZATION OF OS MEMBRANES

A key requirement for efficient visual signaling is a large surface area where the probability of photon capture is maximized and the signal can be amplified—this is what the OS provides the vertebrate photoreceptor. OSs are a type of primary cilia, modified to contain hundreds of tightly packed stacks of membrane disks. This type of organelle was used for light perception in organisms living at least ~550 million years ago, before the protostome and deuterostome lines diverged (Arendt, 2003; Lamb, Collin, & Pugh, 2007).

4.1. Assembling the OS

The development of the OS initially follows that of most cilia (Ishikawa & Marshall, 2011; Pearring, Salinas, Baker, & Arshavsky, 2013; Sedmak & Wolfrum, 2011). When the developing photoreceptor exits the cell cycle, the centrioles move from the periphery of the nucleus to the apical end of the IS where the mother centriole becomes the basal body. As the basal body matures, it is capped by a ciliary vesicle that fuses with the plasma membrane, and nine microtubule doublets extend from the nine triplet microtubules of the basal body forming the axoneme of the cilium. Remarkably, the delineation of OS from IS membranes seems to occur in the earliest stages of ciliogenesis with the formation of the ciliary vesicle from the fusion of intracellular transport vesicles.

As the axoneme extends past the transition zone (connecting cilium), the ensheathing plasma membrane expands and the space between the plasma membrane and the microtubules becomes filled with vesicles and tubules (in warm-blooded species) that subsequently take on the ordered appearance of mature disks (Besharse, Forestner, & Defoe, 1985; Pearring et al., 2013; Sedmak & Wolfrum, 2011). Continuous delivery of new membrane and protein along with the activity of the intraflagellar transport complex fuels the elongation of the OS. The mature disks in rods appear as flattened vesicles of the same diameter all aligned to each other, perpendicular to the axoneme, and separated from the surrounding plasma membrane (closed disks). The disks in cones vary in two ways. The diameter of the disks tapers so that the OS is conical and cone disks are most often continuous infoldings (open disks) of the plasma membrane.

The mechanism of disk biogenesis is not known, though two major models are currently debated. We will briefly discuss some key points of these two models but refer the readers to the following reviews and references therein for a complete discussion (Pearring et al., 2013; Sung & Chuang, 2010). In the evagination model, initially based on observations of electron micrographs, transport vesicles carrying material destined for the OS fuse at the base of the connecting cilium. With this influx of membrane material, the plasma membrane just distal to the connecting cilium evaginates forming an open disk (Steinberg, Fisher, & Anderson, 1980). Repetition of this process explains the formation of cones and the open disks seen at the base of rods, but a remodeling event is needed to form the closed disks making up the majority of rod OS. It has been proposed that the lower edge of an older evagination serves as a template and fuses at the rim with the upper edge of a newer evagination just below it (Corless & Fetter, 1987). Intriguingly, RDS (peripherin-2), a structural protein found just in disk rims, can promote membrane fusion and may be the molecular key to this event (Boesze-Battaglia et al., 2003). In mice lacking RDS, OSs fail to form and this triggers degeneration (Sanyal, De Ruiter, & Hawkins, 1980). OS membranes do form in mice carrying only one allele for RDS, but they form as disorganized whorls of membrane (Hawkins, Jansen, & Sanyal, 1985). Mutations in RDS lead to either retinitis pigmentosa or macular dystrophy, and normal OS disk formation is prevented in animal models expressing some of the most prevalent human RDS mutations (Ding & Naash, 2006; Goldberg, 2006). Rom-1 is a protein similar to RDS that forms complexes with RDS in disk rims. In mice lacking Rom-1, the disks are formed but grow excessively long, further supporting a role for this complex in regulating the formation or maturation of disks (Clarke et al., 2000).

A major challenge for the evagination model is that open disks at the base of rods are not found in all EM preparations of retinas. While this discrepancy can be explained by the use of alternative fixatives leading to artifacts, only closed disks are reported using cryo-EM techniques that do not rely on fixatives and long processing times (Chuang, Zhao, & Sung, 2007; Gilliam et al., 2012; Obata & Usukura, 1992). These observations, along with the finding that SARA, Smad anchor for receptor activation, can link rhodopsin in a complex with syntaxin-3, a target SNARE, gave prominence to the vesicular model of disk biogenesis (Chuang et al., 2007). In this model, rhodopsin-carrying transport vesicles are delivered through the core of the connecting cilium and then fuse with one another using the SNARE machinery common to other membrane fusion events. This occurs just past the connecting cilium forming a series of larger vesicles and tubules that become the closed disks of rods. Note that the vesicular transport model does not automatically negate a need for the RDS/Rom-1 complex in controlling the length or alignment of newly formed disks. In fact, there are likely multiple proteins that participate in shaping the newly forming disks as there are several examples where loss of proteins normally localized to either the connecting cilium or the base of the OS leads to the formation of abnormally sized or aligned disks (Pazour et al., 2002; Rattner et al., 2001; Yang, Chen, et al., 2008; Zacchigna et al., 2009; Zhao et al., 2003). Determining the molecular interplay between these components and any others involved in disk biogenesis will be necessary to build a consensus molecular model for how this important process takes place.

The vesicular model is very appealing because it invokes well-understood mechanisms of membrane fusion used throughout all cells. However, it also raises several questions that will hopefully be addressed by future investigations. For instance, why are vesicles in the connecting cilium that should be supplying the ongoing formation of disks rarely seen? What happens to the SNARE machinery that fuses vesicles to form disks? The components of this machinery should be trapped within a newly formed disk and therefore detectable throughout the length of the OS. Instead, syntaxin-3 is only found at the base of the OS, suggesting that after fusion, there are additional remodeling events that selectively remove the SNARE components. How are the open disks of cones formed? Are rods and cones simply built differently? There are species-specific discrepancies in many of the experiments supporting the differing models of disk formation leading many to suggest that amphibian and mammalian photoreceptors use alterative mechanisms. What evolutionary pressures could have fueled such a split? This is clearly a challenging biophysical puzzle with many missing molecular links.

OS membranes are continuously renewed (Young, 1967). Large amounts of membrane and protein, synthesized in the IS, are selectively trafficked to the base of OS every day to support the ongoing formation of disks. It is estimated that in mice, ~80 rhodopsin molecules per second, throughout the lifetime of the organism, are transported to the connecting cilium for incorporation into the newly forming disks (for the larger frog OS, ~700 rhodopsin molecules per second are transported) (Pearring et al., 2013). Older disks are displaced towards the tip of the OS where they are phagocytized by the adjacent retinal pigment epithelium (RPE). The net length of the OS is thus maintained by balancing the rate of basal disk formation with loss from the tips. Disruptions in either process result in blindness (Kevany & Palczewski, 2010).

4.2. Composition of OS membranes

The protein composition of the OS plasma membrane is distinct from that of disk lamellae and rims. The CNG channels and a Na^+/Ca^{2+} exchanger (NCKX) are the only ion-transporting proteins in the OS, and both are found only in the plasma membrane. CNG channels mediate an influx of Na^+ and Ca^{2+} into the OS and are gated by cGMP. CNG channels are heterotetramers, composed of α- and β-subunits (Peng, Rich, & Varnum, 2004; Shuart, Haitin, Camp, Black, & Zagotta, 2011; Weitz, Ficek, Kremmer, Bauer, & Kaupp, 2002; Zheng, Trudeau, & Zagotta, 2002; Zhong, Molday, Molday, & Yau, 2002). The α-subunits are responsible for ion conduction, while the β-subunits can fine-tune the conductance properties and regulation of the channels (Kaupp & Seifert, 2002; Michalakis et al., 2011; Rebrik, Botchkina, Arshavsky, Craft, & Korenbrot, 2012). Not surprisingly, mutations in human CNG subunits cause blindness. Loss of function of the subunits expressed in rods presents as retinitis pigmentosa, while loss of function of the subunits expressed in cones is the most prevalent cause of achromatopsia (total loss of color vision and cone function). Most frequently, disease-causing mutations prevent proper expression, folding, or trafficking of the channel to the OS (Schon, Biel, & Michalakis, 2013). Loss of CNG channels leads to accumulation of cGMP, toxic changes in OS Ca^{2+} levels, failure to maintain OS structure, and reduced levels of many proteins involved in phototransduction (Huttl et al., 2005; Michalakis et al., 2010; Paquet-Durand et al., 2011; Zhang et al., 2009). The loss in OS signaling is propagated through the cell so that synaptic remodeling is also triggered (Huttl et al., 2005; Michalakis et al., 2013). These studies also revealed that the CNGB1 plays a major role in the trafficking of the channel. Release of the channel from intracellular biosynthetic membranes can depend on masking a retention signal by formation of heterooligomers or the presence of an intact binding site for the membrane adaptor protein, ankyrin-G, in CNGB1 (Kizhatil, Baker, Arshavsky, & Bennett, 2009; Trudeau & Zagotta, 2002). Additionally, a putative ciliary-targeting signal and phosphorylation-dependent interaction with a sorting protein, PACS-1, have been implicated in the ciliary localization of CNG channels (Jenkins et al., 2006; Jenkins, Zhang, Thomas, & Martens, 2009). In the evagination model of disk biogenesis, there is a need for CNG to be further sorted out of the rod disks as they are separated from the plasma membrane, but it is not known how this might happen. Alternatively, in the vesicular model,

CNG would be segregated from disk proteins before different cargo-bearing vesicles reach the OS. Understanding how the stepwise sorting and trafficking of this channel and other OS proteins is coordinated is an exciting area of ongoing work (Pearring et al., 2013).

CNG channels form a complex with dimers of NCKX (Kang et al., 2003). NCKX extrudes Ca^{2+} from the OS at a transport stoichiometry of $1Ca^{2+}$, $1K^{+}$:$4Na^{+}$ ions (Schnetkamp, 2013). In the light when CNG channels close, the continued activity of NCKX depletes intracellular Ca^{2+}. The changing intracellular calcium concentration is used to regulate several aspects of phototransduction so is a key component of light adaptation, which is the ability of photoreceptors to respond to light over a large dynamic range (Arshavsky & Burns, 2012; Fain et al., 2001). A mutation in NCKX1, the isoform expressed in rods, is linked to congenital stationary night blindness, a disorder where night vision is impaired but the cells do not degenerate (Riazuddin et al., 2010).

Rhodopsin, a G protein-coupled receptor bound to chromophore, 11-*cis*-retinal, is the photon receptor that activates the phototransduction cascade. It is the most abundant protein of the OS with an estimated density between 25,000 and 55,000 molecules/μm^2 (Liang et al., 2003; Pugh & Lamb, 1993). It is the only protein documented to segregate into two of the OS membrane subdomains; it is concentrated in disk lamellae and found in the plasma membrane. If the OS is damaged or the membrane barrier between IS and OS membranes breaks down, rhodopsin becomes mislocalized to the IS plasma membrane and this triggers apoptosis (Malanson & Lem, 2009). The trafficking of rhodopsin from biosynthetic membranes in the IS to the OS is the best understood of all the OS-resident proteins. As rhodopsin passes through the Golgi, it recruits ARF4, a small G protein, which in turn recruits a larger signaling complex that promotes the budding of rhodopsin-containing transport vesicles and eventual fusion of the transport vesicles near the connecting cilium (Deretic & Wang, 2012). Failure to transport rhodopsin to the OS causes early and rapid death. Mice completely lacking rhodopsin do not form OSs, and mice with decreased or increased levels of rhodopsin expression have correspondingly smaller or larger OS disks (Humphries et al., 1997; Lem et al., 1999; Liang et al., 2004; Makino et al., 2012; Wen et al., 2009). Therefore, in addition to rhodopsin's well-characterized role in signaling, it participates in generating and maintaining the structure of the OS.

The disk lamellae contain the other components of the phototransduction cascade: transducin, PDE, guanylate cyclase, rhodopsin kinase, and the three components of the transducin GTPase complex (RGS9–Gβ5–R9AP). Despite the extensive studies of phototransduction, there are still some mysterious aspects of the disks. For instance, they contain a protein called progressive rod–cone degeneration (PRCD) with an unknown function whose mutation causes retinal degeneration in dogs and humans (Nevet, Shalev, Zlotogora, Mazzawi, & Ben-Yosef, 2010; Skiba et al., 2013; Zangerl et al., 2006). Another unsolved mystery relates to how the disk proteins are trafficked into this compartment. In an investigation of R9AP trafficking, it was found that in the relatively large OS of frogs, membrane proteins that can exit the ER/Golgi, but lack IS-targeting signals, can accumulate nonspecifically in the OS (Baker et al., 2008; Gospe, Baker, & Arshavsky, 2010). But, work using mice has revealed that R9AP contains trafficking determinants required for OS localization and that R9AP in turn is responsible of the OS localization of its partners, RGS9 and Gβ5 (Gospe et al., 2011; Martemyanov et al., 2003; Pearring et al., 2013). Loss of either functional guanylate cyclase or PDE in OS is particularly damaging as they are involved in controlling intracellular cGMP levels, alteration of which triggers rapid degeneration (Bowes et al., 1990; Fain, 2006; Farber & Lolley, 1974; Sancho-Pelluz et al., 2008). The expression and trafficking of guanylate cyclase requires an interaction with RD3, a protein that binds to the C-terminus of guanylate cyclase and when mutated causes a rapid loss of photoreceptors (Azadi, Molday, & Molday, 2010). The trafficking of PDE also requires a "trafficking chaperone" that regulates the solubility of PDE (Norton et al., 2005; Pearring et al., 2013; Zhang et al., 2007). Dissecting the similarities and differences in the trafficking of OS disk proteins will shed light on how this compartment is maintained.

A lipid transporter found in disks, ATP8a2, is a transporter of phosphatidylserine (PS; Coleman, Kwok, & Molday, 2009). All membranes have an asymmetrical distribution of phospholipids; PS is primarily found in the cytosolic leaflets of membranes, and its translocation to the exterior of apoptotic cells is a well-known signal for phagocytic engulfment (Ravichandran, 2011). An elegant study demonstrated a similar role for PS in photoreceptors. PS becomes exposed on the tips of OS prior to their shedding and engulfment by the RPE, and this process is dependent on signaling events involving the RPE (Ruggiero, Connor, Chen, Langen, & Finnemann, 2012). It is tempting to speculate that ATP8a2 participates in the regulated exposure of PS. Interestingly, an accessory subunit for ATP8a2, CDC50A, is necessary for the efficient trafficking and activity of ATP8a2, but its expression is not limited to OS, as in the case of ATP8a2

(Coleman & Molday, 2011; van der Velden et al., 2010). Perhaps, it interacts with related transporters in IS membranes that could function to maintain the proper symmetry of PS throughout the cell.

A second lipid transporter in the OS is ABCA4. This transporter has received much scrutiny because it is mutated in a number of visual disorders including Stargardt's disease, retinitis pigmentosa, and age-related macular dystrophy (Westerfeld & Mukai, 2008). ABCA4 is responsible for clearing toxic retinoid-acyl adducts from disk membranes (Quazi, Lenevich, & Molday, 2012). 11-*cis*-Retinal is the chromophore bound to opsin, and photon absorption results in its isomerization to all-*trans*-retinal. If all-*trans*-retinal reacts with phosphatidylethanolamine, forming N-ret-PE, it becomes trapped in the luminal side of the disk membrane where it can form other toxic compounds. The transport of N-ret-PE by ABCA4 to the cytoplasmic side of the membrane allows it to be processed and recycled through a series of enzymatic reactions known as the visual cycle (Coleman, Quazi, & Molday, 2013; Pollock & Callaghan, 2011; Tsybovsky, Molday, & Palczewski, 2010).

ABCA4 and the RDS/Rom-1 complex are the disk proteins that are segregated to the rims. One explanation for this distribution is that the extracellular (luminal) domains of these three proteins are simply too large to fit in the lamellar portion of the disk. Across species, the average intradiskal space is ~6nm with the luminal space in the rims being at least twice as large (Gilliam et al., 2012; Molday, 2004; Nickell, Park, Baumeister, & Palczewski, 2007). Another possibility in the case of RDS is that it actually shapes and defines the disk rims. RDS can interact with the beta subunit of CNG that is expressed in rods, and these two proteins likely form the filamentous electron-dense linkages seen between the disk rims and the plasma membrane that stabilize the OS (Zhang et al., 2009). A unique targeting signal in the C-terminus of RDS is required for its precise localization (Salinas, Baker, Gospe, & Arshavsky, 2013; Tam, Moritz, & Papermaster, 2004). It is not yet known how that signal is read, but, since it is different from the targeting signal in rhodopsin, there may be multiple distinct pathways ensuring that OS-resident proteins are efficiently trafficked to this unique domain.

The lipids in OS membranes are primarily phospholipids with some cholesterol. The cholesterol is distributed in a gradient with the largest amount (~30%) found in the basal disks (Albert & Boesze-Battaglia, 2005; Schultz, 2011). This gradient may be a reflection of a role for cholesterol-rich membranes in disk biogenesis. The phospholipids of the OS are enriched in long-chain (C14–C22) and very long-chain (C28–C40) polyunsaturated fatty acids (VLC-PUFAs). Docosahexaenoic acid, C22:6n3 (DHA), is the most abundant PUFA in OS membranes, accounting for ~50% of OS fatty acids (Bush, Malnoe, Reme, & Williams, 1994; Fliesler & Anderson, 1983; Stillwell & Wassall, 2003). This means that the OS membranes are highly fluid, a property that promotes the efficiency of signaling for several components of the phototransduction cascade (Brown, 1994; Gawrisch, Soubias, & Mihailescu, 2008; Litman, Niu, Polozova, & Mitchell, 2001; Mitchell, Niu, & Litman, 2001; Niu, Mitchell, & Litman, 2001). Furthermore, DHA has been observed to have anti-inflammatory properties, and in several studies, increased consumption of fish (which are enriched in omega-3 PUFAs) has been associated with lower risk for development of age-related macular degeneration (Querques, Forte, & Souied, 2011).

VLC-PUFAs are present in nearly 15% of the phosphatidylcholine species making up OS membranes (Rotstein & Aveldano, 1987). ELOVL4 is an elongase that synthesizes these rare lipid species, and mutations in this enzyme cause Stargardt type 3 retinal degeneration (Agbaga et al., 2008; Edwards, Donoso, & Ritter, 2001; Zhang et al., 2001). The Stargardt mutations cause loss of an ER retention signal and inhibit activity (Grayson & Molday, 2005; Logan et al., 2013). Transgenic mice expressing the truncated ELOVL4 form found in Stargardt patients have decreased VLC-PUFA levels and degenerated photoreceptors (Karan et al., 2005). However, conditional knockout of ELOVL4 from rods prevents the accumulation of VLC-PUFAs but unexpectedly does not cause any structural or functional abnormalities (Barabas et al., 2013). This result indicates that the loss of VLC-PUFAs from rods may not be the source of pathogenesis in Stargardt type 3 patients and that VLC-PUFAs are surprisingly not essential for photoreceptors. Clearly, additional work is needed to resolve these differing observations and provide a deeper understanding of how the specific lipid composition of OS membranes contributes to the health of the retina.

5. ORGANIZATION OF IS MEMBRANES

The IS expands soon after retinal progenitor cells become committed to becoming photoreceptors. The composition and function of the IS plasma membrane is distinct from that of the OS membranes but not as well studied, largely because the IS plasma membrane cannot be biochemically purified from the rest of the retina as in the case of OS. One key feature of the IS is that it contains many of the ion exchangers and channels that set the resting membrane potential and shape the light-stimulated responses of the cell.

NKA is an abundant protein in the IS plasma membrane and is necessary for maintenance of the electrochemical gradient (Schneider, Shyjan, & Levenson, 1991; Stahl & Baskin, 1984; Wetzel, Arystarkhova, & Sweadner, 1999). It uses the energy of ATP hydrolysis to pump sodium ions out and potassium ions into the cell, at a stoichiometry of $3Na^{+}:1K^{+}$, thus balancing the inward flow of sodium through the CNG channels in the OS. NKA is composed of a catalytic alpha subunit and an accessory beta subunit (Jorgensen, Hakansson, & Karlish, 2003). NKA can interact with a gamma subunit that modulates its activity (Garty & Karlish, 2006), although the identity of any gamma subunits in photoreceptors has not been described. Expression of the $\alpha_3\beta_2$ isozyme is predominant in photoreceptors (Schneider et al., 1991; Wetzel et al., 1999). Photoreceptors in mice lacking the β_2 subunit develop but quickly degenerate by two and a half weeks of age, emphasizing the importance of this pump for the health of the mature cell (Magyar et al., 1994).

Interestingly, the mobility of NKA within the IS plasma membrane is restricted. This was shown by examining mechanically dissociated frog photoreceptors. In this preparation, the OS and IS can fuse so that the diffusion barrier represented by the connecting cilium can be bypassed. As expected, rhodopsin in the OS abnormally localizes to the IS plasma membrane, but NKA in the IS does not gain access to the OS membranes (Spencer, Detwiler, & Bunt-Milam, 1988). The limited lateral mobility of NKA may reflect mechanisms to ensure the restricted composition of the OS that is optimized for phototransduction and limit competition for energy. In the dark, the major sites of energy consumption are the IS (due in large part to the activity of NKA) and the ST (due to the tonic release of neurotransmitter), while in the light, the OS is the major site of energy consumption (Okawa, Sampath, Laughlin, & Fain, 2008). Interestingly, to prevent NKA from consuming all the ATP in the IS, some of it flows to the ST in the form of phosphocreatine where it is locally converted to ATP by creatine kinase-B (Linton et al., 2010). Failure to localize NKA properly would likely be detrimental to this organized flow of energy. Additionally, NKA can participate in signal transduction and cellular adhesion, and though not explored in the retina, these putative functions would likely be disrupted if NKA was not properly segregated in the IS plasma membrane (Aperia, 2007).

What limits the ability of NKA to laterally diffuse? One possibility is that NKA is tethered to the membrane cytoskeleton by interacting directly with ankyrin, as has been shown in epithelial cells (Devarajan, Scaramuzzino, & Morrow, 1994; Jordan, Puschel, Koob, & Drenckhahn, 1995; Nelson &

Veshnock, 1987). Interestingly, interaction with ankyrin has also been reported to promote the trafficking of NKA out of the ER/Golgi (Stabach, Devarajan, Stankewich, Bannykh, & Morrow, 2008), which could contribute to ensuring NKA is delivered to the proper membrane domain. In photoreceptors, ankyrin-B is found in the IS and shown to colocalize with NKA along the plasma membrane; furthermore, haploinsufficiency of ankyrin-B is accompanied by reduced expression of NKA in the retina (Kizhatil, Sandhu, Peachey, & Bennett, 2009). In cardiomyocytes, ankyrin-B is an essential adaptor that scaffolds NKA in close proximity to the calcium exporter (NCX1) and inositol trisphosphate receptor that is concentrated in a specialization of the membrane known as the T-tubule. However, this complex is specific to cardiomyocytes and was not found in brain (Mohler, Davis, & Bennett, 2005). Immunoprecipitation of NKA from retina pulls down only trace amounts of ankyrin-B (unpublished data). This would be expected if ankyrin-B is primarily involved in early trafficking events that are by nature transient and suggests that additional proteins may be involved in regulating the restricted final localization of NKA in photoreceptors.

One binding partner of NKA unique to the retina that likely contributes to its restricted localization is retinoschisin (Friedrich et al., 2011; Molday, Wu, & Molday, 2007). Retinoschisin is a small protein that forms homooctomeric complexes and is secreted by photoreceptor and bipolar cells in the retina (Gleghorn, Trump, & Bulleid, 2010; Molday, Hicks, Sauer, Weber, & Molday, 2001; Wu, Wong, Kast, & Molday, 2005). Its role in cellular adhesion is demonstrated by the observation that mutations in this protein cause X-linked retinoschisis, a visual disorder where the layers of the retina split (Grayson et al., 2000; Molday, 2007; Tantri et al., 2004; Vijayasarathy, Ziccardi, & Sieving, 2012). It is possible that retinoschisin mediates adhesion by "cross-linking" NKA molecules on adjacent cells within the retina. The NKA–retinoschisin complex also contains SARM1, a putative adaptor protein whose role in the retina has not been determined (Molday et al., 2007). These observations suggest that NKA may contribute to photoreceptor physiology by more than its contribution to setting the electrochemical gradient.

Additional components of the IS plasma membrane are the ion channels that conduct the five principle currents of the IS ($I_{K(v)}$, $I_{K(Ca)}$, I_{Ca}, $I_{Cl(Ca)}$, and I_H); see Fig. 7.2 (Bader, Bertrand, & Schwartz, 1982; Barnes & Hille, 1989). Two of these currents, I_{Ca} and $I_{Cl(Ca)}$, are concentrated in the ST (see Section 6). The other three can all be recorded from photoreceptors lacking terminals that verify their expression in the IS plasma membrane and suggest that their expression in the synapse is lacking or less prominent (MacLeish & Nurse, 2007). Both $I_{K(v)}$ and $I_{K(Ca)}$ carry an outward K^+ current at depolarized potentials supporting the circulating dark current and contributing to the fine-tuning of the membrane potential in light-activated cells. The molecular identity of the K^+ channels expressed in photoreceptors is still being unraveled; some that are expressed include Kv1.2, Kv2.1, Kv4.2, Kv7, and Kv11.1 (Cordeiro, Guseva, Wulfsen, & Bauer, 2011; Klumpp, Song, & Pinto, 1995; Zhang, Yang, & Hughes, 2011). Intermediate and large conductance calcium-activated potassium channels expressed in photoreceptors include SK4 and BK (Pelucchi, Grimaldi, & Moriondo, 2008; Xu & Slaughter, 2005). Another IS potassium current (I_{Kx}) contributes to setting the resting potential and complements I_H to filter the rod light response (Barrow & Wu, 2009; Beech & Barnes, 1989). Ether-a-go-go (EAG) or Kv2.1/Kv8.2 heteromeric channels may be responsible for I_{Kx} (Czirjak, Toth, & Enyedi, 2007; Frings et al., 1998). The importance of Kv8.2 (KCNV2) is highlighted by the finding that its mutation causes defects in both rod and cone signaling in humans (Wu et al., 2006).

I_H is an inward positive current that resets the membrane potential after light-evoked hyperpolarization. It is carried by hyperpolarization-activated and cyclic nucleotide-gated (HCN) channels that are very similar to the CNG channels in the OS. Rods express HCN1, while cones express both HCN1 and HCN3 (Demontis et al., 2002; Fyk-Kolodziej & Pourcho, 2007; Knop et al., 2008; Muller et al., 2003). The activity of these channels is particularly important at medium light intensities when both rods and cones are active. In HCN1 knockout mice, the continued signaling in rods can suppress the cone pathway (Seeliger et al., 2011). HCN1 can interact with several proteins but the best studied is tetratricopeptide repeat-containing Rab8b-interacting protein (TRIP8b), which regulates the expression, localization, and activity of the channel in neurons (Lewis, Estep, & Chetkovich, 2010). However, the extent to which TRIP8b modulates HCN1 can vary. For instance, in the absence of TRIP8b, HCN1 is not selectively trafficked to the distal dendrites of hippocampal pyramidal neurons, but HCN1 is still trafficked to the plasma membrane in cortical pre-STs (Huang et al., 2012; Lewis et al., 2011). In the retina, loss of TRIP8b decreases the amount of HCN1 that is expressed, but the channel still traffics to the surface of retinal neurons (unpublished data). How HCN1 or any of the other channels located in the IS are trafficked specifically to the IS plasma membrane remains unknown.

The composition of the IS plasma membrane is much less characterized than the OS plasma membrane. Here, we have reviewed a few components that contribute to the physiology of light perception. There is a plethora of other proteins that support the needs of the cell including glucose transporters, dopamine receptors, and gap junction complexes (Bloomfield & Volgyi, 2009; Gospe et al., 2010; Witkovsky, 2004). Ongoing studies of the IS should allow us to build a better picture of how these proteins are trafficked to and organized within this domain.

6. ORGANIZATION OF THE SYNAPTIC MEMBRANE

While continuous with the IS, the synaptic membrane is a specialized compartment of the plasma membrane. Unique features include a large accumulation of synaptic vesicles organized around a ribbon and mechanisms to finely tune Ca^{2+} homeostasis. Proteins of the presynaptic plasma membrane play a key role in determining the intricate shape of the pre-ST, adhesion to postsynaptic components, and controlling vesicle fusion.

6.1. Ca^{2+} signaling

One essential component of the synapse is a voltage-gated calcium channel (VGCC). Photoreceptors express an L-type VGCC in contrast to most other CNS neurons that express N- and P/Q-type channels (Ca_v2 family) at the synapse (Catterall, 2011; Doering, Peloquin, & McRory, 2007). The L-type VGCC family consists of four members ($Ca_v1.1$–$Ca_v1.4$), defined by the pore-forming α_1 subunit. $Ca_v1.4$ was discovered through analysis of patients with congenital stationary night blindness, type 2 (CSNB2; Bech-Hansen et al., 1998). CSNB2 is a nonprogressive disease characterized by night blindness and the lack of bipolar responses (b-wave) in electroretinogram (ERG) recordings, indicating lack of synaptic transmission between photoreceptors and bipolar cells. More than 90 different CSNB2-causing mutations have now been identified in the human *CACNA1F* gene (Lodha, Loucks, Beaulieu, Parboosingh, & Bech-Hansen, 2012; Striessnig, Bolz, & Koschak, 2010; Strom et al., 1998; Wutz et al., 2002). $Ca_v1.4$ is present at the membrane as a heterotrimer of the pore-forming α_1 and auxiliary β_2 and $\alpha_2\delta_4$ subunits (Ball, McEnery, Yunker, Shin, & Gregg, 2011; Wycisk et al., 2006), which regulate the trafficking and biophysical properties of the pore-forming subunit (Buraei & Yang, 2013).

Labeling with antibodies specific to $Ca_v1.4$ demonstrates that this channel is localized to a subregion of the synapse known as the ribbon, and elegant single-molecule tracking studies have shown that the mobility of the channel is confined to this domain (Mercer, Chen, & Thoreson, 2011). The ribbon is an elongated flat proteinaceous structure decorated with synaptic vesicles. Currently, there is no agreement of the exact role of the ribbon or the mechanism of its function. In one of the theories, the ribbon acts as a conveyor belt delivering glutamate-filled vesicles to the release sites; in another, the positioning of the vesicles on the ribbon allows them to fuse with each other and with the plasma membrane providing truly continuous release of glutamate into the synaptic cleft (Matthews & Fuchs, 2010). In mouse rods, only one arch-shaped ribbon about 1.5 μm long, 250 nm wide, and only 30 nm thick is present, whereas in cones and rods from amphibian species, tens of ribbons per terminal have been observed (Balkema, Cusick, & Nguyen, 2001; Gabriel & Wilhelm, 2001; Haverkamp, Grunert, & Wassle, 2000). The ribbon is made up of oligomers of RIBEYE (Magupalli et al., 2008; Schmitz, Konigstorfer, & Sudhof, 2000) with many associated proteins, some of which form a structure known as the arciform density that lies between the ribbon proper and the plasma membrane (tom Dieck & Brandstatter, 2006). It is possible that $Ca_v1.4$ is tethered to this region via interactions with components of the arciform density, such as RIM proteins (Han, Kaeser, Sudhof, & Schneggenburger, 2011; Hibino et al., 2002; Kiyonaka et al., 2007).

Similar to CSNB2 patients, $Ca_v1.4$ knockout mouse models have dramatically attenuated b-waves in ERG recordings. Interestingly, these mice have also revealed that $Ca_v1.4$ is required for multiple aspects of synaptic development and maintenance. In all of these mice, the ribbon is very short and often mislocalized. There is also sprouting of horizontal and bipolar cell dendrites into the ONL with formation of ectopic synapses. The entire photoreceptor synaptic layer is thin and very disorganized (Bayley & Morgans, 2007; Chang et al., 2006; Lodha et al., 2010; Mansergh et al., 2005; Raven et al., 2008; Specht et al., 2009; Zabouri & Haverkamp, 2013).

It has been suggested that $Ca_v1.3$ is expressed in photoreceptor synapses, but loss of $Ca_v1.3$ in mice does not result in obvious visual dysfunction (Kersten et al., 2010; Taylor & Morgans, 1998; Wu, Marmorstein, Striessnig, & Peachey, 2007). Also, a recently discovered mutation in $Ca_v1.3$ in humans causing severe deafness did not seem to affect vision (Baig et al., 2011). These results suggest that $Ca_v1.3$ plays only a minimal role in mouse and human photoreceptors.

The primary Ca^{2+} extrusion mechanism in the synapse is via plasma membrane Ca^{2+} ATPase (PMCA1). This pump is found throughout the synaptic plasma membrane (Krizaj & Copenhagen, 1998; Morgans, El Far, Berntson, Wassle, & Taylor, 1998; Wan, Nixon, & Heidelberger, 2012). Ca^{2+} levels inside the synapse are also modulated by calcium-induced calcium release from internal stores, which modulates synaptic output (Babai, Morgans, & Thoreson, 2010; Cadetti, Bryson, Ciccone, Rabl, & Thoreson, 2006; Krizaj, Bao, Schmitz, Witkovsky, & Copenhagen, 1999; Suryanarayanan & Slaughter, 2006). The smooth ER extends into the ST of photoreceptors, and IP3 receptors, ryanodine receptors, and the sarco/endoplasmic reticulum calcium ATPase (SERCA) have all been localized to photoreceptor synapses (Babai et al., 2010; Krizaj, Lai, & Copenhagen, 2003; Mercurio & Holtzman, 1982; Peng, Sharp, Snyder, & Yau, 1991; Ungar, Piscopo, & Holtzman, 1981; Yang et al., 2007).

Ca^{2+} signaling regulates a chloride current in the synapse ($I_{Cl(Ca)}$). The high local Ca^{2+} concentration in the dark activates efflux of chloride. When intracellular Cl^- concentration decreases, the $Ca_v1.4$ channels are inhibited so the net effect of $I_{Cl(Ca)}$ is to provide feedback inhibition and limit the release of neurotransmitter in the dark (Thoreson, Nitzan, & Miller, 1997, 2000; Thoreson & Stella, 2000). $I_{Cl(Ca)}$ is carried by a family of channels known as anoctamin (Ano) or TMEM16 (Caputo et al., 2008; Schroeder, Cheng, Jan, & Jan, 2008; Yang, Cho, et al., 2008). Ano2 (TMEM16B) is expressed in the synapse of rods and is distributed all along the synaptic plasma membrane (Billig, Pal, Fidzinski, & Jentsch, 2011; Dauner, Mobus, Frings, & Mohrlen, 2013; Mercer, Rabl, et al., 2011; Stohr et al., 2009). The molecular identity of the calcium-activated chloride channel in cones is still ambiguous as multiple candidates have been suggested (reviewed in Lalonde, Kelly, & Barnes, 2008).

Another key component of the presynaptic membrane is a glutamate transporter belonging to the excitatory amino acid transporter (EAAT) family (Hasegawa, Obara, Tanaka, & Tachibana, 2006; Rowan, Ripps, & Shen, 2010; Winkler, Kapousta-Bruneau, Arnold, & Green, 1999). There are five EAAT family members found throughout the nervous system in both glia and neurons. Mammalian rods express EAAT5, while cones use a splice variant of EAAT2, Glt-1b (Arriza, Eliasof, Kavanaugh, & Amara, 1997; Eliasof, Arriza, Leighton, Amara, & Kavanaugh, 1998; Eliasof, Arriza, Leighton, Kavanaugh, & Amara, 1998; Fyk-Kolodziej, Qin, Dzhagaryan, & Pourcho, 2004; Pow & Barnett, 2000; Rauen, Wiessner, Sullivan, Lee, & Pow, 2004; Reye, Sullivan, Fletcher, & Pow, 2002; Reye,

Sullivan, & Pow, 2002). Like the chloride channels and PMCA, EAAT5 is expressed throughout the plasma membrane of the ST. The EAATs contribute to photoreceptor signaling in several ways. They clear excess glutamate from the synaptic cleft so that bipolar cells can respond to the light-dependent decreases in glutamate release, which may also prevent excitotoxicity. Glutamate taken up by EAATs can be transported into synaptic vesicles by vesicular glutamate transporters. Finally, there is a Cl^- current associated with glutamate transport by EAATs (Jiang & Amara, 2011) that contributes to the regulation of the synaptic Ca^{2+} current (Grant & Werblin, 1996; Rabl, Bryson, & Thoreson, 2003).

6.2. Scaffold proteins

Maintenance of the intricate shape and organization of the photoreceptor synaptic membrane is due to multiple scaffold and adhesion proteins. They are found both as part of the ribbon complex and lining the membrane. Scaffolding proteins consist of multiple protein–protein interaction domains and form homomeric and heteromeric complexes that link together soluble, cytoskeletal, and membrane proteins (Feng, Long, Fan, Suetake, & Zhang, 2004; Karnak, Lee, & Margolis, 2002). For example, MPP4, PSD95, and Veli3 have all been shown to interact in a complex and loss of MPP4 prevents localization of PSD95 and Veli3 at the photoreceptor synapse. This is accompanied by loss of PMCA and Ano2 that seems to reflect specific targeting or retention of these plasma membrane proteins as some aspects of an organized synapse were intact, such as localization a related scaffold protein, SAP97, and synaptic vesicles (Aartsen et al., 2009, 2006; Stohr et al., 2009; Yang et al., 2007).

Scaffolding proteins often interact with adhesion molecules, providing another way for them to contribute to the organization of the plasma membrane. MAGI-1, an intracellular scaffolding protein with similar protein–protein interaction domains as PSD95, is necessary for the localization of sidekick2. Sidekick2 is a transmembrane cell adhesion molecule that participates in development of specific pre- and postsynaptic contacts in the retina (Sanes & Zipursky, 2010; Yamagata & Sanes, 2010). Neurexin–neuroligin pairs also participate in this process and interact with both PSD95 and SAP97 (Dirks, Thomas, & Montag, 2006). Photoreceptor synapses express components of the dystrophin–glycoprotein complex, which has been most thoroughly characterized at the neuromuscular junction where it functions to link cytoskeletal and signaling proteins with the extracellular matrix (Pilgram, Potikanond, Baines, Fradkin, & Noordermeer, 2010). In photoreceptors, components of this complex including dystrophin, syntrophin, β-dystroglycan, and α-dystroglycan are concentrated near the ribbon and associated with the extracellular protein pikachurin. In pikachurin knockout mice, the connectivity with bipolar cells is altered and photoreceptor synapses display abnormal morphology (Hu, Li, Zhang, & Yu, 2011; Omori et al., 2012; Sato et al., 2008). A major outstanding question in this area is to understand how these various complexes work together to organize the synaptic membrane.

This overview touches on a few proteins involved in developing and maintaining the photoreceptor synapse. Importantly, there must be flexibility or rapid remodeling built into the scaffolding complexes in the synapse to accommodate the changes in membrane volume that occur upon synaptic vesicle exocytosis and endocytosis. Additional studies are needed to understand the complete cast of characters and how specificity is achieved in this highly organized and dynamic membrane domain.

7. PERSPECTIVES

Photoreceptors have been studied for well over 100 years, which has generated a wealth of understanding about their anatomy and physiology. The key to effective signaling is the compartmentalization of the cytoplasm and membrane into various functional units. Here, we have provided an overview of major components of the various photoreceptor plasma membrane domains. The OS, responsible for light detection, is the most extensively studied. Much is known about the protein and lipid composition of these membranes. However, one of the greatest remaining mysteries concerns how new OS membranes are formed on a daily basis. The formation and even the composition of the IS membrane is much less clear although it is recognized that many ion channels localized to this part of the plasma membrane are essential for light perception. It is hoped that future work will allow for a deeper understanding of the composition and organization of photoreceptor membranes.

REFERENCES

Aartsen, W. M., Arsanto, J. P., Chauvin, J. P., Vos, R. M., Versteeg, I., Cardozo, B. N., et al. (2009). PSD95beta regulates plasma membrane Ca2+ pump localization at the photoreceptor synapse. *Molecular and Cellular Neurosciences*, *41*(2), 156–165.

Aartsen, W. M., Kantardzhieva, A., Klooster, J., van Rossum, A. G., van de Pavert, S. A., Versteeg, I., et al. (2006). Mpp4 recruits Psd95 and Veli3 towards the photoreceptor synapse. *Human Molecular Genetics*, *15*(8), 1291–1302.

Agbaga, M. P., Brush, R. S., Mandal, M. N., Henry, K., Elliott, M. H., & Anderson, R. E. (2008). Role of Stargardt-3 macular dystrophy protein (ELOVL4) in the biosynthesis of very long chain fatty acids. *Proceedings of the National Academy of Sciences of the United States of America, 105*(35), 12843–12848.

Albert, A. D., & Boesze-Battaglia, K. (2005). The role of cholesterol in rod outer segment membranes. *Progress in Lipid Research, 44*(2–3), 99–124.

Aperia, A. (2007). New roles for an old enzyme: Na, K-ATPase emerges as an interesting drug target. *Journal of Internal Medicine, 261*(1), 44–52.

Arendt, D. (2003). Evolution of eyes and photoreceptor cell types. *The International Journal of Developmental Biology, 47*(7–8), 563–571.

Arriza, J. L., Eliasof, S., Kavanaugh, M. P., & Amara, S. G. (1997). Excitatory amino acid transporter 5, a retinal glutamate transporter coupled to a chloride conductance. *Proceedings of the National Academy of Sciences of the United States of America, 94*(8), 4155–4160.

Arshavsky, V. Y., & Burns, M. E. (2012). Photoreceptor signaling: Supporting vision across a wide range of light intensities. *The Journal of Biological Chemistry, 287*(3), 1620–1626.

Azadi, S., Molday, L. L., & Molday, R. S. (2010). RD3, the protein associated with Leber congenital amaurosis type 12, is required for guanylate cyclase trafficking in photoreceptor cells. *Proceedings of the National Academy of Sciences of the United States of America, 107*(49), 21158–21163.

Babai, N., Morgans, C. W., & Thoreson, W. B. (2010). Calcium-induced calcium release contributes to synaptic release from mouse rod photoreceptors. *Neuroscience, 165*(4), 1447–1456.

Bader, C. R., Bertrand, D., & Schwartz, E. A. (1982). Voltage-activated and calcium-activated currents studied in solitary rod inner segments from the salamander retina. *The Journal of Physiology, 331*, 253–284.

Baig, S. M., Koschak, A., Lieb, A., Gebhart, M., Dafinger, C., Nurnberg, G., et al. (2011). Loss of Ca(v)1.3 (CACNA1D) function in a human channelopathy with bradycardia and congenital deafness. *Nature Neuroscience, 14*(1), 77–84.

Baker, S. A., Haeri, M., Yoo, P., Gospe, S. M., 3rd., Skiba, N. P., Knox, B. E., et al. (2008). The outer segment serves as a default destination for the trafficking of membrane proteins in photoreceptors. *The Journal of Cell Biology, 183*(3), 485–498.

Balkema, G. W., Cusick, K., & Nguyen, T. H. (2001). Diurnal variation in synaptic ribbon length and visual threshold. *Visual Neuroscience, 18*(5), 789–797.

Ball, S. L., McEnery, M. W., Yunker, A. M., Shin, H. S., & Gregg, R. G. (2011). Distribution of voltage gated calcium channel beta subunits in the mouse retina. *Brain Research, 1412*, 1–8.

Barabas, P., Liu, A., Xing, W., Chen, C. K., Tong, Z., Watt, C. B., et al. (2013). Role of ELOVL4 and very long-chain polyunsaturated fatty acids in mouse models of Stargardt type 3 retinal degeneration. *Proceedings of the National Academy of Sciences of the United States of America, 110*(13), 5181–5186.

Barnes, S., & Hille, B. (1989). Ionic channels of the inner segment of tiger salamander cone photoreceptors. *The Journal of General Physiology, 94*(4), 719–743.

Barrow, A. J., & Wu, S. M. (2009). Complementary conductance changes by IKx and Ih contribute to membrane impedance stability during the rod light response. *Channels, 3*(5), 301–307.

Bayley, P. R., & Morgans, C. W. (2007). Rod bipolar cells and horizontal cells form displaced synaptic contacts with rods in the outer nuclear layer of the nob2 retina. *The Journal of Comparative Neurology, 500*(2), 286–298.

Baylor, D. A., Matthews, G., & Nunn, B. J. (1984). Location and function of voltage-sensitive conductances in retinal rods of the salamander, Ambystoma tigrinum. *The Journal of Physiology, 354*, 203–223.

Bech-Hansen, N. T., Naylor, M. J., Maybaum, T. A., Pearce, W. G., Koop, B., Fishman, G. A., et al. (1998). Loss-of-function mutations in a calcium-channel alpha1-subunit gene in Xp11.23 cause incomplete X-linked congenital stationary night blindness. *Nature Genetics*, *19*(3), 264–267.

Beech, D. J., & Barnes, S. (1989). Characterization of a voltage-gated K+ channel that accelerates the rod response to dim light. *Neuron*, *3*(5), 573–581.

Besharse, J. C., Forestner, D. M., & Defoe, D. M. (1985). Membrane assembly in retinal photoreceptors. III. Distinct membrane domains of the connecting cilium of developing rods. *The Journal of Neuroscience*, *5*(4), 1035–1048.

Billig, G. M., Pal, B., Fidzinski, P., & Jentsch, T. J. (2011). Ca2+-activated Cl- currents are dispensable for olfaction. *Nature Neuroscience*, *14*(6), 763–769.

Bloomfield, S. A., & Volgyi, B. (2009). The diverse functional roles and regulation of neuronal gap junctions in the retina. *Nature Reviews Neuroscience*, *10*(7), 495–506.

Boesze-Battaglia, K., Goldberg, A. F., Dispoto, J., Katragadda, M., Cesarone, G., & Albert, A. D. (2003). A soluble peripherin/Rds C-terminal polypeptide promotes membrane fusion and changes conformation upon membrane association. *Experimental Eye Research*, 77(4), 505–514.

Bowes, C., Li, T., Danciger, M., Baxter, L. C., Applebury, M. L., & Farber, D. B. (1990). Retinal degeneration in the rd mouse is caused by a defect in the beta subunit of rod cGMP-phosphodiesterase. *Nature*, *347*(6294), 677–680.

Breslow, D. K., & Nachury, M. V. (2011). Primary cilia: How to keep the riff-raff in the plasma membrane. *Current Biology*, *21*(11), R434–R436.

Brown, M. F. (1994). Modulation of rhodopsin function by properties of the membrane bilayer. *Chemistry and Physics of Lipids*, *73*(1–2), 159–180.

Buraei, Z., & Yang, J. (2013). Structure and function of the beta subunit of voltage-gated Ca(2+) channels. *Biochimica et Biophysica Acta*, *1828*(7), 1530–1540.

Burns, M. E., & Baylor, D. A. (2001). Activation, deactivation, and adaptation in vertebrate photoreceptor cells. *Annual Review of Neuroscience*, *24*, 779–805.

Bush, R. A., Malnoe, A., Reme, C. E., & Williams, T. P. (1994). Dietary deficiency of N-3 fatty acids alters rhodopsin content and function in the rat retina. *Investigative Ophthalmology & Visual Science*, *35*(1), 91–100.

Cadetti, L., Bryson, E. J., Ciccone, C. A., Rabl, K., & Thoreson, W. B. (2006). Calcium-induced calcium release in rod photoreceptor terminals boosts synaptic transmission during maintained depolarization. *The European Journal of Neuroscience*, *23*(11), 2983–2990.

Caputo, A., Caci, E., Ferrera, L., Pedemonte, N., Barsanti, C., Sondo, E., et al. (2008). TMEM16A, a membrane protein associated with calcium-dependent chloride channel activity. *Science*, *322*(5901), 590–594.

Catterall, W. A. (2011). Voltage-gated calcium channels. *Cold Spring Harbor Perspectives in Biology*, *3*(8), a003947.

Chang, B., Heckenlively, J. R., Bayley, P. R., Brecha, N. C., Davisson, M. T., Hawes, N. L., et al. (2006). The nob2 mouse, a null mutation in Cacna1f: Anatomical and functional abnormalities in the outer retina and their consequences on ganglion cell visual responses. *Visual Neuroscience*, *23*(1), 11–24.

Chuang, J. Z., Zhao, Y., & Sung, C. H. (2007). SARA-regulated vesicular targeting underlies formation of the light-sensing organelle in mammalian rods. *Cell*, *130*(3), 535–547.

Clarke, G., Goldberg, A. F., Vidgen, D., Collins, L., Ploder, L., Schwarz, L., et al. (2000). Rom-1 is required for rod photoreceptor viability and the regulation of disk morphogenesis. *Nature Genetics*, *25*(1), 67–73.

Coleman, J. A., Kwok, M. C., & Molday, R. S. (2009). Localization, purification, and functional reconstitution of the P4-ATPase Atp8a2, a phosphatidylserine flippase in photoreceptor disc membranes. *The Journal of Biological Chemistry, 284*(47), 32670–32679.

Coleman, J. A., & Molday, R. S. (2011). Critical role of the beta-subunit CDC50A in the stable expression, assembly, subcellular localization, and lipid transport activity of the P4-ATPase ATP8A2. *The Journal of Biological Chemistry, 286*(19), 17205–17216.

Coleman, J. A., Quazi, F., & Molday, R. S. (2013). Mammalian P4-ATPases and ABC transporters and their role in phospholipid transport. *Biochimica et Biophysica Acta, 1831*(3), 555–574.

Cordeiro, S., Guseva, D., Wulfsen, I., & Bauer, C. K. (2011). Expression pattern of Kv11 (Ether a-go-go-related gene; erg) K+ channels in the mouse retina. *PLoS One, 6*(12), e29490.

Corless, J. M., & Fetter, R. D. (1987). Structural features of the terminal loop region of frog retinal rod outer segment disk membranes: III. Implications of the terminal loop complex for disk morphogenesis, membrane fusion, and cell surface interactions. *The Journal of Comparative Neurology, 257*(1), 24–38.

Czirjak, G., Toth, Z. E., & Enyedi, P. (2007). Characterization of the heteromeric potassium channel formed by kv2.1 and the retinal subunit kv8.2 in Xenopus oocytes. *Journal of Neurophysiology, 98*(3), 1213–1222.

Dauner, K., Mobus, C., Frings, S., & Mohrlen, F. (2013). Targeted expression of anoctamin calcium-activated chloride channels in rod photoreceptor terminals of the rodent retina. *Investigative Ophthalmology & Visual Science, 54*(5), 3126–3136.

Demontis, G. C., Moroni, A., Gravante, B., Altomare, C., Longoni, B., Cervetto, L., et al. (2002). Functional characterisation and subcellular localisation of HCN1 channels in rabbit retinal rod photoreceptors. *The Journal of Physiology, 542*(Pt 1), 89–97.

Deretic, D., & Wang, J. (2012). Molecular assemblies that control rhodopsin transport to the cilia. *Vision Research, 75*, 5–10.

Devarajan, P., Scaramuzzino, D. A., & Morrow, J. S. (1994). Ankyrin binds to two distinct cytoplasmic domains of Na, K-ATPase alpha subunit. *Proceedings of the National Academy of Sciences of the United States of America, 91*(8), 2965–2969.

Ding, X. Q., & Naash, M. I. (2006). Transgenic animal studies of human retinal disease caused by mutations in peripherin/rds. *Advances in Experimental Medicine and Biology, 572*, 141–146.

Dirks, P., Thomas, U., & Montag, D. (2006). The cytoplasmic domain of NrCAM binds to PDZ domains of synapse-associated proteins SAP90/PSD95 and SAP97. *The European Journal of Neuroscience, 24*(1), 25–31.

Doering, C. J., Peloquin, J. B., & McRory, J. E. (2007). The Ca(v)1.4 calcium channel: More than meets the eye. *Channels, 1*(1), 3–10.

Edwards, A. O., Donoso, L. A., & Ritter, R., 3rd. (2001). A novel gene for autosomal dominant Stargardt-like macular dystrophy with homology to the SUR4 protein family. *Investigative Ophthalmology & Visual Science, 42*(11), 2652–2663.

Eliasof, S., Arriza, J. L., Leighton, B. H., Amara, S. G., & Kavanaugh, M. P. (1998). Localization and function of five glutamate transporters cloned from the salamander retina. *Vision Research, 38*(10), 1443–1454.

Eliasof, S., Arriza, J. L., Leighton, B. H., Kavanaugh, M. P., & Amara, S. G. (1998). Excitatory amino acid transporters of the salamander retina: Identification, localization, and function. *The Journal of Neuroscience, 18*(2), 698–712.

Fain, G. L. (2006). Why photoreceptors die (and why they don't). *Bioessays, 28*(4), 344–354.

Fain, G. L., Matthews, H. R., Cornwall, M. C., & Koutalos, Y. (2001). Adaptation in vertebrate photoreceptors. *Physiological Reviews, 81*(1), 117–151.

Farber, D. B., & Lolley, R. N. (1974). Cyclic guanosine monophosphate: Elevation in degenerating photoreceptor cells of the C3H mouse retina. *Science, 186*(4162), 449–451.

Feng, W., Long, J. F., Fan, J. S., Suetake, T., & Zhang, M. (2004). The tetrameric L27 domain complex as an organization platform for supramolecular assemblies. *Nature Structural & Molecular Biology, 11*(5), 475–480.

Fliesler, S. J., & Anderson, R. E. (1983). Chemistry and metabolism of lipids in the vertebrate retina. *Progress in Lipid Research, 22*(2), 79–131.

Friedrich, U., Stohr, H., Hilfinger, D., Loenhardt, T., Schachner, M., Langmann, T., et al. (2011). The Na/K-ATPase is obligatory for membrane anchorage of retinoschisin, the protein involved in the pathogenesis of X-linked juvenile retinoschisis. *Human Molecular Genetics, 20*(6), 1132–1142.

Frings, S., Brull, N., Dzeja, C., Angele, A., Hagen, V., Kaupp, U. B., et al. (1998). Characterization of ether-a-go-go channels present in photoreceptors reveals similarity to IKx, a K+ current in rod inner segments. *The Journal of General Physiology, 111*(4), 583–599.

Fyk-Kolodziej, B., & Pourcho, R. G. (2007). Differential distribution of hyperpolarization-activated and cyclic nucleotide-gated channels in cone bipolar cells of the rat retina. *The Journal of Comparative Neurology, 501*(6), 891–903.

Fyk-Kolodziej, B., Qin, P., Dzhagaryan, A., & Pourcho, R. G. (2004). Differential cellular and subcellular distribution of glutamate transporters in the cat retina. *Visual Neuroscience, 21*(4), 551–565.

Gabriel, R., & Wilhelm, M. (2001). Structure and function of photoreceptor and second-order cell mosaics in the retina of Xenopus. *International Review of Cytology, 210*, 77–120.

Garty, H., & Karlish, S. J. (2006). Role of FXYD proteins in ion transport. *Annual Review of Physiology, 68*, 431–459.

Gawrisch, K., Soubias, O., & Mihailescu, M. (2008). Insights from biophysical studies on the role of polyunsaturated fatty acids for function of G-protein coupled membrane receptors. *Prostaglandins, Leukotrienes, and Essential Fatty Acids, 79*(3–5), 131–134.

Gilliam, J. C., Chang, J. T., Sandoval, I. M., Zhang, Y., Li, T., Pittler, S. J., et al. (2012). Three-dimensional architecture of the rod sensory cilium and its disruption in retinal neurodegeneration. *Cell, 151*(5), 1029–1041.

Gleghorn, L. J., Trump, D., & Bulleid, N. J. (2010). Wild-type and missense mutants of retinoschisin co-assemble resulting in either intracellular retention or incorrect assembly of the functionally active octamer. *The Biochemical Journal, 425*(1), 275–283.

Goldberg, A. F. (2006). Role of peripherin/rds in vertebrate photoreceptor architecture and inherited retinal degenerations. *International Review of Cytology, 253*, 131–175.

Gospe, S. M., 3rd., Baker, S. A., & Arshavsky, V. Y. (2010). Facilitative glucose transporter Glut1 is actively excluded from rod outer segments. *Journal of Cell Science, 123*(Pt 21), 3639–3644.

Gospe, S. M., 3rd., Baker, S. A., Kessler, C., Brucato, M. F., Winter, J. R., Burns, M. E., et al. (2011). Membrane attachment is key to protecting transducin GTPase-activating complex from intracellular proteolysis in photoreceptors. *The Journal of Neuroscience, 31*(41), 14660–14668.

Grant, G. B., & Werblin, F. S. (1996). A glutamate-elicited chloride current with transporter-like properties in rod photoreceptors of the tiger salamander. *Visual Neuroscience, 13*(1), 135–144.

Grayson, C., & Molday, R. S. (2005). Dominant negative mechanism underlies autosomal dominant Stargardt-like macular dystrophy linked to mutations in ELOVL4. *The Journal of Biological Chemistry, 280*(37), 32521–32530.

Grayson, C., Reid, S. N., Ellis, J. A., Rutherford, A., Sowden, J. C., Yates, J. R., et al. (2000). Retinoschisin, the X-linked retinoschisis protein, is a secreted photoreceptor protein, and is expressed and released by Weri-Rb1 cells. *Human Molecular Genetics, 9*(12), 1873–1879.

Han, Y., Kaeser, P. S., Sudhof, T. C., & Schneggenburger, R. (2011). RIM determines Ca(2)+ channel density and vesicle docking at the presynaptic active zone. *Neuron*, *69*(2), 304–316.

Hasegawa, J., Obara, T., Tanaka, K., & Tachibana, M. (2006). High-density presynaptic transporters are required for glutamate removal from the first visual synapse. *Neuron*, *50*(1), 63–74.

Haverkamp, S., Grunert, U., & Wassle, H. (2000). The cone pedicle, a complex synapse in the retina. *Neuron*, *27*(1), 85–95.

Hawkins, R. K., Jansen, H. G., & Sanyal, S. (1985). Development and degeneration of retina in rds mutant mice: Photoreceptor abnormalities in the heterozygotes. *Experimental Eye Research*, *41*(6), 701–720.

Hibino, H., Pironkova, R., Onwumere, O., Vologodskaia, M., Hudspeth, A. J., & Lesage, F. (2002). RIM binding proteins (RBPs) couple Rab3-interacting molecules (RIMs) to voltage-gated Ca(2+) channels. *Neuron*, *34*(3), 411–423.

Hu, H., Li, J., Zhang, Z., & Yu, M. (2011). Pikachurin interaction with dystroglycan is diminished by defective O-mannosyl glycosylation in congenital muscular dystrophy models and rescued by LARGE overexpression. *Neuroscience Letters*, *489*(1), 10–15.

Huang, Z., Lujan, R., Martinez-Hernandez, J., Lewis, A. S., Chetkovich, D. M., & Shah, M. M. (2012). TRIP8b-independent trafficking and plasticity of adult cortical presynaptic HCN1 channels. *The Journal of Neuroscience*, *32*(42), 14835–14848.

Humphries, M. M., Rancourt, D., Farrar, G. J., Kenna, P., Hazel, M., Bush, R. A., et al. (1997). Retinopathy induced in mice by targeted disruption of the rhodopsin gene. *Nature Genetics*, *15*(2), 216–219.

Huttl, S., Michalakis, S., Seeliger, M., Luo, D. G., Acar, N., Geiger, H., et al. (2005). Impaired channel targeting and retinal degeneration in mice lacking the cyclic nucleotide-gated channel subunit CNGB1. *The Journal of Neuroscience*, *25*(1), 130–138.

Ishikawa, H., & Marshall, W. F. (2011). Ciliogenesis: Building the cell's antenna. *Nature Reviews Molecular Cell Biology*, *12*(4), 222–234.

Jenkins, P. M., Hurd, T. W., Zhang, L., McEwen, D. P., Brown, R. L., Margolis, B., et al. (2006). Ciliary targeting of olfactory CNG channels requires the CNGB1b subunit and the kinesin-2 motor protein, KIF17. *Current Biology*, *16*(12), 1211–1216.

Jenkins, P. M., Zhang, L., Thomas, G., & Martens, J. R. (2009). PACS-1 mediates phosphorylation-dependent ciliary trafficking of the cyclic-nucleotide-gated channel in olfactory sensory neurons. *The Journal of Neuroscience*, *29*(34), 10541–10551.

Jiang, J., & Amara, S. G. (2011). New views of glutamate transporter structure and function: Advances and challenges. *Neuropharmacology*, *60*(1), 172–181.

Jordan, C., Puschel, B., Koob, R., & Drenckhahn, D. (1995). Identification of a binding motif for ankyrin on the alpha-subunit of Na+, K(+)-ATPase. *The Journal of Biological Chemistry*, *270*(50), 29971–29975.

Jorgensen, P. L., Hakansson, K. O., & Karlish, S. J. (2003). Structure and mechanism of Na, K-ATPase: Functional sites and their interactions. *Annual Review of Physiology*, *65*, 817–849.

Kang, K., Bauer, P. J., Kinjo, T. G., Szerencsei, R. T., Bonigk, W., Winkfein, R. J., et al. (2003). Assembly of retinal rod or cone Na(+)/Ca(2+)-K(+) exchanger oligomers with cGMP-gated channel subunits as probed with heterologously expressed cDNAs. *Biochemistry*, *42*(15), 4593–4600.

Karan, G., Yang, Z., Howes, K., Zhao, Y., Chen, Y., Cameron, D. J., et al. (2005). Loss of ER retention and sequestration of the wild-type ELOVL4 by Stargardt disease dominant negative mutants. *Molecular Vision*, *11*, 657–664.

Karnak, D., Lee, S., & Margolis, B. (2002). Identification of multiple binding partners for the amino-terminal domain of synapse-associated protein 97. *The Journal of Biological Chemistry*, *277*(48), 46730–46735.

Kaupp, U. B., & Seifert, R. (2002). Cyclic nucleotide-gated ion channels. *Physiological Reviews*, *82*(3), 769–824.

Kefalov, V. J. (2012). Rod and cone visual pigments and phototransduction through pharmacological, genetic, and physiological approaches. *The Journal of Biological Chemistry*, *287*(3), 1635–1641.

Kersten, F. F., van Wijk, E., van Reeuwijk, J., van der Zwaag, B., Marker, T., Peters, T. A., et al. (2010). Association of whirlin with Cav1.3 (alpha1D) channels in photoreceptors, defining a novel member of the usher protein network. *Investigative Ophthalmology & Visual Science*, *51*(5), 2338–2346.

Kevany, B. M., & Palczewski, K. (2010). Phagocytosis of retinal rod and cone photoreceptors. *Physiology*, *25*(1), 8–15.

Kiyonaka, S., Wakamori, M., Miki, T., Uriu, Y., Nonaka, M., Bito, H., et al. (2007). RIM1 confers sustained activity and neurotransmitter vesicle anchoring to presynaptic Ca2+ channels. *Nature Neuroscience*, *10*(6), 691–701.

Kizhatil, K., Baker, S. A., Arshavsky, V. Y., & Bennett, V. (2009). Ankyrin-G promotes cyclic nucleotide-gated channel transport to rod photoreceptor sensory cilia. *Science*, *323*(5921), 1614–1617.

Kizhatil, K., Sandhu, N. K., Peachey, N. S., & Bennett, V. (2009). Ankyrin-B is required for coordinated expression of beta-2-spectrin, the Na/K-ATPase and the Na/Ca exchanger in the inner segment of rod photoreceptors. *Experimental Eye Research*, *88*(1), 57–64.

Klumpp, D. J., Song, E. J., & Pinto, L. H. (1995). Identification and localization of K+ channels in the mouse retina. *Visual Neuroscience*, *12*(6), 1177–1190.

Knop, G. C., Seeliger, M. W., Thiel, F., Mataruga, A., Kaupp, U. B., Friedburg, C., et al. (2008). Light responses in the mouse retina are prolonged upon targeted deletion of the HCN1 channel gene. *The European Journal of Neuroscience*, *28*(11), 2221–2230.

Krizaj, D., Bao, J. X., Schmitz, Y., Witkovsky, P., & Copenhagen, D. R. (1999). Caffeine-sensitive calcium stores regulate synaptic transmission from retinal rod photoreceptors. *The Journal of Neuroscience*, *19*(17), 7249–7261.

Krizaj, D., & Copenhagen, D. R. (1998). Compartmentalization of calcium extrusion mechanisms in the outer and inner segments of photoreceptors. *Neuron*, *21*(1), 249–256.

Krizaj, D., Lai, F. A., & Copenhagen, D. R. (2003). Ryanodine stores and calcium regulation in the inner segments of salamander rods and cones. *The Journal of Physiology*, *547*(Pt 3), 761–774.

Lalonde, M. R., Kelly, M. E., & Barnes, S. (2008). Calcium-activated chloride channels in the retina. *Channels*, *2*(4), 252–260.

Lamb, T. D., Collin, S. P., & Pugh, E. N., Jr. (2007). Evolution of the vertebrate eye: Opsins, photoreceptors, retina and eye cup. *Nature Reviews Neuroscience*, *8*(12), 960–976.

Lem, J., Krasnoperova, N. V., Calvert, P. D., Kosaras, B., Cameron, D. A., Nicolo, M., et al. (1999). Morphological, physiological, and biochemical changes in rhodopsin knockout mice. *Proceedings of the National Academy of Sciences of the United States of America*, *96*(2), 736–741.

Lewis, A. S., Estep, C. M., & Chetkovich, D. M. (2010). The fast and slow ups and downs of HCN channel regulation. *Channels*, *4*(3), 215–231.

Lewis, A. S., Vaidya, S. P., Blaiss, C. A., Liu, Z., Stoub, T. R., Brager, D. H., et al. (2011). Deletion of the hyperpolarization-activated cyclic nucleotide-gated channel auxiliary subunit TRIP8b impairs hippocampal Ih localization and function and promotes antidepressant behavior in mice. *The Journal of Neuroscience*, *31*(20), 7424–7440.

Liang, Y., Fotiadis, D., Filipek, S., Saperstein, D. A., Palczewski, K., & Engel, A. (2003). Organization of the G protein-coupled receptors rhodopsin and opsin in native membranes. *The Journal of Biological Chemistry*, *278*(24), 21655–21662.

Liang, Y., Fotiadis, D., Maeda, T., Maeda, A., Modzelewska, A., Filipek, S., et al. (2004). Rhodopsin signaling and organization in heterozygote rhodopsin knockout mice. *The Journal of Biological Chemistry*, *279*(46), 48189–48196.

Linton, J. D., Holzhausen, L. C., Babai, N., Song, H., Miyagishima, K. J., Stearns, G. W., et al. (2010). Flow of energy in the outer retina in darkness and in light. *Proceedings of the National Academy of Sciences of the United States of America*, *107*(19), 8599–8604.

Litman, B. J., Niu, S. L., Polozova, A., & Mitchell, D. C. (2001). The role of docosahexaenoic acid containing phospholipids in modulating G protein-coupled signaling pathways: Visual transduction. *Journal of Molecular Neuroscience*, *16*(2–3), 237–242, discussion 279–284.

Lodha, N., Bonfield, S., Orton, N. C., Doering, C. J., McRory, J. E., Mema, S. C., et al. (2010). Congenital stationary night blindness in mice—A tale of two Cacna1f mutants. *Advances in Experimental Medicine and Biology*, *664*, 549–558.

Lodha, N., Loucks, C. M., Beaulieu, C., Parboosingh, J. S., & Bech-Hansen, N. T. (2012). Congenital stationary night blindness: Mutation update and clinical variability. *Advances in Experimental Medicine and Biology*, *723*, 371–379.

Logan, S., Agbaga, M. P., Chan, M. D., Kabir, N., Mandal, N. A., Brush, R. S., et al. (2013). Deciphering mutant ELOVL4 activity in autosomal-dominant Stargardt macular dystrophy. *Proceedings of the National Academy of Sciences of the United States of America*, *110*(14), 5446–5451.

MacLeish, P. R., & Nurse, C. A. (2007). Ion channel compartments in photoreceptors: Evidence from salamander rods with intact and ablated terminals. *Journal of Neurophysiology*, *98*(1), 86–95.

Magupalli, V. G., Schwarz, K., Alpadi, K., Natarajan, S., Seigel, G. M., & Schmitz, F. (2008). Multiple RIBEYE-RIBEYE interactions create a dynamic scaffold for the formation of synaptic ribbons. *The Journal of Neuroscience*, *28*(32), 7954–7967.

Magyar, J. P., Bartsch, U., Wang, Z. Q., Howells, N., Aguzzi, A., Wagner, E. F., et al. (1994). Degeneration of neural cells in the central nervous system of mice deficient in the gene for the adhesion molecule on Glia, the beta 2 subunit of murine Na, K-ATPase. *The Journal of Cell Biology*, *127*(3), 835–845.

Makino, C. L., Wen, X. H., Michaud, N. A., Covington, H. I., DiBenedetto, E., Hamm, H. E., et al. (2012). Rhodopsin expression level affects rod outer segment morphology and photoresponse kinetics. *PLoS One*, 7(5), e37832.

Malanson, K. M., & Lem, J. (2009). Rhodopsin-mediated retinitis pigmentosa. *Progress in Molecular Biology and Translational Science*, *88*, 1–31.

Mansergh, F., Orton, N. C., Vessey, J. P., Lalonde, M. R., Stell, W. K., Tremblay, F., et al. (2005). Mutation of the calcium channel gene Cacna1f disrupts calcium signaling, synaptic transmission and cellular organization in mouse retina. *Human Molecular Genetics*, *14*(20), 3035–3046.

Martemyanov, K. A., Lishko, P. V., Calero, N., Keresztes, G., Sokolov, M., Strissel, K. J., et al. (2003). The DEP domain determines subcellular targeting of the GTPase activating protein RGS9 in vivo. *The Journal of Neuroscience*, *23*(32), 10175–10181.

Matthews, G., & Fuchs, P. (2010). The diverse roles of ribbon synapses in sensory neurotransmission. *Nature Reviews Neuroscience*, *11*(12), 812–822.

Mercer, A. J., Chen, M., & Thoreson, W. B. (2011). Lateral mobility of presynaptic L-type calcium channels at photoreceptor ribbon synapses. *The Journal of Neuroscience*, *31*(12), 4397–4406.

Mercer, A. J., Rabl, K., Riccardi, G. E., Brecha, N. C., Stella, S. L., Jr., & Thoreson, W. B. (2011). Location of release sites and calcium-activated chloride channels relative to calcium channels at the photoreceptor ribbon synapse. *Journal of Neurophysiology*, *105*(1), 321–335.

Mercurio, A. M., & Holtzman, E. (1982). Smooth endoplasmic reticulum and other agranular reticulum in frog retinal photoreceptors. *Journal of Neurocytology*, *11*(2), 263–293.

Michalakis, S., Muhlfriedel, R., Tanimoto, N., Krishnamoorthy, V., Koch, S., Fischer, M. D., et al. (2010). Restoration of cone vision in the CNGA3 −/− mouse model of congenital complete lack of cone photoreceptor function. *Molecular Therapy*, *18*(12), 2057–2063.

Michalakis, S., Schaferhoff, K., Spiwoks-Becker, I., Zabouri, N., Koch, S., Koch, F., et al. (2013). Characterization of neurite outgrowth and ectopic synaptogenesis in response to photoreceptor dysfunction. *Cellular and Molecular Life Sciences*, *70*(10), 1831–1847.

Michalakis, S., Zong, X., Becirovic, E., Hammelmann, V., Wein, T., Wanner, K. T., et al. (2011). The glutamic acid-rich protein is a gating inhibitor of cyclic nucleotide-gated channels. *The Journal of Neuroscience*, *31*(1), 133–141.

Mitchell, D. C., Niu, S. L., & Litman, B. J. (2001). Optimization of receptor-G protein coupling by bilayer lipid composition I: Kinetics of rhodopsin-transducin binding. *The Journal of Biological Chemistry*, *276*(46), 42801–42806.

Mohler, P. J., Davis, J. Q., & Bennett, V. (2005). Ankyrin-B coordinates the Na/K ATPase, Na/Ca exchanger, and InsP3 receptor in a cardiac T-tubule/SR microdomain. *PLoS Biology*, *3*(12), e423.

Molday, R. S. (2004). Molecular organization of rod outer segments. In D. S. Williams (Ed.), *Photoreceptor cell biology and inherited retinal degenerations* (pp. 259–300): Singapore: World Scientific Publishing.

Molday, R. S. (2007). Focus on molecules: Retinoschisin (RS1). *Experimental Eye Research*, *84*(2), 227–228.

Molday, L. L., Hicks, D., Sauer, C. G., Weber, B. H., & Molday, R. S. (2001). Expression of X-linked retinoschisis protein RS1 in photoreceptor and bipolar cells. *Investigative Ophthalmology & Visual Science*, *42*(3), 816–825.

Molday, L. L., Wu, W. W., & Molday, R. S. (2007). Retinoschisin (RS1), the protein encoded by the X-linked retinoschisis gene, is anchored to the surface of retinal photoreceptor and bipolar cells through its interactions with a Na/K ATPase-SARM1 complex. *The Journal of Biological Chemistry*, *282*(45), 32792–32801.

Morgans, C. W., El Far, O., Berntson, A., Wassle, H., & Taylor, W. R. (1998). Calcium extrusion from mammalian photoreceptor terminals. *The Journal of Neuroscience*, *18*(7), 2467–2474.

Muller, F., Scholten, A., Ivanova, E., Haverkamp, S., Kremmer, E., & Kaupp, U. B. (2003). HCN channels are expressed differentially in retinal bipolar cells and concentrated at synaptic terminals. *The European Journal of Neuroscience*, *17*(10), 2084–2096.

Nelson, W. J., & Veshnock, P. J. (1987). Ankyrin binding to (Na+ + K+)ATPase and implications for the organization of membrane domains in polarized cells. *Nature*, *328*(6130), 533–536.

Nevet, M. J., Shalev, S. A., Zlotogora, J., Mazzawi, N., & Ben-Yosef, T. (2010). Identification of a prevalent founder mutation in an Israeli Muslim Arab village confirms the role of PRCD in the aetiology of retinitis pigmentosa in humans. *Journal of Medical Genetics*, *47*(8), 533–537.

Nickell, S., Park, P. S., Baumeister, W., & Palczewski, K. (2007). Three-dimensional architecture of murine rod outer segments determined by cryoelectron tomography. *The Journal of Cell Biology*, *177*(5), 917–925.

Niu, S. L., Mitchell, D. C., & Litman, B. J. (2001). Optimization of receptor-G protein coupling by bilayer lipid composition II: Formation of metarhodopsin II-transducin complex. *The Journal of Biological Chemistry*, *276*(46), 42807–42811.

Norton, A. W., Hosier, S., Terew, J. M., Li, N., Dhingra, A., Vardi, N., et al. (2005). Evaluation of the 17-kDa prenyl-binding protein as a regulatory protein for phototransduction in retinal photoreceptors. *The Journal of Biological Chemistry*, *280*(2), 1248–1256.

Obata, S., & Usukura, J. (1992). Morphogenesis of the photoreceptor outer segment during postnatal development in the mouse (BALB/c) retina. *Cell and Tissue Research*, *269*(1), 39–48.

Okawa, H., Sampath, A. P., Laughlin, S. B., & Fain, G. L. (2008). ATP consumption by mammalian rod photoreceptors in darkness and in light. *Current Biology*, *18*(24), 1917–1921.

Omori, Y., Araki, F., Chaya, T., Kajimura, N., Irie, S., Terada, K., et al. (2012). Presynaptic dystroglycan-pikachurin complex regulates the proper synaptic connection between retinal photoreceptor and bipolar cells. *The Journal of Neuroscience*, *32*(18), 6126–6137.

Palczewski, K. (2012). Chemistry and biology of vision. *The Journal of Biological Chemistry*, *287*(3), 1612–1619.

Paquet-Durand, F., Beck, S., Michalakis, S., Goldmann, T., Huber, G., Muhlfriedel, R., et al. (2011). A key role for cyclic nucleotide gated (CNG) channels in cGMP-related retinitis pigmentosa. *Human Molecular Genetics*, *20*(5), 941–947.

Pazour, G. J., Baker, S. A., Deane, J. A., Cole, D. G., Dickert, B. L., Rosenbaum, J. L., et al. (2002). The intraflagellar transport protein, IFT88, is essential for vertebrate photoreceptor assembly and maintenance. *The Journal of Cell Biology*, *157*(1), 103–113.

Pearring, J. N., Salinas, R. Y., Baker, S. A., & Arshavsky, V. Y. (2013). Protein sorting, targeting and trafficking in photoreceptor cells. *Progress in Retinal and Eye Research*, *36*, 24–51.

Pelucchi, B., Grimaldi, A., & Moriondo, A. (2008). Vertebrate rod photoreceptors express both BK and IK calcium-activated potassium channels, but only BK channels are involved in receptor potential regulation. *Journal of Neuroscience Research*, *86*(1), 194–201.

Peng, C., Rich, E. D., & Varnum, M. D. (2004). Subunit configuration of heteromeric cone cyclic nucleotide-gated channels. *Neuron*, *42*(3), 401–410.

Peng, Y. W., Sharp, A. H., Snyder, S. H., & Yau, K. W. (1991). Localization of the inositol 1,4,5-trisphosphate receptor in synaptic terminals in the vertebrate retina. *Neuron*, *6*(4), 525–531.

Pilgram, G. S., Potikanond, S., Baines, R. A., Fradkin, L. G., & Noordermeer, J. N. (2010). The roles of the dystrophin-associated glycoprotein complex at the synapse. *Molecular Neurobiology*, *41*(1), 1–21.

Pollock, N. L., & Callaghan, R. (2011). The lipid translocase, ABCA4: Seeing is believing. *FEBS Journal*, *278*(18), 3204–3214.

Pow, D. V., & Barnett, N. L. (2000). Developmental expression of excitatory amino acid transporter 5: A photoreceptor and bipolar cell glutamate transporter in rat retina. *Neuroscience Letters*, *280*(1), 21–24.

Pugh, E. N., Jr., & Lamb, T. D. (1993). Amplification and kinetics of the activation steps in phototransduction. *Biochimica et Biophysica Acta*, *1141*(2–3), 111–149.

Quazi, F., Lenevich, S., & Molday, R. S. (2012). ABCA4 is an N-retinylidene-phosphatidylethanolamine and phosphatidylethanolamine importer. *Nature Communications*, *3*, 925.

Querques, G., Forte, R., & Souied, E. H. (2011). Retina and omega-3. *Journal of Nutrition and Metabolism*, 2011:748361, 1–12, .

Rabl, K., Bryson, E. J., & Thoreson, W. B. (2003). Activation of glutamate transporters in rods inhibits presynaptic calcium currents. *Visual Neuroscience*, *20*(5), 557–566.

Rattner, A., Smallwood, P. M., Williams, J., Cooke, C., Savchenko, A., Lyubarsky, A., et al. (2001). A photoreceptor-specific cadherin is essential for the structural integrity of the outer segment and for photoreceptor survival. *Neuron*, *32*(5), 775–786.

Rauen, T., Wiessner, M., Sullivan, R., Lee, A., & Pow, D. V. (2004). A new GLT1 splice variant: Cloning and immunolocalization of GLT1c in the mammalian retina and brain. *Neurochemistry International, 45*(7), 1095–1106.

Raven, M. A., Orton, N. C., Nassar, H., Williams, G. A., Stell, W. K., Jacobs, G. H., et al. (2008). Early afferent signaling in the outer plexiform layer regulates development of horizontal cell morphology. *The Journal of Comparative Neurology, 506*(5), 745–758.

Ravichandran, K. S. (2011). Beginnings of a good apoptotic meal: The find-me and eat-me signaling pathways. *Immunity, 35*(4), 445–455.

Rebrik, T. I., Botchkina, I., Arshavsky, V. Y., Craft, C. M., & Korenbrot, J. I. (2012). CNG-modulin: A novel Ca-dependent modulator of ligand sensitivity in cone photoreceptor cGMP-gated ion channels. *The Journal of Neuroscience, 32*(9), 3142–3153.

Reye, P., Sullivan, R., Fletcher, E. L., & Pow, D. V. (2002). Distribution of two splice variants of the glutamate transporter GLT1 in the retinas of humans, monkeys, rabbits, rats, cats, and chickens. *The Journal of Comparative Neurology, 445*(1), 1–12.

Reye, P., Sullivan, R., & Pow, D. V. (2002). Distribution of two splice variants of the glutamate transporter GLT-1 in the developing rat retina. *The Journal of Comparative Neurology, 447*(4), 323–330.

Riazuddin, S. A., Shahzadi, A., Zeitz, C., Ahmed, Z. M., Ayyagari, R., Chavali, V. R., et al. (2010). A mutation in SLC24A1 implicated in autosomal-recessive congenital stationary night blindness. *American Journal of Human Genetics, 87*(4), 523–531.

Rotstein, N. P., & Aveldano, M. I. (1987). Labeling of phosphatidylcholines of retina subcellular fractions by [1-14C]eicosatetraenoate (20:4(n-6)), docosapentaenoate (22:5(n-3)) and docosahexaenoate (22:6(n-3)). *Biochimica et Biophysica Acta, 921*(2), 235–244.

Rowan, M. J., Ripps, H., & Shen, W. (2010). Fast glutamate uptake via EAAT2 shapes the cone-mediated light offset response in bipolar cells. *The Journal of Physiology, 588*(Pt 20), 3943–3956.

Ruggiero, L., Connor, M. P., Chen, J., Langen, R., & Finnemann, S. C. (2012). Diurnal, localized exposure of phosphatidylserine by rod outer segment tips in wild-type but not Itgb5 −/− or Mfge8 −/− mouse retina. *Proceedings of the National Academy of Sciences of the United States of America, 109*(21), 8145–8148.

Salinas, R. Y., Baker, S. A., Gospe, S. M., 3rd., & Arshavsky, V. Y. (2013). A single valine residue plays an essential role in peripherin/rds targeting to photoreceptor outer segments. *PLoS One, 8*(1), e54292.

Sancho-Pelluz, J., Arango-Gonzalez, B., Kustermann, S., Romero, F. J., van Veen, T., Zrenner, E., et al. (2008). Photoreceptor cell death mechanisms in inherited retinal degeneration. *Molecular Neurobiology, 38*(3), 253–269.

Sanes, J. R., & Zipursky, S. L. (2010). Design principles of insect and vertebrate visual systems. *Neuron, 66*(1), 15–36.

Sanyal, S., De Ruiter, A., & Hawkins, R. K. (1980). Development and degeneration of retina in rds mutant mice: Light microscopy. *The Journal of Comparative Neurology, 194*(1), 193–207.

Sato, S., Omori, Y., Katoh, K., Kondo, M., Kanagawa, M., Miyata, K., et al. (2008). Pikachurin, a dystroglycan ligand, is essential for photoreceptor ribbon synapse formation. *Nature Neuroscience, 11*(8), 923–931.

Schmitz, F., Konigstorfer, A., & Sudhof, T. C. (2000). RIBEYE, a component of synaptic ribbons: A protein's journey through evolution provides insight into synaptic ribbon function. *Neuron, 28*(3), 857–872.

Schneider, B. G., Shyjan, A. W., & Levenson, R. (1991). Co-localization and polarized distribution of Na, K-ATPase alpha 3 and beta 2 subunits in photoreceptor cells. *The Journal of Histochemistry and Cytochemistry, 39*(4), 507–517.

Schnetkamp, P. P. (2013). The SLC24 gene family of Na(+)/Ca(2)(+)-K(+) exchangers: From sight and smell to memory consolidation and skin pigmentation. *Molecular Aspects of Medicine, 34*(2–3), 455–464.

Schon, C., Biel, M., & Michalakis, S. (2013). Gene replacement therapy for retinal CNG channelopathies. *Molecular Genetics and Genomics, 10*, 459–497.

Schroeder, B. C., Cheng, T., Jan, Y. N., & Jan, L. Y. (2008). Expression cloning of TMEM16A as a calcium-activated chloride channel subunit. *Cell, 134*(6), 1019–1029.

Schultz, Z. D. (2011). Raman spectroscopic imaging of cholesterol and docosahexaenoic acid distribution in the retinal rod outer segment. *Australian Journal of Chemistry, 64*(5), 611–616.

Sedmak, T., & Wolfrum, U. (2011). Intraflagellar transport proteins in ciliogenesis of photoreceptor cells. *Biology of the Cell, 103*(10), 449–466.

Seeliger, M. W., Brombas, A., Weiler, R., Humphries, P., Knop, G., Tanimoto, N., et al. (2011). Modulation of rod photoreceptor output by HCN1 channels is essential for regular mesopic cone vision. *Nature Communications, 2*, 532.

Shuart, N. G., Haitin, Y., Camp, S. S., Black, K. D., & Zagotta, W. N. (2011). Molecular mechanism for 3:1 subunit stoichiometry of rod cyclic nucleotide-gated ion channels. *Nature Communications, 2*, 457.

Skiba, N. A., Spencer, W. J., Salinas, R. Y., Lieu, E. C., Thompson, J. W., & Arshavsky, V. Y. (2013). Proteomic identification of unique photoreceptor disc components reveals the presence of PRCD, a protein linked to retinal degeneration. *Journal of Proteome Research, 12*(6), 3010–3018.

Specht, D., Wu, S. B., Turner, P., Dearden, P., Koentgen, F., Wolfrum, U., et al. (2009). Effects of presynaptic mutations on a postsynaptic Cacna1s calcium channel colocalized with mGluR6 at mouse photoreceptor ribbon synapses. *Investigative Ophthalmology & Visual Science, 50*(2), 505–515.

Spencer, M., Detwiler, P. B., & Bunt-Milam, A. H. (1988). Distribution of membrane proteins in mechanically dissociated retinal rods. *Investigative Ophthalmology & Visual Science, 29*(7), 1012–1020.

Stabach, P. R., Devarajan, P., Stankewich, M. C., Bannykh, S., & Morrow, J. S. (2008). Ankyrin facilitates intracellular trafficking of alpha1-Na+−K+−ATPase in polarized cells. *American Journal of Physiology Cell Physiology, 295*(5), C1202–C1214.

Stahl, W. L., & Baskin, D. G. (1984). Immunocytochemical localization of Na+, K+ adenosine triphosphatase in the rat retina. *The Journal of Histochemistry and Cytochemistry, 32*(2), 248–250.

Steinberg, R. H., Fisher, S. K., & Anderson, D. H. (1980). Disc morphogenesis in vertebrate photoreceptors. *The Journal of Comparative Neurology, 190*(3), 501–508.

Stillwell, W., & Wassall, S. R. (2003). Docosahexaenoic acid: Membrane properties of a unique fatty acid. *Chemistry and Physics of Lipids, 126*(1), 1–27.

Stohr, H., Heisig, J. B., Benz, P. M., Schoberl, S., Milenkovic, V. M., Strauss, O., et al. (2009). TMEM16B, a novel protein with calcium-dependent chloride channel activity, associates with a presynaptic protein complex in photoreceptor terminals. *The Journal of Neuroscience, 29*(21), 6809–6818.

Striessnig, J., Bolz, H. J., & Koschak, A. (2010). Channelopathies in Cav1.1, Cav1.3, and Cav1.4 voltage-gated L-type Ca2+ channels. *Pflügers Archiv, 460*(2), 361–374.

Strom, T. M., Nyakatura, G., Apfelstedt-Sylla, E., Hellebrand, H., Lorenz, B., Weber, B. H. F., et al. (1998). An L-type calcium-channel gene mutated in incomplete X-linked congenital stationary night blindness. *Nature Genetics, 19*, 260–263.

Sung, C. H., & Chuang, J. Z. (2010). The cell biology of vision. *The Journal of Cell Biology, 190*(6), 953–963.

Suryanarayanan, A., & Slaughter, M. M. (2006). Synaptic transmission mediated by internal calcium stores in rod photoreceptors. *The Journal of Neuroscience, 26*(6), 1759–1766.

Tam, B. M., Moritz, O. L., & Papermaster, D. S. (2004). The C terminus of peripherin/rds participates in rod outer segment targeting and alignment of disk incisures. *Molecular Biology of the Cell, 15*(4), 2027–2037.

Tantri, A., Vrabec, T. R., Cu-Unjieng, A., Frost, A., Annesley, W. H., Jr., & Donoso, L. A. (2004). X-linked retinoschisis: A clinical and molecular genetic review. *Survey of Ophthalmology, 49*(2), 214–230.

Taylor, W. R., & Morgans, C. (1998). Localization and properties of voltage-gated calcium channels in cone photoreceptors of Tupaia belangeri. *Visual Neuroscience, 15*(3), 541–552.

Thoreson, W. B., Nitzan, R., & Miller, R. F. (1997). Reducing extracellular Cl- suppresses dihydropyridine-sensitive Ca2+ currents and synaptic transmission in amphibian photoreceptors. *Journal of Neurophysiology, 77*(4), 2175–2190.

Thoreson, W. B., Nitzan, R., & Miller, R. F. (2000). Chloride efflux inhibits single calcium channel open probability in vertebrate photoreceptors: Chloride imaging and cell-attached patch-clamp recordings. *Visual Neuroscience, 17*(2), 197–206.

Thoreson, W. B., & Stella, S. L. (2000). Anion modulation of calcium current voltage dependence and amplitude in salamander rods. *Biochimica et Biophysica Acta, 1464*(1), 142–150.

tom Dieck, S., & Brandstatter, J. H. (2006). Ribbon synapses of the retina. *Cell and Tissue Research, 326*(2), 339–346.

Trudeau, M. C., & Zagotta, W. N. (2002). An intersubunit interaction regulates trafficking of rod cyclic nucleotide-gated channels and is disrupted in an inherited form of blindness. *Neuron, 34*(2), 197–207.

Tsybovsky, Y., Molday, R. S., & Palczewski, K. (2010). The ATP-binding cassette transporter ABCA4: Structural and functional properties and role in retinal disease. *Advances in Experimental Medicine and Biology, 703*, 105–125.

Ungar, F., Piscopo, I., & Holtzman, E. (1981). Calcium accumulation in intracellular compartments of frog retinal rod photoreceptors. *Brain Research, 205*(1), 200–206.

van der Velden, L. M., Wichers, C. G., van Breevoort, A. E., Coleman, J. A., Molday, R. S., Berger, R., et al. (2010). Heteromeric interactions required for abundance and subcellular localization of human CDC50 proteins and class 1 P4-ATPases. *The Journal of Biological Chemistry, 285*(51), 40088–40096.

Vijayasarathy, C., Ziccardi, L., & Sieving, P. A. (2012). Biology of retinoschisin. *Advances in Experimental Medicine and Biology, 723*, 513–518.

Wan, Q. F., Nixon, E., & Heidelberger, R. (2012). Regulation of presynaptic calcium in a mammalian synaptic terminal. *Journal of Neurophysiology, 108*(11), 3059–3067.

Weitz, D., Ficek, N., Kremmer, E., Bauer, P. J., & Kaupp, U. B. (2002). Subunit stoichiometry of the CNG channel of rod photoreceptors. *Neuron, 36*(5), 881–889.

Wen, X. H., Shen, L., Brush, R. S., Michaud, N., Al-Ubaidi, M. R., Gurevich, V. V., et al. (2009). Overexpression of rhodopsin alters the structure and photoresponse of rod photoreceptors. *Biophysical Journal, 96*(3), 939–950.

Westerfeld, C., & Mukai, S. (2008). Stargardt's disease and the ABCR gene. *Seminars in Ophthalmology, 23*(1), 59–65.

Wetzel, R. K., Arystarkhova, E., & Sweadner, K. J. (1999). Cellular and subcellular specification of Na, K-ATPase alpha and beta isoforms in the postnatal development of mouse retina. *The Journal of Neuroscience, 19*(22), 9878–9889.

Winkler, B. S., Kapousta-Bruneau, N., Arnold, M. J., & Green, D. G. (1999). Effects of inhibiting glutamine synthetase and blocking glutamate uptake on b-wave generation in the isolated rat retina. *Visual Neuroscience, 16*(2), 345–353.

Witkovsky, P. (2004). Dopamine and retinal function. *Documenta Ophthalmologica Advances in Ophthalmology, 108*(1), 17–40.

Wu, H., Cowing, J. A., Michaelides, M., Wilkie, S. E., Jeffery, G., Jenkins, S. A., et al. (2006). Mutations in the gene KCNV2 encoding a voltage-gated potassium channel subunit cause "cone dystrophy with supernormal rod electroretinogram" in humans. *American Journal of Human Genetics*, *79*(3), 574–579.

Wu, J., Marmorstein, A. D., Striessnig, J., & Peachey, N. S. (2007). Voltage-dependent calcium channel CaV1.3 subunits regulate the light peak of the electroretinogram. *Journal of Neurophysiology*, *97*(5), 3731–3735.

Wu, W. W., Wong, J. P., Kast, J., & Molday, R. S. (2005). RS1, a discoidin domain-containing retinal cell adhesion protein associated with X-linked retinoschisis, exists as a novel disulfide-linked octamer. *The Journal of Biological Chemistry*, *280*(11), 10721–10730.

Wutz, K., Sauer, C., Zrenner, E., Lorenz, B., Alitalo, T., Broghammer, M., et al. (2002). Thirty distinct CACNA1F mutations in 33 families with incomplete type of XLCSNB and Cacna1f expression profiling in mouse retina. *European Journal of Human Genetics*, *10*(8), 449–456.

Wycisk, K. A., Budde, B., Feil, S., Skosyrski, S., Buzzi, F., Neidhardt, J., et al. (2006). Structural and functional abnormalities of retinal ribbon synapses due to Cacna2d4 mutation. *Investigative Ophthalmology & Visual Science*, *47*(8), 3523–3530.

Xu, J. W., & Slaughter, M. M. (2005). Large-conductance calcium-activated potassium channels facilitate transmitter release in salamander rod synapse. *The Journal of Neuroscience*, *25*(33), 7660–7668.

Yamagata, M., & Sanes, J. R. (2010). Synaptic localization and function of Sidekick recognition molecules require MAGI scaffolding proteins. *The Journal of Neuroscience*, *30*(10), 3579–3588.

Yang, Z., Chen, Y., Lillo, C., Chien, J., Yu, Z., Michaelides, M., et al. (2008). Mutant prominin 1 found in patients with macular degeneration disrupts photoreceptor disk morphogenesis in mice. *The Journal of Clinical Investigation*, *118*(8), 2908–2916.

Yang, Y. D., Cho, H., Koo, J. Y., Tak, M. H., Cho, Y., Shim, W. S., et al. (2008). TMEM16A confers receptor-activated calcium-dependent chloride conductance. *Nature*, *455*(7217), 1210–1215.

Yang, J., Pawlyk, B., Wen, X. H., Adamian, M., Soloviev, M., Michaud, N., et al. (2007). Mpp4 is required for proper localization of plasma membrane calcium ATPases and maintenance of calcium homeostasis at the rod photoreceptor synaptic terminals. *Human Molecular Genetics*, *16*(9), 1017–1029.

Yau, K. W., & Hardie, R. C. (2009). Phototransduction motifs and variations. *Cell*, *139*(2), 246–264.

Young, R. W. (1967). The renewal of photoreceptor cell outer segments. *The Journal of Cell Biology*, *33*(1), 61–72.

Zabouri, N., & Haverkamp, S. (2013). Calcium channel-dependent molecular maturation of photoreceptor synapses. *PLoS One*, *8*(5), e63853.

Zacchigna, S., Oh, H., Wilsch-Brauninger, M., Missol-Kolka, E., Jaszai, J., Jansen, S., et al. (2009). Loss of the cholesterol-binding protein prominin-1/CD133 causes disk dysmorphogenesis and photoreceptor degeneration. *The Journal of Neuroscience*, *29*(7), 2297–2308.

Zangerl, B., Goldstein, O., Philp, A. R., Lindauer, S. J., Pearce-Kelling, S. E., Mullins, R. F., et al. (2006). Identical mutation in a novel retinal gene causes progressive rod-cone degeneration in dogs and retinitis pigmentosa in humans. *Genomics*, *88*(5), 551–563.

Zhang, K., Kniazeva, M., Han, M., Li, W., Yu, Z., Yang, Z., et al. (2001). A 5-bp deletion in ELOVL4 is associated with two related forms of autosomal dominant macular dystrophy. *Nature Genetics*, *27*(1), 89–93.

Zhang, H., Li, S., Doan, T., Rieke, F., Detwiler, P. B., Frederick, J. M., et al. (2007). Deletion of PrBP/delta impedes transport of GRK1 and PDE6 catalytic subunits to photoreceptor outer segments. *Proceedings of the National Academy of Sciences of the United States of America*, *104*(21), 8857–8862.

Zhang, Y., Molday, L. L., Molday, R. S., Sarfare, S. S., Woodruff, M. L., Fain, G. L., et al. (2009). Knockout of GARPs and the beta-subunit of the rod cGMP-gated channel disrupts disk morphogenesis and rod outer segment structural integrity. *Journal of Cell Science, 122*(Pt 8), 1192–1200.

Zhang, X., Yang, D., & Hughes, B. A. (2011). KCNQ5/K(v)7.5 potassium channel expression and subcellular localization in primate retinal pigment epithelium and neural retina. *American Journal of Physiology Cell Physiology, 301*(5), C1017–C1026.

Zhao, Y., Hong, D. H., Pawlyk, B., Yue, G., Adamian, M., Grynberg, M., et al. (2003). The retinitis pigmentosa GTPase regulator (RPGR)- interacting protein: Subserving RPGR function and participating in disk morphogenesis. *Proceedings of the National Academy of Sciences of the United States of America, 100*(7), 3965–3970.

Zheng, J., Trudeau, M. C., & Zagotta, W. N. (2002). Rod cyclic nucleotide-gated channels have a stoichiometry of three CNGA1 subunits and one CNGB1 subunit. *Neuron, 36*(5), 891–896.

Zhong, H., Molday, L. L., Molday, R. S., & Yau, K. W. (2002). The heteromeric cyclic nucleotide-gated channel adopts a 3A:1B stoichiometry. *Nature, 420*(6912), 193–198.

CHAPTER EIGHT

The Evolutionary Origin of Epithelial Cell–Cell Adhesion Mechanisms

Phillip W. Miller[*,1], **Donald N. Clarke**[†,1], **William I. Weis**[*,‡], **Christopher J. Lowe**[†], **W. James Nelson**[*,†,2]
[*]Department of Molecular and Cellular Physiology, Stanford University School of Medicine, Stanford, California, USA
[†]Department of Biology, Stanford University, Stanford, California, USA
[‡]Department of Structural Biology, Stanford University School of Medicine, Stanford, California, USA
[2]Corresponding author: e-mail address: wjnelson@stanford.edu

Contents

Abstract

A simple epithelium forms a barrier between the outside and the inside of an organism, and is the first organized multicellular tissue found in evolution. We examine the relationship between the evolution of epithelia and specialized cell–cell adhesion proteins

[1] These authors contributed equally to this work.

Current Topics in Membranes, Volume 72
ISSN 1063-5823
http://dx.doi.org/10.1016/B978-0-12-417027-8.00008-8

comprising the classical cadherin/β-catenin/α-catenin complex (CCC). A review of the divergent functional properties of the CCC in metazoans and non-metazoans, and an updated phylogenetic coverage of the CCC using recent genomic data reveal: (1) The core CCC likely originated before the last common ancestor of unikonts and their closest bikont sister taxa. (2) Formation of the CCC may have constrained sequence evolution of the classical cadherin cytoplasmic domain and β-catenin in metazoa. (3) The α-catenin-binding domain in β-catenin appears to be the favored mutation site for disrupting β-catenin function in the CCC. (4) The ancestral function of the α/β-catenin heterodimer appears to be an actin-binding module. In some metazoan groups, more complex functions of α-catenin were gained by sequence divergence in the non-actin-binding (N-, M-) domains. (5) Allosteric regulation of α-catenin may have evolved for more complex regulation of the actin cytoskeleton.

1. INTRODUCTION

A simple epithelium is a conserved feature of all metazoans and is essential for organized multicellularity. It is comprised of a closed monolayer, often a tube, of polarized cells that surround a luminal space (Fig. 8.1A), thus separating the inside of the organism from its surrounding environment. The cytoskeleton, cytoplasmic organelles, and plasma membrane domains are organized asymmetrically, with the apical plasma membrane facing the luminal space and the basolateral membrane contacting opposing cells and an extracellular matrix (ECM) (Bryant & Mostov, 2008; Gumbiner, 2005; Nelson, Dickinson, & Weis, 2013). Cell–cell adhesion complexes hold epithelial cells together, and an ECM surrounds epithelial tubes. Disruptions in epithelial polarity and cell–cell adhesion cause developmental defects and are found in diseases in adult tissues (Benjamin & Nelson, 2008; Bullions, Notterman, Chung, & Levine, 1997; Kane et al., 1996; Larue et al., 1996; Larue, Ohsugi, Hirchenhain, & Kemler, 1994; Marchiando, Graham, & Turner, 2010; Stepniak, Radice, & Vasioukhin, 2009; Torres et al., 1997; Watabe, Nagafuchi, Tsukita, & Takeichi, 1994).

Simple epithelia predate the origin of metazoans and are the first organized tissues found in evolution. They are present in two distinct lineages of unikonts: the amoebozoans (*Dictyostelium* slime molds and their relatives) and the opisthokonts (eukaryotes) that include metazoans, choanoflagellates, and fungi, which are thought to have developed multicellularity independently. First, the amoebozoan *Dictyostelium discoideum* requires a polarized tip epithelium to form a fruiting body from aggregated amoebae

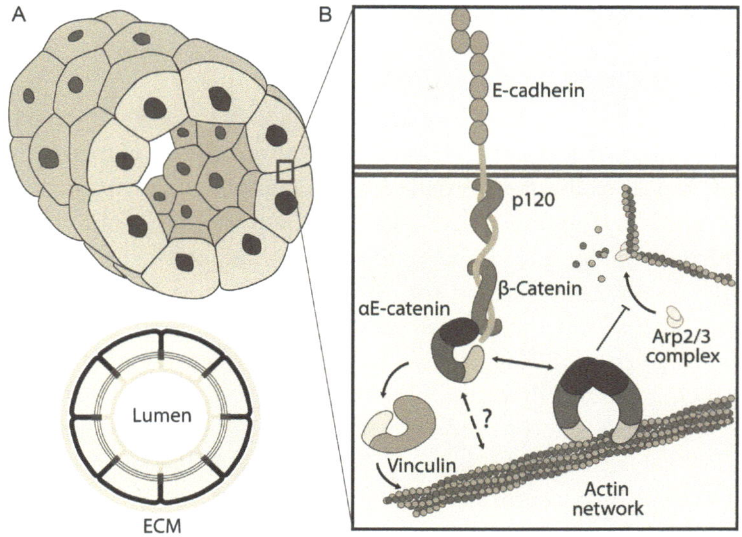

Figure 8.1 *Cadherin/catenin complex at mammalian cell–cell contacts.* (A) A simple, tubular epithelium composed of a closed monolayer of polarized cells. The schematic below demonstrates cell polarity: the apical membrane (yellow) faces the lumen and the basolateral membrane (blue) contacts the underlying ECM and serosa (gray). Cells have a cortical actin belt, which is connected between cells at adherens junctions (red). (B) The cadherin–catenin complex mediates cell–cell adhesion. Classical cadherins contain extracellular repeat domains that mediate trans-interactions with the extracellular domain of cadherins on opposing cells, and a cytoplasmic domain that binds p120 and β-catenin. Monomeric α-catenin localizes to the cadherin–catenin complex by binding β-catenin. The mechanism by which α-catenin associates with the actin cytoskeleton is not well understood. Vinculin binds both α-catenin and actin and may facilitate linkage between the cadherin–catenin complex and actin. Homodimeric α-catenin binds and bundles actin filaments and inhibits Arp2/3 complex-mediated nucleation of actin. (See color plate.)

(Dickinson, Nelson, & Weis, 2011). Second, simple epithelia constitute the core tissues of all metazoans: the feeding chambers of porifera (sponges) are lined with an epithelium (choanoderm) comprising polarized choanocytes that directionally absorb nutrients from seawater (Leys & Hill, 2012), and the placozoan *Trichoplax adhaerens* consists of several thousand cells arranged in an epithelial bilayer of which the ventral layer is required for nutrient absorption (Schierwater, de Jong, & Desalle, 2009; Srivastava et al., 2008). In morphologically complex animals, such as mammals, epithelia define tissue architecture and regulate functionally diverse organs such as the lung, gut, kidney, and epidermis. Thus, formation of a simple polarized epithelium is a principal requirement for the evolution of organized multicellularity and the functional diversification of tissues (Cereijido, Contreras, & Shoshani, 2004).

Classical cadherins are the primary molecules that mediate epithelial cell–cell adhesion in metazoans (Halbleib & Nelson, 2006; Harris & Tepass, 2010). Classical cadherins are defined by a cytoplasmic domain that binds adaptor proteins (catenins) that regulate downstream signaling and actin cytoskeleton dynamics. Classical cadherins establish cell–cell contacts, often in a discrete structure termed the Adherens Junction (AJ) located at the boundary between the apical and basolateral membrane domains (Fig. 8.1A) (Nelson, 2003). AJs are linked to a circumferential actomyosin belt, which generates dynamic forces important for epithelial sheet remodeling and tissue morphogenesis and epithelial tissue integrity (Costa et al., 1998; Nagafuchi, Ishihara, & Tsukita, 1994; Wessells et al., 1971). During cell–cell contact formation, the actin and microtubule cytoskeletons are remodeled to mechanically strengthen contacts, facilitate polarized vesicle trafficking, and maintain cell shape (Adams & Nelson, 1998; Mellman & Nelson, 2008; Nejsum & Nelson, 2007).

The evolution of cadherin-mediated cell–cell adhesion coincided with the formation of different body plans derived from epithelial sheets. Members of the cadherin and catenin protein families (the cadherin–catenin complex, CCC) are present in all metazoans and many premetazoan unikonts (Abedin & King, 2008; Hulpiau, Gul, & van Roy, 2013; Hulpiau & van Roy, 2009; Oda & Takeichi, 2011) (this study). Non-metazoan lineages do not possess a complete set of CCC protein orthologs, and the ancestral function of these cell–cell adhesion proteins in unicellular organisms is unclear. Furthermore, recent functional studies demonstrate divergent properties of the CCC within bilaterians (Desai et al., 2013; Dickinson, Weis, & Nelson, 2011; Drees, Pokutta, Yamada, Nelson, & Weis, 2005; Kwiatkowski et al., 2010; Miller et al., 2013).

The CCC mechanically couples neighboring cells by *trans* interactions between cadherins on opposing cells and linkage to the underlying actin cytoskeletons (Huveneers & de Rooij, 2013; Shapiro & Weis, 2009). In general, the cadherin protein superfamily consists of transmembrane proteins that contain extracellular cadherin repeat domains (CADs) (Boggon et al., 2002; Shapiro & Weis, 2009). Adhesive contacts between classical cadherins require extracellular Ca^{2+} to maintain protein conformation (Koch, Pokutta, Lustig, & Engel, 1997), and are mediated by a strand swap dimer formed between the opposed N-terminal EC1 domains

(Harrison et al., 2011). Classical cadherins have a conserved cytoplasmic domain that binds catenins (Huber & Weis, 2001; Hulpiau & van Roy, 2009). The catenins are responsible for transducing force and molecular signals from the CCC to the actin cytoskeleton (Borghi et al., 2012; Weis & Nelson, 2006; Yonemura, Wada, Watanabe, Nagafuchi, & Shibata, 2010). The armadillo repeat family proteins p120-catenin and β-catenin bind directly to the cadherin cytoplasmic domain; p120-catenin interacts with Rho GTPases that control cytoskeletal dynamics and regulates cadherin endocytosis (for recent reviews see Davis, Ireton, & Reynolds, 2003; Pieters, van Roy, & van Hengel, 2012) and will not be discussed further (for recent reviews see Carnahan, Rokas, Gaucher, & Reynolds, 2010; Menke & Giehl, 2012; Pieters et al., 2012). In turn, β-catenin binds to α-catenin (Herrenknecht et al., 1991) thereby forming the core cytoplasmic protein complex of the CCC (Figs. 8.1B and 8.2A).

α-Catenin is an F (filamentous)-actin-binding protein (Rimm, Koslov, Kebriaei, Cianci, & Morrow, 1995), and is a key protein in the CCC that links cadherin-mediated cell–cell contacts to the underlying actin cytoskeleton. α-Catenin is a paralog of vinculin, which is an F-actin-binding protein at cell–ECM and cell–cell adhesions (Peng, Nelson, Maiers, & DeMali, 2011); we refer to α-catenin/vinculin proteins as VIN-family proteins. Mammalian αE-catenin is composed of a series of four-helix bundles connected to a C-terminal five-helix bundle, and the conformation and accessibility of these domains regulate αE-catenin function (Choi et al., 2012; Ishiyama et al., 2013; Pokutta, Drees, Takai, Nelson, & Weis, 2002; Pokutta & Weis, 2000; Rangarajan & Izard, 2013; Yang, Dokurno, Tonks, & Barford, 2001) (Fig. 8.2A). Mammalian αE-catenin has binding sites for β-catenin and F-actin in the N-terminal and C-terminal domains, respectively (Fig. 8.2A). Mammalian αE-catenin also binds several F-actin-binding proteins, including vinculin (Watabe-Uchida et al., 1998), α-actinin (Knudsen, Soler, Johnson, & Wheelock, 1995), ZO-1 (Itoh, Nagafuchi, Moroi, & Tsukita, 1997; Maiers, Peng, Fanning, & Demali, 2013), l-afadin (Pokutta et al., 2002), and EPLIN (Abe & Takeichi, 2008); whether non-mammalian α-catenin orthologs bind these proteins has not been studied. Mammalian αE-catenin bundles actin filaments (Rimm et al., 1995), inhibits Arp2/3-mediated nucleation of actin filament assembly (Drees et al., 2005), and cofilin severing of actin (Hansen et al., 2013) (Fig. 8.1B).

Here, we focus on evolution of the core components of the CCC—classical cadherins, β-catenin and α-catenin—with particular focus on α-catenin. The evolution of classical cadherins and β-catenin has been

A

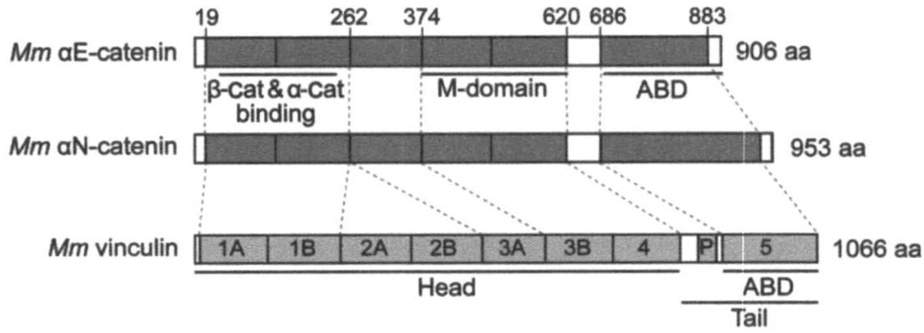

B

Homolog	Dimerization	β-Cat binding	Actin binding	Actin regulation	*In vivo* requirement
Mammalian αE-catenin	Yes	Yes	Homodimer	Bundles actin; inhibits Arp2/3 and cofilin	Adhesion in blastula; epidermal formation
Mammalian αN-catenin	No	Yes	Monomer	?	Neuronal cell migration
Mammalian Vinculin	No	Yes	Autoinhibited	Tail domain bundles actin as a dimer	Heart and brain development beyond E8 mice
G. gallus αN-catenin	?	IP	IP	?	?
G. gallus Vinculin	No	?	Autoinhibited	?	?
X. laevis αE-catenin	?	IP	?	?	Cell–cell adhesion at blastula stage
X. laevis αN-catenin	?	IP	?	?	Gastrulation and neural plate closure
D. rerio αE-catenin	No	Yes	Monomer and α β-cat Heterodimer	Bundles actin	Epiboly and gastrulation
D. melanogaster α-catenin	Yes	IP	Monomer and homodimer	?	Embryogenesis, imaginal discs, oogenesis
C. elegans HMP-1	No	Yes	Autoinhibited	Bundles actin *In vivo*	Ventral closure and body elongation
D. discoideum α-catenin	No	Yes	Monomer	Bundles actin	Fruiting body formation

Figure 8.2 *Functional properties of α-catenin/vinculin family proteins.* (A) Domain organization of mammalian α-catenin/vinculin family proteins. *Mammalian* vinculin is composed of seven four-helix bundles, a proline-rich hinge region, and a C-terminal five-helix bundle. α-Catenins share a similar structure but lack the D2 domain. Head, tail, and actin-binding domains of vinculin as well as β-catenin binding/dimerization, modulation, and F-actin-binding domains in *Mm* αE- and αN-catenin are annotated. Regions of homology are indicated by dashed lines. (B) Functional properties of characterized α-catenin/vinculin family proteins across unikonta. Homodimerization, β-catenin binding, and F-actin binding and regulation *in vitro* using purified proteins is indicated. Indirect evidence of binding by co-immunoprecipitation (IP) is also noted. Developmental and/or *in vivo* requirement for each homolog is summarized on the right. Question marks signify untested properties or inconclusive data. *Panel (A) adapted from Miller et al. (2013).* (See color plate.)

reviewed (Abedin & King, 2008; Hulpiau et al., 2013; Hulpiau & van Roy, 2009; Schneider, Finnerty, & Martindale, 2003). First, we synthesize recent structure–function studies of α-catenin/vinculin family proteins across unikonta (Section 2). Second, we use bioinformatic analysis to identify putative orthologs of the core CCC components by sequence alignment and domain architecture (Section 3). Finally, we combine information about the divergence of α-catenin/vinculin function with bioinformatic observations to provide new insights into how the CCC may have evolved (Section 4).

2. FUNCTIONAL ANALYSIS OF CCC EVOLUTION

2.1. Functional characterization of the α-catenin/vinculin family across unikonta

2.1.1 Mammals

Mammals possess three isoforms of α-catenin, termed αE-, αN-, and αT-catenin, which originated from the same ancestral gene and share the same location on human chromosome 10 (Janssens et al., 2003). αE-, αN-, and αT-catenin are expressed predominantly, but not exclusively, in epithelia, neurons, and heart/testis, respectively (Herrenknecht et al., 1991; Shapiro & Weis, 2009; Uchida et al., 1994). Mammals also express vinculin, which localizes to both integrin-ECM adhesions and AJs (le Duc et al., 2010; Peng et al., 2011; Watabe-Uchida et al., 1998; Ziegler, Liddington, & Critchley, 2006). The functional properties of mammalian αE-catenin have been characterized in detail (Figs. 8.1B and 8.2B; see below). Less is known about αN-catenin, and we are not aware of any detailed biochemical analysis of αT-catenin.

αE-Catenin is essential for the formation of epithelia and morphogenetic cell movements during mammalian development. *Mus musculus* null mutants of αE-catenin (*Mm* αE-catenin) lose cell–cell adhesion in the trophoblast epithelium of the preimplantation embryo, and development is arrested at the blastula stage (Torres et al., 1997). Conditional knockout of *Mm* αE-catenin in keratinocytes disrupts AJ formation and is embryonic lethal due to epithelial hyperproliferation and tumor formation (Vasioukhin, Bauer, Degenstein, Wise, & Fuchs, 2001).

Mm αE-catenin contains three proteolytically defined domains (Fig. 8.2A) homologous to the domain structure of vinculin: an N-terminal domain (NTD) that mediates homodimerization and β-catenin binding, a conformationally flexible M-domain that interacts with several actin-binding proteins (see above), and a C-terminal domain that binds F-actin (ABD) (Aberle, Schwartz, Hoschuetzky, & Kemler, 1996; Herrenknecht et al., 1991; Pokutta & Weis, 2000; Rangarajan & Izard, 2013; Yang et al., 2001).

Mm αE-catenin exists in three oligomeric states: a monomer, an αE-catenin/β-catenin heterodimer, and a homodimer (Drees et al., 2005)

(Fig. 8.2B). All of these oligomeric forms are found in cell extracts from MDCK epithelial cells (Benjamin et al., 2010). The K_d for binding between β-catenin and *Mm* αE-catenin is 25–100 nM (Koslov, Maupin, Pradhan, Morrow, & Rimm, 1997) (Fig. 8.2B; S. Pokutta & W.I.W., unpublished data). The K_d for *Mm* αE-catenin homodimerization is weaker than for αE-catenin–β-catenin binding and is in the single micromolar range (Shapiro & Weis, 2009) (S. Pokutta & W.I.W, unpublished data).

Mammalian αE-catenin binds F-actin (Rimm et al., 1995), but this interaction is regulated by αE-catenin conformation (Fig. 8.2B) (Drees et al., 2005). The *Mm* αE-catenin monomer binds F-actin weakly (Drees et al., 2005; Yamada, Pokutta, Drees, Weis, & Nelson, 2005), whereas homodimerization potentiates actin binding (K_d 0.3 μM) (Drees et al., 2005; Rangarajan & Izard, 2013; Rimm et al., 1995). Since the αE-catenin homodimerization and β-catenin-binding sites overlap (Pokutta & Weis, 2000), αE-catenin can bind β-catenin as a monomer or F-actin as a homodimer, but the E-cadherin/β-catenin/αE-catenin complex binds F-actin weakly in bulk assays *in vitro* (Yamada et al., 2005) (Fig. 8.2A).

The weak binding of *Mm* CCC to F-actin led to a model of αE-catenin function in which clustering of the CCC at cell–cell contacts would produce a high local concentration of actin-binding αE-catenin homodimers, thereby facilitating dynamic interactions between the CCC and actin (Yamada et al., 2005) (Fig. 8.3C). Recent work, however, demonstrated that the CCC is under constitutive actomyosin-generated tension (Borghi et al., 2012), which could regulate αE-catenin conformation at cell–cell adhesions (Yonemura et al., 2010). Thus, an alternative model is that cytoplasmic forces relieve αE-catenin autoinhibition such that the αE-catenin monomer can directly couple the CCC to the cortical actin cytoskeleton. Further work is needed to test this model and define how the CCC interacts with the actin cytoskeleton.

Mm αE-catenin also regulates actin dynamics independently of its role in the CCC. *Mm* αE-catenin homodimers bundle actin filaments (Rimm et al., 1995), and inhibit Arp2/3 complex-mediated nucleation of F-actin and cofilin severing of actin filaments *in vitro* (Drees et al., 2005; Hansen et al., 2013) (Figs. 8.2B and 8.3C). Regulation of actin dynamics by cytoplasmic αE-catenin *in vivo* is important: depletion of the cytosolic pool of αE-catenin homodimers increases actin-dependent membrane dynamics and cell-migration rate of MDCK epithelial cells (Benjamin et al., 2010), and deletion of αE-catenin from keratinocytes *in M. musculus* embryos results in a hyper-migratory phenotype (Vasioukhin et al., 2001).

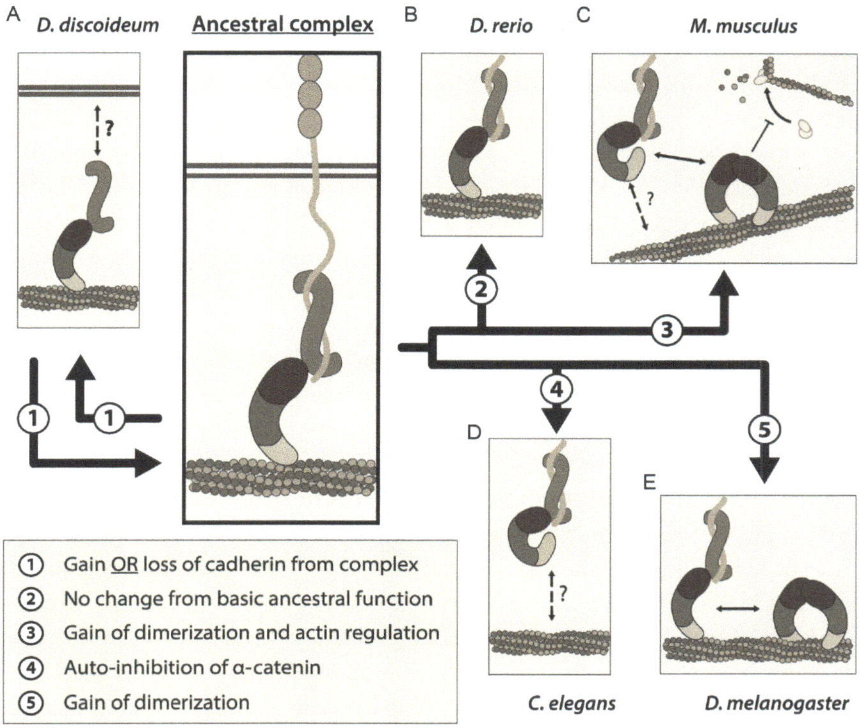

Figure 8.3 *Functional divergence of cadherin–catenin complex.* (A) We propose that ancestral cadherin–catenin complex interactions consisted of monomeric actin-binding α-catenin that coupled the classical cadherin–β-catenin complex to the cortical actin network (center panel). Divergence of the complex in each species from the ancestral complex is depicted in the box on bottom left. *D. discoideum* α-catenin is a monomer that forms a heterodimer complex with Aardvark (β-catenin homolog), and localizes to cell–cell contacts and the cortical actin cytoskeleton. *Dd* α-catenin bundles actin, but the mechanism by which the *Dd* α-catenin–Aardvark complex associates with the cell membrane is not known. *D. discoideum* does not possess cadherin homologs and the adhesion proteins to which the heterodimer complex is attached at cell–cell contacts have not been identified. (B) *D. rerio* αE-catenin is a monomer that is not autoinhibited, and can simultaneously bind β-catenin and F-actin, but does not regulate actin dynamics via Arp2/3 inhibition. (C) The mammalian cadherin–catenin complex possesses divergent regulatory properties from the ancestral complex. Monomeric αE-catenin associates with the E-cadherin–β-catenin complex, but is allosterically regulated and binds actin weakly. The mechanism by which the cadherin–catenin complex is linked to the cortical actin cytoskeleton is poorly understood. Homodimerization of αE-catenin potentiates bundling of actin and inhibition of Arp2/3 complex-mediated actin polymerization. (D) *C. elegans* HMP-1 forms a ternary complex with HMR-1 (classical cadherin homolog) and HMP-2 (β-catenin homolog), but is autoinhibited and cannot bind F-actin *in vitro*. The mechanism by which actin binding is activated *in vivo* is not known. (E) *D. melanogaster* α-catenin exists as monomer and homodimer species, both of which bind actin. *D. melanogaster* α-catenin forms a ternary complex with DE-cadherin and Armadillo. (See color plate.)

Less is known about the functional and biochemical properties of αN-catenin. Two splice variants of αN-catenin have been identified (Uchida et al., 1994), but their significance is unknown. *M. musculus* null mutants of αN-catenin have defects in proper cell layering in the cerebellum and hippocampus (Park, Falls, Finger, Longo-Guess, & Ackerman, 2002). This phenotype may be due to defects in neuronal cell migration and cell–cell contacts, since depletion of *Mm* αN-catenin from dendritic spines increases membrane activity while αN-catenin over-expression reduces membrane dynamics (Abe, Chisaka, Van Roy, & Takeichi, 2004). Although *Mm* αN-catenin appears to regulate cell–cell adhesion and cell migration like αE-catenin, initial studies indicate that it has different biochemical properties. Multi-angle light scattering and small-angle X-ray scattering show that *Mm* αN-catenin is monomeric with a less compact conformation than *Mm* αE-catenin (Ishiyama et al., 2013). However, it remains unclear if *Mm* αN-catenin homodimerizes at higher concentrations or whether it is allosterically regulated like *Mm* αE-catenin. *Mm* αN-catenin binds β-catenin and F-actin (Ishiyama et al., 2013) (S. Pokutta & WIW, unpublished data), but it is not known whether it binds simultaneously to β-catenin and the actin cytoskeleton. Therefore, there are insufficient data to propose a model of *Mm* αN-catenin function in cadherin-mediated cell–cell adhesion.

The function of mammalian vinculin in cadherin-mediated adhesion is not well understood, compared to its well-known roles in integrin-mediated cell adhesion to the ECM (Bakolitsa et al., 2004; Plotnikov, Pasapera, Sabass, & Waterman, 2012; Ziegler et al., 2006). Vinculin knockout mice exhibit gross defects in neural tube closure and heart development, and reduced cell-substrate adhesion to the ECM, but no specific defects in AJ formation or cell–cell contacts were described (Xu, Baribault, & Adamson, 1998) (Fig. 8.2B). Nevertheless, vinculin has biochemical properties similar to *Mm* αE-catenin. The N-terminal head domain of mammalian vinculin has binding sites for β-catenin and αE-catenin, and the C-terminal tail domain binds actin (Bakolitsa, de Pereda, Bagshaw, Critchley, & Liddington, 1999; Craig & Johnson, 1996; Janssen et al., 2006; Peng, Cuff, Lawton, & DeMali, 2010) (Fig. 8.2). Vinculin can bundle actin filaments through homodimerization of the C-terminal tail domain (Janssen et al., 2006), but the full-length protein is autoinhibited by a strong intramolecular head–tail interaction (Bakolitsa et al., 2004; Choi et al., 2012; Rangarajan & Izard, 2012; Ziegler et al., 2006) (Fig. 8.2B). It has been proposed that vinculin mediates binding between αE-catenin at the CCC and the cortical actin cytoskeleton (Huveneers & de Rooij, 2013;

Peng et al., 2011; Yonemura et al., 2010), and *Mm* αE-catenin can activate vinculin binding to actin *in vitro* (Choi et al., 2012). Further work is needed to define how αE-catenin and vinculin autoinhibition is relieved, and how these paralogs bind the CCC to the actin cytoskeleton *in vivo*.

2.1.2 Gallus gallus

Like mammals and other amniotes, *G. gallus* (chicken) expresses αE-, αN-, and αT-catenin isoforms, and a vinculin. There are no published functional or *in vivo* studies of *Gg* αE- or αT-catenin. *Gg* αN-catenin and vinculin have been used in place of their mammalian orthologs in several studies and exhibit conserved functional properties. *Gg* αN-catenin co-immunoprecipitates with β-catenin and actin (Fig. 8.2B) (Hirano & Takeichi, 1994) and can rescue cadherin-mediated cell–cell adhesion in human lung adenocarcinoma cells, which express E-cadherin and β-catenin but not α-catenin (Hirano & Takeichi, 1994). Functional properties of *Gg* vinculin are also conserved with mammals: *Gg* vinculin binds αE-catenin and actin *in vitro*, and is autoinhibited by strong head–tail domain interactions (Fig. 8.2B) (Chen, Choudhury, & Craig, 2006; Choi et al., 2012).

2.1.3 Xenopus laevis

Amphibians and other anamniote vertebrates possess two isoforms of α-catenin, αE-catenin and αN-catenin, and a vinculin. The function of both α-catenin isoforms has been investigated in *X. laevis* following α-catenin depletion with morpholino antisense oligonucleotides, but direct biochemical studies of interactions between α-catenin and other CCC members have not been performed. *Xl* vinculin has not been studied.

Xl αE-catenin co-localizes with E-cadherin at cell–cell contacts in early embryos and is redistributed to the cytoplasm upon depletion of E-cadherin (Kurth et al., 1999). *Xl* αE-catenin morphants have defects in cell–cell adhesion at the blastula stage and are arrested in further development (Kofron, Spagnuolo, Klymkowsky, Wylie, & Heasman, 1997) (Fig. 8.2B). *Xl* αE-catenin and αN-catenin co-immunoprecipitate with β-catenin and E-cadherin in whole embryo lysates (Kurth et al., 1999; Nandadasa, Tao, Shoemaker, Cha, & Wylie, 2012) (Fig. 8.2B). *Xl* αN-catenin morphants and null mutants have defects in neural plate closure and gastrulation, respectively (Nandadasa et al., 2012; Sehgal, Gumbiner, & Reichardt, 1997). Whether *Xl* αE-catenin or αN-catenin binds actin or regulates cytoskeletal dynamics has not been studied.

2.1.4 Danio rerio

D. rerio (zebrafish) contains orthologs of αE- and αN-catenin, and vinculin. *D. rerio* αE-catenin (*Dr* αE-catenin) has been characterized *in vivo* by depletion with morpholinos (Schepis, Sepich, & Nelson, 2012), and *in vitro* using purified recombinant proteins (Miller et al., 2013). Biochemical properties and cellular functions of *D. rerio* αN-catenin and vinculin have not been described.

Dr αE-catenin is expressed at cell–cell contacts between enveloping layer cells (EVL) and deep cells at the mid-blastula transition in wild-type embryos. αE-catenin morphants are delayed in epiboly due to defects in cell migration and adhesion in both the EVL and deep cells (Fig. 8.2B) (Schepis et al., 2012). EVL cells become elongated at 80% epiboly in control embryos, whereas they retain a spherical morphology in αE-catenin morphants (Schepis et al., 2012). At 50% epiboly, deep cells in αE-catenin morphants display reverse radial intercalation at a rate approximately equal to normal radial intercalation (Schepis et al., 2012). Defects in radial intercalation may be related to increased plasma membrane blebbing and loss of cadherin-mediated adhesion in αE-catenin-depleted deep cells (Schepis & Nelson, 2012; Schepis et al., 2012). αE-catenin is required to maintain cortical tension and normal adhesive forces (Maitre et al., 2012). Taken together, these data indicate that αE-catenin regulates cell–cell adhesion and membrane dynamics by anchoring E-cadherin to the cortical actin cytoskeleton. Interestingly, *Mm* αE-catenin only partially rescues epiboly in *Dr* αE-catenin morphants (Schepis et al., 2012). This suggests that functional and/or regulatory properties of *D. rerio* and mammalian αE-catenin have diverged (see below).

Despite 90% amino acid sequence identity, *D. rerio* and *Mm* αE-catenin have surprisingly different biochemical properties (Miller et al., 2013). *Dr* αE-catenin is monomeric in solution (Fig. 8.2B), and binds F-actin as a monomer or as a αE-catenin/β-catenin heterodimer and, therefore, is not autoinhibited (Fig. 8.2B). However, *Dr* αE-catenin binds F-actin >20× more weakly when bound to β-catenin, similar to the *Mm* αE-catenin/β-catenin heterodimer (Miller et al., 2013). *Dr* αE-catenin bundles F-actin poorly compared to *Mm* αE-catenin and does not inhibit Arp2/3 complex nucleation of F-actin (Miller et al., 2013) (Fig. 8.2B). Thus, *Dr* αE-catenin may directly link the CCC to the cortical actin cytoskeleton, but have a limited effect on actin dynamics (Fig. 8.3B). This is consistent with *in vivo* data, in which depletion of *Dr* αE-catenin disrupts cortical tension (Maitre et al., 2012) and induces protracted membrane blebbing (Schepis & Nelson, 2012; Schepis et al., 2012) that does not involve the Arp2/3 complex (Charras & Paluch, 2008).

2.1.5 Drosophila melanogaster

D. melanogaster, like all other invertebrates so far examined, expresses one homolog of α-catenin and one homolog of vinculin. Sequence analysis indicates that invertebrate α-catenin is most similar to vertebrate αN-catenin (Hulpiau & van Roy, 2009), although this has not been confirmed biochemically. *D. melanogaster* α-catenin (*Dm* α-catenin) has been studied extensively *in vivo*, but detailed biochemical experiments *in vitro* of the DE-cadherin/Armadillo (β-catenin)/α-catenin complex have not been performed. Deletion of vinculin, by inversion of the X-chromosome, results in viable and fertile adults indicating that it is not essential (Alatortsev, Kramerova, Frolov, Lavrov, & Westphal, 1997).

Although the binding affinities between proteins in the *Dm* CCC have not been determined, it appears that *Dm* DE-cadherin, Armadillo, and α-catenin form a ternary complex at cell–cell junctions. *Dm* α-catenin localizes to cell–cell contacts throughout embryogenesis, and co-immunoprecipitates with Armadillo and DE-cadherin (Oda et al., 1993) (Fig. 8.2B). Depletion of α-catenin disrupts embryogenesis, oogenesis and imaginal disc development (Sarpal et al., 2012), and causes the collapse of epithelia (Cavey, Rauzi, Lenne, & Lecuit, 2008). *Dm* α-catenin function in cell–cell adhesion requires binding to Armadillo (Desai et al., 2013), indicating that it may function as an anchor between DE-cadherin/Armadillo and the actin cytoskeleton (Sarpal et al., 2012). A DE-cadherin:α-catenin chimera can rescue cell–cell adhesion in an α-catenin weak allele background but the embryos die during larval stages (Sarpal et al., 2012), suggesting that DE-cadherin:α-catenin fusion proteins cannot replace all functions of endogenous α-catenin. *Dm* α-catenin mutants with deletions of either the NTD or F-actin-binding domain (ABD) do not rescue embryos with a weak α-catenin allele (Desai et al., 2013). Interestingly, Desai et al. suggest that *Mm* αN-catenin can rescue a weak *Dm* α-catenin phenotype (Desai et al., 2013). *Dm* α-catenin also appears to regulate the dynamics of the actin cytoskeleton: RNAi-mediated depletion of α-catenin inhibits the accumulation of Rho GTPase near the apical membrane (Magie et al., 2002), and disruption of the Arp2/3 complex or its activator SCAR ameliorates the hypermigratory phenotype of an α-catenin weak allele (Sarpal et al., 2012).

Dm α-catenin is a monomer or a homodimer (Desai et al., 2013), but unlike *Mm* α-catenin, both monomeric and homodimeric forms of *Dm* α-catenin bind actin (Figs. 8.2B and 8.3E) (Desai et al., 2013). While these results and the rescue experiments (above) indicate that *Dm* α-catenin directly links the CCC to the actin cytoskeleton (Fig. 8.3E) this has not been confirmed biochemically. That *Dm* α-catenin homodimerizes suggests that it has a cytoplasmic function similar to *Mm* αE-catenin homodimer, since *Mm* αE-catenin homodimer cannot interact with β-catenin in the CCC.

2.1.6 Caenorhabditis elegans

C. elegans possesses α-catenin (HMP-1) and vinculin (DEB-1) orthologs that are expressed in different cell types (Hardin, Lynch, Loveless, & Pettitt, 2013). HMP-1 has been investigated in loss-of-function mutants by biochemical characterization *in vitro* and through genetic dissection of actin binding. Little is known about the biochemical or cellular functions of DEB-1, although null mutations are lethal due to defects in body elongation and muscle development (Barstead & Waterston, 1989, 1991).

The *C. elegans* CCC comprises HMR-1 (E-cadherin), HMP-2 (β-catenin), and HMP-1, localizes to cell–cell contacts and the actin cytoskeleton, and is required for proper morphogenetic cell shape changes in early embryogenesis (Costa et al., 1998; Hardin et al., 2013). During epidermal morphogenesis, actomyosin-driven contractile forces are transmitted along circumferential actin filament bundles to cell–cell junctions and are required for cell shape changes and elongation of the embryo (Costa et al., 1998). In HMP-1 loss of function mutants, circumferential actin filament bundles are detached from the plasma membrane, which results in dorsal folding of the epidermis (the Humpback phenotype) (Costa et al., 1998) (Fig. 8.2B). Strong loss of function HMP-1 mutants also display defects in closure of the ventral midline between ventral hypodermal cells and the posterior of the body (Costa et al., 1998).

Purified HMP-1 is a monomer even at high concentrations, but the full-length protein does not bind actin *in vitro* (Fig. 8.2B), although the C-terminal ABD binds F-actin directly with an affinity similar to *Mm* αE-catenin homodimer (Kwiatkowski et al., 2010). This indicates that the ABD is inaccessible in the full-length protein, perhaps due to autoinhibitory interactions between the NTD and ABD. HMP-1 forms a ternary complex with HMP-2 and Casein kinase1-phosphorylated HMR-1, but this *in vitro* reconstituted CCC also does not bind actin (Kwiatkowski et al., 2010) (Figs. 8.2B and 8.3D). *In vivo*, however, HMP-1 must bind both HMP-2 and actin directly as deletion of either the HMP-2-binding region or ABD recapitulates the Humpback phenotype (Kwiatkowski et al., 2010). HMP-1 mutants lacking residues 687–742, 802, or 826–927 in the C-terminal ABD have a phenotype similar to embryos lacking HMP-1 or HMP-2 (Maiden et al., 2013), indicating that these C-terminal amino acids are critical to HMP-1 interactions with the actin cytoskeleton. However, the regulatory mechanism involved in activating actin binding by HMP-1 *in vivo* remains unknown (Fig. 8.2B and 8.3D).

2.1.7 Dictyostelium discoideum

The genome of the amoebozoan *D. discoideum* encodes two VIN proteins. One is a divergent sequence with a ~1000-amino acid insertion at the C-terminus that has not been characterized. The other is a putative homolog of α-catenin with equal sequence identity to both metazoan α-catenin and vinculin that has been characterized; this homolog is more like α-catenin than vinculin since it localizes to cell–cell contacts and not to cell–ECM contacts, and *in vitro* characterization of recombinant protein shows that it does not bind talin or form strong NTD–ABD interactions (Dickinson, Nelson, et al., 2011).

D. discoideum lives as a unicellular amoeba, but upon starvation develops into a multicellular fruiting body through a process called culmination (Grimson et al., 2000; Schaap et al., 2006). The fruiting body comprises a spore head supported by a rigid vertical stalk and surrounded by a single layer of cells called the tip epithelium (Dickinson, Nelson, et al., 2011). The tip epithelium has actin-associated intercellular junctions reminiscent of metazoan AJs (Grimson et al., 2000), and *Dd* α-catenin localizes to cell–cell contacts (Dickinson, Nelson, et al., 2011). *D. discoideum* possesses a protein with multiple armadillo repeats termed Aardvark (Grimson et al., 2000) that binds to *Dd* α-catenin through a sequence similar to *Mm* β-catenin (Dickinson, Nelson, et al., 2011). Knockdown of either Aardvark or *Dd* α-catenin inhibits the organization of cells in culminants, resulting in the disruption of stalk and spore head formation (Dickinson, Nelson, et al., 2011) (Fig. 8.2B). Aardvark appears to mediate the association of *Dd* α-catenin and intercellular junctions, since *Dd* α-catenin does not localize to cell–cell contacts in the absence of Aardvark (Dickinson, Nelson, et al., 2011). Thus in the context of binding *Dd* α-catenin, Aardvark has properties of a β-catenin homolog (Dickinson, Nelson, et al., 2011). The presence of a polarized epithelium in *D. discoideum* that is organized by α-catenin and β-catenin homologs suggests that multicellularity may be a more ancient evolutionary development than previously thought (Dickinson, Nelson, & Weis, 2012).

D. discoideum is the only premetazoan organism in which a VIN protein has been characterized. *Dd* α-catenin is a monomer, and binds either Aardvark or *Mm* β-catenin *in vitro* indicating that this interaction is evolutionarily conserved (Dickinson, Nelson, et al., 2011) (Fig. 8.2B). *Dd* α-catenin is not autoinhibited, as the monomer binds and bundles F-actin, but it does not inhibit the Arp2/3 complex (Dickinson, Nelson, et al., 2011) (Fig. 8.2B). Currently, it is unclear how the *Dd* α-catenin/Aardvark complex couples the cortical actin cytoskeleton to the plasma membrane, since homologs of classical cadherins are absent in *D. discoideum* (Fig. 8.3A). Nevertheless, like the monomeric *Dr* αE-catenin homolog, *Dd* α-catenin appears to be an actin-binding protein that links adhesive junctions and the actin cytoskeleton (Dickinson, Nelson, et al., 2011).

2.2. Summary: An evolutionary perspective of how the CCC formed

The CCC has properties essential for organized multicellular development, tissue organization, and physiology of metazoans and, to date, at least one non-metazoan (an amoebozoan). The core CCC components—classical cadherins, β-catenin, and α-catenin—are present in all metazoans, and it might be expected that there would be significant functional conservation between homologs. The biochemical properties of classical cadherins and β-catenin have not been studied in detail across a phylogenetically diverse range of organisms. Nevertheless, existing data indicate high conservation of homotypic interactions between cadherin extracellular CADs, binding between the cadherin cytoplasmic domain and the β-catenin armadillo repeat domain, and the formation of heterodimers between β-catenin and α-catenin (Abedin & King, 2008; Dickinson, Nelson, et al., 2011; Grimson et al., 2000; Nichols, Dirks, Pearse, & King, 2006; Nichols, Roberts, Richter, Fairclough, & King, 2012; Schneider et al., 2003). In contrast, α-catenin homologs have surprisingly divergent functional properties across metazoans and a non-metazoan, which raises two questions: (1) Did the CCC as a whole evolve different functions that were tailored to specific requirements in different organisms; and (2) Is α-catenin functionally more variable than other CCC components, and hence α-catenin variability evolved to regulate CCC interactions with the actin cytoskeleton?

Answers to these questions require analysis of CCC protein sequences during evolution. Therefore we screened for CCC proteins from phylogenetically diverse genomes to begin to reconstruct the ancient origins of the core components of the CCC, investigate whether domain organization was preserved between diverse homologs, and speculate whether molecular interactions between CCC members restricted or expanded their evolution.

3. GENOMIC ANALYSIS OF CCC EVOLUTION

3.1. Ancient origins of core CCC components

Each core member of the CCC has characteristic domains: the extracellular cadherin repeat (CAD) of classical cadherins; the armadillo repeat domain of β-catenin comprising multiple ARM repeats; and several helical bundle domains comprising the vinculin/α-catenin (VIN) family proteins (Fig. 8.2A). Protein sequences containing these domains are found in a diverse range of eukaryotes, bacteria, and archaea, but outside the metazoa their presence seems to vary independently between clades (Fig. 8.4A). Recent genomic evidence indicates that the core components of the CCC predate the origin of metazoa.

The presence or absence of these characteristic CCC protein domains within a proteome does not necessarily indicate they mediate protein–protein interactions and functions of the CCC (see Section 2), but it does provide evidence for the potential evolutionary point of origin of each CCC component. To determine evolutionary relationships more definitively, bioinformatics analysis must be extended to the identification of putative orthologs, structure-based sequence alignment and domain architecture, and ultimately functional characterization. Here we report the presence of CCC protein domains across cellular organisms, with a number of putative orthologs found within opisthokonts and their near ancestors. We used non-exhaustive Hidden-Markov Model (HMM)-based searches for the pFam profiles for CAD, ARM, and VIN (PF00028, PF00514, and PF01044, respectively) (Punta et al., 2012).

The CAD and ARM repeat domains are found in a large number of proteins in highly variable contexts. CADs are adhesive domains responsible for hetero- and homotypic interactions in many transmembrane and secreted proteins, and are dependent on calcium for their conformation and adhesive function (Harrison et al., 2011; Ivanov, Philippova, & Tkachuk, 2001; Koch et al., 1997). Bacterial CAD proteins have calcium-dependent homotypic and heterotypic interactions *in vitro*, and bind to the cell surface when added exogenously to bacterial cultures (Fraiberg, Borovok, Weiner, & Lamed, 2010). Within metazoans, CADs are essential for cell–cell adhesion and specificity of cell–cell contacts (Hulpiau & van Roy, 2011; Oda & Takeichi, 2011; Wheelock & Johnson, 2003). An ARM domain comprises multiple copies of a three-helix ARM repeat motif that associate to form a rigid protein-binding structure (Huber, Nelson, & Weis, 1997) involved in

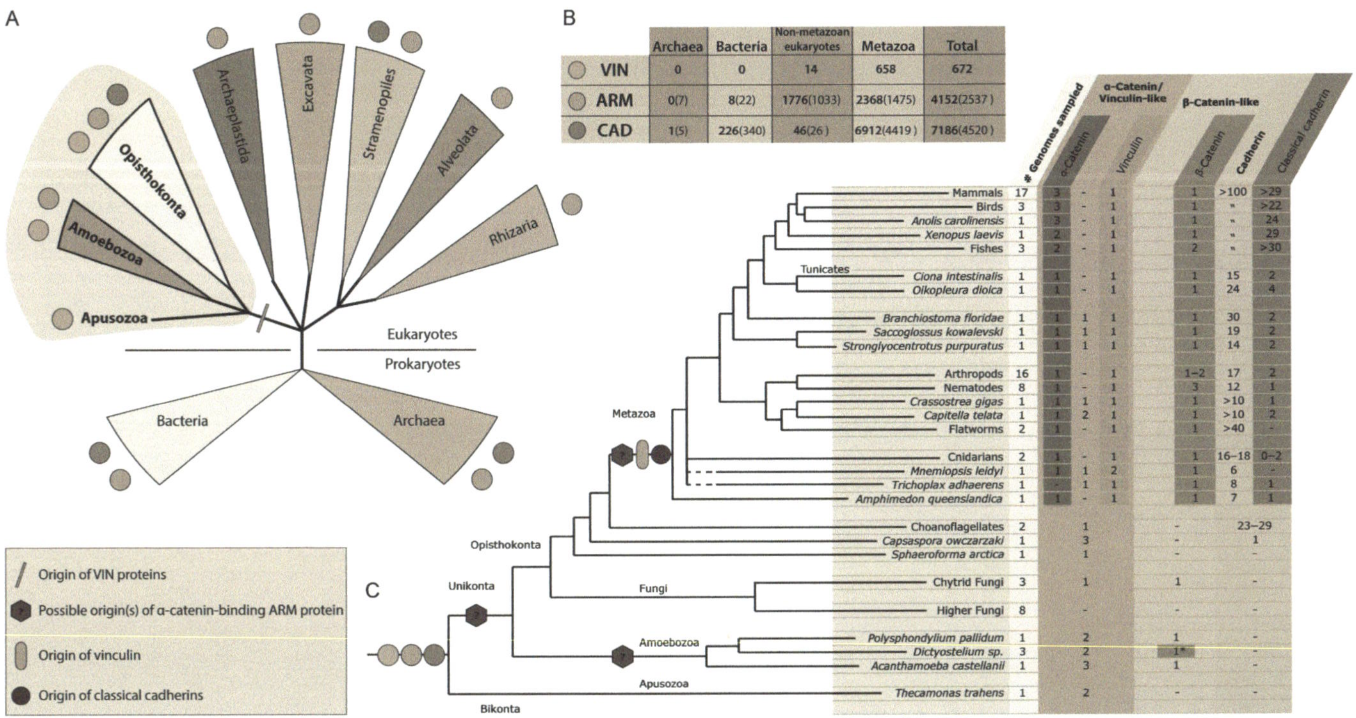

Figure 8.4 *Distribution of cadherin/catenin complex proteins in cellular organisms.* (A) A cladogram indicating the presence of the VIN, ARM, and CAD proteins (blue, purple, and green circles, respectively) in bacteria, archaea, and major eukaryotic lineages. The blue line indicates the lineage-specific origin of VIN proteins. (B) Numbers of CCC proteins found within each domain of life found in the pFam (in bold) and SMART (in parentheses) online databases (Letunic, Doerks, & Bork, 2009; Punta et al., 2012). (C) Distribution of CCC proteins within metazoans and opisthokont and unikont outgroups, deliniating between true orthologs and similar sequences. *$D.$ *discoideum* Aardvark is referred to here as a β-Catenin due to its observed interaction with *Dd* α-Catenin, but it should be noted that other canonical functions of metazoan β-Catenin have not yet been confirmed for *Dd* Aardvark. (See color plate.)

different cellular processes, including cytoskeleton organization, growth factor signaling, nuclear import, and gene transcription (Nelson & Nusse, 2004). ARM domain proteins contain a variable number of ARM repeats (Coates, 2003; Tewari, Bailes, Bunting, & Coates, 2010). Multi-CAD and multi-ARM repeat proteins are found in small numbers in both archaea and bacteria, indicating that the domains themselves likely evolved prior to the divergence of the last common ancestor of eukaryotes and prokaryotes. Thus multi-CAD and multi-ARM repeat proteins were part of the ancestral proteome of cellular organisms (Fig. 8.4A).

The vinculin-homology family (VIN) comprises a series of four-helix bundles connected to a C-terminal five-helix bundle. VIN proteins bind F-actin, and link actin filaments to proteins at the plasma membrane (see Section 2). Multiple VIN-containing proteins are detected in the proteomes of all metazoans, and are found in some opisthokont relatives of metazoa (choanoflagellates, chytridomyctes), as well as in the nearest eukaryotic sister clades, apusozoa and amoebozoa (Abedin & King, 2008; Dickinson, Nelson, et al., 2011; Grimson et al., 2000) (Fig. 8.4C). A VIN has not been detected in organisms outside of this larger clade of opisthokonts and their near relatives, indicating a lineage-specific origin near the base of the eukaryotic branch leading to metazoans (Fig. 8.4A).

The basic protein repertoire necessary for the assembly of the core CCC—a multi-CAD protein, a multi-ARM protein, and a VIN protein—likely originated sometime before the last common ancestor of unikonts and their closest bikont sister taxa. Therefore, it can be inferred that in opisthokont clades, such as fungi and choanoflagellates, the absence of some or all of these proteins represents a loss of ancestral sequence diversity. It also indicates that the evolution of the CCC was not driven by the emergence of novel domains, but rather through domain rearrangement and the accumulation of functional interactions between CCC proteins within certain taxa.

3.2. Premetazoan assembly of a functional CCC

Three changes in the CCC coincided with the advent of multicellularity: (1) Modification of the unstructured cytoplasmic domain of cadherin to contain a binding motif for β-catenin; (2) Gain of α-catenin- and cadherin-binding sites in an ARM protein (β-catenin); (3) Duplication of an ancestral VIN-containing protein resulting in separate α-catenin and vinculin protein families. When compared to the nearest outgroup to metazoa, the choanoflagellates, each of these changes appears to be an evolutionary gain of function, but their novelty becomes uncertain when examined in a broader context. First, a more complete search reveals that most non-metazoans have two or more VIN proteins, and that choanoflagellates are an exception (Fig. 8.4C). Second, *D. discoideum* has a multi-ARM repeat protein, Aardvark, that binds to a α-catenin ortholog (Dickinson, Nelson, et al., 2011). The presence of the β-catenin-binding motif in classical cadherins, however, appears to be a metazoan novelty. Still, the unexpected functional diversity of VIN-containing proteins (see Section 2) and the observation of functional CCC protein interactions in an amoebozoan support the hypothesis that the core actin-binding functions of the CCC complex arose before the advent of metazoan multicellularity (Fig. 8.4). Below, we discuss these trends in light of more complete genomic evidence.

3.2.1 Cadherins

An interesting trend can be observed in the abundance of cadherins across the opisthokont-metazoan boundary (Figs. 8.4 and 8.5). A larger diversity of cadherins is found in choanoflagellates than in any invertebrate phylum outside deuterostomes and platyhelminthes, but none of the choanoflagellate cadherins contain the characteristic catenin-binding motif found in metazoan classical cadherins (Fig. 8.5). As this motif is observed in the classical cadherins of sponges, which are currently thought to be the most basally branching metazoan group, a β-catenin-binding cadherin should be considered an ancestral character of the metazoa (Fahey & Degnan, 2010; Nichols et al., 2012). Interestingly, the cadherins in two invertebrate phyla, ctenophora and platyhelminthes, do not have a discernable β-catenin-binding domain. It is possible that the lack of a β-catenin-binding site in these cadherins is an artifact due to low genomic coverage in the available sequence data for these groups. However, it is also possible that ancestors of these phyla secondarily lost a classical cadherin. If these phyla truly do not possess a classical cadherin, determination of the transmembrane component of the CCC (or the functional substitute for the entire complex) in flatworms and comb jellies could be informative. Nevertheless, if we exclude these two phyla, either on grounds of insufficient data or the principle of parsimony, we can conclude that members of the metazoan lineage inherited a classical cadherin with a conserved cytoplasmic domain that bound β-catenin.

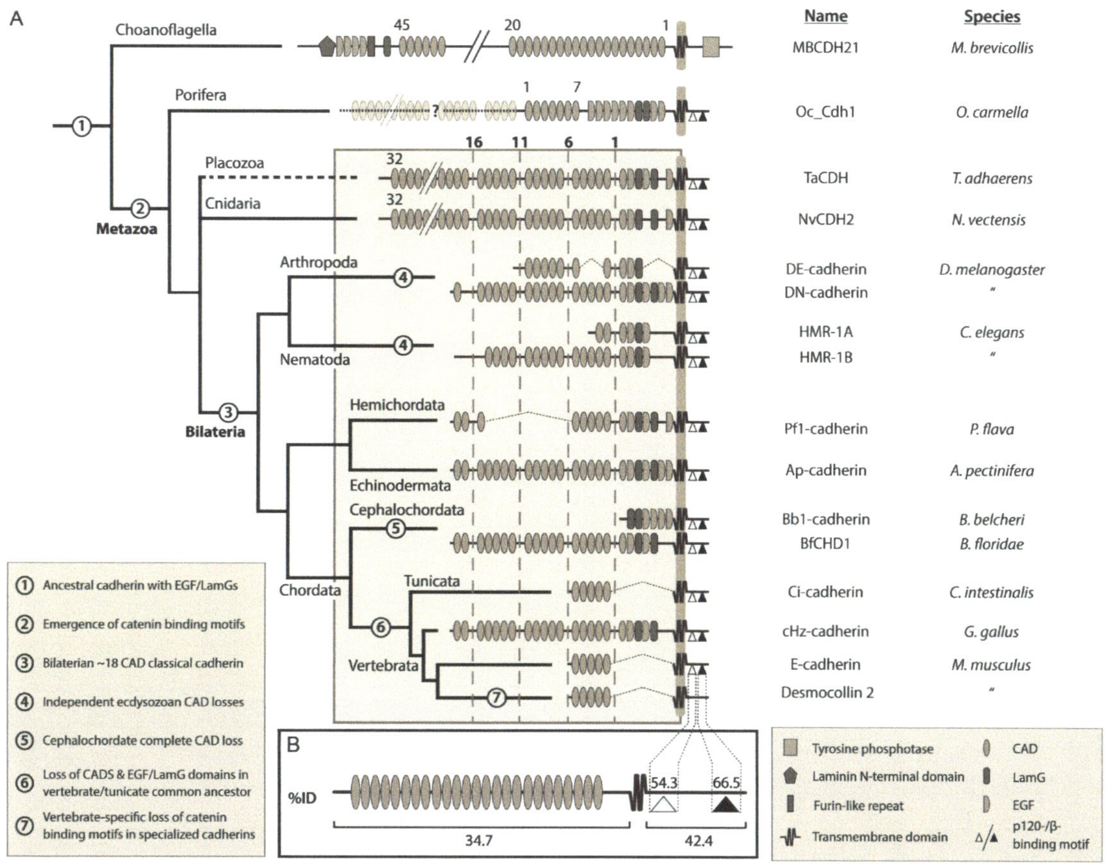

Figure 8.5 *Evolution of classical cadherin domain architecture from porifera to vertebrates.* (A) A phylogeny of representative classical cadherin proteins from multiple animal phyla, indicating the modification of the extracellular region by loss and/or rearrangement of conserved domains. Key transitions are noted by number in the legend in the lower left. (B) Schematic representation of a classical cadherin, indicating the average percent amino acid sequence identity of the catenin-binding motifs and the cytoplasmic and extracellular regions of the protein. (See color plate.)

3.2.2 β-Catenin

Analysis of non-metazoan ARM proteins reveals a lack of clear orthologs of β-catenin, although β-catenin-like proteins are present (Fig. 8.4). All three available chytridomycte genomes (*Allomyces macrogynus*, *Batrachochytrium dendrobatidis*, and *Spizellomyces punctatus*) each have a 12–13 ARM repeat protein that is ~20% identical to *Mm* β-catenin (19%, 20%, and 20% identity, respectively), but each is more similar to the mammalian protein Armadillo-repeat Protein 4 (48%, 22%, and 51% identical, respectively). The *D. discoideum* protein Aardvark has a single ortholog in each of the other amoebozoans analyzed. With the exception of *D. discoideum* Aardvark, none of these ARM proteins have been assayed for binding to their corresponding VIN protein (α-catenin), nor analyzed to determine their evolutionary relationships, so inferring anything about their role in a putative CCC is impossible. Without further functional evidence to guide us, these data produce two equally probable hypotheses: (1) An α-catenin-binding ARM protein evolved twice independently (i.e., *D. discoideum* Aardvark is convergent with metazoan β-catenin); or (2) A single gain-of-function event occurred in an ancestral opisthokont, followed by domain rearrangement or loss in some clades. Additional sequence data may be necessary to determine the evolutionary relationships between identified β-catenin-like ARM proteins, but the hypothesis that the known sequences interact with their corresponding VIN protein is readily testable.

3.2.3 α-Catenin

Previous studies of α-catenin evolution have examined only a sparse phylogeny of metazoans (Zhao, Reynolds, & Gaucher, 2011). This resulted in the perspective that only a single α-catenin existed prior to evolution of the chordate lineage, and that there were two subsequent duplication events that gave rise to the three known isoforms of mammalian α-catenin (αE, αN, and αT; see above) and vinculin. Expanding the analysis to include additional invertebrate phyla confirms this initial finding in terms of direct α-catenin orthologs (Fig. 8.4). However, deeper analysis shows that one or more additional α-catenin-like proteins are found in many metazoan phyla, including ctenophores, mollusks, annelids, and basal deuterostome phyla (Fig. 8.4). Interestingly, the basal metazoan *Trichoplax* does not have a direct α-catenin ortholog, but it does have a single vinculin protein and an additional α-catenin-like protein with ambiguous orthology (19% and 17% identity to α-catenin and vinculin, respectively). Another previously unexamined basal metazoan, the ctenophore *Mnemiopsis leiydi*, also has a α-catenin-like protein and two vinculins. The discovery of additional α-catenin-like proteins raises questions about our current understanding of their evolutionary relationships—this will be addressed briefly in the following sections, but should also be considered an important question for further investigation.

3.3. Classical cadherins: Variation and constraint mediated by catenin interactions

Of the three core components of the CCC, classical cadherins display the most diversity in terms of domain composition and organization (Oda & Takeichi, 2011). Much is known about cadherin evolution in terms of the relatedness of the different classes and of the architectural diversity of the extracellular region (Hulpiau et al., 2013; Hulpiau & van Roy, 2011; Oda & Takeichi, 2011). Here, we focus on functional variation between metazoan classical cadherins in relation to the CCC, and make a general comparison between variability in the extracellular and cytoplasmic domains.

A general trend in classical cadherin evolution is that organismal complexity is positively correlated with cadherin diversity and inversely correlated with cadherin complexity—nearly all bilaterians have a larger number of classical cadherins, each with fewer extracellular domains than non-bilaterians (Fig. 8.5A). For example, the placozoan classical cadherin (*T. adhaerens* only has one discernable isoform) contains 32 CADs, while the classical cadherins of bilaterians contain between 2 and 17 CADs (Hulpiau & van Roy, 2009) (Fig. 8.5). The length of classical cadherins in sponge is unclear, as all published sequences are annotated as fragments; however, sponge is thought to have only one classical cadherin with at least 10 or more CADs (Fahey & Degnan, 2010; Nichols et al., 2012).

Additionally, non-chordate classical cadherins contain multiple repeats of epidermal growth factor-like domains (EGFs), laminin A globular domains (LamGs), and a region homologous to the extracellular proteolytic cleavage site of *Dm* DE-cadherin in the C-terminus of the extracellular domain proximal to the plasma membrane (Oda & Takeichi, 2011; Oda & Tsukita, 1999). The function of these additional domains is poorly understood, but initial *in vivo* evidence from *D. melanogaster* indicates that they may play a signaling or regulatory role in development (Haruta, Warrior, Yonemura, & Oda, 2010). Thus non-chordate classical cadherins may have multiple functions beyond cell–cell adhesion.

From this analysis, we posit that the specialized cell/tissue-type classical cadherins of bilaterians evolved by duplication and sub-functionalization of a ubiquitously expressed classical cadherin with a multi-functional extracellular region. However, studies of expression patterns and functional interactions of large classical cadherins in basal metazoans are needed to test this hypothesis.

Within the highly variable cadherin superfamily, some elements of the protein sequence remained static in order to maintain necessary functions at cell–cell contacts. In order to participate in the CCC, a classical cadherin in a given species must maintain over evolutionary time a β-catenin-binding motif within the cytoplasmic domain. There are two binding motifs in the cytoplasmic domain of classical cadherins for p120-catenin and β-catenin, respectively (Hulpiau & van Roy, 2011). Using a heuristic method suitable for general comparison, we can state that the average percent amino acid sequence identity within the β-catenin-binding motif is nearly twofold higher than across the entire protein (Fig. 8.5B). The p120-binding motif is also more conserved than the whole protein, though to a lesser extent than the β-catenin-binding motif. An important caveat is that the algorithms used to calculate pair-wise identity are designed to return an optimal alignment, and they ignore large gaps and terminal regions in order to optimize the alignment score. Therefore, the numbers reported here should not be considered indicative of conservation within a region, but rather a means of holistic comparison between regions of a protein. Nevertheless, it is clear that the β-catenin-binding motif is more conserved than the rest of cadherin sequence, indicating that in spite of large rearrangements of the extracellular domain of classical cadherins, the sequence in cadherins required for participation in the CCC (i.e., binding to β-catenin) is evolutionarily constrained to maintain this critical function.

3.4. β-Catenin evolution: Domain homology, evidence for evolutionary constraint by α-catenin interaction, and a consensus α-catenin-binding motif

β-Catenin sequence and function appear to be highly conserved within metazoans (Fig. 8.6). β-Catenin has a characteristic structure consisting of unstructured N- and C-terminal regions involved in the regulation of its degradation and transcriptional activity, respectively, a central region consisting of 12 ARM repeats, and an α-catenin-binding domain located partially N-terminal to the first ARM repeat (Huber et al., 1997; Shapiro & Weis, 2009) (Fig. 8.6A). The order and identity of the ARM repeats within β-catenin are highly conserved. When sequences of metazoan β-catenin orthologs are subdivided based upon their individual ARM repeats and

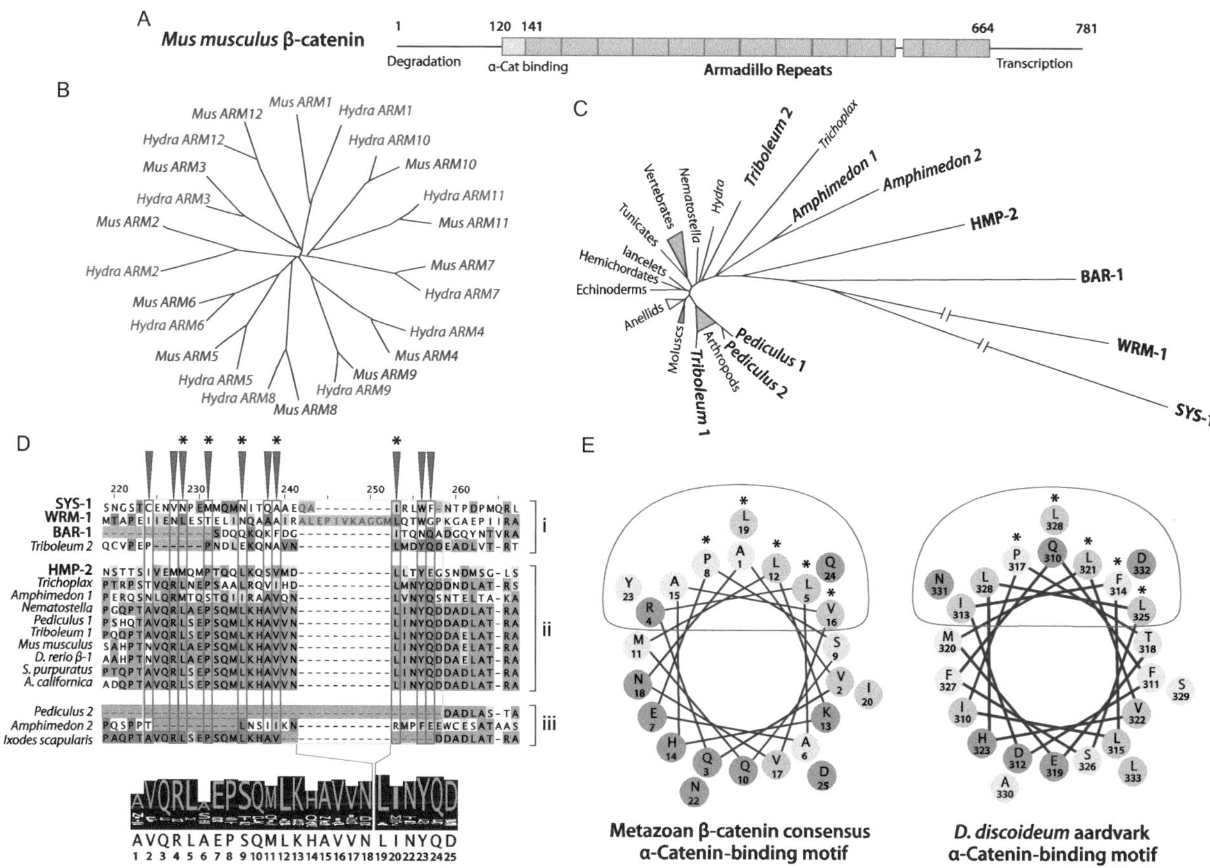

Figure 8.6 For legend see next page.

clustered algorithmically according to sequence similarity, it is clear that each ARM repeat groups exclusively, even between distantly related taxa (Schneider et al., 2003) (Fig. 8.6B).

In spite of this strong sequence conservation across the metazoa, there are a small number of β-catenin orthologs that are quite divergent. On a neighbor-joining phylogeny in which branch length corresponds to the number of amino-acid substitutions, several proteins stand apart as clear outliers. Interestingly, most of these occur in species in which there has been a lineage-specific duplication of β-catenin: *A. queenslandica, T. casteneum,* and most notably in the three additional *C. elegans* proteins BAR-1, SYS-1, and WRM-1 (Liu, Phillips, Amaya, Kimble, & Xu, 2008; Zhao et al., 2011) (Fig. 8.6C). In *A. queenslandica*, the two β-catenin paralogs have different mRNA expression patterns, but neither their subcellular localization nor function have been examined (Adamska et al., 2010). The *C. elegans* β-catenin paralogs HMP-2, BAR-1, SYS-1, and WRM-1 also exhibit sub-functionalization. Of the four, only HMP-2 binds to mammalian α-catenin and HMP-1 (Kwiatkowski et al., 2010; Natarajan, Witwer, & Eisenmann, 2001), while BAR-1, SYS-1, and WRM-1 are involved in Wnt signaling and gene transcription (Liu et al., 2008; Natarajan et al., 2001). In *T. castaneum*, β-catenin paralogs exhibit partial sub-functionalization in cell–cell adhesion (β-1) and centrosomal regulation (β-2) (Bao, Fischer, Bolognesi, Brown, & Friedrich, 2012), which is also a function of mammalian β-catenin (Mbom, Nelson, & Barth, 2013).

Figure 8.6 *Evolutionary constraint of β-catenin by the α/β-catenin interaction.* (A) A schematic representation of *Mm* β-catenin, indicating the position of the α-catenin-binding region and Armadillo repeats. (B) A phylogeny of *Hydra* and *Mus* Armadillo repeats 1–12 indicating homology of each repeat. (C) A neighbor-joining phylogeny of metazoan β-catenins, indicating divergent paraogs in bold. Branch length correlates to number of amino acid substitutions between proteins. (D) An alignment generated using MUSCLE (Edgar, 2004), visualized in JalView (Waterhouse, Procter, Martin, Clamp, & Barton, 2009), highlighting the α-catenin-binding motif (yellow box), conserved binding surface (red boxes/arrows), and structurally important residues (asterisks). Paralogs that fail to bind α-catenin (i) and untested paralogs and orthologs with abberant binding motifs (iii) are segregated from α-catenin-binding paralogs and single orthologs (ii) for clarity. A consensus sequence generated from the proteins in group ii is displayed below. (E) Helical wheel representations of the consensus α-catenin-binding helix from D, and that of *D. discoideum*. Charged residues are colored orange, and hydrophobic residues are colored cyan. Red ovals and asterisks indicate the α-catenin-binding surface and structurally important residues as in D. *Panel (B) after Schneider et al. (2003), panel (E) after Dickinson, Nelson, et al. (2011).* (See color plate.)

Additionally, *T. castaneum* β-2 has a 6-amino-acid deletion within the conserved α-catenin-binding motif, but maintains a high level of sequence conservation in residues important for E-cadherin binding (Bao et al., 2012; Pai et al., 1996). These data indicate that β-catenin participation in the CCC may be a form of evolutionary constraint that is partially responsible for the high level of sequence conservation within β-catenin. Release from this selective pressure might have been achieved through disruption of α-catenin-binding in β-catenin paralogs.

A multiple alignment of single and duplicated β-catenin paralogs across metazoa reveals that all highly divergent paralogs possess significant insertions or deletions within the α-catenin-binding domain (Fig. 8.6D). The paralogs for which there is evidence supporting a loss of α-catenin-binding function (Fig. 8.6D, top) all possess insertions or deletions in this region. In contrast, α-catenin-binding paralogs align closely with those organisms that have only one single β-catenin ortholog (Fig. 8.6D, center). *A. queenslandica* and *P. humanus* possess β-catenin paralogs with significant alterations in the α-catenin-binding domain (Fig. 8.6D, bottom), indicating that these proteins may have lost the capacity to bind α-catenin. Interestingly, *I. scapularis* has a single β-catenin ortholog with a deletion within the α-catenin-binding domain. One hypothesis is that the α-catenin-binding site in β-catenin is the favored mutation site for disrupting β-catenin function in the CCC, as this would produce only an adhesive phenotype (Brembeck et al., 2004; Hoffmans & Basler, 2007; Huber & Weis, 2001; Roura, Miravet, Piedra, Garcia de Herreros, & Dunach, 1999; Taurin, Sandbo, Qin, Browning, & Dulin, 2006); in contrast, a mutation of the central ARM repeat region would have multiple effects on adhesion and Wnt signaling due to the significant overlap of binding sites for E-cadherin (Huber & Weis, 2001), TCF (Graham, Ferkey, Mao, Kimelman, & Xu, 2001; Graham, Weaver, Mao, Kimelman, & Xu, 2000), Axin (Xing, Clements, Kimelman, & Xu, 2003) and Adenomatous Polyposis Coli (Ha, Tonozuka, Stamos, Choi, & Weis, 2004; Spink, Fridman, & Weis, 2001; Xing et al., 2004).

From this sequence alignment, we are also able to derive a consensus α-catenin-binding motif in β-catenin (depicted on a helical wheel in Fig. 8.6E). Excluding *C. elegans* sequences and other divergent paralogs, the β-catenin residues shown to form structurally important contacts with α-catenin (consensus residues 5, 8, 12, 16, and 19) are entirely conserved within bilaterians, and mostly conserved across metazoans (Aberle et al., 1996; Pokutta & Weis, 2000). These key residues are also conserved in the α-catenin-binding motif of *D. discoideum* Aardvark, demonstrating their importance for α/β heterodimer interaction (Dickinson, Nelson, et al., 2011) (Fig. 8.6E). A more detailed comparison of these two motifs may yield a consensus search sequence usable for the detection of putative α-catenin-binding ARM proteins in divergent non-metazoan eukaryotes. This α-catenin-binding region should also be considered a candidate region for studies of intermolecular co-evolution between α- and β-catenin.

3.5. Evolutionary history of the α-catenin/vinculin family

In light of recent studies that have identified unexpected diversity in functional properties of α-catenins (Figs. 8.1–8.3) and the sequences reported above that were not described previously, a more thorough review of VIN protein phylogeny is merited. Phylogenetic analysis of distantly related sequences is often inconclusive, but when combined with functional evidence and domain architecture we can draw conclusions about the ancestry of α-catenin and related proteins. Here we describe the presence of a previously unknown group of metazoan α-catenin-like proteins that may be representative of an ancestral metazoan α-catenin, and discuss the origin of the vinculin family as a duplication event from an α-catenin-like protein in an ancestral metazoan.

To determine the relatedness of metazoan α-catenins and vinculins and non-metazoan VIN proteins, a maximum-likelihood phylogeny based on the sequence of 96 VIN proteins was made from diverse metazoan and non-metazoan phyla (Fig. 8.7A). Known α-catenins and vinculins group as distinct clades, but non-metazoan sequences do not group in either of these clades. The chytridomycete and amoebozoan VIN proteins also group as independent clades. These clades arise from a polytomy of indeterminate relatedness that also contains sequences from choanoflagellates and their near relative, the apusozoan *Thecamonas trahens*, and all the metazoan sequences that were not described previously. These metazoan sequences group together as a clade (0.66 ML posterior probability), but their relatedness to the described α-catenins cannot be accurately determined as this clade groups unreliably at the base of the α-catenin branch (0.28 ML posterior probability, depicted in Fig. 8.7A as collapsed to the polytomy). This clade has both protostome and deuterostome VIN members, and also contains the only α-catenin-like sequence found in *Trichoplax,* indicating that it is both widely distributed and ancestral in the metazoa. As this clade was not described until now, the function of these α-catenin-like proteins is not

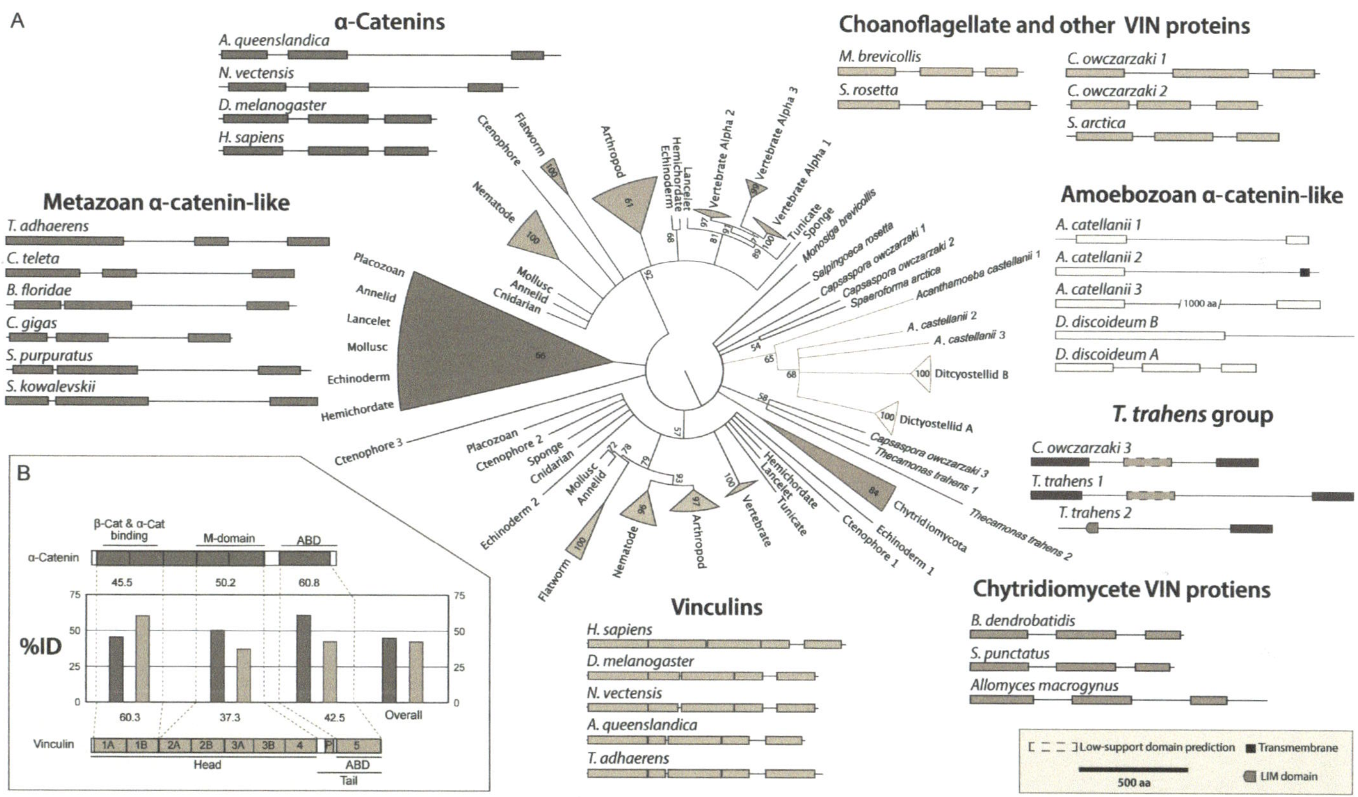

Figure 8.7 *Evolutionary history of VIN families.* (A) A maximum likelihood phylogeny of 96 VIN-containing proteins generated using RAxML (Stamatakis, 2006) with 1000 rapid bootstrap iterations and best-fit model parameters as determined by ProtTest3 (PROTGAMMALGF; Darriba, Taboada, Doallo, & Posada, 2011) on a trimmed alignment (trimAl; Capella-Gutierrez et al., 2009). Schematics of domain architecture are colored and grouped to correspond with clades on the phylogeny. Ungrouped sequences are colored gray. (B) A general schematic of the domain architecture of α-catenin and vinculin, with a bar graph indicating the average percent amino acid sequence identity of the highlighted regions within each protein family. (See color plate.)

known; a comparison between organisms that possess both α-catenin and a member of this clade with organisms possessing only one or the other may provide important data for understanding the evolution of the α-catenin protein family. This phylogeny demonstrates that the α-catenin and vinculin families are distinct metazoan lineages that diverged at the base of the animal tree. However, the relatedness of divergent ancestral groups to these two clades cannot be readily determined from the phylogeny alone.

As indicated by previous studies of the α-catenin-related protein in *D. discoideum*, general domain architecture may be more indicative of functional relatedness than strict sequence comparison—*Dd* α-catenin is equally related to *H. sapiens* αN-catenin and vinculin by sequence (17% identity), but the arrangement of its functional domains and biochemical properties corresponds more closely to α-catenin than to vinculin (Dickinson, Nelson, et al., 2011). Similar comparison of domain architecture between the groups represented here reveals both a surprising diversity within divergent VIN sequences, as well as the presence of a conserved architectural pattern. Within opisthokonts, all VIN proteins contain only helical bundle domains characteristic of the VIN family and no other recognizable functional domains.

Outside of opisthokonts, other protein motifs are included such as a LIM domain in *T. trahens VIN 2* and the inclusion of a transmembrane domain in *A. castellani VIN* 2. The LIM domain is a tandem zinc-finger structure that functions as a protein-binding interface and is associated with cytoskeletal organization and signal transduction from the plasma membrane to the nucleus (Kadrmas & Beckerle, 2004). Many known LIM proteins have been shown to bind F-actin or actin-binding proteins (Khurana, Khurana, Khaire, & Noegel, 2002), but none have been described that contain a VIN domain. *T. trahens VIN 2* appears to be a novel VIN–LIM-containing protein that represents a class of eukaryotic LIM proteins that were not described previously (Koch, Ryan, & Baxevanis, 2012). The inclusion of a transmembrane domain in *A. castellani VIN* 2 is also novel—no other described VIN protein has been observed to have a transmembrane domain.

There is significant variation in domain architecture within previously described VIN proteins. Even among VIN proteins, there are clear outliers in which we can observe domain loss or the addition of large insertions (Fig. 8.7A). The loss of one or more VIN domains is observed in several amoebozoans, including *Dictyostelium* sp. and *A. castellani,* and putatively within the opisthokont *C. owczarzaki* and the apusozoan *T. trahens*, where a third VIN domain is identified with unreliable significance (Fig. 8.7A, dashed outline). Insertion sequences are observed in α-catenins and α-catenin-like sequences of many basal metazoans. For instance, both

N. vectensis and *T. adhaerens* have large insertions (~400 and 600 amino acids, respectively) between the M- and actin-binding domains. There are multiple phosphorylation sites in this region in *Mm* αE-catenin (Huttlin et al., 2010) that are conserved within the insertions of basal metazoans, but the functional significance of these sites is not understood. Nothing is known of how these inserted sequences affect the conformation or functional properties of α-catenin such as actin binding, homodimerization capacity, or autoinhibition. However, the location of these insertions provides a natural study system in which to test hypotheses of the roles of internal residues in regulating α-catenin structure and function.

As observed with classical cadherins and β-catenin, we would expect to find signs of constraint within α-catenin sequences resulting from its necessary function as a link between the CCC and the actin cytoskeleton. However, as α-catenin plays multiple roles at the plasma membrane in the mammalian CCC (see Section 2), it is unclear which regions would be expected to have high sequence conservation. In a comparison of α-catenin sequences, we observe that the C-terminal ABD is more conserved relative to the rest of the protein, whereas the N-terminal (homo- and hetero-) dimerization domain is more variable (Fig. 8.7B). This observation indicates that the actin-binding domain may be under selection to maintain the capacity to bind actin, whereas the N-terminal dimerization and M domains have varied more significantly over evolutionary time.

Several related hypotheses can be posited to explain why the interaction between the β-catenin/cadherin complex and α-catenin varies significantly between metazoan groups. (1) The interaction with the actin cytoskeleton is the more ancestral function of an α/β-catenin heterodimer. (2) Greater sequence divergence occurred in the N- and M-domains as more functions at the plasma membrane are gained in some metazoan groups. (3) Regulation of actin binding may have evolved by autoinhibition/allostery rather than loss of function mutations due to overlapping functional regions/binding surfaces of VIN domains. These hypotheses are not mutually exclusive nor are they the only interpretation, but they stand out in light of recent functional evidence in diverse organisms.

4. CONCLUSION AND SYNTHESIS

4.1. Functional divergence within a highly conserved protein complex

It is generally accepted that homologous genes and proteins have conserved functions between different organisms, and molecular evolution occurs through changes in gene regulatory elements (Carroll, 2008). A common assumption made in sequence comparison is that shifts in protein function will correspond to large differences in sequence, but at the molecular level the importance of a single residue cannot be underestimated. Few studies have tested whether protein homology, as determined by sequence comparison, is indicative of conservation of biochemical, cellular, or developmental properties, and those that have tried had a false positive rate of over 50% when calling orthologs with equivalent functions (Ponting, 2001; Watson, Laskowski, & Thornton, 2005; Yu et al., 2012). Indeed similarities in sequence are not predictive of conserved protein function throughout the cadherin superfamily, hedgehog signaling pathway components, and vertebrate glucocorticoid receptor homologs (Dickinson, Weis, et al., 2011; Ortlund, Bridgham, Redinbo, & Thornton, 2007). As demonstrated here, *in vitro* biochemical experiments make it possible to directly test hypotheses of conservation of protein function and, by careful selection of a set of homologs for analysis, can reveal how proteins acquired novel functions in evolution.

The CCC is highly conserved throughout metazoans, where evolutionary assembly of the complex likely occurred in unicellular ancestors. Molecular interactions between the cytoplasmic domain of classical cadherins and ARM repeats of β-catenin have not been investigated over as diverse a set of organisms as α-catenin, but cadherin/β-catenin complex formation appears to be highly conserved in animals and positively regulated by phosphorylation of the cadherin cytoplasmic domain (Kwiatkowski et al., 2010; Nichols et al., 2006, 2012; Schneider et al., 2003) (this study). Direct binding between α-catenin and β-catenin is conserved throughout metazoans (Aberle et al., 1996; Hirano & Takeichi, 1994; Koslov et al., 1997; Kwiatkowski et al., 2010; Miller et al., 2013; Schneider et al., 2003).

While metazoan α-catenin orthologs share >20% sequence identity and conserved domain architecture, regulatory mechanisms of actin binding are divergent among unikonts, and novel functional properties have developed in some homologs. Autoinhibition reduces the actin-binding affinity of mammalian αE-catenin, *C. elegans* HMP-1, and vertebrate vinculin. However, autoinhibition is absent from *Dd* α-catenin, *Dm* α-catenin, and *Dr* αE-catenin (see Section 2). *Dm* α-catenin and *Mm* αE-catenin dimerize in solution. β-Catenin binding also negatively regulates *Dr* and *Mm* αE-catenin binding to F-actin *in vitro* (Miller et al., 2013; Yamada et al., 2005), but it is not known whether this occurs in other species. Future *in vivo* experiments should determine how these distinct biochemical properties cater to specific developmental or physiological requirements. Between unikonts, α-catenin appears to have sequence divergence as well as variable function. Functional implications of changes in domain structure of non-bilaterian and premetazoan VIN homologs in organisms that do not contain all core components of the CCC will require biochemical characterization of purified proteins.

4.2. Sequence versus function of α-catenin/vinculin family proteins

With available bioinformatics tools, it is not possible to determine the specific biochemical and functional properties of α-catenin/vinculin family proteins from sequence alone. For instance, it is not possible to use a comparison of sequence identity to classify premetazoan VIN proteins as members of either the α-catenin or vinculin protein families. *Dd* α-catenin, which has an equal level of sequence identity to metazoan α-catenin and vinculin, possesses biochemical and functional properties of α-catenin and not vinculin (Dickinson, Nelson, et al., 2011; Dickinson, Weis, et al., 2011). Furthermore, as described above, a large polytomy of α-catenin-like proteins results from phylogenetic analysis of VIN-family proteins (Fig. 8.7A). While domain organization may make stronger predictions of protein function and relatedness than strict sequence identity, functional predictions based on domain organization can be confounded by deletion, convergent evolution, or duplication of domains. As a result, in the absence of functional characterization one cannot predict specific functions of these proteins in cell–cell or cell-substrate adhesion.

In a sequence-based phylogeny of biochemically characterized α-catenin orthologs, mammalian and *D. rerio* αE-catenin cluster together with 90% sequence identity (Fig. 8.8A). *Dm* α-catenin clusters with mammalian αE-catenin with ~60% sequence identity, while *C. elegans* HMP-1 and *Dd* α-catenin are outgroups with <40% and <20% identity, respectively, to the mammalian orthologs (Fig. 8.8A). To directly juxtapose sequence-based methods with functional comparisons of α-catenin family proteins, we generated an alternative cladogram in which orthologs are clustered together based on common functional properties (Fig. 8.8B). Based upon functional properties, we would group *Mm* αE-catenin and *Dm* α-catenin due to their shared character of dimerization, and place *Dd* α-catenin and *Dr* αE-catenin together in a separate group because they are monomers that bind and bundle actin constitutively, with *C. elegans* HMP-1 forming a divergent outgroup to these due to its autoinhibition of actin binding.

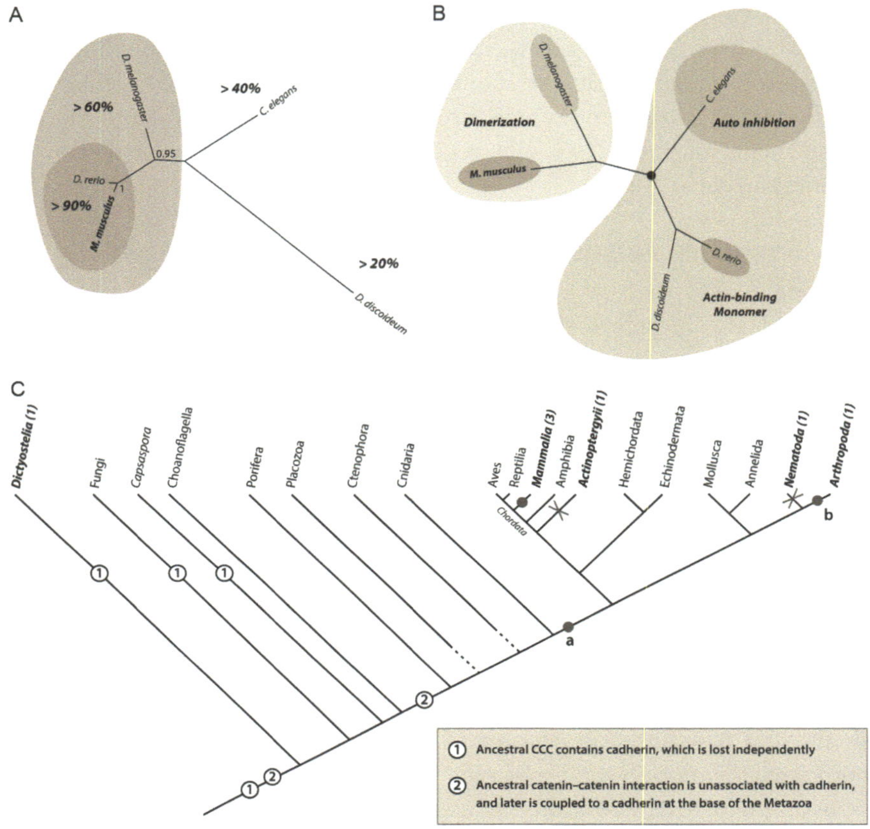

Figure 8.8 *Sequence identity versus functional equivalence, and CCC evolutionary hypotheses.* (A) A neighbor-joining tree of the five characterized metazoan α-catenin orthologs. Percent identity to *Mm* αE-catenin is indicated in bold. Orthologs sharing >90% identity are grouped in blue, >60% in green, and <60% left uncolored. (B) A cladogram of the same proteins, as grouped by function. The red oval groups actin-binding monomers, the yellow groups proteins capable of homodimerization, and the purple indicates autoinhibition of actin binding. (C) A phylogeny of unikonts, indicating two evolutionary hypotheses of CCC evolution (numbered circles), and two hypotheses for the evolution of dimerization in metazoan α-catenins. See text for detailed explanations of hypotheses. (See color plate.)

Another possible functional grouping might place mammalian and *C. elegans* α-catenins together because they both appear to be inhibited in their capacity to bind actin as monomers, but autoinhibition of actin binding in mammalian α-catenin is less well supported. Regardless of how the tree is drawn, one distinction is clear—when grouped functionally, the two proteins most similar by sequence, *Mm* and *Dr* α-catenin, cannot be grouped together, and must be grouped with proteins to which they have much lower sequence identity (Fig. 8.8B). Thus, sequence similarity is a poor indicator of functional similarity between studied α-catenin orthologs. These examples highlight the necessity of biochemical characterization as a means of assessing functional equivalence between orthologs. Current bioinformatic methods can identify putative orthologs with remarkable speed and accuracy, but our capacity to predict tertiary structure and biochemical function based on protein sequence is quite limited. Only through *in vitro* characterization and *in vivo* observation can we describe in detail the subtleties of protein activity. As demonstrated here, such observations can vastly increase the power of bioinformatic analysis: knowledge of binding sites, structurally important residues, and other functional properties, when synthesized with sequenced-based methods, can yield deeper inferences about protein evolution. By using known biochemical properties and domain boundaries to inform sequence analysis, it may be possible to understand the evolution of the CCC at a deeper level.

4.3. Evolutionary context of the CCC

Classical cadherins are clearly unique to the metazoa, but as the last common ancestor of all eukaryotes necessarily had some form of CAD protein, the role of cadherins during the formation of the ancestral CCC is unclear. The loss of cadherins from amoebozoans, apusozoans, and fungi can be seen either as a loss of cadherin from an ancestral complex, or as an indication that cadherins did not constitute the original membrane component of the CCC. Evidence from *D. discoideum* indicates that the α-catenin/Aardvark heterodimer links actin to the plasma membrane at cell contacts, but with only a single data point we are unable to say if this is an amoebazoan-specfic (or species-specific) loss of the cadherin component of the CCC due to derivation of other novel junction complexes, or if cadherin was absent from the ancestral CCC altogether. Further investigation of the CCC components of choanoflagellates and *C. owczarzaki,* the only non-metazoan opthistokonts known to have proteins of all three CCC protein families, is necessary to elucidate the evolution of cadherin–catenin interactions.

The evolution of β-catenin appears to be due to gain of function. Based on the homology of α-catenin-binding motifs between metazoan β-catenins and *D. discoideum* Aardvark (Fig. 8.6E), the capacity to bind α-catenin, and thereby potentially link the actin cytoskeleton to the plasma membrane, was the ancestral function of the common ancestor of these proteins. The role of β-catenin in Wnt signaling is a metazoan novelty acquired through cooption following the evolution of the Wnt pathway. Based on this evidence, we can infer that an α-catenin-binding ARM protein was part of the shared inheritance of opthistokonts, which has either been lost in multiple groups (*capsaspora*, ichthyosporeans, choanflagellates, fungi, etc.), or remains to be discovered. As previous search attempts have focused on finding full β-catenin orthologs, and have not utilized a strategy based on identification of α-catenin-binding motifs, the latter is likely true, but only further analysis of the ARM proteins of unicellular organisms can determine how ancient this interaction truly is.

We propose a model of α-catenin evolution in which the ancestral homolog was a constitutively active F-actin-binding monomer that gained β-catenin binding in a unicellular ancestor to amoebozoa and opisthokonta (Fig. 8.8C). In unicellular organisms, the α/β-catenin heterodimer may have supported cell–cell contact formation during mating and feeding behavior. Duplication of an α-catenin-like protein in an ancestral unikont resulted in multiple VIN-family proteins, which developed distinct functions in cell–cell and cell–ECM adhesion in metazoans. Innovation of vinculin-like proteins likely facilitated cell-substrate adhesion during crawling in unicellular organisms, and perhaps strengthened cell–cell contacts in early multicellular organisms.

α-Catenin has not been characterized in a basal, non-bilaterian metazoan and we can only speculate about its regulatory properties. Nevertheless, we posit that it was likely a monomeric, β-catenin and F-actin-binding protein that gained additional functions for complex development and tissue construction through association with classical cadherins. Given the limited phylogenetic coverage of currently characterized orthologs (groups in bold, Fig. 8.8C), there are two equally probable evolutionary histories of bilaterian α-catenin: (1) α-Catenin developed the ability to homodimerize in an ancestor to the bilaterian clade, and this function was subsequently lost in *C. elegans* and *D. rerio* (Fig. 8.8C: red dot and X, respectively). (2) The innovation of homodimerization is convergent between *D. melanogaster* and mammals (Fig. 8.8C; blue dots). The fact that homodimerization in mammals, but not in *Drosophila,* is necessary to potentiate F-actin binding by relieving autoinhibition supports a hypothesis of convergent evolution. Similarly autoinhibition may have originated in a common ancestor to bilateria and was subsequently disrupted in *D. melanogaster* and *D. rerio*, or may have been independent innovations of *C. elegans* and mammals. Future characterization of orthologs from other bilaterians will allow a fuller explanation of the evolutionary origin of these regulatory properties. In addition, while αE-, αN-, and αT-catenin isoforms originated from a single gene in higher vertebrates, the significance of three isoforms is not well understood, and future biochemical and cell biological studies will reveal the properties underlying their distinct developmental functions.

In light of evidence reviewed here, the assembly of the CCC can be viewed under two alternative hypotheses: (1) The ancestral CCC included a cadherin, which was lost in some non-metazoan lineages (1 on Fig. 8.8C); or (2) An ancestral interaction between an ARM protein and a VIN protein linked the actin cytoskeleton to the plasma membrane by way of a currently unknown protein, and a cadherin was co-opted to replace the membrane portion of this complex at the base of the metazoa (2 on Fig. 8.8C). Both hypotheses assert that the core α-catenin/β-catenin interaction of the CCC is ancestral and homologous within unikonts, but differs in how they regulate cell–cell adhesion. The first hypothesis stipulates that the capacity to form cell–cell contacts mediated by the CCC was ancestral to unikonts, and therefore that extant noncolonial unicellular lineages may be secondarily simplified. The second hypothesis requires fewer evolutionary transitions.

Interpreting the validity of these hypotheses based on only parsimony belies the evolutionary complexity of cell–cell junction formation. An ancestral capacity to form cell–cell contacts is in line with the prevalence and diversity of colonial, aggregate, and multicellular organisms within the Unikonta, but without additional data points from fungi, choanoflagellates, and other non-metazoan opthistokonts we can only speculate about the importance of the ancestral CCC in the evolution of these organisms. Given the presence of CCC complex in all metazoans, it is likely that conserved residues and domains at binding interfaces mediated refinement of the ancestral CCC into the complex seen in bilaterians, and thus constrained the independent evolution of its constituent proteins. Studies of co-evolution of classical cadherins, β-catenin, and α-catenin homologs will provide further understanding of CCC function within metazoans, and investigation of distant homologs in non-metazoans will provide insight into the evolution of the CCC and the origins of multicellularity.

ACKNOWLEDGMENTS

This work was supported by NIH Ruth L. Kirschstein National Research Service Award GM007276 (P. W. M.), NSF Pre-doctoral Fellowship (D. N. C.), NIH GM035527 (W. J. N.), NIH GM56169 (W. I. W.), and NIH U01 GM094663 (W. J. N., W. I. W.).

REFERENCES

Abe, K., Chisaka, O., Van Roy, F., & Takeichi, M. (2004). Stability of dendritic spines and synaptic contacts is controlled by alpha N-catenin. *Nature Neuroscience*, *7*, 357–363.

Abe, K., & Takeichi, M. (2008). EPLIN mediates linkage of the cadherin catenin complex to F-actin and stabilizes the circumferential actin belt. *Proceedings of the National Academy of Sciences of the United States of America*, *105*, 13–19.

Abedin, M., & King, N. (2008). The premetazoan ancestry of cadherins. *Science*, *319*, 946–948.

Aberle, H., Schwartz, H., Hoschuetzky, H., & Kemler, R. (1996). Single amino acid substitutions in proteins of the armadillo gene family abolish their binding to alpha-catenin. *The Journal of Biological Chemistry*, *271*, 1520–1526.

Adams, C. L., & Nelson, W. J. (1998). Cytomechanics of cadherin-mediated cell-cell adhesion. *Current Opinion in Cell Biology*, *10*, 572–577.

Adamska, M., Larroux, C., Adamski, M., Green, K., Lovas, E., Koop, D., et al. (2010). Structure and expression of conserved Wnt pathway components in the demosponge Amphimedon queenslandica. *Evolution and Development*, *12*, 494–518.

Alatortsev, V. E., Kramerova, I. A., Frolov, M. V., Lavrov, S. A., & Westphal, E. D. (1997). Vinculin gene is non-essential in Drosophila melanogaster. *FEBS Letters*, *413*, 197–201.

Bakolitsa, C., Cohen, D. M., Bankston, L. A., Bobkov, A. A., Cadwell, G. W., Jennings, L., et al. (2004). Structural basis for vinculin activation at sites of cell adhesion. *Nature*, *430*, 583–586.

Bakolitsa, C., de Pereda, J. M., Bagshaw, C. R., Critchley, D. R., & Liddington, R. C. (1999). Crystal structure of the vinculin tail suggests a pathway for activation. *Cell*, *99*, 603–613.

Bao, R., Fischer, T., Bolognesi, R., Brown, S. J., & Friedrich, M. (2012). Parallel duplication and partial subfunctionalization of beta-catenin/armadillo during insect evolution. *Molecular Biology and Evolution*, *29*, 647–662.

Barstead, R. J., & Waterston, R. H. (1989). The basal component of the nematode dense-body is vinculin. *The Journal of Biological Chemistry*, *264*, 10177–10185.

Barstead, R. J., & Waterston, R. H. (1991). Vinculin is essential for muscle function in the nematode. *The Journal of Cell Biology*, *114*, 715–724.

Benjamin, J. M., Kwiatkowski, A. V., Yang, C., Korobova, F., Pokutta, S., Svitkina, T., et al. (2010). AlphaE-catenin regulates actin dynamics independently of cadherin-mediated cell-cell adhesion. *The Journal of Cell Biology*, *189*, 339–352.

Benjamin, J. M., & Nelson, W. J. (2008). Bench to bedside and back again: Molecular mechanisms of alpha-catenin function and roles in tumorigenesis. *Seminars in Cancer Biology*, *18*, 53–64.

Boggon, T. J., Murray, J., Chappuis-Flament, S., Wong, E., Gumbiner, B. M., & Shapiro, L. (2002). C-cadherin ectodomain structure and implications for cell adhesion mechanisms. *Science*, *296*, 1308–1313.

Borghi, N., Sorokina, M., Shcherbakova, O. G., Weis, W. I., Pruitt, B. L., Nelson, W. J., et al. (2012). E-cadherin is under constitutive actomyosin-generated tension that is increased at cell-cell contacts upon externally applied stretch. *Proceedings of the National Academy of Sciences of the United States of America*, *109*, 12568–12573.

Brembeck, F. H., Schwarz-Romond, T., Bakkers, J., Wilhelm, S., Hammerschmidt, M., & Birchmeier, W. (2004). Essential role of BCL9-2 in the switch between beta-catenin's adhesive and transcriptional functions. *Genes & Development*, *18*, 2225–2230.

Bryant, D. M., & Mostov, K. E. (2008). From cells to organs: Building polarized tissue. *Nature Reviews Molecular Cell Biology*, *9*, 887–901.

Bullions, L. C., Notterman, D. A., Chung, L. S., & Levine, A. J. (1997). Expression of wild-type alpha-catenin protein in cells with a mutant alpha-catenin gene restores both growth regulation and tumor suppressor activities. *Molecular and Cellular Biology*, *17*, 4501–4508.

Capella-Gutierrez, S., Silla-Martinez, J. M., & Gabaldon, T. (2009). trimAl: a tool for automated alignment trimming in large-scale phylogenetic analyses. *Bioinformatics*, *25*, 1972–1973.

Carnahan, R. H., Rokas, A., Gaucher, E. A., & Reynolds, A. B. (2010). The molecular evolution of the p120-catenin subfamily and its functional associations. *PLoS One*, *5*, e15747.

Carroll, S. B. (2008). Evo-devo and an expanding evolutionary synthesis: A genetic theory of morphological evolution. *Cell*, *134*, 25–36.

Cavey, M., Rauzi, M., Lenne, P.-F., & Lecuit, T. (2008). A two-tiered mechanism for stabilization and immobilization of E-cadherin. *Nature*, *453*, 751–756.

Cereijido, M., Contreras, R. G., & Shoshani, L. (2004). Cell adhesion, polarity, and epithelia in the dawn of metazoans. *Physiological Reviews*, *84*, 1229–1262.

Charras, G., & Paluch, E. (2008). Blebs lead the way: How to migrate without lamellipodia. *Nature Reviews Molecular Cell Biology*, *9*, 730–736.

Chen, H., Choudhury, D. M., & Craig, S. W. (2006). Coincidence of actin filaments and talin is required to activate vinculin. *The Journal of Biological Chemistry*, *281*, 40389–40398.

Choi, H.-J., Pokutta, S., Cadwell, G. W., Bobkov, A. A., Bankston, L. A., Liddington, R. C., et al. (2012). αE-catenin is an autoinhibited molecule that coactivates vinculin. *Proceedings of the National Academy of Sciences of the United States of America*, *109*, 8576–8581.

Coates, J. C. (2003). Armadillo repeat proteins: Beyond the animal kingdom. *Trends in Cell Biology*, *13*, 463–471.

Costa, M., Raich, W., Agbunag, C., Leung, B., Hardin, J., & Priess, J. R. (1998). A putative catenin-cadherin system mediates morphogenesis of the Caenorhabditis elegans embryo. *The Journal of Cell Biology*, *141*, 297–308.

Craig, S. W., & Johnson, R. P. (1996). Assembly of focal adhesions: Progress, paradigms, and portents. *Current Opinion in Cell Biology*, *8*, 74–85.

Darriba, D., Taboada, G. L., Doallo, R., & Posada, D. (2011). ProtTest 3: Fast selection of best-fit models of protein evolution. *Bioinformatics*, *27*, 1164–1165.

Davis, M. A., Ireton, R. C., & Reynolds, A. B. (2003). A core function for p120-catenin in cadherin turnover. *The Journal of Cell Biology*, *163*, 525–534.

Desai, R., Sarpal, R., Ishiyama, N., Pellikka, M., Ikura, M., & Tepass, U. (2013). Monomeric α-catenin links cadherin to the actin cytoskeleton. *Nature Cell Biology*, *15*, 261–273.

Dickinson, D. J., Nelson, W. J., & Weis, W. I. (2011). A polarized epithelium organized by beta- and alpha-catenin predates cadherin and metazoan origins. *Science*, *331*, 1336–1339.

Dickinson, D. J., Nelson, W. J., & Weis, W. I. (2012). An epithelial tissue in Dictyostelium challenges the traditional origin of metazoan multicellularity. *Bioessays*, *34*, 833–840.

Dickinson, D. J., Weis, W. I., & Nelson, W. J. (2011). Protein evolution in cell and tissue development: Going beyond sequence and transcriptional analysis. *Developmental Cell*, *21*, 32–34.

Drees, F., Pokutta, S., Yamada, S., Nelson, W. J., & Weis, W. I. (2005). Alpha-catenin is a molecular switch that binds E-cadherin-beta-catenin and regulates actin-filament assembly. *Cell*, *123*, 903–915.

Edgar, R. C. (2004). MUSCLE: Multiple sequence alignment with high accuracy and high throughput. *Nucleic Acids Research*, *32*, 1792–1797.

Fahey, B., & Degnan, B. M. (2010). Origin of animal epithelia: Insights from the sponge genome. *Evolution and Development, 12*, 601–617.
Fraiberg, M., Borovok, I., Weiner, R. M., & Lamed, R. (2010). Discovery and characterization of cadherin domains in Saccharophagus degradans 2-40. *Journal of Bacteriology, 192*, 1066–1074.
Graham, T. A., Ferkey, D. M., Mao, F., Kimelman, D., & Xu, W. (2001). Tcf4 can specifically recognize beta-catenin using alternative conformations. *Nature Structural Biology, 8*, 1048–1052.
Graham, T. A., Weaver, C., Mao, F., Kimelman, D., & Xu, W. (2000). Crystal structure of a beta-catenin/Tcf complex. *Cell, 103*, 885–896.
Grimson, M. J., Coates, J. C., Reynolds, J. P., Shipman, M., Blanton, R. L., & Harwood, A. J. (2000). Adherens junctions and beta-catenin-mediated cell signalling in a non-metazoan organism. *Nature, 408*, 727–731.
Gumbiner, B. M. (2005). Regulation of cadherin-mediated adhesion in morphogenesis. *Nature Reviews Molecular Cell Biology, 6*, 622–634.
Ha, N. C., Tonozuka, T., Stamos, J. L., Choi, H. J., & Weis, W. I. (2004). Mechanism of phosphorylation-dependent binding of APC to beta-catenin and its role in beta-catenin degradation. *Molecular Cell, 15*, 511–521.
Halbleib, J. M., & Nelson, W. J. (2006). Cadherins in development: Cell adhesion, sorting, and tissue morphogenesis. *Genes & Development, 20*, 3199–3214.
Hansen, S. D., Kwiatkowski, A. V., Ouyang, C. Y., Liu, H., Pokutta, S., Watkins, S. C., et al. (2013). Alpha-E-catenin actin binding domain alters actin filament conformation and regulates binding of nucleation and disassembly factors. *Molecular Biology of the Cell, in press*.
Hardin, J., Lynch, A., Loveless, T., & Pettitt, J. (2013). Cadherins and their partners in the nematode worm Caenorhabditis elegans. *Progress in Molecular Biology and Translational Science, 116*, 239–262.
Harris, T. J. C., & Tepass, U. (2010). Adherens junctions: From molecules to morphogenesis. *Nature Reviews Molecular Cell Biology, 11*, 502–514.
Harrison, O. J., Jin, X., Hong, S., Bahna, F., Ahlsen, G., Brasch, J., et al. (2011). The extracellular architecture of adherens junctions revealed by crystal structures of type I cadherins. *Structure, 19*, 244–256.
Haruta, T., Warrior, R., Yonemura, S., & Oda, H. (2010). The proximal half of the Drosophila E-cadherin extracellular region is dispensable for many cadherin-dependent events but required for ventral furrow formation. *Genes to Cells, 15*, 193–208.
Herrenknecht, K., Ozawa, M., Eckerskorn, C., Lottspeich, F., Lenter, M., & Kemler, R. (1991). The uvomorulin-anchorage protein alpha catenin is a vinculin homologue. *Proceedings of the National Academy of Sciences of the United States of America, 88*, 9156–9160.
Hirano, S., & Takeichi, M. (1994). Differential expression of alpha N-catenin and N-cadherin during early development of chicken embryos. *The International Journal of Developmental Biology, 38*, 379–384.
Hoffmans, R., & Basler, K. (2007). BCL9-2 binds Arm/beta-catenin in a Tyr142-independent manner and requires Pygopus for its function in Wg/Wnt signaling. *Mechanisms of Development, 124*, 59–67.
Huber, A. H., Nelson, W. J., & Weis, W. I. (1997). Three-dimensional structure of the armadillo repeat region of beta-catenin. *Cell, 90*, 871–882.
Huber, A. H., & Weis, W. I. (2001). The structure of the beta-catenin/E-cadherin complex and the molecular basis of diverse ligand recognition by beta-catenin. *Cell, 105*, 391–402.
Hulpiau, P., Gul, I. S., & van Roy, F. (2013). New insights into the evolution of metazoan cadherins and catenins. *Progress in Molecular Biology and Translational Science, 116*, 71–94.
Hulpiau, P., & van Roy, F. (2009). Molecular evolution of the cadherin superfamily. *The International Journal of Biochemistry & Cell Biology, 41*, 349–369.

Hulpiau, P., & van Roy, F. (2011). New insights into the evolution of metazoan cadherins. *Molecular Biology and Evolution, 28*, 647–657.

Huttlin, E. L., Jedrychowski, M. P., Elias, J. E., Goswami, T., Rad, R., Beausoleil, S. A., et al. (2010). A tissue-specific atlas of mouse protein phosphorylation and expression. *Cell, 143*, 1174–1189.

Huveneers, S., & de Rooij, J. (2013). Mechanosensitive systems at the cadherin-F-actin interface. *Journal of Cell Science, 126*, 403–413.

Ishiyama, N., Tanaka, N., Abe, K., Yang, Y. J., Abbas, Y. M., Umitsu, M., et al. (2013). An autoinhibited structure of α-catenin and its implications for vinculin recruitment to adherens junctions. *The Journal of Biological Chemistry, 288*, 15913–15925.

Itoh, M., Nagafuchi, A., Moroi, S., & Tsukita, S. (1997). Involvement of ZO-1 in cadherin-based cell adhesion through its direct binding to alpha catenin and actin filaments. *The Journal of Cell Biology, 138*, 181–192.

Ivanov, D. B., Philippova, M. P., & Tkachuk, V. A. (2001). Structure and functions of classical cadherins. *Biochemistry (Mosc), 66*, 1174–1186.

Janssen, M. E. W., Kim, E., Liu, H., Fujimoto, L. M., Bobkov, A., Volkmann, N., et al. (2006). Three-dimensional structure of vinculin bound to actin filaments. *Molecular Cell, 21*, 271–281.

Janssens, B., Mohapatra, B., Vatta, M., Goossens, S., Vanpoucke, G., Kools, P., et al. (2003). Assessment of the CTNNA3 gene encoding human alpha T-catenin regarding its involvement in dilated cardiomyopathy. *Human Genetics, 112*, 227–236.

Kadrmas, J. L., & Beckerle, M. C. (2004). The LIM domain: From the cytoskeleton to the nucleus. *Nature Reviews Molecular Cell Biology, 5*, 920–931.

Kane, D. A., Maischein, H. M., Brand, M., van Eeden, F. J., Furutani-Seiki, M., Granato, M., et al. (1996). The zebrafish early arrest mutants. *Development, 123*, 57–66.

Khurana, B., Khurana, T., Khaire, N., & Noegel, A. A. (2002). Functions of LIM proteins in cell polarity and chemotactic motility. *The EMBO Journal, 21*, 5331–5342.

Knudsen, K. A., Soler, A. P., Johnson, K. R., & Wheelock, M. J. (1995). Interaction of alpha-actinin with the cadherin/catenin cell-cell adhesion complex via alpha-catenin. *The Journal of Cell Biology, 130*, 67–77.

Koch, A. W., Pokutta, S., Lustig, A., & Engel, J. (1997). Calcium binding and homoassociation of E-cadherin domains. *Biochemistry, 36*, 7697–7705.

Koch, B. J., Ryan, J. F., & Baxevanis, A. D. (2012). The diversification of the LIM superclass at the base of the metazoa increased subcellular complexity and promoted multicellular specialization. *PLoS One, 7*, e33261.

Kofron, M., Spagnuolo, A., Klymkowsky, M., Wylie, C., & Heasman, J. (1997). The roles of maternal alpha-catenin and plakoglobin in the early Xenopus embryo. *Development, 124*, 1553–1560.

Koslov, E. R., Maupin, P., Pradhan, D., Morrow, J. S., & Rimm, D. L. (1997). Alpha-catenin can form asymmetric homodimeric complexes and/or heterodimeric complexes with beta-catenin. *The Journal of Biological Chemistry, 272*, 27301–27306.

Kurth, T., Fesenko, I. V., Schneider, S., Munchberg, F. E., Joos, T. O., Spieker, T. P., et al. (1999). Immunocytochemical studies of the interactions of cadherins and catenins in the early Xenopus embryo. *Developmental Dynamics, 215*, 155–169.

Kwiatkowski, A. V., Maiden, S. L., Pokutta, S., Choi, H.-J., Benjamin, J. M., Lynch, A. M., et al. (2010). In vitro and in vivo reconstitution of the cadherin-catenin-actin complex from Caenorhabditis elegans. *Proceedings of the National Academy of Sciences of the United States of America, 107*, 14591–14596.

Larue, L., Antos, C., Butz, S., Huber, O., Delmas, V., Dominis, M., et al. (1996). A role for cadherins in tissue formation. *Development, 122*, 3185–3194.

Larue, L., Ohsugi, M., Hirchenhain, J., & Kemler, R. (1994). E-cadherin null mutant embryos fail to form a trophectoderm epithelium. *Proceedings of the National Academy of Sciences of the United States of America, 91*, 8263–8267.

le Duc, Q., Shi, Q., Blonk, I., Sonnenberg, A., Wang, N., Leckband, D., et al. (2010). Vinculin potentiates E-cadherin mechanosensing and is recruited to actin-anchored sites within adherens junctions in a myosin II-dependent manner. *The Journal of Cell Biology*, *189*, 1107–1115.

Letunic, I., Doerks, T., & Bork, P. (2009). SMART 6: Recent updates and new developments. *Nucleic Acids Research*, *37*, D229–D232.

Leys, S. P., & Hill, A. (2012). The physiology and molecular biology of sponge tissues. *Advances in Marine Biology*, *62*, 1–56.

Liu, J., Phillips, B. T., Amaya, M. F., Kimble, J., & Xu, W. (2008). The C. elegans SYS-1 protein is a bona fide beta-catenin. *Developmental Cell*, *14*, 751–761.

Magie, C. R., Pinto-Santini, D., & Parkhurst, S. M. (2002). Rho1 interacts with p120ctn and alpha-catenin, and regulates cadherin-based adherens junction components in Drosophila. *Development*, *129*, 3771–3782.

Maiden, S. L., Harrison, N., Keegan, J., Cain, B., Lynch, A. M., Pettitt, J., et al. (2013). Specific conserved C-terminal amino acids of Caenorhabditis elegans HMP-1/α-catenin modulate F-actin binding independently of vinculin. *The Journal of Biological Chemistry*, *288*, 5694–5706.

Maiers, J. L., Peng, X., Fanning, A. S., & Demali, K. A. (2013). ZO-1 recruitment to alpha-catenin: A novel mechanism for coupling the assembly of tight junctions to adherens junctions. *Journal of Cell Science*, *126*, 3904–3905.

Maitre, J.-L., Berthoumieux, H., Krens, S. F. G., Salbreux, G., Julicher, F., Paluch, E., et al. (2012). Adhesion functions in cell sorting by mechanically coupling the cortices of adhering cells. *Science*, *338*, 253–256.

Marchiando, A. M., Graham, W. V., & Turner, J. R. (2010). Epithelial barriers in homeostasis and disease. *Annual Review of Pathology*, *5*, 119–144.

Mbom, B. C., Nelson, W. J., & Barth, A. (2013). Beta-catenin at the centrosome: Discrete pools of beta-catenin communicate during mitosis and may co-ordinate centrosome functions and cell cycle progression. *Bioessays*, *35*, 804–809.

Mellman, I., & Nelson, W. J. (2008). Coordinated protein sorting, targeting and distribution in polarized cells. *Nature Reviews Molecular Cell Biology*, *9*, 833–845.

Menke, A., & Giehl, K. (2012). Regulation of adherens junctions by Rho GTPases and p120-catenin. *Archives of Biochemistry and Biophysics*, *524*, 48–55.

Miller, P. W., Pokutta, S., Ghosh, A., Almo, S. C., Weis, W. I., Nelson, W. J., et al. (2013). Danio rerio αE-catenin is a monomeric F-actin binding protein with distinct properties from Mus musculus αE-catenin. *The Journal of Biological Chemistry*, *288*, 22324–22332.

Nagafuchi, A., Ishihara, S., & Tsukita, S. (1994). The roles of catenins in the cadherin-mediated cell adhesion: Functional analysis of E-cadherin-alpha catenin fusion molecules. *The Journal of Cell Biology*, *127*, 235–245.

Nandadasa, S., Tao, Q., Shoemaker, A., Cha, S.-W., & Wylie, C. (2012). Regulation of classical cadherin membrane expression and F-actin assembly by alpha-catenins, during Xenopus embryogenesis. *PLoS One*, *7*, e38756.

Natarajan, L., Witwer, N. E., & Eisenmann, D. M. (2001). The divergent Caenorhabditis elegans beta-catenin proteins BAR-1, WRM-1 and HMP-2 make distinct protein interactions but retain functional redundancy in vivo. *Genetics*, *159*, 159–172.

Nejsum, L. N., & Nelson, W. J. (2007). A molecular mechanism directly linking E-cadherin adhesion to initiation of epithelial cell surface polarity. *The Journal of Cell Biology*, *178*, 323–335.

Nelson, W. J. (2003). Adaptation of core mechanisms to generate cell polarity. *Nature*, *422*, 766–774.

Nelson, W. J., Dickinson, D. J., & Weis, W. I. (2013). Roles of cadherins and catenins in cell-cell adhesion and epithelial cell polarity. *Progress in Molecular Biology and Translational Science*, *116*, 3–23.

Nelson, W. J., & Nusse, R. (2004). Convergence of Wnt, beta-catenin, and cadherin pathways. *Science*, *303*, 1483–1487.
Nichols, S. A., Dirks, W., Pearse, J. S., & King, N. (2006). Early evolution of animal cell signaling and adhesion genes. *Proceedings of the National Academy of Sciences of the United States of America*, *103*, 12451–12456.
Nichols, S. A., Roberts, B. W., Richter, D. J., Fairclough, S. R., & King, N. (2012). Origin of metazoan cadherin diversity and the antiquity of the classical cadherin/β-catenin complex. *Proceedings of the National Academy of Sciences of the United States of America*, *109*, 13046–13051.
Oda, H., & Takeichi, M. (2011). Evolution: Structural and functional diversity of cadherin at the adherens junction. *The Journal of Cell Biology*, *193*, 1137–1146.
Oda, H., & Tsukita, S. (1999). Nonchordate classic cadherins have a structurally and functionally unique domain that is absent from chordate classic cadherins. *Developmental Biology*, *216*, 406–422.
Oda, H., Uemura, T., Shiomi, K., Nagafuchi, A., Tsukita, S., & Takeichi, M. (1993). Identification of a Drosophila homologue of alpha-catenin and its association with the armadillo protein. *The Journal of Cell Biology*, *121*, 1133–1140.
Ortlund, E. A., Bridgham, J. T., Redinbo, M. R., & Thornton, J. W. (2007). Crystal structure of an ancient protein: Evolution by conformational epistasis. *Science*, *317*, 1544–1548.
Pai, L. M., Kirkpatrick, C., Blanton, J., Oda, H., Takeichi, M., & Peifer, M. (1996). Drosophila alpha-catenin and E-cadherin bind to distinct regions of Drosophila Armadillo. *The Journal of Biological Chemistry*, *271*, 32411–32420.
Park, C., Falls, W., Finger, J. H., Longo-Guess, C. M., & Ackerman, S. L. (2002). Deletion in Catna2, encoding alpha N-catenin, causes cerebellar and hippocampal lamination defects and impaired startle modulation. *Nature Genetics*, *31*, 279–284.
Peng, X., Cuff, L. E., Lawton, C. D., & DeMali, K. A. (2010). Vinculin regulates cell-surface E-cadherin expression by binding to beta-catenin. *Journal of Cell Science*, *123*, 567–577.
Peng, X., Nelson, E. S., Maiers, J. L., & DeMali, K. A. (2011). New insights into vinculin function and regulation. *International Review of Cell and Molecular Biology*, *287*, 191–231.
Pieters, T., van Roy, F., & van Hengel, J. (2012). Functions of p120ctn isoforms in cell-cell adhesion and intracellular signaling. *Frontiers in Bioscience*, *17*, 1669–1694.
Plotnikov, S. V., Pasapera, A. M., Sabass, B., & Waterman, C. M. (2012). Force fluctuations within focal adhesions mediate ECM-rigidity sensing to guide directed cell migration. *Cell*, *151*, 1513–1527.
Pokutta, S., Drees, F., Takai, Y., Nelson, W. J., & Weis, W. I. (2002). Biochemical and structural definition of the l-afadin- and actin-binding sites of alpha-catenin. *The Journal of Biological Chemistry*, *277*, 18868–18874.
Pokutta, S., & Weis, W. I. (2000). Structure of the dimerization and beta-catenin-binding region of alpha-catenin. *Molecular Cell*, *5*, 533–543.
Ponting, C. P. (2001). Issues in predicting protein function from sequence. *Briefings in Bioinformatics*, *2*, 19–29.
Punta, M., Coggill, P. C., Eberhardt, R. Y., Mistry, J., Tate, J., Boursnell, C., et al. (2012). The Pfam protein families database. *Nucleic Acids Research*, *40*, D290–D301.
Rangarajan, E. S., & Izard, T. (2012). The cytoskeletal protein alpha-catenin unfurls upon binding to vinculin. *The Journal of Biological Chemistry*, *287*, 18492–18499.
Rangarajan, E. S., & Izard, T. (2013). Dimer asymmetry defines α-catenin interactions. *Nature Structural & Molecular Biology*, *20*, 188–193.
Rimm, D. L., Koslov, E. R., Kebriaei, P., Cianci, C. D., & Morrow, J. S. (1995). Alpha 1 (E)-catenin is an actin-binding and -bundling protein mediating the attachment of

F-actin to the membrane adhesion complex. *Proceedings of the National Academy of Sciences of the United States of America, 92*, 8813–8817.
Roura, S., Miravet, S., Piedra, J., Garcia de Herreros, A., & Dunach, M. (1999). Regulation of E-cadherin/Catenin association by tyrosine phosphorylation. *The Journal of Biological Chemistry, 274*, 36734–36740.
Sarpal, R., Pellikka, M., Patel, R. R., Hui, F. Y. W., Godt, D., & Tepass, U. (2012). Mutational analysis supports a core role for Drosophila α-catenin in adherens junction function. *Journal of Cell Science, 125*, 233–245.
Schaap, P., Winckler, T., Nelson, M., Alvarez-Curto, E., Elgie, B., Hagiwara, H., et al. (2006). Molecular phylogeny and evolution of morphology in the social amoebas. *Science, 314*, 661–663.
Schepis, A., & Nelson, W. J. (2012). Adherens junction function and regulation during zebrafish gastrulation. *Cell Adhesion & Migration, 6*, 173–178.
Schepis, A., Sepich, D., & Nelson, W. J. (2012). αE-catenin regulates cell-cell adhesion and membrane blebbing during zebrafish epiboly. *Development, 139*, 537–546.
Schierwater, B., de Jong, D., & Desalle, R. (2009). Placozoa and the evolution of Metazoa and intrasomatic cell differentiation. *The International Journal of Biochemistry & Cell Biology, 41*, 370–379.
Schneider, S. Q., Finnerty, J. R., & Martindale, M. Q. (2003). Protein evolution: Structure-function relationships of the oncogene beta-catenin in the evolution of multicellular animals. *Journal of Experimental Zoology Part B, Molecular and Developmental Evolution, 295*, 25–44.
Sehgal, R. N., Gumbiner, B. M., & Reichardt, L. F. (1997). Antagonism of cell adhesion by an alpha-catenin mutant, and of the Wnt-signaling pathway by alpha-catenin in Xenopus embryos. *The Journal of Cell Biology, 139*, 1033–1046.
Shapiro, L., & Weis, W. I. (2009). Structure and biochemistry of cadherins and catenins. *Cold Spring Harbor Perspectives in Biology, 1*, a003053.
Spink, K. E., Fridman, S. G., & Weis, W. I. (2001). Molecular mechanisms of β-catenin recognition by Adenomatous Polyposis Coli revealed by the structure of an APC/β-catenin complex. *The EMBO Journal, 20*, 6203–6212.
Srivastava, M., Begovic, E., Chapman, J., Putnam, N. H., Hellsten, U., Kawashima, T., et al. (2008). The Trichoplax genome and the nature of placozoans. *Nature, 454*, 955–960.
Stamatakis, A. (2006). RAxML-VI-HPC: Maximum likelihood-based phylogenetic analyses with thousands of taxa and mixed models. *Bioinformatics, 22*, 2688–2690.
Stepniak, E., Radice, G. L., & Vasioukhin, V. (2009). Adhesive and signaling functions of cadherins and catenins in vertebrate development. *Cold Spring Harbor Perspectives in Biology, 1*, a002949.
Taurin, S., Sandbo, N., Qin, Y., Browning, D., & Dulin, N. O. (2006). Phosphorylation of beta-catenin by cyclic AMP-dependent protein kinase. *The Journal of Biological Chemistry, 281*, 9971–9976.
Tewari, R., Bailes, E., Bunting, K. A., & Coates, J. C. (2010). Armadillo-repeat protein functions: Questions for little creatures. *Trends in Cell Biology, 20*, 470–481.
Torres, M., Stoykova, A., Huber, O., Chowdhury, K., Bonaldo, P., Mansouri, A., et al. (1997). An alpha-E-catenin gene trap mutation defines its function in preimplantation development. *Proceedings of the National Academy of Sciences of the United States of America, 94*, 901–906.
Uchida, N., Shimamura, K., Miyatani, S., Copeland, N. G., Gilbert, D. J., Jenkins, N. A., et al. (1994). Mouse alpha N-catenin: Two isoforms, specific expression in the nervous system, and chromosomal localization of the gene. *Developmental Biology, 163*, 75–85.
Vasioukhin, V., Bauer, C., Degenstein, L., Wise, B., & Fuchs, E. (2001). Hyperproliferation and defects in epithelial polarity upon conditional ablation of alpha-catenin in skin. *Cell, 104*, 605–617.
Watabe, M., Nagafuchi, A., Tsukita, S., & Takeichi, M. (1994). Induction of polarized cell-cell association and retardation of growth by activation of the E-cadherin-

catenin adhesion system in a dispersed carcinoma line. *The Journal of Cell Biology, 127,* 247–256.
Watabe-Uchida, M., Uchida, N., Imamura, Y., Nagafuchi, A., Fujimoto, K., Uemura, T., et al. (1998). Alpha-catenin-vinculin interaction functions to organize the apical junctional complex in epithelial cells. *The Journal of Cell Biology, 142,* 847–857.
Waterhouse, A. M., Procter, J. B., Martin, D. M., Clamp, M., & Barton, G. J. (2009). Jalview Version 2—A multiple sequence alignment editor and analysis workbench. *Bioinformatics, 25,* 1189–1191.
Watson, J. D., Laskowski, R. A., & Thornton, J. M. (2005). Predicting protein function from sequence and structural data. *Current Opinion in Structural Biology, 15,* 275–284.
Weis, W. I., & Nelson, W. J. (2006). Re-solving the cadherin-catenin-actin conundrum. *The Journal of Biological Chemistry, 281,* 35593–35597.
Wessells, N. K., Spooner, B. S., Ash, J. F., Bradley, M. O., Luduena, M. A., Taylor, E. L., et al. (1971). Microfilaments in cellular and developmental processes. *Science, 171,* 135–143.
Wheelock, M. J., & Johnson, K. R. (2003). Cadherins as modulators of cellular phenotype. *Annual Review of Cell and Developmental Biology, 19,* 207–235.
Xing, Y., Clements, W. K., Kimelman, D., & Xu, W. (2003). Crystal structure of a beta-catenin/axin complex suggests a mechanism for the beta-catenin destruction complex. *Genes & Development, 17,* 2753–2764.
Xing, Y., Clements, W. K., Le Trong, I., Hinds, T. R., Stenkamp, R., Kimelman, D., et al. (2004). Crystal structure of a beta-catenin/APC complex reveals a critical role for APC phosphorylation in APC function. *Molecular Cell, 15,* 523–533.
Xu, W., Baribault, H., & Adamson, E. D. (1998). Vinculin knockout results in heart and brain defects during embryonic development. *Development, 125,* 327–337.
Yamada, S., Pokutta, S., Drees, F., Weis, W. I., & Nelson, W. J. (2005). Deconstructing the cadherin-catenin-actin complex. *Cell, 123,* 889–901.
Yang, J., Dokurno, P., Tonks, N. K., & Barford, D. (2001). Crystal structure of the M-fragment of alpha-catenin: Implications for modulation of cell adhesion. *The EMBO Journal, 20,* 3645–3656.
Yonemura, S., Wada, Y., Watanabe, T., Nagafuchi, A., & Shibata, M. (2010). Alpha-catenin as a tension transducer that induces adherens junction development. *Nature Cell Biology, 12,* 533–542.
Yu, D. S., Lee, D. H., Kim, S. K., Lee, C. H., Song, J. Y., Kong, E. B., et al. (2012). Algorithm for predicting functionally equivalent proteins from BLAST and HMMER searches. *Journal of Microbiology and Biotechnology, 22,* 1054–1058.
Zhao, Z. M., Reynolds, A. B., & Gaucher, E. A. (2011). The evolutionary history of the catenin gene family during metazoan evolution. *BMC Evolutionary Biology, 11,* 198.
Ziegler, W. H., Liddington, R. C., & Critchley, D. R. (2006). The structure and regulation of vinculin. *Trends in Cell Biology, 16,* 453–460.

CHAPTER NINE

"Cell Biology Meets Physiology: Functional Organization of Vertebrate Plasma Membranes"—The Immunological Synapse

Silvia Curado[*,†,‡,1], **Sudha Kumari**[*,†,‡], **Michael L. Dustin**[*,†,‡,1]

[*]Skirball Institute of Biomolecular Medicine, New York University School of Medicine, New York, USA
[†]Department of Pathology, New York University School of Medicine, New York, USA
[‡]Kennedy Institute of Rheumatology, NDORMS, University of Oxford, Headington, Oxfordshire, Oxford, UK
[1]Corresponding authors: e-mail address: silvia.curado@med.nyu.edu; michael.dustin@kennedy.ox.ac.uk

Contents

Abstract

The immunological synapse (IS) is an excellent example of cell–cell communication, where signals are exchanged between two cells, resulting in a well-structured line of defense during adaptive immune response. This process has been the focus of several studies that aimed at understanding its formation and subsequent events and has led to the realization that it relies on a well-orchestrated molecular program that only occurs when specific requirements are met. The development of more precise and controllable T cell activation systems has led to new insights including the role of mechanotransduction in the process of formation of the IS and T cell activation. Continuous advances in our understanding of the IS formation, particularly in the context of T cell activation and differentiation, as well the development of new T cell activation systems are being applied to the establishment and improvement of immune therapeutical approaches.

Current Topics in Membranes, Volume 72
ISSN 1063-5823
http://dx.doi.org/10.1016/B978-0-12-417027-8.00009-X

1. OVERVIEW

Our body responds to pathogenic challenges (caused by viruses, bacteria, fungi, or parasites) not only by efficiently clearing invading pathogens but also by developing long-lasting immunity against infectious diseases. This defense against infection can be divided into innate and adaptive immune response on the basis of specificity of recognition (Janeway & Medzhitov, 2002). Whereas innate response is readily available to fight a range of pathogens in an evolutionarily preprogrammed manner, the adaptive immune response results from a selection of novel specificities during infection with a pathogen and often leads to long-term protection to reinfection by that specific pathogen. These two arms of immune response work hand in hand to fight infections, prevent autoimmunity, and also to resolve the immune response when infection has cleared. The main cellular players of the adaptive immune response include antigen-presenting cells (APCs) which process and display on their surface foreign antigens that subsequently interact with B cells or T cells, leading to an adaptive immune response (Trombetta & Mellman, 2005). While APC sounds like a general term, this is actually highly specific to a group of cells that express high levels of histocompatibility complex (MHC) class II, such as B cells and dendritic cells (DCs). Almost all nucleated cells express MHC class I, but they are not generally called APC, or professional APC, if they lack MHC class II expression. Whereas B cell activation leads to the production of antibodies that fight extracellular pathogens, T cells are crucial in cell-mediated immune responses for pathogens replicating inside cells (Banchereau & Steinman, 1998). In the latter, the antigen presented by the APC triggers the T cell receptor (TCR) and leads to the formation of the "immunological synapse" (IS) (Fooksman et al., 2010; Grakoui et al., 1999), the interface between the APC and T cell, establishing the communication between the two cells and ultimately leading to the T cell activation through a signaling network. The stable IS serves a variety of functions including sustaining contact-dependent receptor engagement with appropriate topology, polarization of molecules in the partner cells, and confinement of secreted products (Angus & Griffiths, 2013; Choudhuri, Wiseman, Brown, Gould, & van der Merwe, 2005). Numerous studies have contributed to our understanding of how the IS is formed at the cellular and molecular level and how this reorganization and molecular architecture of the membrane interface between two cells impacts the response to an invading pathogen (Egen et al., 2008; Harris et al., 2012; Waite et al., 2011; Zinselmeyer et al., 2013). The study of the IS provides opportunities for basic investigations in cell biology and valuable insights into immunopathologies and immunotherapy.

2. THE IS

Among the APCs, DCs are the main players in the initiation of the adaptive immune response because they coordinate antigen processing and presentation capacity with migration. Multiple likes of myeloid development converge on the DC phenotypes, making definitive fate mapping a challenge. "Immature" DCs reside near barriers and in the marginal zone of the spleen where they sample self-antigens and benign foreign antigens as part of maintaining self-tolerance and prevention of inappropriate response to foreign antigens such as allergies (Steinman, Hawiger, & Nussenzweig, 2003). Upon encounter with foreign antigens in combination with innate activation, the DCs initiate a program of migration into lymphatics and into lymph nodes or across the marginal zone into white pulp (Edelson et al., 2011; Ohl et al., 2004). The migration program is coordinated with "maturation" of the DCs, which involves upregulation of major MHC class II complexes and costimulatory molecules in which type I interferons play an important role (Longhi et al., 2009). Within the T cell zones of lymph nodes or the spleen white pulp, the DCs scan thousands of naïve T cells per hour to bring MHC–peptide complexes together with the one in one million T cells that has appropriate specificity (Miller, Hejazi, Wei, Cahalan, & Parker, 2004).

During this initial key step, the digested antigenic peptides, associated with the MHC on the surface of the DCs, together with costimulatory molecules, trigger the cell receptors on the naïve T cell, leading to cytoskeleton remodeling and the formation of the IS at the interface of these two cells, ultimately leading to activation of the T cell (Celli, Lemaitre, & Bousso, 2007). CD4+ T cells that are activated by DCs can then go on to help B cells through synapse formation at the junction between the T and B cell zones and later in germinal centers (Allen, Okada, Tang, & Cyster, 2007; Victora et al., 2010).

2.1. Types of IS

There is a remarkable diversity of synaptic interactions in immune cells. Synapse-like contact interfaces are observed in many immune cells, and these cells communicate with a variety of cellular partners in the steady state and during an immune response. Although the contact between T cell–APC and cytotoxic T cell and their targets remains the most characterized form of synapse, recent studies in other immune cell types have also characterized stable contacts that involve B cells, NK-T cells, NK cells, and even phagocytes (Fooksman et al., 2010; Goodridge et al., 2011). In addition, "synapse"-like contacts have also been described between T cell subtypes such as between CD4 and CD8 T cells. While most of the studies characterizing the higher order structure of non-T cell synapses have been carried out *in vitro* using controlled activation conditions, *in vivo* dynamics of these interactions has also been investigated in specific tissue contexts, and has confirmed the specificities of these interactions (Breart, Lemaitre, Celli, & Bousso, 2008; Mempel, Henrickson, & Von Andrian, 2004; Ruocco et al., 2012). Diverse synapse types between different immune cells, although exhibiting variations in the overall organization of molecules at the cell–cell interface, follow some common themes as studied in T cell–APC synapses. This includes the activation and spatial segregation of immunoreceptors at the cell surface, regulation of this process by coreceptors and adhesion molecules, phosphorylation of intracellular immunoreceptor tyrosine-based motifs (ITAM), and activation of intracellular signaling and eventually Ca^{2+} flux (Skokos et al., 2007). Most forms of synapse also exhibit cytoskeleton remodeling and specialized organization of actin cytoskeleton, which is required for cell polarization and optimal synapse function (Burkhardt, Carrizosa, & Shaffer, 2008).

B cells form synapses with various APCs—macrophages, DCs, and follicular DCs (Harwood & Batista, 2011). The antigen displayed on the APC interacts with the B cell receptor (BCR) on the B cell surface, and this engagement process triggers the formation of BCR microclusters and B cell–APC synapse. Activated B cells internalize BCR–antigen complexes and process the antigen to present it in the context of MHC to T cells. Thus, B cells relay the antigen from APCs to T cells via synapse formation (Batista, Iber, & Neuberger, 2001). Actin cytoskeleton is thought to play a negative role in B cell synapse, where plasma membrane BCR is restricted in cortical actin meshwork present underneath the membrane, and antigen engagement releases this diffusion trap (Harwood & Batista, 2011). Further actin-dependent spreading of B cell facilitates formation of a larger contact area and BCR microclusters. A stable synapse is established when the B cell interface contracts following expansion, mobilizing BCR microclusters to the center of the contact. The process of BCR internalization after the contraction phase is currently under investigation in search of precise molecular players that regulate the endocytic process.

T cell–APC synapse is the most extensively characterized form of immune cell synapse. T cells can form synapses with antigen-loaded B cells or antigen-loaded DCs. The TCR recognizes peptides bound to MHC on APCs and undergoes activation. TCR triggering then activates a series of signaling events inside the T cell leading to its activation. TCR-stimulated T cells then serve a variety of functions such as cytolysis of target cells by CD8 T cells and production of effector cytokines by CD4 T cells. Though less characterized, T cells can form synapses with non-APCs. A good example of this class of interaction is the synapse between T cells. Antigen CD8 T cells can form stable conjugates with CD4 T cells and this interaction can lead to amplification of the CD8 T cell response (Chaudhri et al., 2009). Alternatively, a synapse between activated CD4 T cells can also form, which can be utilized for sharing of paracrine cytokines such as IL-2 and IFN-γ, thus consolidating CD4 T cell activation response at population level (Gerard et al., 2013; Sabatos et al., 2008). The molecular players involved in these synaptic interactions are not known; however, a critical role for integrin lymphocyte function associated antigen-1 (LFA-1) has been established in homotypic T cell synapses. In addition to the above-mentioned synapse, effector T cells can also form conjugates with regulatory T cells and NK cells; however, the precise cell biology of these interactions has not been characterized.

Unlike T cells where the presence of cognate antigen–MHC on the APC is sufficient to trigger activation, NK cells utilize a combination of a positive and negative surface receptor signal on the DC to regulate their cytolytic response. NK cells are innate cells that express germline-encoded activating and inhibitory receptors. The activating receptors, mainly NKG2D and NCRS, recognize a variety of molecules expressed on infected or transformed target cells. The inhibitory receptors CD94/NKG2A and KIRs recognize nonclassical and classical MHC molecules (Barreira da Silva & Munz, 2011). The relative signaling of these two functionally distinct types of signaling receptors decides the nature of the NK cell synapse, that is, activating or inhibitory. Actin cytoskeleton plays a crucial role in the strength of the contact and has been proposed to mediate polarized and localized secretion of cytolytic granules (Rak, Mace, Banerjee, Svitkina, & Orange, 2011).

Another interesting variation of typical recognition-specificity-based synapse is the synapse formed by the NK-T cells. These cells are innate-like cells and the majority of them express a semi-invariant TCR consisting of invariant α-chain along with diverse β chain, and exhibit features of both NK cells and T cells. These cells are thought to represent a link between the innate and adaptive immune system and, likewise, can interact with a variety of innate and adaptive immune cells. The TCR on the NK-T cell recognizes self and non-self lipid antigens presented on the CD1d molecule. CD1d is an MHC-like molecule expressed by a variety of APCs, including macrophages, DCs, B cells, CD4–CD8 thymocytes, and epithelial cells (Joyce, Girardi, & Zajonc, 2011). Upon recognition of lipid antigen/CD1d on the APC, NK-T cells establish a synapse with the APC. Stable synapse formation and polarization of lytic granules accompanies the optimal effector response.

2.2. T cell IS

The T cell IS has been the best studied as having a close contact is indisputably critical for antigen recognition by a T cell (Grakoui et al., 1999).

The naïve T cell is initially primed by a pathogen-activated DC in the lymph node. The formation of an IS between these two cells results in the activation of the T cell which clonally expands ≥ 10 generations for each T cell that is activated by a DC (Moon et al., 2007). These progeny differentiate into effector cells that exit the lymph node or spleen and in turn patrol the whole body scanning for matching MHC peptide complexes on the surface of APCs.

Once a T cell encounters an APC, several conditions have to be met for an IS to form (Shaw & Dustin, 1997). These requirements were formalized over 15 years ago, but new appreciation of the function of actin networks necessitates updating these:

(1) Cell membrane proximity and cell–cell adhesion

The APC and T cell have to be close enough to enable the molecules on each cell's membrane to interact efficiently. One obstacle in reaching proximity between the two cells is the layer of glycocalyx (consisting of large glycoproteins) that covers both the T cell and APC and that creates repulsion between the two cells (Cyster, Shotton, & Williams, 1991). This repulsion effect is partially overcome by the integrin-based adhesion between the ICAM-1, on the APC and (LFA-1, a member of the integrin family of receptors), on the T cell, an antigen-nonspecifc interaction that is driven in part by chemokines (Alon & Dustin, 2007). Although LFA-1–ICAM-1-based adhesion brings the two cells' membranes within ~45 nm of each other (Chen, Lou, Evans, & Zhu, 2012), the distance between the two cells

(APC and T cell) has to be brought to less than 15 nm so that the pMHC complex on the APC can bind to the TCR on the T cell. To achieve the optimal intercellular distance, and reduce it from 45 to 15 nm, actin polymerization-based protrusions (invasive pseudopodia) of the T cell reach the APC, allowing the T cell to sample a large number of pMHC complexes at the surface of each APC to scan for a specific peptide. Such "invadosome-like protusions" (ILP) have been recently visualized at high resolution using endothelial cells to provide a thin, flat APC with excellent properties for imaging (Sage et al., 2012).

(2) Recognition of matching TCR–pMHC

If the TCR (on the T cell) does not recognize a specific matching antigen (pMHC complex), the T cell will separate from the APC to continue sampling. Conversely, upon recognition and binding of the specific pMHC by the TCR, the 15-nm optimal distance for interaction between the two cell membranes is sustained, allowing formation of a microcluster over a period of several seconds (Campi, Varma, & Dustin, 2005; Varma, Campi, Yokosuka, Saito, & Dustin, 2006).

On a longer time scale, binding of the TCR to the specific antigen (pMHC complex) leads to the formation of an IS composed of supramolecular activation clusters (SMACs) (Grakoui et al., 1999; Monks, Freiberg, Kupfer, Sciaky, & Kupfer, 1998).

(3) Costimulatory system

In addition to the TCR–pMHC and adhesion (LFA-1–ICAM-1) systems, the interaction of CD28 (on the surface of the T cell) with a B7 ligand on the APC (CD80 or CD86) provides a costimulatory signal that enhances TCR triggering of naïve T cells (Levine et al., 1997).

Several cell surface molecules are therefore required for T cell and APC to interact:

T Cell::APC

TCR::pMHC (class I, CD8 T cells; class II, CD4 T cells)

CD28::B7 molecules (CD80, CD86)

LFA-1::ICAM-I

2.3. Organizational architecture of the IS

The IS is a well-organized structure that consists of three concentric rings which differ in their receptor composition and density (Fig. 9.1). The different surface receptors that are part of the IS are segregated into distinct areas in the form of clusters. The TCR is initially present at the T cell surface

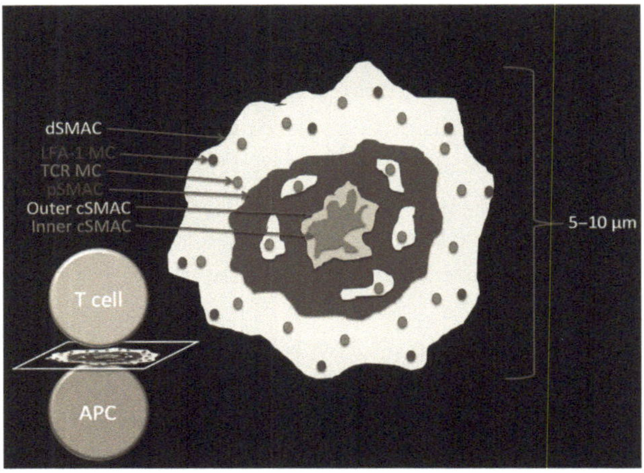

Figure 9.1 Components of the immunological synapse. The mature immunological synapse is composed of three SMACs and at least two types of microclusters (MC). The dSMAC (outer yellow) is F-actin rich and is the location in which TCR (green) and LFA-1 (red) microclusters form. The pSMAC is a perforated network of engaged LFA-1 (red) that contains some TCR microclusters in the perforations. The outer cSMAC (pale green) contains a small amount of TCR and accumulates all the CD28 (over time). It is active in signaling. The inner cSMAC (green) is dependent upon ubiquitin recognition and is not active in signaling. There is inward movement of MCs, inward movement of LFA-1 in the pSMAC (unpublished observations), and inward movement of TCR from the outer to the inner cSMAC. (See color plate.)

in clusters of 5–10 molecules of TCR in "confinement zones" or "protein islands," possibly held by an underlying actin meshwork and/or anchored to the cortical actin (Lillemeier et al., 2010). Upon engagement of the TCR, microclusters start forming in small contact areas between the T cell and APC membranes, and their number increases as the contact area between the two cells increases with T cell spreading. The TCR microclusters form in the periphery of the synapse in the distal supramolecular activation cluster (dSMAC) and within a few minutes of cluster formation, they translocate centripetally, through the peripheral supramolecular activation cluster (pSMAC) eventually fusing into the central supramolecular activation cluster (cSMAC) zone. It is at the cSMAC, where the TCR clusters colocalize with markers of protein degradation (Varma et al., 2006) and components of the ubiquitin pathway (Vardhana, Choudhuri, Varma, & Dustin, 2010), that signaling is thought to be terminated, with the TCR–pMHC being internalized possibly by phagocytosis (Alarcon, Mestre, & Martinez-Martin, 2011; Coombs, Kalergis, Nathenson, Wofsy, & Goldstein, 2002; Liu, Rhodes,

Wiest, & Vignali, 2000). As the TCR microclusters form, in the pSMAC, they associate with the signaling molecules that will eventually trigger T cell activation (described below); however, as the microclusters migrate toward the cSMAC, these associations are lost, thus, despite the large number of TCRs present in the cSMAC, signaling does not occur in this central part of the IS.

Contrary to TCR clusters, LFA-1 clusters are organized in the pSMAC—a highly contractile zone—and do not cross the cSMAC–pSMAC border. As molecules engage and TCR microclusters are formed at the pSMAC, these start signaling and continue to do so as they translocate to the cSMAC at the center of the IS. After the TCR microclusters are delivered to the cSMAC, TCR signaling is downregulated, possibly as a way to control signaling strength at the synapse (Varma et al., 2006).

Studies that aimed at investigating the involvement of actin polymerization and remodeling in the formation of the IS and segregation of its main components have shown that actin is dynamically remodeled during this process. The observation that inhibition of actin polymerization inhibits T cell activation implies that actin is a key player. In an initial phase, actin rapidly polymerizes as the T cell spreads, with expansion of lamellipodia-like edges. At this stage the IS presents three distinct actin organization zones: an actin-enriched area at the periphery of the IS (dSMAC), an actomyosin-rich area (pSMAC), and a central F-actin poor zone (cSMAC). Following this stage, the T cell lamellipodia start contracting, leading to a retrograde flow of actin toward the cSMAC at the same time as the TCR translocates toward the cSMAC. Several studies have suggested that actin is indeed coupled to the translocation of TCR microclusters (the fact that sometimes translocation is slower than actin flow has been suggested to occur by occasional slippage). Another model suggests that the coupling between microclusters and actin is transient ("frictional coupling mechanism") through weak links that form and break at different times producing a frictional force (DeMond, Mossman, Starr, Dustin, & Groves, 2008; Smoligovets, Smith, Wu, Petit, & Groves, 2012; Yu, Wu, Kaizuka, Vale, & Groves, 2010). Such studies on the role of actin in microclusters translocation and organization of the IS include the measurement of the velocity of TCR microclusters and actin speckles when tracking actin-GFP in Jurkat T cells (Kaizuka, Douglass, Varma, Dustin, & Vale, 2007) and the use of chrome barriers that obstruct TCR–pMHC translocation but not the actin flow in supported lipid bilayers (SLBs) (DeMond et al., 2008). The mechanism by which microclusters are coupled to F-actin is, however, still not clear. Experiments using myosin II activity inhibitors or depletion of myosin II also resulted in defective microcluster translocation, suggesting myosin II activity to be involved in TCR migration toward the cSMAC in an additive manner to actin polymerization (Babich et al., 2012; Ilani, Vasiliver-Shamis, Vardhana, Bretscher, & Dustin, 2009; Kumari et al., 2012; Yi, Wu, Crites, & Hammer, 2012). This myosin II activity appears not to be required for TCR translocation in motile T cells, whereas actin depolymerization is (Beemiller, Jacobelli, & Krummel, 2012). Recent studies have also implied the involvement of microtubule cytoskeleton in the TCR microcluster motility toward the cSMAC as TCR microclusters seem to be able to interact with the microtulule motor dynein (Hashimoto-Tane et al., 2011).

The final step of the IS, that is the separation of T cell and APC, is as critical, as in order for the T cell to move on to scan for its cognate pMHC it must disengage from the IS. This stage of the IS is referred to as "kinapse" (Dustin, 2008). The mechanism underlying the synapse–kinapse transition is not yet fully understood.

2.4. Functional consequences of the IS

Upon antigen presentation to the T cell and recognition of the pMHC and TCR engagement, downstream signaling pathways are triggered, leading to a series of events that can result in different outcomes.

2.4.1 TCR signaling

As the TCR recognizes and binds the cognate antigen on the APC, the ITAM motifs of the cytosolic tail of the TCR-associated CD3 chains—that is thought to be associated with the lipids in the inner leaflet of the plasma membrane in the inactive conformation—undergo phosphorylation by the lymphocyte-specific protein tyrosine kinase (Lck).

It is not clear whether the release of the cytosolic TCR-associated CD3 zeta chains, protected from Lck, is the cause ("safety model") (Gagnon, Schubert, Gordo, Chu, & Wucherpfennig, 2012; Shi et al., 2013; Xu et al., 2008) or the consequence (Zhang, Cordoba, Dushek, & van der Merwe, 2011) of the ITAM motifs' phosphorylation.

The "safety model," which proposes that the TCR/CD3 cytoplasmic domains are associated with the membrane to prevent phosphorylation of the ITAM motifs, has been recently questioned, mainly due to the observation that inhibition of the association of the CD3 cytoplasmic domain with the membrane did not enhance CD3 phosphorylation, and reduced it instead, and that the ITAM motifs' phosphorylation was required for the TCR/CD3 chains' dissociation from the membrane (DeFord-Watts et al., 2011; Zhang et al., 2011). At the IS, the phosphorylated TCR-associated CD3 chains and the inhibitory phosphatase CD45 (which localizes to the pSMAC and dSMAC) are mutually exclusive, which indicates that this is a requirement for the receptor phosphorylation (Leupin, Zaru, Laroche, Muller, & Valitutti, 2000; Varma et al., 2006). In fact, ITAM motifs can have a low level of phosphorylation in resting T cells, as they are constantly being phosphorylated, by Lck, and dephosphorylated by CD45; given the short distance between T cell and APC when TCR and pMHC bind (15 nm), glycoproteins of larger size are excluded from the TCR contact area and are localized to the pSMAC, allowing prolonged phosphorylation of the TCR-CD3 ITAM motifs and subsequent TCR triggering (Choudhuri et al., 2009). The phosphorylation of the ITAM motifs allows the recruitment of the Syk-kinase zeta chain-associated protein of 70 kDa (ZAP70) (Wang et al., 2010), which is then also phosphorylated, and activated, by Lck. The phosphorylated ZAP70 (pZAP70) phosphorylates the transmembrane protein Linker of Activated T cells (LAT). Whereas some have proposed that LAT and TCR are segregated in sub-micrometer "protein islands" surrounded by a protein-poor "lipid sea" (Lillemeier et al., 2010)—a barrier that would have to be overcome for signaling to occur—others have suggested that LAT molecules are not laterally recruited but that there is an intracellular pool of LAT sufficient to drive signaling (Williamson et al., 2011). LAT has been shown to exist in sub-synaptic vesicles that could either be phosphorylated in trans or instead, tethered or fused with the plasma membrane. The LAT "protein island" and "LAT in vesicles" models are not mutually exclusive and may reflect different experimental conditions, such as time points, or may be related to different signaling outcomes. Once phosphorylated, LAT phosphorylates and docks SH2 domain-containing leukocyte protein of 76 kDa (SLP76), phospholipase Cγ (PLCγ), phosphatidylinositol 3-kinase (PI3K), growth factor receptor-bound protein 2 (Gb2), and Grb2-homologous adaptor (GADS) (Cruz-Orcutt & Houtman, 2009; Liu, Berry, & McGlade, 2001; Wu & Koretzky, 2004; Zhang et al., 2011). The phosphorylation of LAT (4 s after TCR–pMHC engagement) and subsequent docking of PLCγ regulate Ca^{2+} influx (6–7 s post-TCR–pMHC engagement). PLCγ activity leads to an increase in cytosolic Ca^{2+} through the generation of IP3, which releases Ca^{2+} from endoplasmic reticulum (store-operated Ca^{2+} entry, SOCE), then triggering the influx of Ca^{2+} across the plasma membrane (Ca^{2+} release-activated Ca^{2+}, CRAC) channels (reviewed in Shaw, Qu, Hoth, & Feske, 2013). One of the main consequences of the Ca^{2+} influx is the relocation of the transcription factor nuclear factor of activated T cells (NFAT) from the cytosol to the nucleus, through dephosphorylation by the Ca^{2+}-dependent phosphatase calcineurin (Hogan, Chen, Nardone, & Rao, 2003), eventually resulting in cytokine transcription and secretion (IL-2, IL-4, IL-17, IFN-γ, and TNF) by the T cell. SLP76, phosphorylated by LAT, plays a role as scaffold for several actin effectors. Both the guanine–nucleotide-exchange factor (GEF) Vav1 (that appears to associate with the LAT–SLP76 complex; Pauker, Hassan, Noy, Reicher, & Barda-Saad, 2012; Sylvain, Nguyen, & Bunnell, 2011) and Wiskott–Aldrich syndrome protein WASp, F-actin nucleation and remodeling regulators have been reported to be recruited to the TCR–pMHC sites and regulate actin polymerization (Barda-Saad et al., 2005; Miletic et al., 2006, 2009; Sasahara et al., 2002; Tybulewicz, 2005; Yokosuka et al., 2005; Zeng et al., 2003).

The formation of the IS can have different functional consequences to provide an appropriate response in a context-dependent manner. The role of the IS is multifaceted: to aid the interaction of T cells with APC, efficient antigen–MHC sampling, to form stable conjugate with cognate pMHC, and to activate and downregulate TCR signaling (which is as important to achieve an optimal and controlled T cell response). The IS allows for the development of the appropriate response ranging from cytokine secretion, proliferation, and effector functions. Though cytokine secretion and proliferation have been thought of as a singular pathway, initiated by TCR triggering, ITAM phosphorylation, Ca^{2+} influx, and transcription activation, Guy et al. have recently shown that proliferation and cytokine secretion are nonoverlapping pathways (Guy et al., 2013). The TCR–CD3 complex has 10 functional ITAM motifs—more than the two motifs that have been believed to be required to trigger TCR signaling. Guy et al. have found that even though low multiplicity of TCR–CD3 ITAM motifs (only two to four functional ITAM motifs) is indeed sufficient to engage canonical TCR signaling events that lead to cytokine secretion, such as IL-2 and IFN-γ, a high multiplicity of TCR–CD3 ITAM motifs is required for TCR-induced proliferation (Guy et al., 2013). The proliferation pathway was shown to rely on the formation of a compact cSMAC, on Notch1 and upregulation of c-Myc after ligation of TCR–CD3 to the pMHC (Guy et al., 2013). Having distinct pathways driving different functional responses, such as secretion of certain cytokines and proliferation, may be biologically relevant as it may enable the T cell to respond differently under different conditions. Indeed, Guy et al. have shown that T cells presented with superagonists have an enhanced expression and translocation of c-Myc and a greater proliferation than when presented with a weak agonist.

2.4.2 Polarized secretion

Formation of the IS and subsequent induction of cell polarization are required for directed secretion by CD8 cytotoxic T cells (CTL) of specialized secretory lisosomes (or lytic granules) containing lytic pore-forming perforin, which enables entry of granzymes into virally-infected target cells or tumor cells, ultimately triggering apoptosis (reviewed in Stinchcombe & Griffiths, 2007). Movement of the centrosome to the membrane at the edge of the cSMAC is a key step in IS-induced cell polarization, which has been observed in CTLs (Stinchcombe, Majorovits, Bossi, Fuller, & Griffiths, 2006) as well as in CD4 cells (Ueda, Morphew, McIntosh, & Davis, 2011), NK and NKT cells (Stinchcombe et al., 2011), indicating that docking of the centrosome to the plasma membrane is a common event in direct secretion. The cytoskeleton and motor proteins such as dynein, as well as diacyl-glycerol (DAG) accumulation and restriction to the IS and other TCR signaling players, have been suggested to play a role in inducing the movement of the centrosome (Gharbi et al., 2011; Quann, Merino, Furuta, & Huse, 2009). Several candidate proteins have been suggested to also be involved in the centrosome polarization (reviewed in Angus & Griffiths, 2013). Other TCR signaling players, such as Lck, have been implicated specifically in the docking of the centrosome to the plasma membrane (Tsun et al., 2011), essential for granule delivery to the IS. The movement and docking of the centrosome—the microtubule organizing center (MTOC) of the cell—to the IS reorganizes the intracellular organizing cytoskeleton and is thought to allow polarized and precise secretion of lytic granules to the IS, inducing cell death of the target cell. Though there have been advances in the study of polarized secretion in T cells regarding centrosome movement and exocytosis of the secretory vesicles at the IS (reviewed in Griffiths, Tsun, & Stinchcombe, 2010), much remains to be understood. Besides directing secretion of lytic granules, the IS can also induce polarized secretion of cytokines (such as IL-2, IL-4, IL-5, and IFN-γ) toward the APC (Kupfer, Mosmann, & Kupfer, 1991; Poo, Conrad, & Janeway, 1988). In addition, it has been proposed that also the cytokine receptors can be selectively recruited to the synapse, adding a new regulation layer (Maldonado et al., 2009).

2.4.3 T cell differentiation

One important aspect of the immune response is the ability to generate the appropriate type of T cell for a suitable response. The IS plays a role in T cell differentiation by leading to changes in transcription factor programs initiated by TCR triggering and ultimately the activation of transcription factors such as NFAT, activator protein 1 (AP-1), and nuclear factor-kB (NF-kB), which in turn bind to regulatory elements in genes that encode cytokines and other relevant transcription factors (reviewed in Padhan & Varma, 2010).

Depending on whether an immediate response or long-lasting memory is needed, the organism can generate two different subsets of cells: effector or memory T cells, respectively. The mechanism underlying T cell differentiation into effector or memory fate remains controversial and is still to be fully understood. While some studies support the hypothesis that memory cells can differentiate in a linear developmental fashion directly from effector cells (Bannard, Kraman, & Fearon, 2009), it has also been proposed that the progeny of activated T cells is not homogenous compositionally leading to differences in fate potential. The latter hypothesis is supported by the evidence that following antigen presentation by the APC, the mitotic T cell shows an asymmetric distribution of polarity proteins—such as aPKC and Par3—resulting in its asymmetric division and an uneven segregation of molecular cell fate determinants between the daughter cells, ultimately leading to two daughter cells with two different fate potentials: effector and memory (Chang et al., 2007; Oliaro et al., 2010).

The IS therefore plays a role in T cell differentiation by modulating the trafficking of transcription factors between nucleus and cytoplasm and by regulating the asymmetric division of the T cell.

2.4.4 Signal modulation and termination

Though it has been proposed that the cSMAC plays a role in the temporal regulation of the synapse (Varma et al., 2006), with TCR signaling down-regulation and termination taking place at the cSMAC through internalization of the TCR–pMHC, it has also been suggested that the cSMAC may enhance the stimulatory effect of a weak TCR agonist, and thus that the balance between signaling and termination is dependent on the quality of the specific antigen (Cemerski et al., 2008; Lee et al., 2003). A potential pathway to resolution of this controversy is the observation that there are actually two compartments within the cSMAC, a TCR-enriched inner cSMAC compartment that is dependent upon ubiquitin recognition for signal termination (Vardhana et al., 2010) and a CD28-enriched outer cSMAC compartment that is active in TCR and CD28 signaling (Tseng, Waite, Liu, Vardhana, & Dustin, 2008; Yokosuka et al., 2008). It is possible that the use of weak TCR ligands results in generation of an outer cSMAC without an inner cSMAC due to low ubiquitination.

The continued TCR signaling through new microcluster formation or an active, outer cSMAC compartment is important for T cell "deceleration" or the "stop" signal of the motile T cell, to ensure the engagement of a longer duration with the APC (Dustin, Bromley, Kan, Peterson, & Unanue, 1997; Sims et al., 2007; Skokos et al., 2007). The increase in cytoplasmic Ca^{2+} is required for rapid deceleration in some conditions *in vivo* (Skokos et al., 2007; Waite et al., 2013).

2.5. Mechanotransduction

Other factors appear to play a role in T cell activation. Mechanical forces have been shown to play an important role in several biological processes such as cell differentiation, mitosis, and migration of certain cell types (reviewed in Eyckmans, Boudou, Yu, & Chen, 2011), and the T cell appears to be no exception (Kim et al., 2012). Several groups have shown that the T cell is responsive to mechanical forces that can trigger T cell activation (Husson, Chemin, Bohineust, Hivroz, & Henry, 2011; Judokusumo, Tabdanov, Kumari, Dustin, & Kam, 2012; Kim et al., 2009; Li et al., 2010). The application of lateral, but not normal, forces to the TCR using a monoclonal anti-CD3 antibody that is poorly stimulatory even when presented on a particle, resulted in an increase in cytoplasmic Ca^{2+} levels, suggesting that the TCR receptor is an anisotropic mechanosensor (Kim et al., 2009). Li et al. have used an APC system expressing an elongated ligand against CD3 that in spite of binding to the TCR does not induce signaling under static conditions, making it a good model to test whether physical forces can trigger TRC signaling (Li et al., 2010). They observed an increase in intracellular Ca^{2+} levels when a force was applied normal to the T cell surface. The direction of activation of force is distinct in this study compared to the Kim et al. study above. The same effect was observed when T cells were pulled away from the APC (Li et al., 2010). Husson et al. have also shown, using a biomembrane force probe, that T cells do generate pushing and pulling forces when their TCR is engaged by anti-CD3 (Husson et al., 2011). In a physiological setting, such mechanical forces could easily be exerted on the TCR as it binds pMHC during migration and scanning of

APC surfaces in search for the cognate pMHC and T cell detachment, which can create tensile mechanical forces as the T cell moves in an actin-dependent manner with respect to the APC (receptor deformation model); however, there is no experimental support for this (Ma & Finkel, 2010; Ma, Sharp, Janmey, & Finkel, 2008). It is possible that mechanical forces applied to the TCR may play a role in dislodging the ITAM motifs from the inner leaflet of the plasma membrane making the ITAM motifs more accessible to be phosphorylated, initiating the signaling cascade, with or without the participation of actin (Xu et al., 2008). More studies will be required to address these hypotheses. The lateral transport of microclusters toward the center of the synapse may also generate force on the TCR and LFA-1 (Kaizuka et al., 2007; Zhu et al., 2008).

Additional recent studies have aimed at integrating mechanical stimuli in T cell culture, by manipulating the substrate elastic modulus (stiffness), as a way to direct T cell response, using mouse (Judokusumo et al., 2012) and human (O'Connor et al., 2012) T cells. These groups have shown that there is indeed a correlation between the substrate stiffness and the T cell response. Mouse CD4+ T cells cultured on polyacrylamide gels with increasing elastic modulus coated with antibodies against CD3 and CD28 (mimicking the APC) secreted more IL-2. The observed difference in levels of IL-2 secretion by T cells cultured in gels of different stiffness was abolished by blebbistatin, which suggests that actomyosin cytoskeleton is involved in the T cell response to mechanical stimuli. In this study it was also reported that when the anti-CD28 antibody was tethered to the polyacrylamide substrate of different elastic modulus and the anti-CD3 antibody was immobilized onto beads (of constant rigidity), the difference in IL-2 secretion was no longer observed (unlike the inverse situation), supporting Li et al.'s hypothesis that the mechanical force acts through the TCR. Consistently with the increased levels of IL-2 produced by T cells cultured on stiffer substrates, also clusters of phosphorylated ZAP70 (pZAP70), phosphorylated Src family kinase proteins (Lck and Fyn) and phosphorylated proline-rich tyrosine kinase-2 (pPYK2) were observed at the T cell–substrate interface, contrary to what was observed in cells cultured in softer substrates where no (pZAP70 and pSFK) or only a small number of clusters (pPYK2) was detected. The observation that blebbistatin induced a small decrease in levels of pPYK2 across substrates of different rigidities suggests that PYK2 responds to cell contractility and may contribute to T cell mechanosensing (Judokusumo et al., 2012). One important aspect to take into account is that a different cellular response may not necessarily imply a mechanical response, but may result from a difference in chemistry of ligand presentation among substrates of different elastic modulus (Trappmann et al., 2012). Interestingly, PKY2, which is structurally related to the mechanotransducer focal adhesion kinase (FAK), appears to be critical for CD8 T cell short-lived effector fate, as its deficiency results in the impairment of T cell activation and the loss of this type of cell CD8+ effector cells, supporting mechanotransduction as an important player in T cell signaling (Beinke et al., 2010).

An approach similar to the ones developed by Judokusumo et al. and O'Connor et al. was used to investigate the effect of cell culture antigen-presenting substrates of different stiffness (polyacrylamide gels) on B cells and led to the conclusion that B cells are also affected by the substrate stiffness: mouse B cells cultured in stiffer substrates showed a stronger activation (Wan et al., 2013). Another study on B cells also suggested that the B cells mechanically test the strength of antigen binding discriminating antigen affinities for development of high-affinity antibodies (Natkanski et al., 2013). Besides substrate stiffness, other parameters such as topology and protein distribution and molecular motility remain to be accessed regarding their role on the IS and T cell activation.

2.6. T cell activation systems and imaging tools

A good understanding of the spatial organization of the signaling molecules and how they are distributed in the plasma membrane and connected to downstream pathways is crucial to understand the underlying mechanisms of T cell activation.

A number of tools have been developed and optimized and have been increasingly used to study the IS.

The development of planar TCR activation systems has allowed for the visualization of the dynamic protein distribution during the IS formation and T cell activation.

2.6.1 Supported lipid bilayers

SLBs consist of a self-assembled lipid bilayers on a solid substrate reconstituted with an adhesion molecule like ICAM-1, antigenic pMHC, or with anti-CD3 antibody, aiming at mimicking the phospholipid bilayer of the APC membrane (Dustin, Starr, Varma, & Thomas, 2007; Groves & Dustin, 2003). SLBs can be formed on planar substrates, which is ideal for imaging, or on glass beads, which is useful for functional assays (Gay, Coeshott, Golde, Kappler, & Marrack, 1986). This system was utilized sparingly through the 1980s for studies on Fc receptors and the TCR

(McConnell, Watts, Weis, & Brian, 1986), but the use of lipid-anchored proteins that are mobile in bilayers with fluorescence microscopy has led to expanded use (Dustin, Ferguson, Chan, Springer, & Golan, 1996). SLBs capture the lateral mobility of natural ligands that are involved in the IS, allowing T cells to conjugate with activating proteins, as if recognizing an APC, and forming an IS. The mobility of this system allows the reproduction of the IS reorganization as it enables the translocation of ligated protein clusters. SLBs provide not only a planar orientation for better visualization of the synapse but also allow for the control of antigen density for TCR triggering and study of its effect in IS formation.

Others have aimed at mimicking T cell–APC interactions by seeding T cells on planar multicomponent protein substrates with lithographically defined patterns of tethered TCR ligands (anti-CD3 antibody) and the adhesion molecule ICAM-1 attempting to reproduce the microscale organization of the IS (Doh & Irvine, 2006). This approach offers the possibility of gaining control over the IS geometry and has been expanded so that in addition to the distribution of anti-CD3, also that of the costimulatory CD28 ligand could be controlled, against a background of ICAM-1 through multiple rounds of microcontact printing. This approach has allowed the study of how T cells respond differently to different patterns and how the costimulatory signal can affect this response (Shen, Thomas, Dustin, & Kam, 2008).

The mobility of proteins on the SLB, or restriction of their mobility when immobilized onto substrates, may not represent the physiological conditions of the APC/T cell interface and may influence signaling events considerably. In order to control lateral mobility of molecules in the supported membrane, some studies have used fabricated barriers onto the underlying substrate. These engineered supported membranes can lead to new insights into our understanding of the IS. Mossman et al. have imposed geometric constraints on the formation of the IS by using supported bilayer membranes with nanometer-scale structures onto the underlying substrate and have shown that restriction of the motion of TCR clusters to the center of the IS enhanced TCR signaling, that is, TCR microclusters mechanically trapped in the periphery of the IS resulted in prolonged signaling from the TCR clusters, which is consistent with the hypothesis that translocation of the TCR clusters toward the cSMAC acts as a signal regulation mechanism (Mossman, Campi, Groves, & Dustin, 2005). Lemond et al. have used SLB-reconstituted substrates patterned with molecular mazes of chromium, which enabled the study of TCR dynamics. In this system, microclusters were not permanently trapped by the barriers and displayed long-range movement on the cell surface, leading to insightful results described above. Conversely, the use of grids instead of maze lines also affected the mobility of TCR but caused TCR to be confined in a fixed position (DeMond et al., 2008). Other groups have tried to engineer supported membranes introducing patterns to control molecule diffusion (reviewed in Yu & Groves, 2010).

Other model systems that aim at mimicking the APCs include nonplanar surfaces such as biomimetic droplets—liquid colloids in the form of droplets grafted with specific molecules. This system enabled the visualization of the binding molecules on the T cell and the droplet and has been shown to trigger the T cell signaling cascade (Bourouina, Husson, Hivroz, & Henry, 2012).

The introduction into nonimmune cells of genes encoding the TCR and other proteins required for regulation of its phosphorylation and conjugation of these cells with an APC (James & Vale, 2012) and red blood cells directly coated with pMHC via biotin–streptavidin coupling, serving as an artificial APC (Huang et al., 2010), have also been used as reconstituted T cell activation systems to study initial TCR-triggering events (and also as force sensors in the latter case).

Because the IS is based on the communication between two cells and localized to their contact interface, visualization of the IS and subsequent events in the cells relies on the use of specific imaging techniques.

2.6.2 Total internal reflection fluorescence microscopy

Total internal reflection fluorescence microscopy (TIRF) offers the advantage of a much reduced background fluorescence by illuminating only $100<200$ nm into the cell, enabling the generation of high-resolution images of the plane between the T cell and APC, and has thus been used to study the organization of the IS. TIRF is routinely used to study the T cell interface with SLBs, where the well-defined planar interface facilitates high-resolution imaging. TIRF can also be combined with the techniques described below.

Other imaging techniques that allow for visualization at a molecular nanometer scale have been developed and extremely useful in gaining new insights in the field; these include photoactivated localization microscopy (PALM) (Betzig et al., 2006), stochastic optical reconstruction microscopy (STORM) (Rust, Bates, & Zhuang, 2006), which can be used for detection and precise localization of single molecules, and fluorescent resonance energy transfer (FRET) (Joo, Balci, Ishitsuka, Buranachai, & Ha,

2008; Moerner, 2007) that enables single molecule tracking and provides high-resolution information on dynamic molecular processes.

Advances in super-resolution fluorescence microscopy applied to the IS have been key in gaining further insights into several aspects of TCR signaling. This tool has been particularly relevant in discerning between different models for (1) LAT distribution upon TCR engagement, in membrane "protein islands" (Lillemeier et al., 2010) versus cytoplasmic pool (Williamson et al., 2011) and arrangement of other signaling molecules (Hsu & Baumgart, 2011; Sherman et al., 2011); (2) role of ITAM phosphorylation in the coordination of distinct TRC-driven pathways (Guy et al., 2013); (3) clustering mechanism of Lck driven by conformational changes of Lck following TCR activation (Rossy, Owen, Williamson, Yang, & Gaus, 2013); and (4) mechanism for release of intracellular domains of the CD3 from the inner leaflet of the plasma membrane (Gagnon et al., 2012; Shi et al., 2013; Zhang et al., 2011) (reviewed in Rossy, Pageon, Davis, & Gaus, 2013).

One of the challenges in imaging the IS is to have it positioned in the horizontal imaging plane so that the microcluster localization can be properly resolved. To manipulate the T cell–APC conjugates and to allow for vertical cell orientation, laser tweezers or laser trap have been successfully used (Oddos et al., 2008). This system, however, presents some limitations such as cellular phototoxicity complications and/or thermotoxicity due to heating of the medium, and does not allow for high-throughput analysis due to its laborious nature.

To facilitate high-resolution imaging of the IS overcoming these limitations, Biggs et al. developed microfabricated substrates (micropit arrays in polydimethylsiloxane, PDMS) to sequester single T cell–APC conjugates, not substrate-adhered or influenced by interaction with adjacent cells, so that the IS forms in the horizontal imaging plane. This experimental design made use of K562-based APCs (K562 cells transduced to express an array of stimulatory and costimulatory molecules used to activate and expand T cells) that were initially sequestered in the fabricated micropits, followed by loading of T cells. This system allowed for large-scale imaging analysis of the IS interface of T cell–APC conjugates both in fixed and live samples, as multiple cell conjugates can be analyzed within a single imaging field (Biggs, Milone, Santos, Gondarenko, & Wind, 2011).

The development of artificial substrates to mimic the T cell–APC interface has become a main focus in the IS field, with the generation of minimal systems with greater control and high resolution. The use of such substrates will provide a system that allows dissection of the molecular mechanism underlying the IS, in particular the requirements regarding the surface molecules' density, precise distribution, and mechanical and topological characteristics of the surface where the APC surface molecules lay. Future approaches will aim at improving the antigen-presenting substrate regarding the resolution of surface molecules as well as the substrate's fluidity so that the dynamics of the IS can be accurately reproduced.

Molecular imaging techniques continue to improve and will undoubtedly lead to a better understanding of the IS—its formation and subsequent T cell activation downstream consequences. The ability to analyze single molecules will enable a more accurate capture of the dynamics throughout these events not only in fixed but also living cells.

3. CONCLUDING REMARKS AND PERSPECTIVES

3.1. IS evolution

The two broad arms of the immune system, that is innate and adaptive, appear to have arisen at different timescales during evolution (Cooper & Alder, 2006). While all multicellular organisms have innate defense mechanisms and innate-like cells, the adaptive system is found primarily in vertebrates. Adaptive immune cell antigen receptors exhibit selective reactivity to various antigens and these utilize somatic recombination mechanisms to generate a diversity of receptor recognition specificities. Needless to say that the cell–cell conjugate formation involving the mechanism of recognition or the classical "IS" has thus been studied extensively in vertebrate systems. Although theoretically the IS can exist between any two immune cells, the antigen sampling by an adaptive immune cell via interaction with another immune cell has been the focus of extensive research during the past decade (Flajnik & Kasahara, 2010).

The components of the innate immune system such as innate cells expressing pattern recognition receptors are found in the entire animal kingdom. The evidence of adaptive immune system emerged first in jawless vertebrates, such as lamprey and hagfish, when the immunization with specific antigens resulted in production of specific agglutinins. Interestingly, instead of Immunoglobulin (Ig)-domain-rich immunoreceptors, as found in lymphocytes of higher vertebrates, lymphoid cells in these organisms have variable lymphocyte receptors (VLRs) rich in leucine-containing repeats (Boehm et al., 2012). Lamprey has two VLR genes—VLRA and VLRB—which exhibit allelic exclusion and encode membrane-bound and secreted forms of the

VLRs, respectively. The adaptive immune system in higher vertebrates includes cells expressing BCR, TCR, and MHC molecules, where these molecules participate in specialized mechanisms of antigen recognition and synapse formation. It will be interesting to know if mechanotransduction is common to both the LRR and Ig-superfamily antigen receptors. This could make sense in that lymphocytes in both systems are freely migrating cells derived from mesenchyme. Cells from this tissue lineage utilize integrins to measure the physical properties of the environment and apply forces that are important in formation of solid tissues. Thus, a key mechanism in immune triggering may be a common sensory module handed down to immune cells through evolution (Dobereiner et al., 2006).

3.2. IS and disease

Cellular-based immunotherapy relies on providing a robust source of responsive cells to target the disease, adoptive transfer being an important approach (June, 2007a). One advantage of applying adoptive T cell therapy in cancer is that T cells can have the ability to specifically target tumor cells through recognition of tumor proteins presented on the target cell surface. Provided that relevant tumor-specific antigens are identified, antigen-specific T cells can be generated and expanded *ex vivo* in a bioreactor system. This approach, though specific, can be costly and labor intensive. An alternative approach is based on the activation and expansion of polyclonal T cells and relies on the presence of tumor-specific T cells within the polyclonal pool—technically more rapid and feasible but depends on the number of T cells within the expanded population that are tumor-specific and responsive to tumor-antigens presented by the APCs *in vivo* (reviewed in June, 2007b). Adoptive immunotherapy can be based on allogeneic donor leukocytes, which through graft-versus-leukemic effects can be successful in patients with myeloid leukemia (Loren & Porter, 2006). To resolve the complications associated with graft-versus-host disease, autologous immunotherapy is a promising approach. This approach relies on the use of patient cells that need to be expanded in culture systems *ex vivo* to allow the T cells to proliferate before being reintroduced into the same patient. Though APCs could be used for this purpose, they may not be controllable enough to achieve optimal differentiation of the patient T cells, whose long-term persistence and efficacy seem to be the key. Therefore, the current protocol for expansion of patient T cells *ex vivo*, used in early-phase clinical trials, is based on the use of cell-size magnetic beads coated with antibodies against

CD3 and CD28 (Brentjens et al., 2011; Kalos et al., 2011; Levine et al., 1997; Rapoport et al., 2005; Scholler et al., 2012). Efforts to replace the natural APC and reproduce the IS details in a carefully designed APC–mimetic system are being made not only to gain a better understanding of the IS but also to better control *ex vivo* T cell expansion, directing or enhancing T cell function of patients under certain pathological conditions. Engineered surfaces have already been used to show that T cells can respond to microscale geometry of the IS—using controlled ligand localization (Shen et al., 2008)—and to stiffness (Judokusumo et al., 2012; O'Connor et al., 2012), and appear to be superior to the currently used anti-CD3/CD28 antibody-coated beads, suggesting that these fabricated surfaces may mimic the natural T cell–APC interface more closely.

T cells also have the capability of being genetically engineered during *ex vivo* manipulation using TCRs or chimeric antigen receptors (CARs) to confer tumor specificity. This is achieved by linking a tumor recognition domain, most commonly the antigen-specific domain of antibodies, to molecules involved in signaling T cell effector function, leading to T cell granule release and cytotoxicity, cytokine production and proliferation. It is crucial that CAR-engineered T cells selectively and efficiently target the malignant tissue without harming normal cells *in vivo*. One major advantage of this approach is that CARs are not MHC-restricted and can be applied to different patients expressing the same target molecule. Lentiviral and retroviral transduction has been used to introduce CARs with costimulatory domains (to increase antitumor activity *in vivo*) into T cells that are expanded for adoptive transfer. Important factors for the optimization of CAR design include costimulation, density of target molecules in the tumor cell for effective T cell signaling, accessibility, and affinity of the engineered T cell and CAR molecules to the target tumor cell membrane (reviewed in Riddell, Jensen, & June, 2013). This approach has proven promising in pilot clinical trials for treatment of patients with chronic myeloid leukemia (CLL), where T cells are removed from the patient, modified to express the CAR, expanded, and reinfused into the patient (Kalos et al., 2011; Porter, Levine, Kalos, Bagg, & June, 2011). Some of the current studies on CARs include the development of safety strategies through incorporation of a conditional suicide gene or regulated expression of the tumor-targeting receptor, and the study of the effect of composition of CAR T cell products (also reviewed in Riddell et al., 2013).

A better understanding of the IS and how it is formed in healthy individuals can be extremely insightful in the study and design of appropriate treatment of certain pathologies, where the IS may show defects. T cells of CLL patients show defects in the formation of the IS with APCs—this synapse dysfunction appears to result both from CLL cells having poor APC function and defective actin polymerization in T cells from patients with CLL (Ramsay et al., 2008). Ramsay et al. observed, using cells of CLL patients, that the conjugation of T cells with APCs and subsequent polarization of F-actin, and recruitment of key regulatory proteins to the IS contact site (TCRs, adhesion molecules, and actin cytoskeleton proteins) were inhibited. These defects were observed both in T cells from CLL patients and T cells from healthy individuals that had been in contact with CLL cells; these defects could be reduced using the immunomodulating drug lenalidomide that appears to repair F-actin polymerization and signaling at the IS. This finding can have implications in the treatment of CLL patients both through autologous and allogeneic immunotherapy approaches and indicates that the repair of IS defects can be an important step in improving cancer immunotherapy.

As mentioned above, granule exocytosis-mediated cytotoxicity is the major effector function of CD8 T cells in adaptive immunity. Upon association of the CTL with the target cell and TCR engagement, lytic granules move to the IS and ultimately fuse with the effector cell plasma membrane to release cytolytic proteins. This lytic activity has been shown to be defective in human cancer and mouse human models (Frey & Monu, 2008). Lee et al. have reported that tumor-derived gangliosides inhibit lytic function in CD8 CTLs by impeding the process of TCR-induced lytic granule release (Lee et al., 2012).

In the context of HIV infection, components of the IS may facilitate latency as well as active infection. In addition, the asymmetric cell division in response to T cell stimulation by the IS could be a mechanism by which HIV infection can be maintained in a latent state and at the same time generating progeny that actively express HIV virus. This may suggest that manipulation of the IS may lead to establishment and maintenance of latency of the virus (reviewed in Kulpa et al., 2013).

Several immune diseases—autoimmune diseases—result from the recognition of self-pMHC by the organism's T cells. These T cells, with self-reactive TCRs, manage to escape negative selection and yet are able to respond to self-antigen with enough strength in the target organ. ISs formed by self-reactive T cells of patients with active autoimmune pathologies (such as multiple sclerosis and type 1 diabetes), though unusual and showing a reduced formation of the cSMAC and reduced pMHC recruitment, still appear to be able to sustain active TCR signaling (Schubert et al., 2012), possibly due to sustained motility of the T cells, and similarly to ISs formed with a mutant weak pMHC agonist (Vardhana et al., 2010). The observation that T cells with altered IS can have pathogenic potential strongly supports the need for a better understanding of the IS, both in healthy individuals and immune disease patients.

3.3. Future perspectives

There are still a number of fundamental questions about all levels of organization in the IS that remain answered. It is unclear how TCR triggering works and it is likely that there are multiple ways that it can be initiated. It will be interesting to know if TCR microclusters are a common pathway that are formed regardless of the triggering mechanism. More studies will be needed to address questions such as the relevance of synapse stability in the T cell response, the molecular determinants involved, the cross-talk in TCR signaling pathways that will lead to diversification of T cell functional responses and other open questions. Moreover, the combination of both the understanding of the molecular players and the development of new T cell activation systems will enable the generation of platforms that can ultimately be used to manipulate T cell response, to optimize T cell expansion and differentiation into a phenotype of choice. The ability to manipulate the immune response is particularly important in designing effective immunotherapy strategies. Cellular immunotherapy is becoming a preferred approach in the treatment of patients with cancer; thus, all efforts that may improve our understanding of the events that occur from the formation of the IS to the T cell functional response will be key in the development and refinement of these therapeutic strategies.

REFERENCES

Alarcon, B., Mestre, D., & Martinez-Martin, N. (2011). The immunological synapse: A cause or consequence of T-cell receptor triggering? *Immunology*, *133*(4), 420–425.

Allen, C. D., Okada, T., Tang, H. L., & Cyster, J. G. (2007). Imaging of germinal center selection events during affinity maturation. *Science*, *315*(5811), 528–531.

Alon, R., & Dustin, M. L. (2007). Force as a facilitator of integrin conformational changes during leukocyte arrest on blood vessels and antigen-presenting cells. *Immunity*, *26*(1), 17–27.

Angus, K. L., & Griffiths, G. M. (2013). Cell polarisation and the immunological synapse. *Current Opinion in Cell Biology*, *25*(1), 85–91.

Babich, A., Li, S., O'Connor, R. S., Milone, M. C., Freedman, B. D., & Burkhardt, J. K. (2012). F-actin polymerization and retrograde flow drive sustained PLCgamma1 signaling during T cell activation. *The Journal of Cell Biology*, *197*(6), 775–787 [Research Support, N.I.H., Extramural].

Banchereau, J., & Steinman, R. M. (1998). Dendritic cells and the control of immunity. *Nature*, *392*(6673), 245–252.

Bannard, O., Kraman, M., & Fearon, D. T. (2009). Secondary replicative function of CD8+ T cells that had developed an effector phenotype. *Science*, *323*(5913), 505–509.

Barda-Saad, M., Braiman, A., Titerence, R., Bunnell, S. C., Barr, V. A., & Samelson, L. E. (2005). Dynamic molecular interactions linking the T cell antigen receptor to the actin cytoskeleton. *Nature Immunology*, *6*(1), 80–89.

Barreira da Silva, R., & Munz, C. (2011). Natural killer cell activation by dendritic cells: Balancing inhibitory and activating signals. *Cellular and Molecular Life Sciences*, *68*(21), 3505–3518.

Batista, F. D., Iber, D., & Neuberger, M. S. (2001). B cells acquire antigen from target cells after synapse formation. *Nature*, *411*(6836), 489–494.

Beemiller, P., Jacobelli, J., & Krummel, M. F. (2012). Integration of the movement of signaling microclusters with cellular motility in immunological synapses. *Nature Immunology*, *13*(8), 787–795 [Research Support, N.I.H., Extramural].

Beinke, S., Phee, H., Clingan, J. M., Schlessinger, J., Matloubian, M., & Weiss, A. (2010). Proline-rich tyrosine kinase-2 is critical for CD8 T-cell short-lived effector fate. *Proceedings of the National Academy of Sciences of the United States of America*, *107*(37), 16234–16239.

Betzig, E., Patterson, G. H., Sougrat, R., Lindwasser, O. W., Olenych, S., Bonifacino, J. S., et al. (2006). Imaging intracellular fluorescent proteins at nanometer resolution. *Science*, *313*(5793), 1642–1645.

Biggs, M. J., Milone, M. C., Santos, L. C., Gondarenko, A., & Wind, S. J. (2011). High-resolution imaging of the immunological synapse and T-cell receptor microclustering through microfabricated substrates. *Journal of the Royal Society Interface*, *8*(63), 1462–1471.

Boehm, T., McCurley, N., Sutoh, Y., Schorpp, M., Kasahara, M., & Cooper, M. D. (2012). VLR-based adaptive immunity. *Annual Review of Immunology*, *30*, 203–220 [Research Support, N.I.H., Extramural Research Support, Non-U.S. Gov't Review].

Bourouina, N., Husson, J., Hivroz, C., & Henry, N. (2012). Biomimetic droplets for artificial engagement of living cell surface receptors: The specific case of the T-cell. *Langmuir*, *28*(14), 6106–6113 [Research Support, Non-U.S. Gov't].

Breart, B., Lemaitre, F., Celli, S., & Bousso, P. (2008). Two-photon imaging of intratumoral CD8+ T cell cytotoxic activity during adoptive T cell therapy in mice. *The Journal of Clinical Investigation*, *118*(4), 1390–1397.

Brentjens, R. J., Riviere, I., Park, J. H., Davila, M. L., Wang, X., Stefanski, J., et al. (2011). Safety and persistence of adoptively transferred autologous CD19-targeted T cells in patients with relapsed or chemotherapy refractory B-cell leukemias. *Blood*, *118*(18), 4817–4828.

Burkhardt, J. K., Carrizosa, E., & Shaffer, M. H. (2008). The actin cytoskeleton in T cell activation. *Annual Review of Immunology*, *26*, 233–259.

Campi, G., Varma, R., & Dustin, M. L. (2005). Actin and agonist MHC-peptide complex-dependent T cell receptor microclusters as scaffolds for signaling. *The Journal of Experimental Medicine*, *202*(8), 1031–1036.

Celli, S., Lemaitre, F., & Bousso, P. (2007). Real-time manipulation of T cell-dendritic cell interactions in vivo reveals the importance of prolonged contacts for CD4+ T cell activation. *Immunity*, *27*(4), 625–634.

Cemerski, S., Das, J., Giurisato, E., Markiewicz, M. A., Allen, P. M., Chakraborty, A. K., et al. (2008). The balance between T cell receptor signaling and degradation at the center of the immunological synapse is determined by antigen quality. *Immunity*, *29*(3), 414–422.

Chang, J. T., Palanivel, V. R., Kinjyo, I., Schambach, F., Intlekofer, A. M., Banerjee, A., et al. (2007). Asymmetric T lymphocyte division in the initiation of adaptive immune responses. *Science*, *315*(5819), 1687–1691.

Chaudhri, G., Quah, B. J., Wang, Y., Tan, A. H., Zhou, J., Karupiah, G., et al. (2009). T cell receptor sharing by cytotoxic T lymphocytes facilitates efficient virus control. *Proceedings of the National Academy of Sciences of the United States of America, 106*(35), 14984–14989.

Chen, W., Lou, J., Evans, E. A., & Zhu, C. (2012). Observing force-regulated conformational changes and ligand dissociation from a single integrin on cells. *The Journal of Cell Biology, 199*(3), 497–512 [Research Support, N.I.H., Extramural Research Support, Non-U.S. Gov't].

Choudhuri, K., Parker, M., Milicic, A., Cole, D. K., Shaw, M. K., Sewell, A. K., et al. (2009). Peptide-major histocompatibility complex dimensions control proximal kinase-phosphatase balance during T cell activation. *The Journal of Biological Chemistry, 284*(38), 26096–26105.

Choudhuri, K., Wiseman, D., Brown, M. H., Gould, K., & van der Merwe, P. A. (2005). T-cell receptor triggering is critically dependent on the dimensions of its peptide-MHC ligand. *Nature, 436*(7050), 578–582.

Coombs, D., Kalergis, A. M., Nathenson, S. G., Wofsy, C., & Goldstein, B. (2002). Activated TCRs remain marked for internalization after dissociation from pMHC. *Nature Immunology, 3*(10), 926–931.

Cooper, M. D., & Alder, M. N. (2006). The evolution of adaptive immune systems. *Cell, 124*(4), 815–822.

Cruz-Orcutt, N., & Houtman, J. C. (2009). PI3 kinase function is vital for the function but not formation of LAT-mediated signaling complexes. *Molecular Immunology, 46*(11–12), 2274–2283.

Cyster, J. G., Shotton, D. M., & Williams, A. F. (1991). The dimensions of the T lymphocyte glycoprotein leukosialin and identification of linear protein epitopes that can be modified by glycosylation. *The EMBO Journal, 10*(4), 893–902.

DeFord-Watts, L. M., Dougall, D. S., Belkaya, S., Johnson, B. A., Eitson, J. L., Roybal, K. T., et al. (2011). The CD3 zeta subunit contains a phosphoinositide-binding motif that is required for the stable accumulation of TCR-CD3 complex at the immunological synapse. *The Journal of Immunology, 186*(12), 6839–6847.

DeMond, A. L., Mossman, K. D., Starr, T., Dustin, M. L., & Groves, J. T. (2008). T cell receptor microcluster transport through molecular mazes reveals mechanism of translocation. *Biophysical Journal, 94*(8), 3286–3292 [Research Support, U.S. Gov't, Non-P.H.S.].

Dobereiner, H. G., Dubin-Thaler, B. J., Hofman, J. M., Xenias, H. S., Sims, T. N., Giannone, G., et al. (2006). Lateral membrane waves constitute a universal dynamic pattern of motile cells. *Physical Review Letters, 97*(3), 038102.

Doh, J., & Irvine, D. J. (2006). Immunological synapse arrays: Patterned protein surfaces that modulate immunological synapse structure formation in T cells. *Proceedings of the National Academy of Sciences of the United States of America, 103*(15), 5700–5705.

Dustin, M. L. (2008). Hunter to gatherer and back: Immunological synapses and kinapses as variations on the theme of amoeboid locomotion. *Annual Review of Cell and Developmental Biology, 24*, 577–596.

Dustin, M. L., Bromley, S. K., Kan, Z., Peterson, D. A., & Unanue, E. R. (1997). Antigen receptor engagement delivers a stop signal to migrating T lymphocytes. *Proceedings of the National Academy of Sciences of the United States of America, 94*(8), 3909–3913.

Dustin, M. L., Ferguson, L. M., Chan, P. Y., Springer, T. A., & Golan, D. E. (1996). Visualization of CD2 interaction with LFA-3 and determination of the two-dimensional dissociation constant for adhesion receptors in a contact area. *The Journal of Cell Biology, 132*(3), 465–474.

Dustin, M. L., Starr, T., Varma, R., & Thomas, V. K. (2007). Supported planar bilayers for study of the immunological synapse. *Current Protocols in Immunology,* [Chapter 18, Unit 18 13].

Edelson, B. T., Bradstreet, T. R., Hildner, K., Carrero, J. A., Frederick, K. E., Kc, W., et al. (2011). CD8alpha(+) Dendritic cells Are an obligate cellular entry point for productive infection by Listeria monocytogenes. *Immunity*, *35*(2), 236–248.

Egen, J. G., Rothfuchs, A. G., Feng, C. G., Winter, N., Sher, A., & Germain, R. N. (2008). Macrophage and T cell dynamics during the development and disintegration of mycobacterial granulomas. *Immunity*, *28*(2), 271–284.

Eyckmans, J., Boudou, T., Yu, X., & Chen, C. S. (2011). A hitchhiker's guide to mechanobiology. *Developmental Cell*, *21*(1), 35–47.

Flajnik, M. F., & Kasahara, M. (2010). Origin and evolution of the adaptive immune system: Genetic events and selective pressures. *Nature Reviews Genetics*, *11*(1), 47–59.

Fooksman, D. R., Vardhana, S., Vasiliver-Shamis, G., Liese, J., Blair, D. A., Waite, J., et al. (2010). Functional anatomy of T cell activation and synapse formation. *Annual Review of Immunology*, *28*, 79–105.

Frey, A. B., & Monu, N. (2008). Signaling defects in anti-tumor T cells. *Immunological Reviews*, *222*, 192–205.

Gagnon, E., Schubert, D. A., Gordo, S., Chu, H. H., & Wucherpfennig, K. W. (2012). Local changes in lipid environment of TCR microclusters regulate membrane binding by the CD3{varepsilon} cytoplasmic domain. *The Journal of Experimental Medicine*, *209*(13), 2423–2439.

Gay, D., Coeshott, C., Golde, W., Kappler, J., & Marrack, P. (1986). The major histocompatibility complex-restricted antigen receptor on T cells. IX. Role of accessory molecules in recognition of antigen plus isolated IA. *The Journal of Immunology*, *136*(6), 2026–2032.

Gerard, A., Khan, O., Beemiller, P., Oswald, E., Hu, J., Matloubian, M., et al. (2013). Secondary T cell-T cell synaptic interactions drive the differentiation of protective CD8+ T cells. *Nature Immunology*, *14*(4), 356–363.

Gharbi, S. I., Rincon, E., Avila-Flores, A., Torres-Ayuso, P., Almena, M., Cobos, M. A., et al. (2011). Diacylglycerol kinase zeta controls diacylglycerol metabolism at the immunological synapse. *Molecular Biology of the Cell*, *22*(22), 4406–4414.

Goodridge, H. S., Reyes, C. N., Becker, C. A., Katsumoto, T. R., Ma, J., Wolf, A. J., et al. (2011). Activation of the innate immune receptor Dectin-1 upon formation of a 'phagocytic synapse'. *Nature*, *472*(7344), 471–475.

Grakoui, A., Bromley, S. K., Sumen, C., Davis, M. M., Shaw, A. S., Allen, P. M., et al. (1999). The immunological synapse: A molecular machine controlling T cell activation. *Science*, *285*(5425), 221–227.

Griffiths, G. M., Tsun, A., & Stinchcombe, J. C. (2010). The immunological synapse: A focal point for endocytosis and exocytosis. *The Journal of Cell Biology*, *189*(3), 399–406.

Groves, J. T., & Dustin, M. L. (2003). Supported planar bilayers in studies on immune cell adhesion and communication. *Journal of Immunological Methods*, *278*(1–2), 19–32.

Guy, C. S., Vignali, K. M., Temirov, J., Bettini, M. L., Overacre, A. E., Smeltzer, M., et al. (2013). Distinct TCR signaling pathways drive proliferation and cytokine production in T cells. *Nature Immunology*, *14*(3), 262–270.

Harris, T. H., Banigan, E. J., Christian, D. A., Konradt, C., Tait Wojno, E. D., Norose, K., et al. (2012). Generalized Levy walks and the role of chemokines in migration of effector CD8+ T cells. *Nature*, *486*(7404), 545–548 [Research Support, N.I.H., Extramural Research Support, Non-U.S. Gov't Research Support, U.S. Gov't, Non-P.H.S.].

Harwood, N. E., & Batista, F. D. (2011). The cytoskeleton coordinates the early events of B-cell activation. *Cold Spring Harbor Perspectives in Biology*, *3*(2).

Hashimoto-Tane, A., Yokosuka, T., Sakata-Sogawa, K., Sakuma, M., Ishihara, C., Tokunaga, M., et al. (2011). Dynein-driven transport of T cell receptor microclusters regulates immune synapse formation and T cell activation. *Immunity*, *34*(6), 919–931 [Research Support, Non-U.S. Gov't].

Hogan, P. G., Chen, L., Nardone, J., & Rao, A. (2003). Transcriptional regulation by calcium, calcineurin, and NFAT. *Genes & Development*, *17*(18), 2205–2232.

Hsu, C. J., & Baumgart, T. (2011). Spatial association of signaling proteins and F-actin effects on cluster assembly analyzed via photoactivation localization microscopy in T cells. *PLoS One*, *6*(8), e23586 [Research Support, N.I.H., Extramural Research Support, U.S. Gov't, Non-P.H.S.].

Huang, J., Zarnitsyna, V. I., Liu, B., Edwards, L. J., Jiang, N., Evavold, B. D., et al. (2010). The kinetics of two-dimensional TCR and pMHC interactions determine T-cell responsiveness. *Nature*, *464*(7290), 932–936 [Research Support, N.I.H., Extramural Research Support, Non-U.S. Gov't].

Husson, J., Chemin, K., Bohineust, A., Hivroz, C., & Henry, N. (2011). Force generation upon T cell receptor engagement. *PLoS One*, *6*(5), e19680.

Ilani, T., Vasiliver-Shamis, G., Vardhana, S., Bretscher, A., & Dustin, M. L. (2009). T cell antigen receptor signaling and immunological synapse stability require myosin IIA. *Nature Immunology*, *10*(5), 531–539 [Research Support, N.I.H., Extramural Research Support, Non-U.S. Gov't].

James, J. R., & Vale, R. D. (2012). Biophysical mechanism of T-cell receptor triggering in a reconstituted system. *Nature*, *487*(7405), 64–69 [Research Support, Non-U.S. Gov't].

Janeway, C. A., Jr., & Medzhitov, R. (2002). Innate immune recognition. *Annual Review of Immunology*, *20*, 197–216.

Joo, C., Balci, H., Ishitsuka, Y., Buranachai, C., & Ha, T. (2008). Advances in single-molecule fluorescence methods for molecular biology. *Annual Review of Biochemistry*, *77*, 51–76.

Joyce, S., Girardi, E., & Zajonc, D. M. (2011). NKT cell ligand recognition logic: Molecular basis for a synaptic duet and transmission of inflammatory effectors. *The Journal of Immunology*, *187*(3), 1081–1089.

Judokusumo, E., Tabdanov, E., Kumari, S., Dustin, M. L., & Kam, L. C. (2012). Mechanosensing in T lymphocyte activation. *Biophysical Journal*, *102*(2), L5–L7.

June, C. H. (2007a). Adoptive T cell therapy for cancer in the clinic. *The Journal of Clinical Investigation*, *117*(6), 1466–1476.

June, C. H. (2007b). Principles of adoptive T cell cancer therapy. *The Journal of Clinical Investigation*, *117*(5), 1204–1212.

Kaizuka, Y., Douglass, A. D., Varma, R., Dustin, M. L., & Vale, R. D. (2007). Mechanisms for segregating T cell receptor and adhesion molecules during immunological synapse formation in Jurkat T cells. *Proceedings of the National Academy of Sciences of the United States of America*, *104*(51), 20296–20301.

Kalos, M., Levine, B. L., Porter, D. L., Katz, S., Grupp, S. A., Bagg, A., et al. (2011). T cells with chimeric antigen receptors have potent antitumor effects and can establish memory in patients with advanced leukemia. *Science Translational Medicine*, *3*(95), 95ra73.

Kim, S. T., Shin, Y., Brazin, K., Mallis, R. J., Sun, Z. Y., Wagner, G., et al. (2012). TCR mechanobiology: Torques and tunable structures linked to early T cell signaling. *Frontiers in Immunology*, *3*, 76.

Kim, S. T., Takeuchi, K., Sun, Z. Y., Touma, M., Castro, C. E., Fahmy, A., et al. (2009). The alphabeta T cell receptor is an anisotropic mechanosensor. *The Journal of Biological Chemistry*, *2009*(284), 31028–31037.

Kulpa, D. A., Brehm, J. H., Fromentin, R., Cooper, A., Cooper, C., Ahlers, J., et al. (2013). The immunological synapse: The gateway to the HIV reservoir. *Immunological Reviews*, *254*(1), 305–325.

Kumari, S., Vardhana, S., Cammer, M., Curado, S., Santos, L., Sheetz, M. P., et al. (2012). T lymphocyte myosin IIA is required for maturation of the immunological synapse. *Frontiers in Immunology*, *3*, 230.

Kupfer, A., Mosmann, T. R., & Kupfer, H. (1991). Polarized expression of cytokines in cell conjugates of helper T cells and splenic B cells. *Proceedings of the National Academy of Sciences of the United States of America*, *88*(3), 775–779.

Lee, K. H., Dinner, A. R., Tu, C., Campi, G., Raychaudhuri, S., Varma, R., et al. (2003). The immunological synapse balances T cell receptor signaling and degradation. *Science*, *302*(5648), 1218–1222.

Lee, H. C., Wondimu, A., Liu, Y., Ma, J. S., Radoja, S., & Ladisch, S. (2012). Ganglioside inhibition of CD8+ T cell cytotoxicity: Interference with lytic granule trafficking and exocytosis. *The Journal of Immunology*, *189*(7), 3521–3527.

Leupin, O., Zaru, R., Laroche, T., Muller, S., & Valitutti, S. (2000). Exclusion of CD45 from the T-cell receptor signaling area in antigen-stimulated T lymphocytes. *Current Biology*, *10*(5), 277–280.

Levine, B. L., Bernstein, W. B., Connors, M., Craighead, N., Lindsten, T., Thompson, C. B., et al. (1997). Effects of CD28 costimulation on long-term proliferation of CD4+ T cells in the absence of exogenous feeder cells. *The Journal of Immunology*, *159*(12), 5921–5930.

Li, Y. C., Chen, B. M., Wu, P. C., Cheng, T. L., Kao, L. S., Tao, M. H., et al. (2010). Cutting Edge: Mechanical forces acting on T cells immobilized via the TCR complex can trigger TCR signaling. *The Journal of Immunology*, *184*(11), 5959–5963.

Lillemeier, B. F., Mortelmaier, M. A., Forstner, M. B., Huppa, J. B., Groves, J. T., & Davis, M. M. (2010). TCR and Lat are expressed on separate protein islands on T cell membranes and concatenate during activation. *Nature Immunology*, *11*(1), 90–96 [Research Support, N.I.H., Extramural Research Support, Non-U.S. Gov't Research Support, U.S. Gov't, Non-P.H.S.].

Liu, S. K., Berry, D. M., & McGlade, C. J. (2001). The role of Gads in hematopoietic cell signalling. *Oncogene*, *20*(44), 6284–6290.

Liu, H., Rhodes, M., Wiest, D. L., & Vignali, D. A. (2000). On the dynamics of TCR:CD3 complex cell surface expression and downmodulation. *Immunity*, *13*(5), 665–675.

Longhi, M. P., Trumpfheller, C., Idoyaga, J., Caskey, M., Matos, I., Kluger, C., et al. (2009). Dendritic cells require a systemic type I interferon response to mature and induce CD4+ Th1 immunity with poly IC as adjuvant. *The Journal of Experimental Medicine*, *206*(7), 1589–1602 [Research Support, N.I.H., Extramural Research Support, Non-U.S. Gov't].

Loren, A. W., & Porter, D. L. (2006). Donor leukocyte infusions after unrelated donor hematopoietic stem cell transplantation. *Current Opinion in Oncology*, *18*(2), 107–114.

Ma, Z., & Finkel, T. H. (2010). T cell receptor triggering by force. *Trends in Immunology*, *31*(1), 1–6.

Ma, Z., Sharp, K. A., Janmey, P. A., & Finkel, T. H. (2008). Surface-anchored monomeric agonist pMHCs alone trigger TCR with high sensitivity. *PLoS Biology*, *6*(2), e43 [Research Support, N.I.H., Extramural Research Support, Non-U.S. Gov't].

Maldonado, R. A., Soriano, M. A., Perdomo, L. C., Sigrist, K., Irvine, D. J., Decker, T., et al. (2009). Control of T helper cell differentiation through cytokine receptor inclusion in the immunological synapse. *The Journal of Experimental Medicine*, *206*(4), 877–892.

McConnell, H. M., Watts, T. H., Weis, R. M., & Brian, A. A. (1986). Supported planar membranes in studies of cell-cell recognition in the immune system. *Biochimica et Biophysica Acta*, *864*(1), 95–106.

Mempel, T. R., Henrickson, S. E., & Von Andrian, U. H. (2004). T-cell priming by dendritic cells in lymph nodes occurs in three distinct phases. *Nature*, *427*(6970), 154–159.

Miletic, A. V., Graham, D. B., Sakata-Sogawa, K., Hiroshima, M., Hamann, M. J., Cemerski, S., et al. (2009). Vav links the T cell antigen receptor to the actin cytoskeleton and T cell activation independently of intrinsic Guanine nucleotide exchange activity. *PLoS One*, *4*(8), e6599.

Miletic, A. V., Sakata-Sogawa, K., Hiroshima, M., Hamann, M. J., Gomez, T. S., Ota, N., et al. (2006). Vav1 acidic region tyrosine 174 is required for the formation of T cell receptor-induced microclusters and is essential in T cell development and activation. *The Journal of Biological Chemistry, 281*(50), 38257–38265.
Miller, M. J., Hejazi, A. S., Wei, S. H., Cahalan, M. D., & Parker, I. (2004). T cell repertoire scanning is promoted by dynamic dendritic cell behavior and random T cell motility in the lymph node. *Proceedings of the National Academy of Sciences of the United States of America, 101*(4), 998–1003.
Moerner, W. E. (2007). New directions in single-molecule imaging and analysis. *Proceedings of the National Academy of Sciences of the United States of America, 104*(31), 12596–12602.
Monks, C. R., Freiberg, B. A., Kupfer, H., Sciaky, N., & Kupfer, A. (1998). Three-dimensional segregation of supramolecular activation clusters in T cells. *Nature, 395*(6697), 82–86.
Moon, J. J., Chu, H. H., Pepper, M., McSorley, S. J., Jameson, S. C., Kedl, R. M., et al. (2007). Naive CD4(+) T cell frequency varies for different epitopes and predicts repertoire diversity and response magnitude. *Immunity, 27*(2), 203–213.
Mossman, K. D., Campi, G., Groves, J. T., & Dustin, M. L. (2005). Altered TCR signaling from geometrically repatterned immunological synapses. *Science, 310*(5751), 1191–1193.
Natkanski, E., Lee, W. Y., Mistry, B., Casal, A., Molloy, J. E., & Tolar, P. (2013). B cells use mechanical energy to discriminate antigen affinities. *Science, 340*(6140), 1587–1590.
O'Connor, R. S., Hao, X., Shen, K., Bashour, K., Akimova, T., Hancock, W. W., et al. (2012). Substrate rigidity regulates human T cell activation and proliferation. *The Journal of Immunology, 189*(3), 1330–1339.
Oddos, S., Dunsby, C., Purbhoo, M. A., Chauveau, A., Owen, D. M., Neil, M. A., et al. (2008). High-speed high-resolution imaging of intercellular immune synapses using optical tweezers. *Biophysical Journal, 95*(10), L66–L68.
Ohl, L., Mohaupt, M., Czeloth, N., Hintzen, G., Kiafard, Z., Zwirner, J., et al. (2004). CCR7 governs skin dendritic cell migration under inflammatory and steady-state conditions. *Immunity, 21*(2), 279–288.
Oliaro, J., Van Ham, V., Sacirbegovic, F., Pasam, A., Bomzon, Z., Pham, K., et al. (2010). Asymmetric cell division of T cells upon antigen presentation uses multiple conserved mechanisms. *The Journal of Immunology, 185*(1), 367–375.
Padhan, K., & Varma, R. (2010). Immunological synapse: A multi-protein signalling cellular apparatus for controlling gene expression. *Immunology, 129*(3), 322–328.
Pauker, M. H., Hassan, N., Noy, E., Reicher, B., & Barda-Saad, M. (2012). Studying the dynamics of SLP-76, Nck, and Vav1 multimolecular complex formation in live human cells with triple-color FRET. *Science Signaling, 5*(221), rs3.
Poo, W. J., Conrad, L., & Janeway, C. A., Jr. (1988). Receptor-directed focusing of lymphokine release by helper T cells. *Nature, 332*(6162), 378–380.
Porter, D. L., Levine, B. L., Kalos, M., Bagg, A., & June, C. H. (2011). Chimeric antigen receptor-modified T cells in chronic lymphoid leukemia. *The New England Journal of Medicine, 365*(8), 725–733.
Quann, E. J., Merino, E., Furuta, T., & Huse, M. (2009). Localized diacylglycerol drives the polarization of the microtubule-organizing center in T cells. *Nature Immunology, 10*(6), 627–635.
Rak, G. D., Mace, E. M., Banerjee, P. P., Svitkina, T., & Orange, J. S. (2011). Natural killer cell lytic granule secretion occurs through a pervasive actin network at the immune synapse. *PLoS Biology, 9*(9), e1001151.
Ramsay, A. G., Johnson, A. J., Lee, A. M., Gorgun, G., Le Dieu, R., Blum, W., et al. (2008). Chronic lymphocytic leukemia T cells show impaired immunological synapse formation that can be reversed with an immunomodulating drug. *The Journal of Clinical Investigation, 118*(7), 2427–2437.

Rapoport, A. P., Stadtmauer, E. A., Aqui, N., Badros, A., Cotte, J., Chrisley, L., et al. (2005). Restoration of immunity in lymphopenic individuals with cancer by vaccination and adoptive T-cell transfer. *Nature Medicine*, *11*(11), 1230–1237.

Riddell, S. R., Jensen, M. C., & June, C. H. (2013). Chimeric antigen receptor-modified T cells: Clinical translation in stem cell transplantation and beyond. *Biology of Blood and Marrow Transplantation*, *19*(1 Suppl), S2–S5.

Rossy, J., Owen, D. M., Williamson, D. J., Yang, Z., & Gaus, K. (2013). Conformational states of the kinase Lck regulate clustering in early T cell signaling. *Nature Immunology*, *14*(1), 82–89.

Rossy, J., Pageon, S. V., Davis, D. M., & Gaus, K. (2013). Super-resolution microscopy of the immunological synapse. *Current Opinion in Immunology*, *25*(3), 307–312.

Ruocco, M. G., Pilones, K. A., Kawashima, N., Cammer, M., Huang, J., Babb, J. S., et al. (2012). Suppressing T cell motility induced by anti-CTLA-4 monotherapy improves antitumor effects. *The Journal of Clinical Investigation*, *122*(10), 3718–3730.

Rust, M. J., Bates, M., & Zhuang, X. (2006). Sub-diffraction-limit imaging by stochastic optical reconstruction microscopy (STORM). *Nature Methods*, *3*(10), 793–795.

Sabatos, C. A., Doh, J., Chakravarti, S., Friedman, R. S., Pandurangi, P. G., Tooley, A. J., et al. (2008). A synaptic basis for paracrine interleukin-2 signaling during homotypic T cell interaction. *Immunity*, *29*(2), 238–248.

Sage, P. T., Varghese, L. M., Martinelli, R., Sciuto, T. E., Kamei, M., Dvorak, A. M., et al. (2012). Antigen recognition is facilitated by invadosome-like protrusions formed by memory/effector T cells. *The Journal of Immunology*, *188*(8), 3686–3699.

Sasahara, Y., Rachid, R., Byrne, M. J., de la Fuente, M. A., Abraham, R. T., Ramesh, N., et al. (2002). Mechanism of recruitment of WASP to the immunological synapse and of its activation following TCR ligation. *Molecular Cell*, *10*(6), 1269–1281.

Scholler, J., Brady, T. L., Binder-Scholl, G., Hwang, W. T., Plesa, G., Hege, K. M., et al. (2012). Decade-long safety and function of retroviral-modified chimeric antigen receptor T cells. *Science Translational Medicine*, *4*(132), 132ra153.

Schubert, D. A., Gordo, S., Sabatino, J. J., Jr., Vardhana, S., Gagnon, E., Sethi, D. K., et al. (2012). Self-reactive human CD4 T cell clones form unusual immunological synapses. *The Journal of Experimental Medicine*, *209*(2), 335–352.

Shaw, A. S., & Dustin, M. L. (1997). Making the T cell receptor go the distance: A topological view of T cell activation. *Immunity*, *6*(4), 361–369.

Shaw, P. J., Qu, B., Hoth, M., & Feske, S. (2013). Molecular regulation of CRAC channels and their role in lymphocyte function. *Cellular and Molecular Life Sciences*, *70*, 2637–2656.

Shen, K., Thomas, V. K., Dustin, M. L., & Kam, L. C. (2008). Micropatterning of costimulatory ligands enhances CD4+ T cell function. *Proceedings of the National Academy of Sciences of the United States of America*, *105*(22), 7791–7796.

Sherman, E., Barr, V., Manley, S., Patterson, G., Balagopalan, L., Akpan, I., et al. (2011). Functional nanoscale organization of signaling molecules downstream of the T cell antigen receptor. *Immunity*, *35*(5), 705–720 [Research Support, N.I.H., Intramural].

Shi, X., Bi, Y., Yang, W., Guo, X., Jiang, Y., Wan, C., et al. (2013). Ca2+ regulates T-cell receptor activation by modulating the charge property of lipids. *Nature*, *493*(7430), 111–115 [Research Support, Non-U.S. Gov't].

Sims, T. N., Soos, T. J., Xenias, H. S., Dubin-Thaler, B., Hofman, J. M., Waite, J. C., et al. (2007). Opposing effects of PKCtheta and WASp on symmetry breaking and relocation of the immunological synapse. *Cell*, *129*(4), 773–785.

Skokos, D., Shakhar, G., Varma, R., Waite, J. C., Cameron, T. O., Lindquist, R. L., et al. (2007). Peptide-MHC potency governs dynamic interactions between T cells and dendritic cells in lymph nodes. *Nature Immunology*, *8*(8), 835–844.

Smoligovets, A. A., Smith, A. W., Wu, H. J., Petit, R. S., & Groves, J. T. (2012). Characterization of dynamic actin associations with T-cell receptor microclusters in primary T cells. *Journal of Cell Science*, *125*(Pt 3), 735–742.

Steinman, R. M., Hawiger, D., & Nussenzweig, M. C. (2003). Tolerogenic dendritic cells. *Annual Review of Immunology*, *21*, 685–711.

Stinchcombe, J. C., & Griffiths, G. M. (2007). Secretory mechanisms in cell-mediated cytotoxicity. *Annual Review of Cell and Developmental Biology*, *23*, 495–517.

Stinchcombe, J. C., Majorovits, E., Bossi, G., Fuller, S., & Griffiths, G. M. (2006). Centrosome polarization delivers secretory granules to the immunological synapse. *Nature*, *443*(7110), 462–465.

Stinchcombe, J. C., Salio, M., Cerundolo, V., Pende, D., Arico, M., & Griffiths, G. M. (2011). Centriole polarisation to the immunological synapse directs secretion from cytolytic cells of both the innate and adaptive immune systems. *BMC Biology*, *9*, 45.

Sylvain, N. R., Nguyen, K., & Bunnell, S. C. (2011). Vav1-mediated scaffolding interactions stabilize SLP-76 microclusters and contribute to antigen-dependent T cell responses. *Science Signaling*, *4*(163), ra14.

Trappmann, B., Gautrot, J. E., Connelly, J. T., Strange, D. G., Li, Y., Oyen, M. L., et al. (2012). Extracellular-matrix tethering regulates stem-cell fate. *Nature Materials*, *11*(7), 642–649 [Research Support, Non-U.S. Gov't].

Trombetta, E. S., & Mellman, I. (2005). Cell biology of antigen processing in vitro and in vivo. *Annual Review of Immunology*, *23*, 975–1028.

Tseng, S. Y., Waite, J. C., Liu, M., Vardhana, S., & Dustin, M. L. (2008). T cell-dendritic cell immunological synapses contain TCR-dependent CD28-CD80 clusters that recruit protein kinase Ctheta. *The Journal of Immunology*, *181*(7), 4852–4863.

Tsun, A., Qureshi, I., Stinchcombe, J. C., Jenkins, M. R., de la Roche, M., Kleczkowska, J., et al. (2011). Centrosome docking at the immunological synapse is controlled by Lck signaling. *The Journal of Cell Biology*, *192*(4), 663–674.

Tybulewicz, V. L. (2005). Vav-family proteins in T-cell signalling. *Current Opinion in Immunology*, *17*(3), 267–274.

Ueda, H., Morphew, M. K., McIntosh, J. R., & Davis, M. M. (2011). CD4+ T-cell synapses involve multiple distinct stages. *Proceedings of the National Academy of Sciences of the United States of America*, *108*(41), 17099–17104.

Vardhana, S., Choudhuri, K., Varma, R., & Dustin, M. L. (2010). Essential role of ubiquitin and TSG101 protein in formation and function of the central supramolecular activation cluster. *Immunity*, *32*(4), 531–540.

Varma, R., Campi, G., Yokosuka, T., Saito, T., & Dustin, M. L. (2006). T cell receptor-proximal signals are sustained in peripheral microclusters and terminated in the central supramolecular activation cluster. *Immunity*, *25*(1), 117–127.

Victora, G. D., Schwickert, T. A., Fooksman, D. R., Kamphorst, A. O., Meyer-Hermann, M., Dustin, M. L., et al. (2010). Germinal center dynamics revealed by multiphoton microscopy with a photoactivatable fluorescent reporter. *Cell*, *143*(4), 592–605.

Waite, J. C., Leiner, I., Lauer, P., Rae, C. S., Barbet, G., Zheng, H., et al. (2011). Dynamic imaging of the effector immune response to listeria infection in vivo. *PLoS Pathogens*, 7(3), e1001326.

Waite, J. C., Vardhana, S., Shaw, P. J., Jang, J.-E., McCarl, C. A., Cameron, T. O., et al. (2013). Interference with Ca2+ release activated Ca2+ (CRAC) channel function delays T-cell arrest in vivo. *European Journal of Immunology*. http://dx.doi.org/10.1002/eji.201243255.

Wan, Z., Zhang, S., Fan, Y., Liu, K., Du, F., Davey, A. M., et al. (2013). B cell activation is regulated by the stiffness properties of the substrate presenting the antigens. *The Journal of Immunology*, *190*(9), 4661–4675.

Wang, H., Kadlecek, T. A., Au-Yeung, B. B., Goodfellow, H. E., Hsu, L. Y., Freedman, T. S., et al. (2010). ZAP-70: An essential kinase in T-cell signaling. *Cold Spring Harbor Perspectives in Biology*, *2*(5), a002279.

Williamson, D. J., Owen, D. M., Rossy, J., Magenau, A., Wehrmann, M., Gooding, J. J., et al. (2011). Pre-existing clusters of the adaptor Lat do not participate in early T cell signaling events. *Nature Immunology*, *12*(7), 655–662.

Wu, J. N., & Koretzky, G. A. (2004). The SLP-76 family of adapter proteins. *Seminars in Immunology*, *16*(6), 379–393.

Xu, C., Gagnon, E., Call, M. E., Schnell, J. R., Schwieters, C. D., Carman, C. V., et al. (2008). Regulation of T cell receptor activation by dynamic membrane binding of the CD3epsilon cytoplasmic tyrosine-based motif. *Cell*, *135*(4), 702–713.

Yi, J., Wu, X. S., Crites, T., & Hammer, J. A., 3rd. (2012). Actin retrograde flow and acto-myosin II arc contraction drive receptor cluster dynamics at the immunological synapse in Jurkat T cells. *Molecular Biology of the Cell*, *23*(5), 834–852.

Yokosuka, T., Kobayashi, W., Sakata-Sogawa, K., Takamatsu, M., Hashimoto-Tane, A., Dustin, M. L., et al. (2008). Spatiotemporal regulation of T cell costimulation by TCR-CD28 microclusters and protein kinase C theta translocation. *Immunity*, *29*(4), 589–601.

Yokosuka, T., Sakata-Sogawa, K., Kobayashi, W., Hiroshima, M., Hashimoto-Tane, A., Tokunaga, M., et al. (2005). Newly generated T cell receptor microclusters initiate and sustain T cell activation by recruitment of Zap70 and SLP-76. *Nature Immunology*, *6*(12), 1253–1262.

Yu, C. H., & Groves, J. T. (2010). Engineering supported membranes for cell biology. *Medical & Biological Engineering & Computing*, *48*(10), 955–963.

Yu, C. H., Wu, H. J., Kaizuka, Y., Vale, R. D., & Groves, J. T. (2010). Altered actin centripetal retrograde flow in physically restricted immunological synapses. *PLoS One*, *5*(7), e11878.

Zeng, R., Cannon, J. L., Abraham, R. T., Way, M., Billadeau, D. D., Bubeck-Wardenberg, J., et al. (2003). SLP-76 coordinates Nck-dependent Wiskott-Aldrich syndrome protein recruitment with Vav-1/Cdc42-dependent Wiskott-Aldrich syndrome protein activation at the T cell-APC contact site. *The Journal of Immunology*, *171*(3), 1360–1368.

Zhang, H., Cordoba, S. P., Dushek, O., & van der Merwe, P. A. (2011). Basic residues in the T-cell receptor zeta cytoplasmic domain mediate membrane association and modulate signaling. *Proceedings of the National Academy of Sciences of the United States of America*, *108*(48), 19323–19328.

Zhu, J., Luo, B. H., Xiao, T., Zhang, C., Nishida, N., & Springer, T. A. (2008). Structure of a complete integrin ectodomain in a physiologic resting state and activation and deactivation by applied forces. *Molecular Cell*, *32*(6), 849–861.

Zinselmeyer, B. H., Heydari, S., Sacristan, C., Nayak, D., Cammer, M., Herz, J., et al. (2013). PD-1 promotes immune exhaustion by inducing antiviral T cell motility paralysis. *The Journal of Experimental Medicine*, *210*(4), 757–774.

INDEX

Note: Page numbers followed by "*f*" indicate figures and "*t*" indicate tables.

N

O

P

T

U

V

W

X

Z

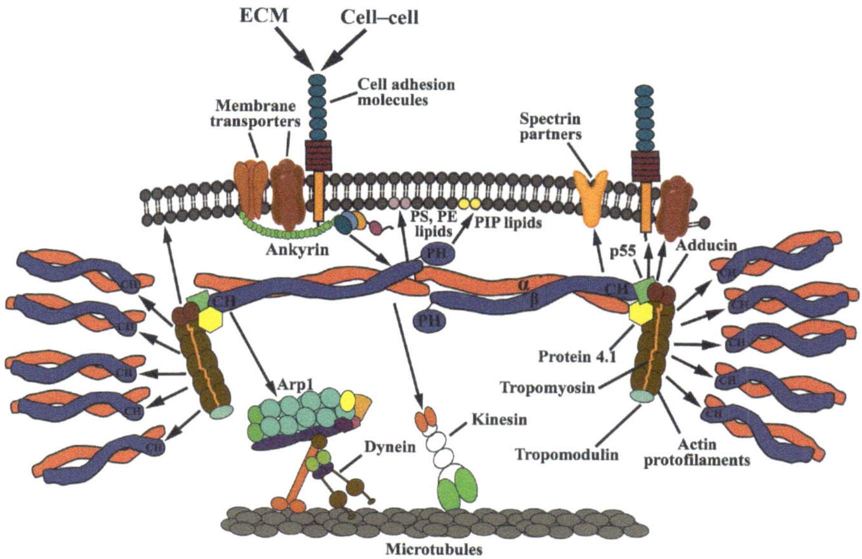

Please refer to Fig. 1.1 in text for figure legend.

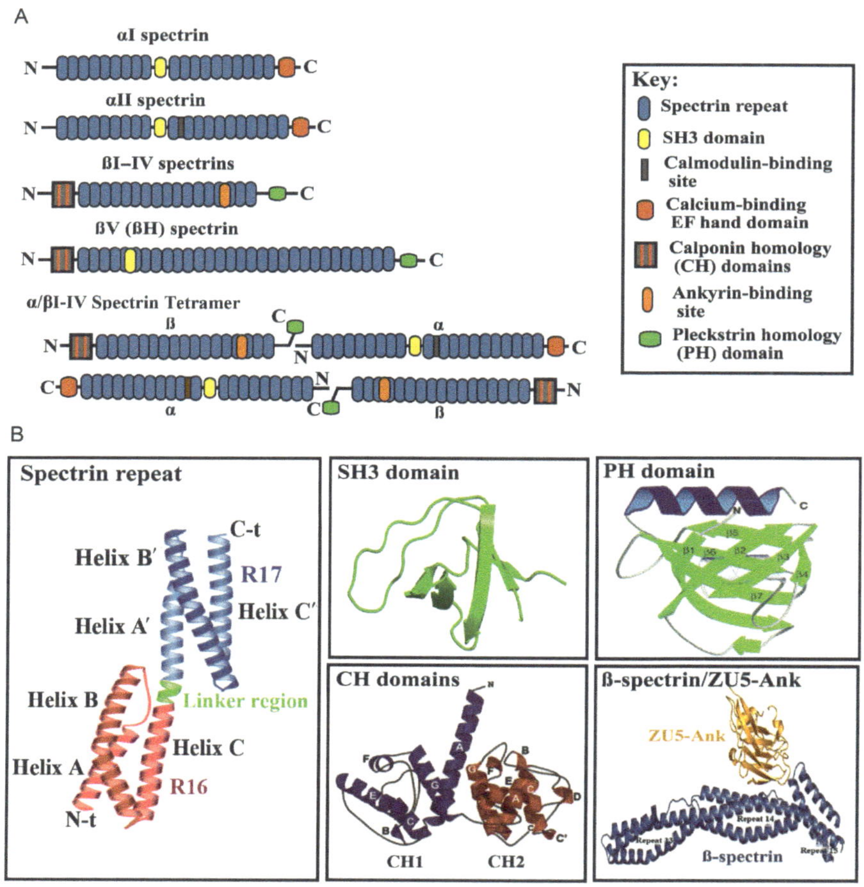

Please refer to Fig. 1.2 in text for figure legend.

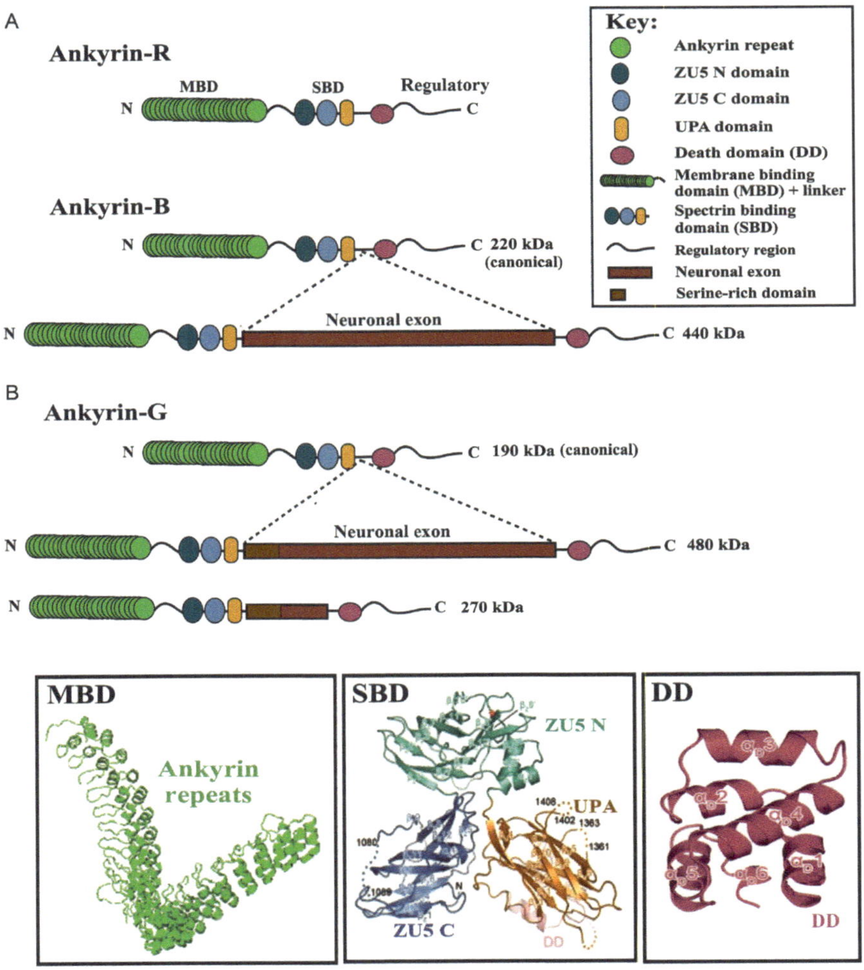

Please refer to Fig. 1.3 in text for figure legend.

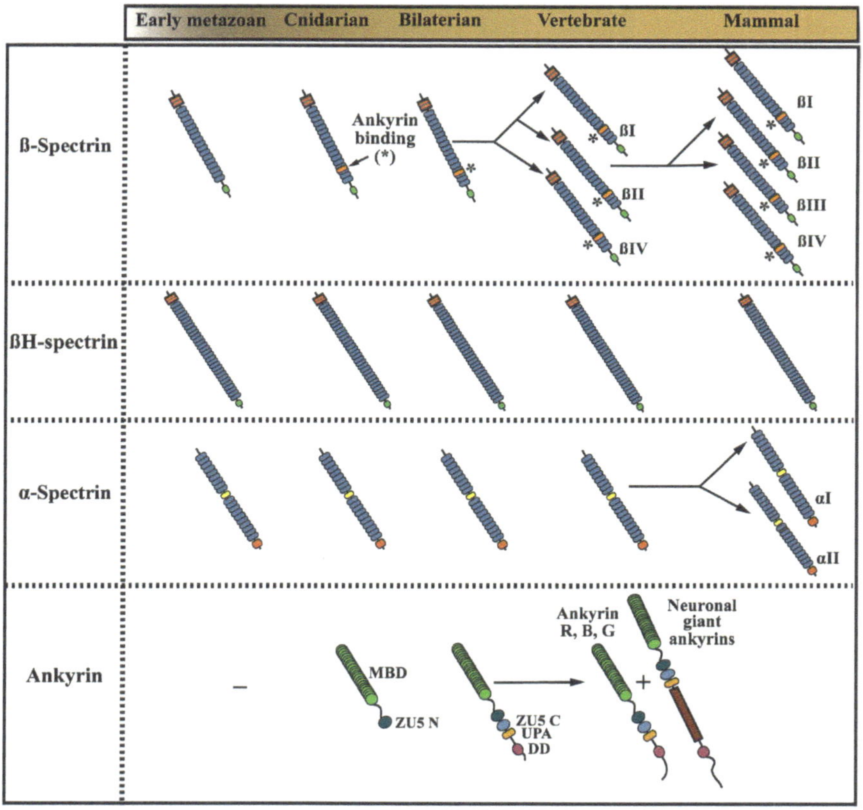

Please refer to Fig. 1.4 in text for figure legend.

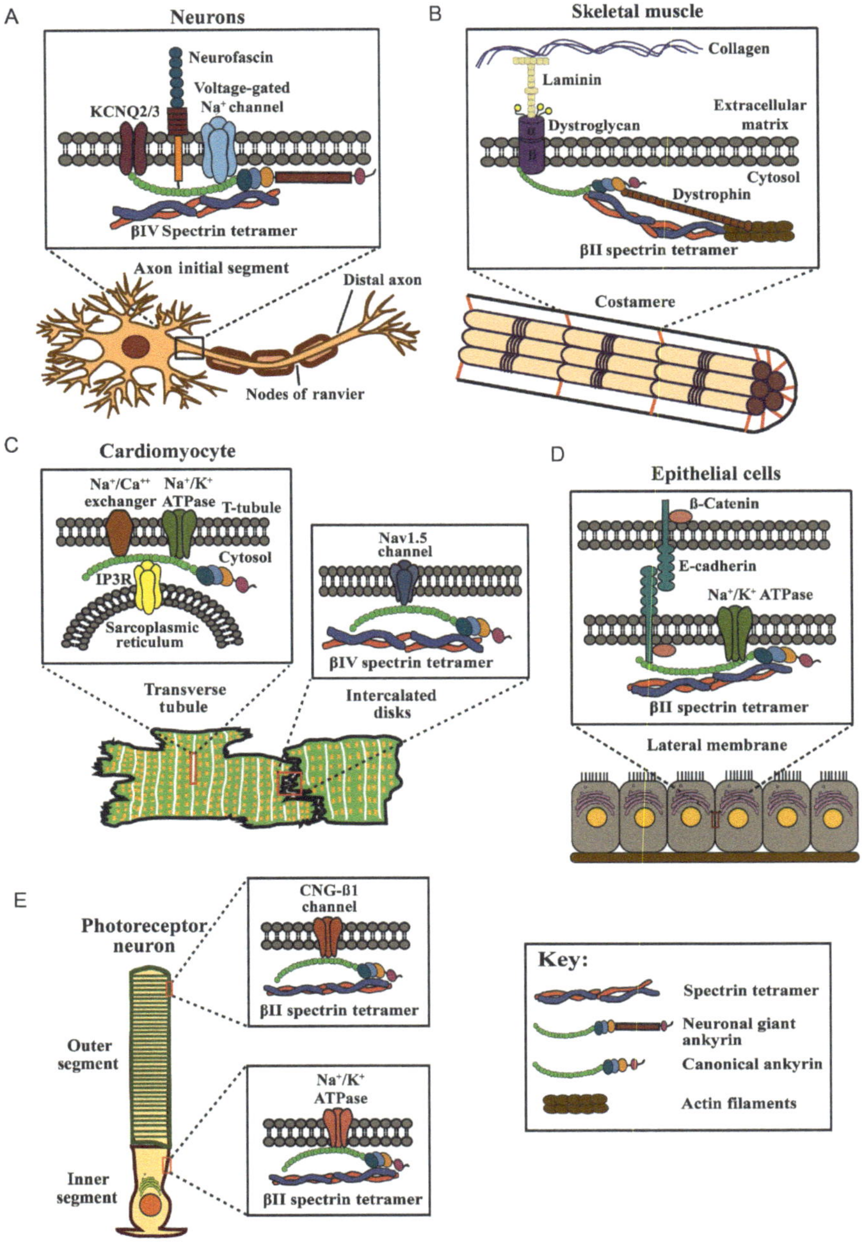

Please refer to Fig. 1.5 in text for figure legend.

Family	*Ankyrin Partner*	*Ankyrin Binding Motif*	*Present*	*~ mybp*
Cell adhesion molecules	**Dystroglycan**	**GVPIIFADELDDSK**	**Cnidarian**	**580**
	NF/L1CAMs	**QFNEDGSFIGQY**	**Bilaterian**	**555**
	E-cadherin	**KEPLLPPEDDTRDNVYYYDEE**	**Urochordate**	**520**
Membrane Transporters	**NaV**	**VPIAVGESDFE**	**Cephalochordate**	**500**
	KCNQ2	**PYIAEGESDTDSD**	**Jawed fish**	**440**
	Kir6.2	**VPIVAEED**	**Jawed fish**	**440**
	CNG beta	**PEPGEQILSVKMPE**	**Jawed fish**	**440**
	AE1	**ELVMDEKNQEL...PAVLTRSGDPS**	**Mammals**	**200**
	RhBG	**KLPFLDSPP**	**Mammals**	**200**

Please refer to Table 1.1 in text for Table legend.

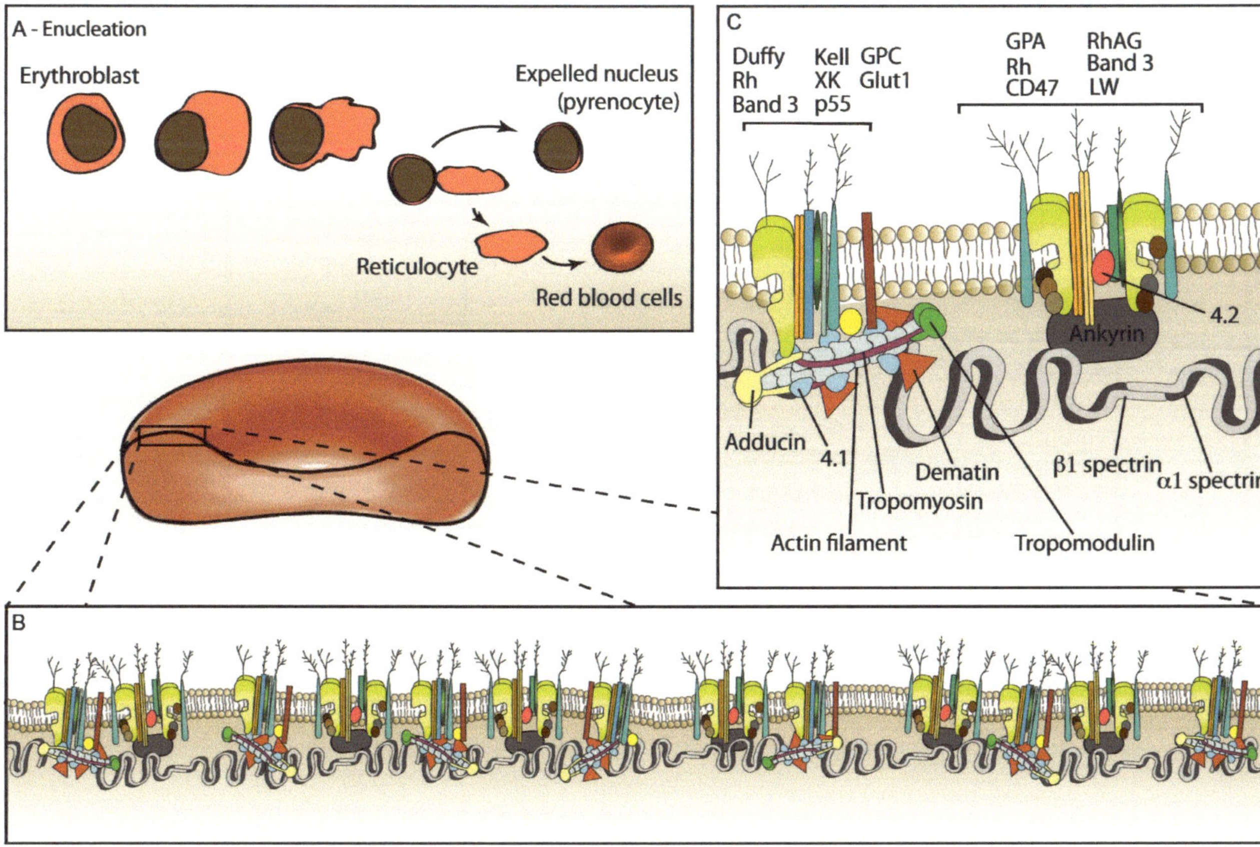

Please refer to Fig. 2.1 in text for figure legend.

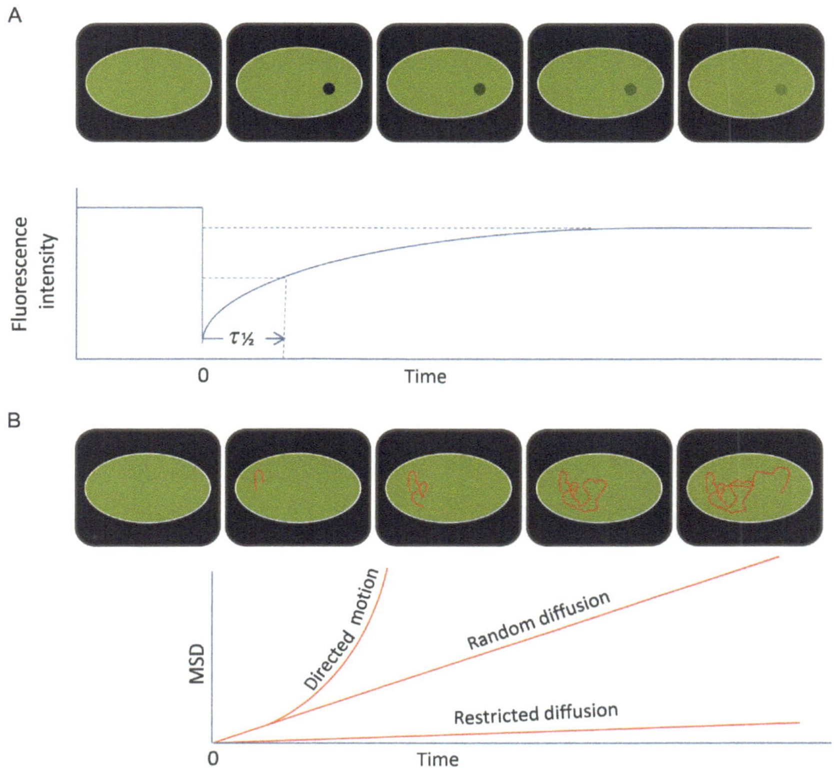

Please refer to Fig. 3.2 in text for figure legend.

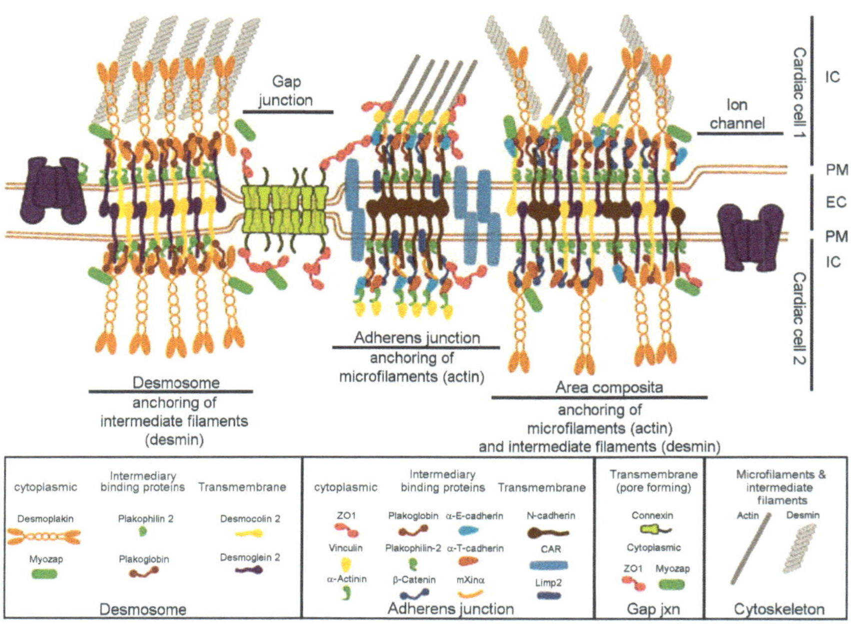

Please refer to Fig. 4.3 in text for figure legend.

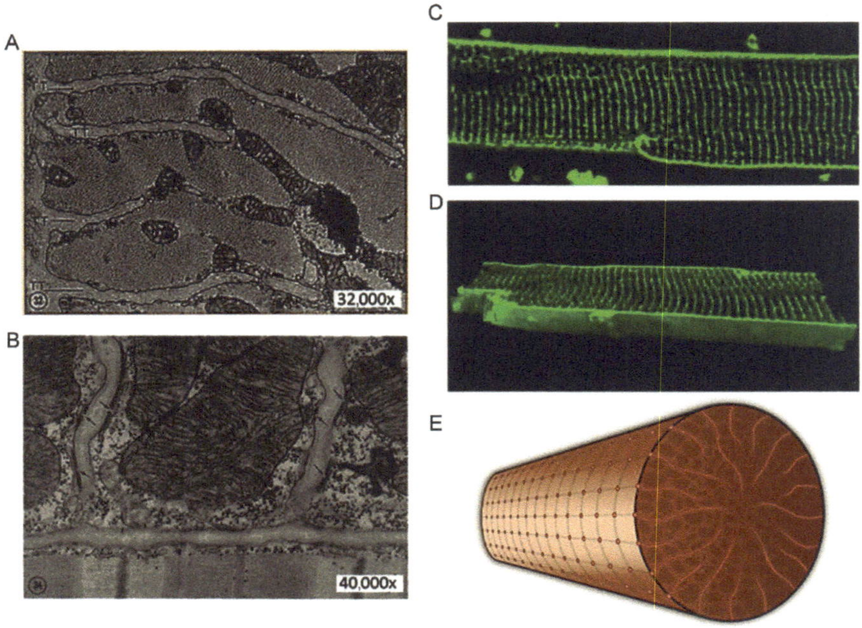

Please refer to Fig. 4.4 in text for figure legend.

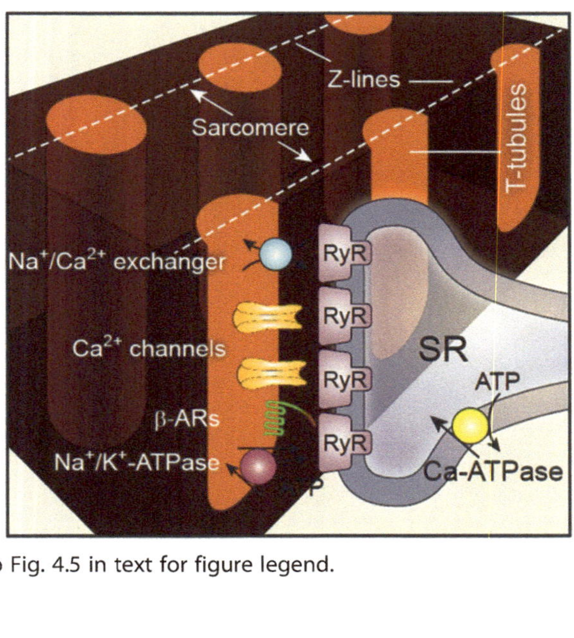

Please refer to Fig. 4.5 in text for figure legend.

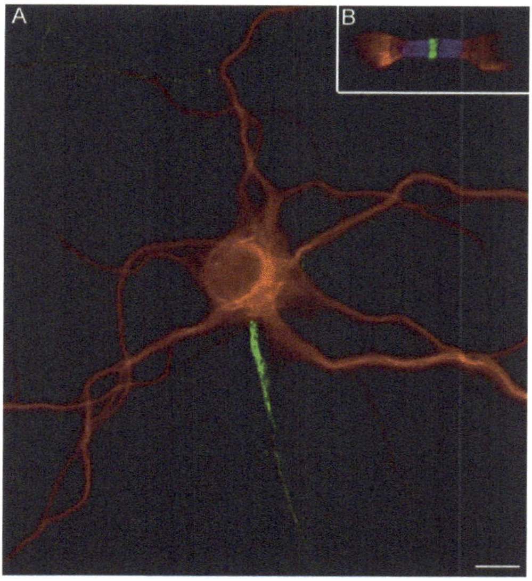

Please refer to Fig. 5.2 in text for figure legend.

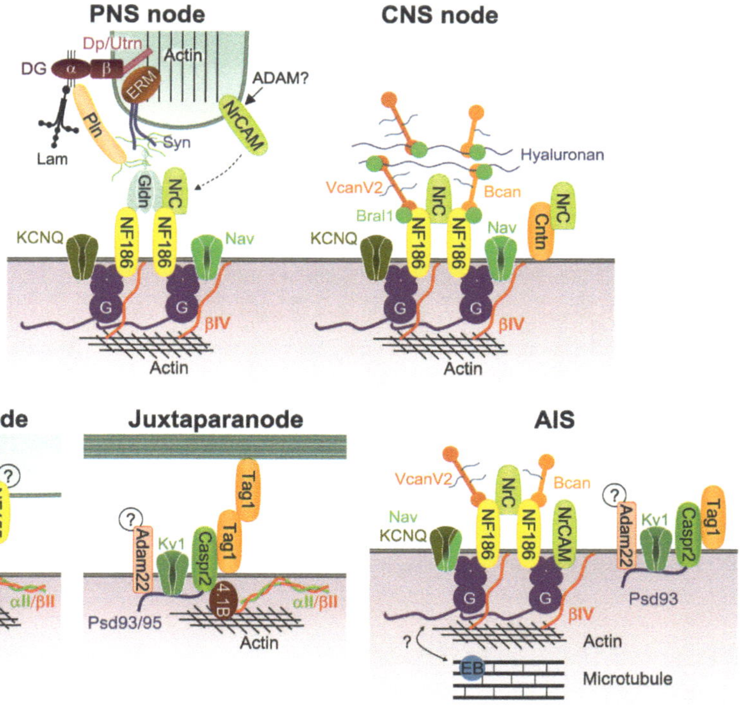

Please refer to Fig. 5.3 in text for figure legend.

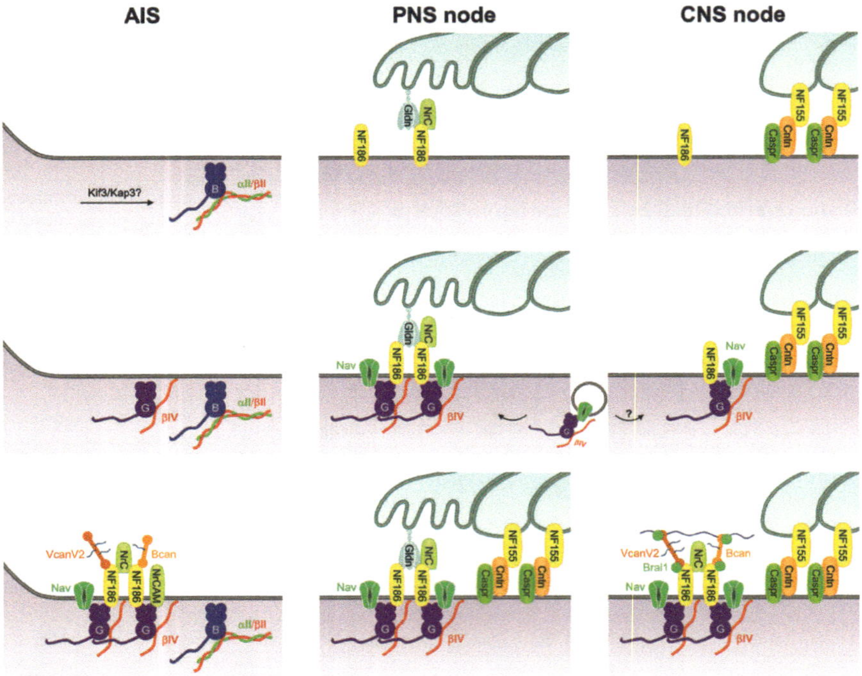

Please refer to Fig. 5.4 in text for figure legend.

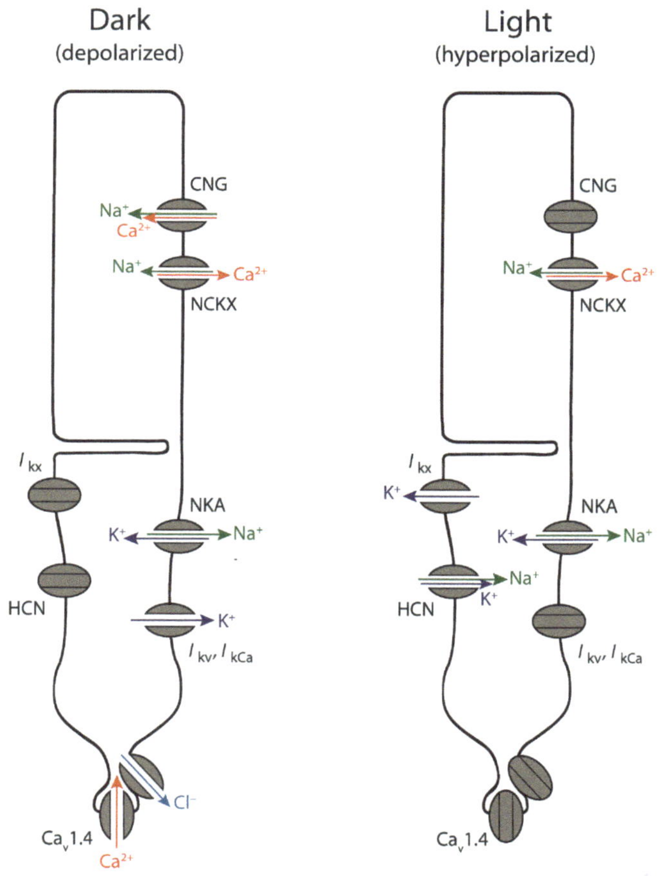

Please refer to Fig. 7.2 in text for figure legend.

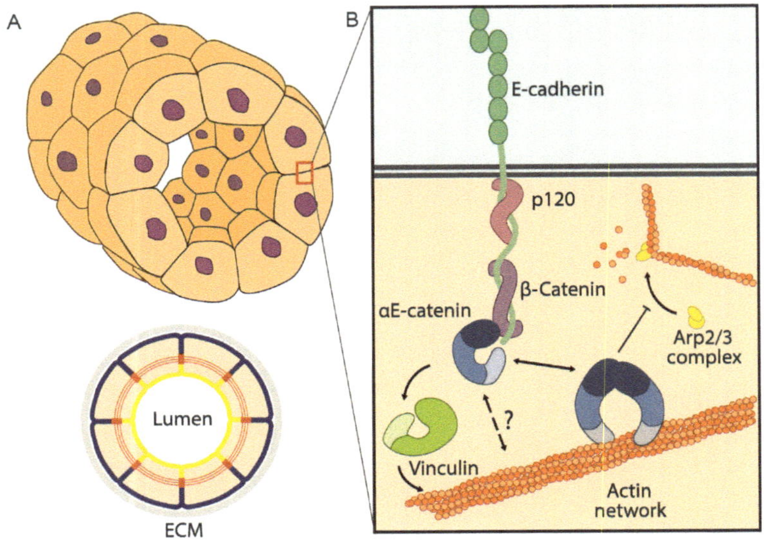

Please refer to Fig. 8.1 in text for figure legend.

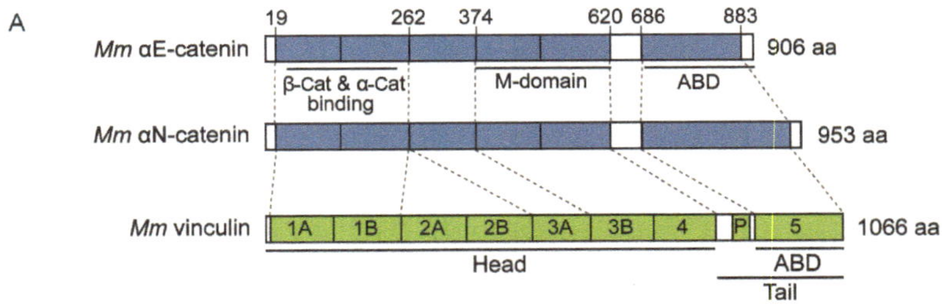

B

Homolog	Dimerization	β-Cat binding	Actin binding	Actin regulation	*In vivo* requirement
Mammalian αE-catenin	Yes	Yes	Homodimer	Bundles actin; inhibits Arp2/3 and cofilin	Adhesion in blastula; epidermal formation
Mammalian αN-catenin	No	Yes	Monomer	?	Neuronal cell migration
Mammalian vinculin	No	Yes	Autoinhibited	Tail domain bundles actin as a dimer	Heart and brain development beyond E8 mice
G. gallus αN-catenin	?	IP	IP	?	?
G. gallus Vinculin	No	?	Autoinhibited	?	?
X. laevis αE-catenin	?	IP	?	?	Cell–cell adhesion at blastula stage
X. laevis αN-catenin	?	IP	?	?	Gastrulation and neural plate closure
D. rerio αE-catenin	No	Yes	Monomer and α β-cat Heterodimer	Bundles actin	Epiboly and gastrulation
D. melanogaster α-catenin	Yes	IP	Monomer and homodimer	?	Embryogenesis, imaginal discs, oogenesis
C. elegans HMP-1	No	Yes	Autoinhibited	Bundles actin *In vivo*	Ventral closure and body elongation
D. discoideum α-catenin	No	Yes	Monomer	Bundles actin	Fruiting body formation

Please refer to Fig. 8.2 in text for figure legend.

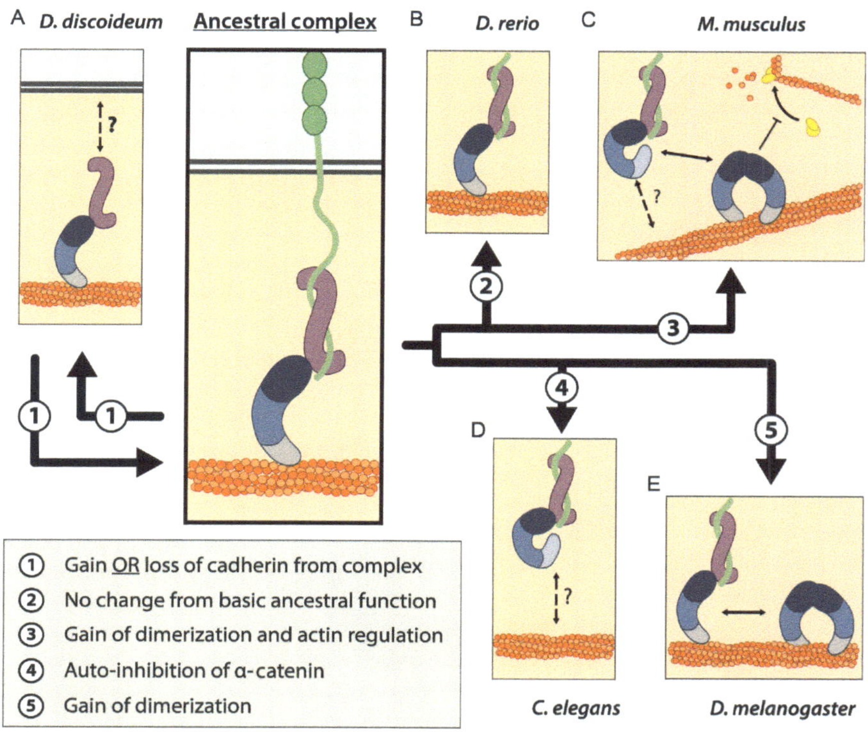

Please refer to Fig. 8.3 in text for figure legend.

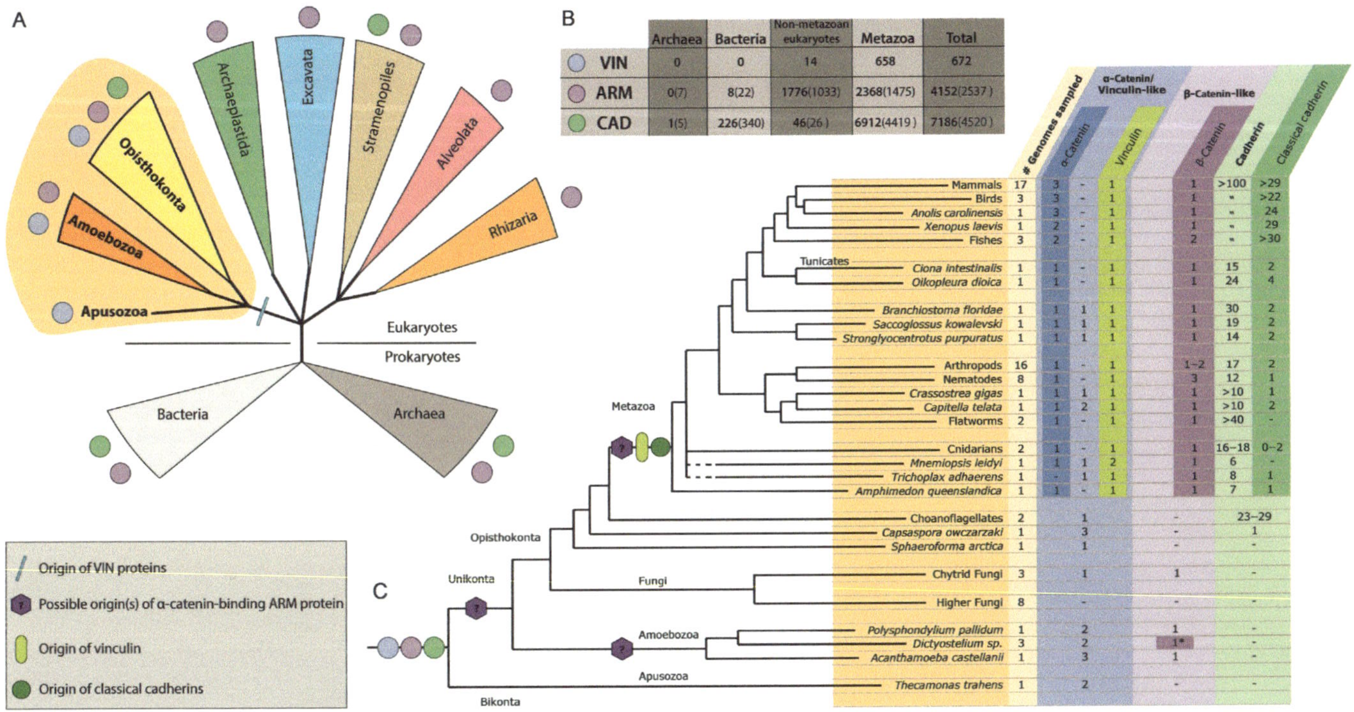

Please refer to Fig. 8.4 in text for figure legend.

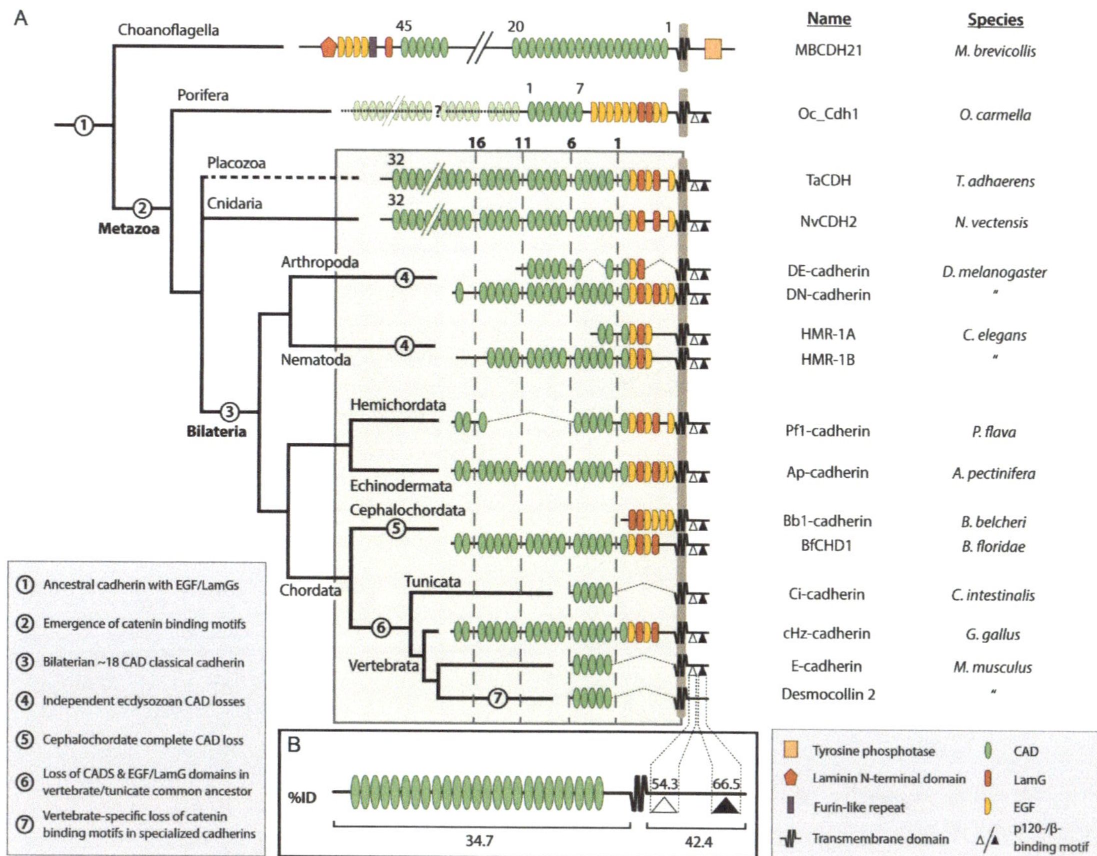

Please refer to Fig. 8.5 in text for figure legend.

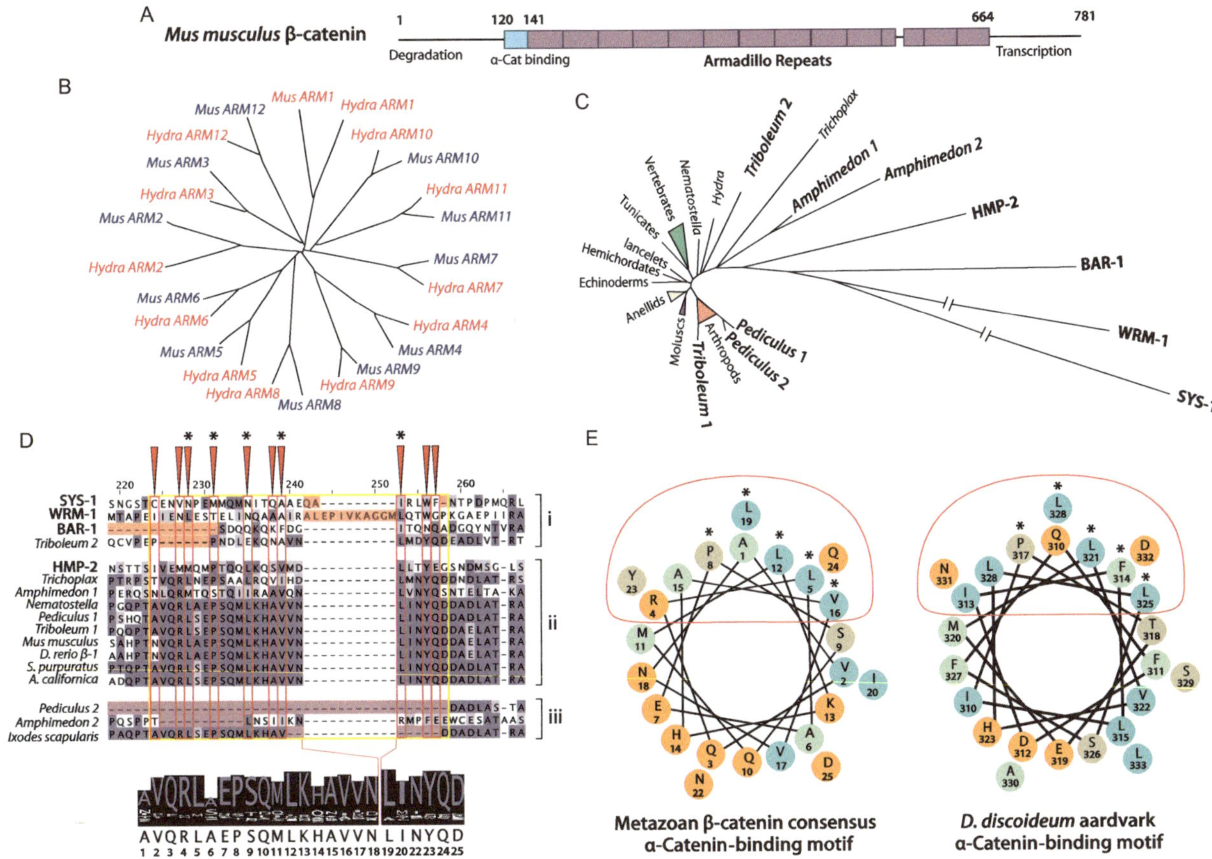

Please refer to Fig. 8.6 in text for figure legend.

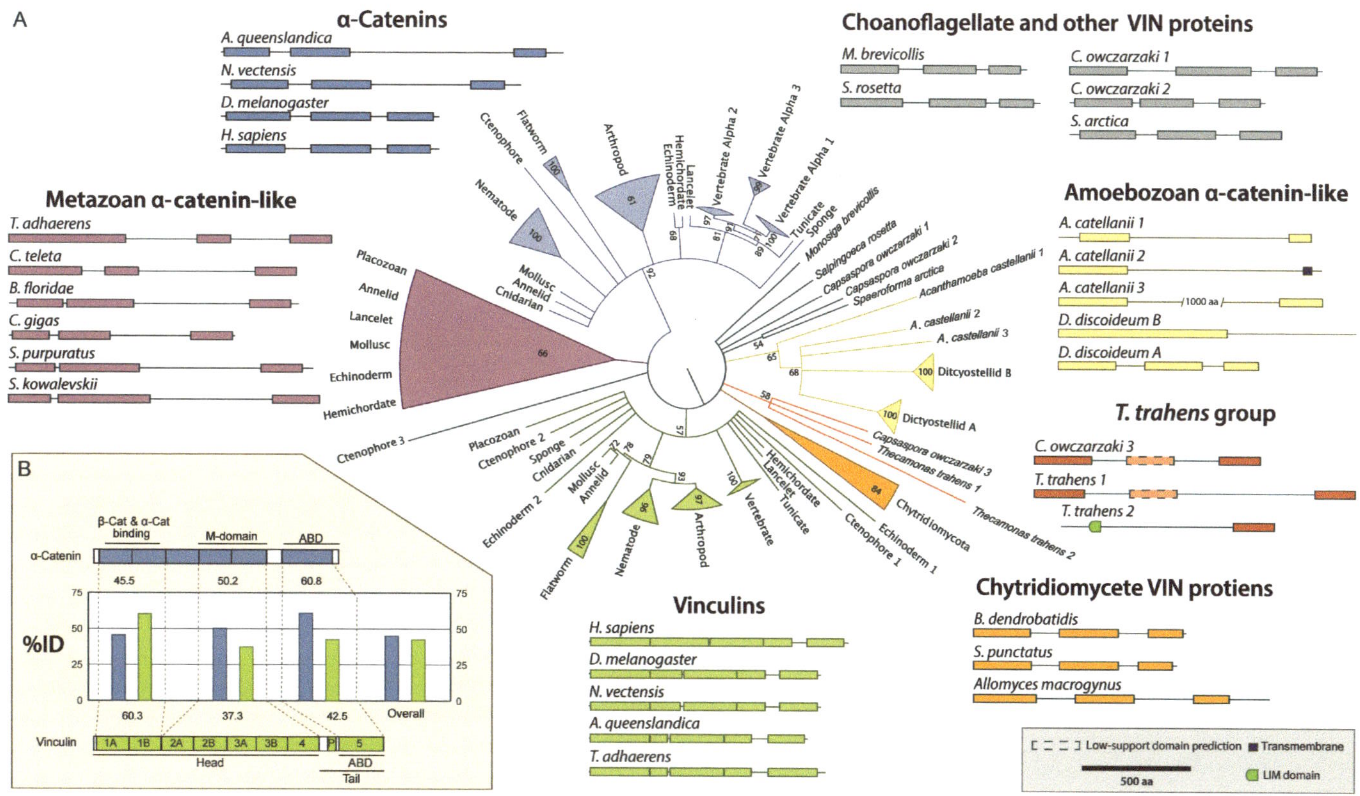

Please refer to Fig. 8.7 in text for figure legend.

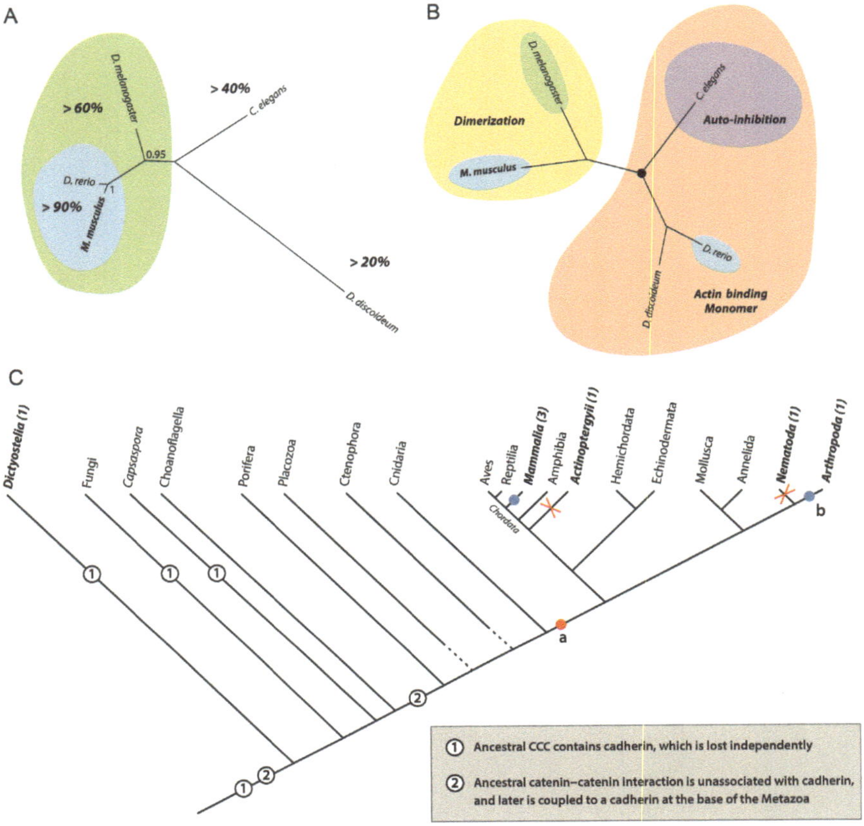

Please refer to Fig. 8.8 in text for figure legend.

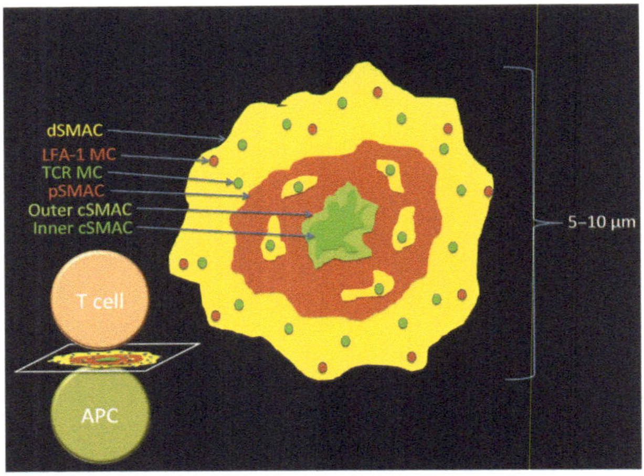

Please refer to Fig. 9.1 in text for figure legend.

FSC
www.fsc.org
MIX
Papier | Fördert gute Waldnutzung
FSC® C083411

Kolibri 360 GmbH
Ferdinand-Jühlke-Straße 7
99095 Erfurt, Deutschland
produktsicherheit@kolibri360.de